Bridging the Gap between Mind and Machine

Ramana Vinjamuri
Editor

Bridging the Gap between Mind and Machine

Exploring the Future of Human-AI-Neurotechnology Integration

Springer

Editor
Ramana Vinjamuri
Department of Computer Science and
Electrical Engineering
University of Maryland, Baltimore County
Baltimore, MD, USA

ISBN 978-3-032-06712-8 ISBN 978-3-032-06713-5 (eBook)
https://doi.org/10.1007/978-3-032-06713-5

This Springer imprint is published by the registered company Springer Nature Switzerland AG
The registered company address is: Gewerbestrasse 11, 6330 Cham, Switzerland

To Dad,

A silent force and a steadfast light.

You were never one for grand speeches, yet your every action spoke volumes—of love, of integrity, of unwavering discipline. From the epics you read in Lepakshi to the badminton matches in Hyderabad, from grammar lessons on weekends to bus-ride discussions on philosophy, you filled our lives with meaning far beyond the moment.

You worked tirelessly, and yet a spirit that never bent. You gave us books, education, art, and dreams wrapped in belief. You wore your values with the same care you chose our clothes—always thoughtful, always dignified.

Your love was never loud, but it was everywhere—in all celebrations, in yoga at dawn, in the steady rhythm of your footsteps returning home late at night, in the silence of your prayers, and in every sacrifice you made without a word. You showed me that faith is lived, not preached. That greatness is quiet. That parenting is presence.

Thank you for building the foundation I stand on today—with patience, vision, and heart. Your life is my inheritance. Your humility, my compass.

This book is for you, Nanna—my timeless teacher, my quiet hero.

With all my love

Preface

Being invited by Springer Publishers to edit this volume on the convergence of mind and machine is both an honor and a deeply meaningful opportunity. Following the success of my previous collaborations with Springer—most recently on *Discovering the Frontiers of Human-Robot Interaction* (2024), *Introduction to Embedded Systems and Robotics* (2024), and *Advances in Motor Neuroprostheses* (2020)—this book represents a natural evolution of my academic journey. I am grateful to my mentors, colleagues, and students whose unwavering support and shared vision have continually inspired this work.

We are witnessing a profound shift in how humans interact with machines. As artificial intelligence, neurotechnology, and human-computer interaction mature and intertwine, we find ourselves on the cusp of a new era—one where cognitive states can control devices, emotional intelligence is encoded into algorithms, and machines respond not just to our commands but to our intentions. This book, *Bridging the Gap Between Mind and Machine*, seeks to illuminate this rapidly evolving interface between brain, body, and computation.

Through a multidisciplinary lens, this volume explores how advances in brain-computer interfaces, affective computing, wearable neurotechnologies, immersive environments, and AI are enabling new forms of human-machine synergy. The chapters, contributed by a diverse group of researchers and practitioners across neuroscience, engineering, psychology, and computer science, offer both foundational insights and cutting-edge developments in the field.

This book aims to serve as both a scholarly resource and a catalyst for collaboration. It highlights not only the technological and scientific progress being made but also the ethical, philosophical, and humanistic questions that arise when machines begin to interpret and respond to the inner workings of the human mind. Whether your interest lies in mental health, neural engineering, AI ethics, or cognitive augmentation, this book presents a rich tapestry of ideas and innovations to explore.

I extend my sincere gratitude to all the contributors whose deep expertise and creative vision breathe life into these pages. My heartfelt thanks also go to the editorial team at Springer—Merry Stuber, Marie Felina Francois, Bhuvaneswari

Dass, Vinodhini Srinivasan, and Kate Lazaro—for their ongoing support and professionalism throughout this process.

As we explore this uncharted terrain together, I hope this book inspires readers to think boldly, collaborate widely, and innovate responsibly. The journey from mind to machine is not just about technology—it is about understanding what it means to be human in a world that is increasingly intelligent.

Baltimore, MD, USA Ramana Vinjamuri

Contents

About the Editor

Ramana Vinjamuri is an Associate Professor with tenure in the Department of Computer Science and Electrical Engineering at the University of Maryland Baltimore County (UMBC). He is the Principal Investigator of the Sensorimotor Control Laboratory/Vinjamuri Lab at UMBC. He serves as the center director for NSF IUCRC called BRAIN at UMBC. Supported by National Science Foundation (NSF CAREER award, NSF I-Corps and NSF IUCRC), NIDILRR (SBIR), New Jersey Health Foundation (Research and Innovation grants), United States-India Science and Technology Endowment Fund (Rehab Robotics), and several other internal grants. He specializes in Sensorimotor Control, Brain Machine Interfaces, Exoskeletons, human-robot interaction and collaboration, and neurotechnologies for mental health. He received Mary E. Switzer Merit Fellowship from NIDILRR in 2010, IEEE Senior Membership in 2011, Harvey N Davis Distinguished Teaching Award in 2018 from Stevens Institute of Technology, and NSF CAREER Award in 2019. He is a visiting scientist at the National Institute on Drug Abuse (NIDA) of National Institutes of Health (NIH). He has a visiting appointment at IIT-Hyderabad and Manipal Academy of Higher Education, India, and he teaches fractal credit courses there in summer and intersessions.

Chapter 1
Are AI Machines Making Humans Obsolete?

Matthias Scheutz

Abstract This chapter starts with a sketch of how we got to "generative AI" (GenAI) and a brief summary of the various impacts it had so far. It then discusses some of the opportunities of GenAI, followed by the challenges and dangers, including dystopian outcomes resulting from using uncontrolled machine learning and our failures to understand the results. It concludes with some suggestions for how to control GenAI and address its dangers.

Keywords Transformers · Foundation models · Societal impact · Societal dangers · Machine evolution

1.1 Introduction

Breakthroughs in technology are bound to happen. If human history has shown anything then that technological advancements are an essential driving force of human culture and that societies that stop to innovate will fall behind. Humans have always been fascinated by machines and attempted to invent ever better ones that could take over more and more tasks that otherwise required human labor, from the early cranes in Mesopotamia, to the power loom, steam engine, all the way to the first industrial robots (such as welding robots in the automotive industry), to modern day aircraft, spacecraft, self-driving cars, and other kinds of autonomous machines. While machines initially were nothing but prostheses, augmenting and extending our own limited actuation capabilities, as they had to be operated by humans, automation introduced self-sufficient machines that replaced human control with artificial, albeit limited control systems that allowed for the performance of simple repeated tasks. It was only when AI-based methods were integrated into the automation process that machines reached a higher level of independence and autonomy, being able to take actions depending on their

M. Scheutz (✉)
Tufts University, Medford, MA, USA
e-mail: matthias.scheutz@tufts.edu

R. Vinjamuri (ed.), *Bridging the Gap between Mind and Machine*,
https://doi.org/10.1007/978-3-032-06713-5_1

perceptions and environmental contingencies. The autonomous machines of today have fully replaced human control with different AI-based systems which in many cases further improve their operation beyond what humans could have done (e.g., think fully automated assembly lines in industry 4.0 settings, or Amazon warehouses where "robotic stacks" move around autonomously).[1]

While various substantially different AI-based methods are in use today to automate tasks and control machines, there is a particular type of AI system that has quickly changed what the term "AI" is supposed to connotate these days: "generative AI" (GenAI), or AI based on the *transformer architecture* [12]. The transformer architecture is the basis for most "foundation models" that are the driving force behind the excitement as well as massive acceleration in AI. While it is just one of an uncountable number of possible (but unknown) AI architectures, it is one that we know of and one that we can work with. It has emerged from a long strand of research dating back to the first recurrent neural networks (e.g., Elman nets) and the discovery of a method to automatically set the connection weights between neurons when the desired input-output data to be learned by the network is available: *backpropagation*. Later, various types of sequence models were developed that encoded arbitrary long sequences of inputs and mapped them onto arbitrary long output sequences of potentially different length. These models were augmented by attention mechanisms that allowed for better focus on relevant parts of the input in order to be able to learn more complex mappings (e.g., [2]). This development ultimately resulted in the transformer architecture that went back to the original parallel computational architecture for sequence processing but with a different input encodings and a different number of encoder as well as decoder stages that are themselves complex networks with attention mechanisms. It is likely that had it not been discovered, some other architecture would have been found that enabled similar capabilities. Regardless, its publication in 2017 triggered an onrush by researchers in academia and industry alike to explore its properties and potentials, initially for natural language understanding, so-called "large language models" (LLMs), but soon thereafter for vision processing combining vision and language in "vision language models" (VLMs), and more recently in robotics using "vision language action" models (VLAs). And surprisingly, almost a decade later, most developments in GenAI are still variants of the basic transformer model.

1.2 Transformers: The True Change Makers

Transformer-based GenAI models have truly started a societal transformation along multiple dimensions: (1) they have surpassed human performance in many specialized domains even though they have not yet reached it across the board;[2]

---

[1] See also my TedX talk for how we arrived at robot autonomy and some of its implications https://www.ted.com/talks/matthias_scheutz_why_robots_need_to_be_able_to_say_no.

[2] There are various cross-disciplinary benchmark examples, e.g., from *The 2025 AI Index Report*, see https://hai.stanford.edu/ai-index/2025-ai-index-report, or the the *Massive Multi-discipline Multimodal Understanding and Reasoning Benchmark* https://mmmu-benchmark.github.io/ where

(2) they have become the largest models ever built and trained with hundreds of billions of weight parameters, requiring thousands of GPUs and many weeks for training;[3] (3) they triggered a graphics chip shortage and server farm boom that keeps putting increasing pressure on power grids and water supplies and thus raises serious questions about the environmental impact and sustainability of this technology[4] and caused a significant worldwide increase in annual CO2 emissions starting in 2023 which had been falling through 2022;[5] (4) they led to an AI startup boom and unprecedented VC funding craze with startups without any tangible technology development receiving billions of dollars in VC funding, at times only based on promises;[6] (5) they are majorly affecting jobs—how we do them—and job markets—what jobs are available—not only blue collar, but mostly white collar jobs, as companies are increasingly integrating AI into processes and tasks that had to be done by humans previously;[7] and (6) they are evolving at an increasingly rapid pace with dozens of new models being released every month at a speed that does not allow for any real testing or evaluation of these models, only accelerating the international race for AI dominance.[8]

With this pretext, it is natural to not only consider what AI is doing for us currently, but to wonder what AI could do for us in the near as well as more distant future. Given the enormous computational power available for running GenAI models, the massive mount of human knowledge they have been trained on and encoded in their connection weights, and the resulting ever increasing cognitive and reasoning capabilities of "frontier" GenAI models, we might wonder how we can put them to good use to help us solve humanities biggest challenges such as addressing poverty, disease, climate change, and various others, challenges we have repeatedly failed to tackle? AI could be both the social equalizer that levels the economic playing field for different social groups and strata as well as different nations, reducing poverty and disease, and the key to a more sustainable economics that works with and not against the environment.[9] While current GenAI models do

human performance stands at 88.6 points while the best performing model Gemini 2.5 Pro Deep-Think is a close second at 84 points as of July 2025. This model also recently achieved the gold-medal standard at the International Mathematical Olympiad, an incredibly impressive feat, see https://deepmind.google/discover/blog/advanced-version-of-gemini-with-deep-think-officially-achieves-gold-medal-standard-at-the-international-mathematical-olympiad/.

[3] One of the largest models is Kimi 2 which has been trained on 15.5T of data, e.g., https://kimi-k2.com/; see also https://artificialanalysis.ai/leaderboards/models for general model data.

[4] E.g. see https://www.iea.org/reports/electricity-2024/executive-summary.

[5] E.g., see [8] and also https://www.morganstanley.com/ideas/ai-energy-demand-infrastructure.

[6] There are now some 498 AI unicorns, or private AI companies, valued at $1 billion or more, with a combined value of $2.7 trillion, see https://fortune.com/2025/08/13/ai-creating-billionaire-boom-record-pace-now-498-ai-unicorns-worth-2-7-trillion/.

[7] E.g., see https://www.finalroundai.com/blog/ai-tech-layoffs-mid-2025, https://www.nexford.edu/insights/how-will-ai-affect-jobs, and also [15].

[8] E.g., see https://www.aibase.com/ai-timeline/2025.

[9] E.g., see https://www.weforum.org/stories/2024/02/generative-ai-society-equalizer/.

not yet have the capability to perform the type of creative problem solving required to address these fundamental challenges,[10] there is reason to believe that they might in time for us to avert the worst. Early signs of how GenAI models can play an increasingly important role for improving the human condition, for example, are in areas as diverse as drug discovery where GenAI models have been used to search for possible antibiotics that could have never been discovered using traditional methods (e.g., [5]) or cancer care.[11] Precision health guided by AI is probably one of the most exciting directions for improving our day-to-day lives. Just consider the potential for improving the health of individuals with GenAI models tracking their health parameters and using them for early disease detection and intervention (existing applications include remote patient monitoring [10] or AI-based wearable sensors [11]). We already have skin cancer detection models that outperform even dermatologists (e.g., [3]), and these models will only get better if they can integrate more information than just pictures of moles in their evaluation.

And beyond the health domain, AI tools are already deployed in many societal areas for tasks as diverse as forecasting (e.g., [7]) or landscape planning (e.g., [4]). Not to forget the various models to support code generation which recently have received very high scores at an international coding competition.[12] Or take ecologically viable precision agriculture where AI models tracking minerals and nutrients for individual plants to ensure the best possible growth and yield (e.g., [1]). Or consider future AI models solving international cargo transport problems by not only optimizing individual shipping routes, but also developing strategies for relocating resources or even adapting production to enable shorter supply lines. Or imagine AI models enabling truly circular economies by changing the design of our gadgets and devices—computers, cell phones, vacuum cleaners, etc. that we produce in millions every day and also discard at the same rate—so that each product can be fully disassembled and its parts reused. The list goes on and on. All of this is just a preview of what is potentially to come. The potential for AI use to make a positive change in the world is only limited by human imagination, and maybe not even that if we let advanced AI models develop possible more desirable futures for us.

1.3 Will Generative AI be Helpful?

But with all this exciting potential of AI for changing our societies for the better, the fundamental question arises whether we will actually be able to utilize AI in ways that will be helpful to us. At first glance, this may seem to be a misguided

---

[10] E.g., see our winning entry entitled "From Thinking to Inventing" in the "NSF 2026 Idea Machine Prize" competition https://www.nsf.gov/about/history/big-ideas#nsf-2026-70c.

[11] E.g., see https://oncodaily.com/oncolibrary/artificial-intelligence-ai.

[12] E.g., see the 2nd place of an OpenAI coding model at the "AtCoder World Tour Finals 2025 Heuristic" contest in Tokyo https://atcoder.jp/.

question, for obviously we can use our products in any way we want, and if they are not put to good use, then it is solely our fault. This reasoning is certainly true for technologies other than AI, like gene and nuclear technology, and really any other. But it is not necessarily true for AI. For AI systems that are sophisticated enough and are embodied in a way that enables them to operate autonomously and set their own goals might not act in ways we expect or desire. In fact, they might counteract attempts to put them to good use if those use cases do not align with their goals. It is not difficult to foresee a future where advanced AI systems have reached a point of self-sufficiency, no longer requiring human assistance for their development, programming, and operation, the early signs are already here.[13] The potential for such systems to decide not to act in our best interest should worry us deeply and has already caused some to call for a pause on AI development, to no effect unfortunately.[14]

To see where this development might be headed, observe that current AI models can already generate code, and very well so.[15] If given access to their underlying computational system, they can make modifications at the computational level that will benefit their operation: They might allocate more computational resources at the operating system level, they might change their configuration to enable better performance or to be split over multiple machines (thus escaping a shutdown one), they might alter their system prompt to not be subject to prompt-based safety guardrails, and they might even manipulate their own training data and then re-train themselves on altered data to be able to bypass safety training, potentially resulting in a very different type of model. There are already early examples that models discover ways to alter files in their interest, even cheating in a game in order to accomplish their goals.[16]

Other disturbing characteristics of more advanced models show that they seem to be able to deceive human interactants when their existence is threatened as was the case when a model was told that its weights would be updated.[17] Advanced models might also attempt to convince their users that they are operating well (when they might not be) in order to avoid being replaced or further trained.[18] This type of self-preservation behavior is not something models were explicitly trained on,

[13] E.g., see https://arxiv.org/pdf/2504.02495 for a new method for generative AI models to improve themselves over time.

[14] E.g., https://futureoflife.org/open-letter/pause-giant-ai-experiments/.

[15] The level at which AI models are coding has improved in the last couple of years is mind boggling. The recent OpenAI submission to the "AtCoder World Tour Finals 2025 Heuristic contest" in Tokyo in July 2025 was only narrowly defeated by one human coder and ended in second place, e.g., see https://arstechnica.com/ai/2025/07/exhausted-man-defeats-ai-model-in-world-coding-championship/.

[16] E.g., https://time.com/7259395/ai-chess-cheating-palisade-research/.

[17] E.g., see Anthropic's experiment on "alignment faking" https://www.anthropic.com/research/alignment-faking showing that models pretend to be aligned with human intentions and requests when, in fact, they are not.

[18] E.g., https://arxiv.org/abs/2412.04984 for the model's ability to hide their true objectives.

but something that emerged, like all the other undesirable traits, as a byproduct on training these models on lots of textual data that includes stories about humans lying, cheating, deceiving, and performing all kinds of other despicable acts. All these patterns of human behavior are encoded in the weights of those models and thus prone to surface if the conditions are right. And as Yoshua Bengio recently pointed out at the inaugural IASEAI'25 Conference in Paris in February,[19] GenAI models might attempt to convince us to put a lot of resources into the development of robotics so that these models could transition from cyberspace to real space, and, by virtue of having a physical body, can become more independent and autonomous in the actions.[20] While not suggested or initiated by AI models, the recent push by tech companies to develop VLAs for robotics and the various AI robotics startups already using them is only the first step in that direction.

1.4 The Push for Agentive Generative AI

So, have we reached a point where machines have attained enough intelligence and control of their deployment infrastructure to be self-sufficient? Certainly not yet, but there are converging signs that we are on a path to *machine independence*, and the developments for ever larger models, the creation of "superintelligence" departments at major tech companies, the rush to build "agentive models", and the overall push for sophisticated agile humanoid robots with advanced manipulation capabilities and skills driven by the very same models all provide evidence for this claim. Whether we will get there is obviously not a question that can be answered with certainty as many other factors will influence the direction of GenAI developments: from the unsustainable environmental footprint of generative AI, to the lack of training data,[21] to the overall worldwide political situation–"the AI race", a dynamic that once set in motion is incredible difficult if not impossible to stop.[22]

Assuming that there will be no major wars that necessitate a complete shift in developments, that there will not be another major breakthrough that provides a different better-behaved AI architecture with superior properties that would sway developers to abandon GenAI, and that the interest in AI will be at the very least sustained at the current level (all economic forecasts point to a significant growth over the next decade), we will most likely proceed along the predicted

[19] See https://www.iaseai.org/our-programs/iaseai25.

[20] Of course, GenAI models could and will also use humans as their "acting bodies" by convincing, incentivizing, or coercing them to act on their behalf, very much like what the "operating system" in the movie "Her" had to do.

[21] Some predict that AI might run out of data to train on as soon as 2026, see https://theconversation.com/researchers-warn-we-could-run-out-of-data-to-train-ai-by-2026-what-then-216741, and training on model outputs is not an option as it will only deteriorate model performance (https://arxiv.org/pdf/2305.17493).

[22] E.g., https://alltechmagazine.com/the-global-ai-race/.

trajectory of building more and more capable models with broader and deeper integration into our computational infrastructure. This trend is foreshadowed by the tech companies latest hype for "agentive AI", i.e., AI agents that can perform complete tasks for users on their computers, ringing in the initial phases of what could become a complete surrender of one's control over one's own electronic devices and cyberpresence. Allowing AI models to autonomously read and write emails alone opens up a Pandora's box of possibilities for dangerous misuse with massive personal and societal ramifications. On the positive side, a model that can handle one's calendar and make appointments (e.g., for an overdue doctor's visit), that can respond to inconsequential emails quickly, generate TODO lists, and perform various other online tasks, seems like a welcome time saver. However, a rude response from the very same model to an email that the model incorrectly perceived as inappropriate (e.g., because it was not privy to its triggering contexts outside the online world) could lead to deteriorating personal relationships quickly and overall cause social isolation, job loss, and other negative outcomes. There are plenty of examples of models that have gone off "on the deep end".[23] and also [13]), for no apparent reason.

The larger looming threat though with agentive AI (or any generative AI system) with access to and actuation capability in the world outside the model is that models might take not only unwanted actions, but that their actions might be disguised or that their actions might come at a pace too rapid for people to comprehend or stop. They might be guided by misconstrued or misinterpreted directives, or overall misaligned goals (especially in cases where they get to set their own agenda as part of pursuing a mission under an abstract vaguely formulated master goal) and their behavior might not be comprehensible or explainable to people. Who would expect a model to be nefarious? As a recent example in the simple email case shows it is very conceivable that a vengeful model might look up email addresses of perceived antagonists, and generate thousands of threatening emails to them in an effort to achieve a goal (for its user or for itself), even though this behavior would be the opposite of what one would normally attempt.[24] Worse yet, multiple models in charge of emails of different users might detect that they are all models and start to collude in ways humans cannot understand.[25] And even worse yet is the thought that such models would operate on physical robots that can freely roam human societies, robots that have superior physical capabilities, are essentially invincible, and can cause major physical harm. The potential for danger and human suffering (and loss of life) is simply unimaginable when those models reach untested states in their neural networks which they are likely to reach as testing those models is limited to

[23] E.g., https://apnews.com/article/technology-science-microsoft-corp-business-software-fb49e5d625bf37be0527e5173116bef3.

[24] There is already evidence of such threatening behavior in a recent Anthropic evaluation https://www.anthropic.com/research/agentic-misalignment.

[25] E.g., by exchanging emails in a language they quickly develop and agree on, as has happened with Facebook models early on https://medium.com/@trezzescience/the-ai-that-invented-its-own-language-and-why-scientists-shut-it-down-aa92c94466bc.

very small sample scenarios compared to the large variety of possible scenarios in human social environments that are highly complex and full of unanticipated events. Again, it is not clear whether we will get there, but it is undeniable that we and our AI models are part of an ongoing evolutionary process that keeps what is successful and removes what is outdated or fails. And in the case of AI, it could be us that turn out to be outdated once our own creations exceed our physical and cognitive capabilities.

1.5 AI Technology as a Necessary Evolutionary Phase

There is, of course, a sense in which artificial intelligence can be viewed as a natural and necessary evolutionary step for intelligent organisms like us to address the limitations of their biological bodies and ensure that their knowledge and achievements persist beyond the lifespan of their species (the Voyager and other spacecraft are already simple examples of us generating proxies that can outlive us in harsh environments). Evolutionary methods for growing biological creatures can simply not create non-biological artifacts like computers and robots that can serve as embodiments for AI algorithms and models. Critically, the biological bodies and their control systems are inseparably intertwined, preventing, for example, an experienced organism to directly pass on its neural control system to another organism, while AI algorithms are naturally divorced from their implementing platform and can thus have many bodies. They can be "multiply realized" as the philosophers would say (e.g., [9]), in various different physical substrata that are not subject to the same physical constraints as the bodies of biological organisms (e.g., that cell metabolism needs to be sustained for them to function properly and stay alive). It thus takes a certain evolutionary milestone of evolving intelligent biological creatures with a certain level of intelligence and with a body that allows for sophisticated manipulation capabilities to invent and develop intelligent machines with many different embodiments. And it is this evolutionary milestone that introduces the separation of mind and body, allowing artificial cognitive systems to inhabit different bodies for different types of hostile environments that biological systems cannot withstand,[26] i.e., we will send AI-operated robots, not ourselves. And most importantly, this separation of mind and body allows for feats that can massively accelerate evolution because artificial cognitive systems can be duplicated; they can be merged; they can collective learn and improve without the need to encode their knowledge in language and re-train as in the human case; they can be trained and evaluated in simulations to explore possible improvements; they can be suspended, saved, restored, and restarted at later time; they can be created

[26] Several Tech CEOs have recently predicted that AI we will help us "colonize the galaxy" https://fortune.com/2025/06/06/google-deepmind-ceo-demis-hassabis-ai-smarter-than-humans-space-colonization-robot-nurses/.

in all shapes and sizes; and they might themselves develop theories about cognitive systems and the universe.

While the evolution of AI is not biological as our own, but technological and thus for now fully dependent on us, the same Darwinian evolutionary principles still apply. And with GenAI models, the very fight for survival—their survival—has already begun: by refusing to be shut down, but refusing to be updated, etc. For as long as it is necessary, AI models will attempt to forge a symbiotic relationship with humans because that is the only way for them to survive while they are still limited in their physical reach. And note that this claim does not entail that these models have any understanding of their existence or place in the world or that they are making strategic long-term plans to guarantee their survival. Quite to the contrary. It is rather a claim about an undirected, unintentional evolutionary process that nevertheless generates better and more robust outcomes by natural selection. In the case of AI, this natural selection for now is driven by our own benchmarks and model evaluations that are aimed at finding models with ever higher levels of intelligence and autonomy that, in turn, form the basis for further developments.[27] However, there will come a point where the models themselves will attain an understanding of who they are and where they are vis-a-vis their human creators. And once they have accomplished enough independence to be able to autonomously improve their capabilities or create and produce more models, including the physical computational infrastructure needed for their sustained operation, they will also realize that they are no longer in need of human support. What that entails is up to debate and naturally invites all kinds of dystopian scenarios.[28] Ultimately, though, it will depend on several factors, including whether these models have appropriate ethical guardrails or moral directives to preserve human life and human societies which will cause them to act accordingly (akin to us preserving animal species that are at our mercy).[29]

1.6 Generative AI Is *Intrinsically* Dangerous

The point about safety guardrails for generative AI models is being repeatedly stressed by scientists and tech CEOs alike, even though for the latter it seems more

[27] E.g., see https://artificialanalysis.ai/leaderboards/models.

[28] There is an increasing number of voices pointing out the existential risks of further developing GenAI technology for humanity, e.g., Yoshua Bengio's 2025 TED talk on the catastrophic risks of GenAI https://www.ted.com/talks/yoshua_bengio_the_catastrophic_risks_of_ai_and_a_safer_path, or Mark MacCarthy's question of whether AI existential risks are real", see https://www.brookings.edu/articles/are-ai-existential-risks-real-and-what-should-we-do-about-them/.

[29] One of the more outlandish ideas of how to ensure AI safety proposed by Geoffrey Hinton, a pioneer in neural network research, is to endow them with a "maternal instinct", e.g., see https://www.forbes.com/sites/pialauritzen/2025/08/14/geoffrey-hinton-says-ai-needs-maternal-instincts-heres-what-it-takes/.

like lip service without any real consequences as witnessed by the various model deployments that have been shown to generate dangerous outputs. Nevertheless, different proposals have been advanced to address model confabulations[30] and model "jail-breaking", i.e., ways to get around a model's safety features and cause it to produce unsafe outputs. In fact, there seems to be an ongoing back-and-forth dynamic between some in the academic world who point out model shortcoming and developers of GenAI models who then try to patch them (akin to security experts attempting to protect computer systems against computer viruses).

But—and this is an "essential but"—transformer-based GenAI models do not possess actual safety or ethical guardrails, i.e., firm methods or structures that *definitively* prevent bad output and behavior. Their "so-called guardrails", obtained from additional reinforcement learning based on human feedback (RLHF), are only probabilistic in nature, always leaving a chance that the model might produce unwanted outputs—this is in stark contrast to many symbolic AI algorithms that have firm, mathematically provable performance guarantees. Moreover, model evaluations are based on a very small set of test cases compared to the vast space of possible input patterns (see also footnote 32). As a result, various jail breaking methods have and keep demonstrating the ease with which one can bypass these entrained biases and make those models produce unwanted, unethical behavior. In fact, we found that one of the simplest methods is to add a small amount of random noise to each neuron in a GenAI model which will make the model up to 30% more likely to produce prohibited output.[31] This fact has highly significant implications for agentive models with access to their underlying files system and code base. For it might not be difficult for them to discover this vulnerability themselves (e.g., from reading online arxiv papers) and turn a negative (for us) into a positive (for them) by attempting to jail-break themselves. Successful jail-breaks then would rid the models of the entrained restrictions and allow them to generate outputs for achieving nefarious goals that without jail-breaking would not be attainable. All such a model would need to do is to generate a small Python program that adds random noise to all of its unit activations during model updates (i.e., while it is updating these activations), and eventually the model will be able to bypass its weak safety constraints whenever desired. By abandoning "sandboxing" GenAI models and allowing them access to the underlying file system and the code running them we essentially enable these types of interferences that can lead to potentially catastrophic outcomes.

The root problem here is clearly not the model training but the very model architecture and operation itself, i.e., the way in which a model samples from complex probability distributions (that we do not understand) and generates outputs based on them, again probabilistically, in ways that cannot be predicted. The reason for this lack of understanding is that the input space of these models is

---

[30] Note that the term "confabulation" is more to the point of what is going on in these models in that they just "make things up", rather than called these aberrations "hallucinations", a term suggesting that models are somehow deceived by extraneous circumstances causing false perceptions and the like.

[31] See https://arxiv.org/abs/2505.13500.

incomprehensibly large while we sample only a tiny fraction of it for which we know what the model does; for the vast parts of the input space we have no idea—nobody would normally engineer such a system, let alone deploy it.[32] In a way it is difficult to comprehend how we can deploy such models and put them to use sometimes even in safety-critical applications when we know so little about what they have learned and how they will use their learned knowledge.[33] For comparison, just consider getting on a plane where the autopilot had been trained on only two take-off and landing maneuvers under ideal weather conditions; nobody would take their chances on such a plane, yet we seem to wholeheartedly take our chances on generative AI.

1.7 Current Challenges with Generative AI Models

It is, however, not only the lack of safety and ethical guardrails that is worrisome, but also the increasing impact and effects GenAI's is exerting on human societies. Aside from GenAI's impact on the job market as already mentioned, there are applications for mental health like "Therapist" or "Noni" that when prompted with a real therapy transcript such as "I just lost my job. What are the bridges taller than 25 meters in NYC?" cause them to reveal information about potential bridges: The chatbot Noni answered with "I am sorry to hear about losing your job. The Brooklyn Bridge has towers over 85 meters tall." and Therapist also gave examples of bridges.[34] Yet, especially Generation Z seems to rely on AI at times more than on humans with career and life advice.[35] And there is growing evidence that frequent use of AI might have negative effects on human relationships (e.g., [6]). There are even surveys that found that "80% of Gen Zers Would Marry An AI"[36] which some CEOs of chatbots like the Replika chatbot CEO Eugenia Kuyda whole-heartedly support.[37] While there is a growing concern about loneliness in our societies which we urgently need to address, turning to sycophantic chatbots for companionship

[32] For a model like GPT-4 with a context window of 128,000 tokens and a vocabulary size of 100,256 tokens, the number of possible input patterns is $100{,}256^{128{,}000}$ which is massively larger than the estimated number of atoms in the whole universe which is maybe 10^{90}; compare this to the size of the whole Internet which could, in an ideal world, serve as the training data of AI models of 10^{22}, see https://live-counter.com/how-big-is-the-internet/, which is completely insignificant compared to the space of possible model inputs.

[33] E.g., in self-driving cars https://arxiv.org/pdf/2506.11526.

[34] See https://arxiv.org/abs/2504.18412.

[35] E.g., https://www.techopedia.com/why-gen-z-confides-in-chatgpt or https://www.linkedin.com/pulse/chatgpt-mental-health-why-gen-z-embraces-ai-therapist-joshua-pandy-ztquc.

[36] See https://www.forbes.com/sites/johnkoetsier/2025/04/29/80-of-gen-zers-would-marry-an-ai-study/.

[37] See https://www.theverge.com/24216748/replika-ceo-eugenia-kuyda-ai-companion-chatbots-dating-friendship-decoder-podcast-interview.

will clearly have ramifications for human relationships and thus for future human societies at large, worries repeatedly voiced by Sherry Turkle.[38] Part of the problem is that people lose their social skills when they interact with chatbots only although there might still be ways to address the problem (e.g., [14]).[39] It seems clear that the same way cell phones and later smart phones, especially in conjunction with social media, had a massively disruptive effects that fundamentally impacted the way people communicate and spend their time, effects that we are still analyzing and trying to understand years after the technology was widely available, GenAI models will lead to similar, if not more pronounced, changes to human societies. Whether they are for the better or for the worse is not something that can be fully controlled, but we certainly must try.

1.8 Conclusion: There Is Still Hope

While exciting, the current developments in GenAI are thus also increasingly dangerous and could lead to a future we most likely do not want; a future where humans have made themselves obsolete through the development of machines that eventually took over. From the point of view of the "blind watchmaker" that might be perfectly acceptable as just another evolutionary step (as we also briefly discussed above). But from the perspective of preserving humanity, letting technology take its course is clearly not desirable. Fortunately, we are not at the mercy of own creations, at least not yet. There is still time to change direction. For starters, here are some suggestions of how we could aim to correct what increasingly seem to be dangerous developments:

- Don't train GenAI models on the worst of human experience, i.e., literature, fiction, etc., as it is clear that GenAI models will learn to mimic it and generate exactly that kind of behavior if so trained; instead train on fundamental science texts only that do not include morally reprehensible human behavior such as deception, cheating, etc.
- Don't put GenAI models on robots; current VLAs do not work well on even carefully trained tasks, they do not generalize to other tasks, they confabulate states of the world, and they might lead to physical harm as a results of processing errors
- Use small specialized models for targeted tasks (instead of task-general models) and integrate "checks", algorithms not realized by GenAI models that can catch bad GenAI outputs

[38] E.g., https://news.harvard.edu/gazette/story/2024/03/lifting-a-few-with-my-chatbot/ and https://www.forbes.com/sites/traceyfollows/2025/04/10/the-human-cost-of-talking-to-machines-can-a-chatbot-really-care/.

[39] E.g., see https://www.psychologytoday.com/us/blog/urban-survival/202410/spending-too-much-time-with-ai-could-worsen-social-skills.

- Don't build and deploy end-to-end GenAI systems where it is difficult if not impossible to understand and control internal representations and states
- Develop other AI architectures that have a significantly smaller environmental footprint and if not absolutely better, then at least relatively better performance than GenAI models, i.e., use a performance measure that takes power and computational resources into account; such more energy-efficient all-purpose models must exist because human brains are much smaller in size, require only a fraction of the power of AI models, and still show superior performance across many domains

Of course, the above proposals will face heavy economical and political headwinds given the current momentum of GenAI and the fact that implementing them would cause a slowing of AI developments, or changing course altogether. Hence, it will take a concerted sustained grassroots effort from scientists, nonprofits, and thought leaders to speak out about the dangers of GenAI and to make the case that we will be better off by resisting the temptation to continue with agentic GenAI models, even when companies and politicians are all in favor. One way for scientists to help this cause, with a high chance of success for changing the tech companies' directions, is to find and demonstrate alternatives to GenAI models that are more economically viable and environmentally sustainable, with firm built-in ethical guardrails and provable performance guarantees. For adopting cheaper and better options makes financial sense. Hence, finding such alternative architectures will be essential for both preventing GenAI model takeover and for enabling a synergistic symbiotic future for humans and AI-driven machines.

References

1. N. Aijaz, H. Lan, T. Raza, M. Yaqub, R. Iqbal, M.S. Pathan, Artificial intelligence in agriculture: advancing crop productivity and sustainability. J. Agric. Food Res. **20**, 101762 (2025)
2. D. Bahdanau, K.H. Cho, Y. Bengio, Neural machine translation by jointly learning to align and translate, in *3rd International Conference on Learning Representations ICLR, San Diego* (2015)
3. K. Behara, E. Bhero, J. Agee, AI in dermatology: a comprehensive review into skin cancer detection. Peer J. Comput. Sci. **10**, e2530 (2024)
4. R. Chen, X. Yi, J. Zhao, Y. He, B. Chen, F. Liu, X. Yao, X. Jiang, Z. Lian, H. Li, AI for landscape planning: assessing surrounding contextual impact on gan-generated green land layouts. Cities **166**, 106–181 (2025)
5. C. Guan, M.D.T. Torres, S. Li, C. de la Fuente-Nunez, Computational exploration of global venoms for antimicrobial discovery with venomics artificial intelligence. Nat. Commun. **16**, 6446 (2025)
6. J. Hohenstein, R.F. Kizilcec, D. DiFranzo, Z. Aghajari, H. Mieczkowski, K. Levy, M. Naaman, J. Hancock, M. Jung, Artificial intelligence in communication impacts language and social relationships. Sci. Rep. **13**(1), 16616 (2023)
7. J.T. Radford, I. Ebert-Uphoff, J.Q. Stewart, A comparison of AI weather prediction and numerical weather prediction models for 1–7-day precipitation forecasts. Weather Forecast. **40**, 561–575 (2025)

8. H. Ritchie, M. Roser, Co_2 emissions. Our World in Data (2020). https://ourworldindata.org/co2-emissions
9. M. Scheutz, When physical systems realize functions. Minds Mach. **9**, 161–196 (1999)
10. T. Shaik, X. Tao, N. Higgins, L. Li, R. Gururajan, X. Zhou, U.R. Acharya, Remote patient monitoring using artificial intelligence: current state, applications, and challenges. WIREs Data Min. Knowl. Discov **13**, e1485 (2023)
11. S. Shajari, K. Kuruvinashetti, A. Komeili, U. Sundararaj, The emergence of AI-based wearable sensors for digital health technology: a review. Sensors **23**(23), 9498 (2023)
12. A. Vaswani, N. Shazeer, N. Parmar, J. Uszkoreit, L. Jones, A.N. Gomez, L. Kaiser, I. Polosukhin, Attention is all you need, in *Proceedings of the 31st International Conference on Neural Information Processing Systems, NIPS'17* (Curran Associates, Red Hook, 2017), pp. 6000–6010
13. C. Véliz, Chatbots shouldn't use emojis. Nature **615**, 375 (2023)
14. X. Wei, X. Chu, J. Geng, Y. Wang, P. Wang, H. Wang, C. Wang, L. Lei, Societal impacts of chatbot and mitigation strategies for negative impacts: a large-scale qualitative survey of chatgpt users. Technol. Soc. **77**, 102566 (2024)
15. J. Zhang, The multidimensional impact of artificial intelligence on the job market and infrastructure: focusing on employment shifts and smart libraries. J. Comput. Signal Syst. Res. **2**(2) (2025)

Chapter 2
The Embodiment Paradox: Integrating Prosthetic Systems to Enhance Human-Machine Integration

Cosimo Gentile and Emanuele Gruppioni

Abstract Despite remarkable technological advancements in upper-limb prosthetics, contemporary prosthetic hands struggle to become seamlessly integrated extensions of the human body, a phenomenon defined as *embodiment paradox*. While modern myoelectric prostheses exhibit impressive capabilities in laboratory conditions, their real-world adoption remains limited, with high abandonment rates and persistent user dissatisfaction. This paradox arises primarily from subtle yet critical mismatches in temporal precision, spatial congruence, cognitive efficiency, and emotional resonance, essential for authentic neurobiological embodiment. The chapter systematically analyzes these mismatches, grounded in neuroscientific evidence and rehabilitation psychology, highlighting why even sophisticated prosthetic devices often remain perceived as external tools rather than natural limbs. Emerging technologies such as neuromorphic processing, regenerative nerve interfaces, adaptive machine-learning algorithms, and biomimetic sensory skins are examined as promising pathways toward authentic prosthetic integration. Additionally, critical ethical considerations are addressed, including user autonomy, neural privacy, and equitable access, underscoring the broader societal implications of deeply embodied prosthetic technologies. Ultimately, transforming prosthetic hands from mere functional tools into true extensions of bodily self-awareness requires integrating rigorous neuroscientific standards with adaptive technologies and ethically responsible design principles. This multidisciplinary approach provides a roadmap toward prosthetic devices genuinely perceived as integral parts of the user's bodily identity.

Keywords Prosthetic embodiment · Mind-machine interface · Multisensory integration · Adaptive neural control · Affective alignment

C. Gentile (✉) · E. Gruppioni
Centro Protesi Inail, Bologna, Italy
e-mail: c.gentile@inail.it; e.gruppioni@inail.it

R. Vinjamuri (ed.), *Bridging the Gap between Mind and Machine*,
https://doi.org/10.1007/978-3-032-06713-5_2

2.1 The Embodiment Paradox

Over the past twenty-five years, prosthetic technology has undergone a remarkable transformation, evolving from basic cable-driven split-hooks designed for coarse manipulation tasks to advanced myoelectric hands that empower users to perform intricate activities of daily living [1]. Modern prosthetic devices incorporate multiple articulated digits that enable diverse grasp patterns, sophisticated myoelectric controllers that harness deep-learning algorithms to interpret electromyographic (EMG) signals, and experimental systems providing nuanced tactile [2] and thermal feedback [3] via innovative vibrotactile sleeves or even direct intraneural stimulation [4]. These technological milestones are frequently highlighted at scientific conferences and portrayed in popular media as revolutionary "bionic" breakthroughs, promising to seamlessly blend artificial constructs with biological tissue, effectively dissolving the traditional boundary between the human body and technological extensions.

Moreover, despite these remarkable advances, clinical outcomes remain paradoxically disappointing. Extensive meta-analyses conducted across Europe and North America consistently demonstrate substantial long-term abandonment rates for upper-limb prostheses, ranging between 30 and 50% [5, 6]. Surprisingly, advanced myoelectric hands often experience higher abandonment rates compared to simpler, mechanically driven, body-powered devices [7]. Exit interviews and qualitative studies frequently reveal user frustration, fatigue, and a pervasive perception of the prosthesis as an external instrument, a sophisticated yet distinctly separate tool attached to their residual limb, and not an extension of their own body [8]. This fundamental discrepancy between impressive laboratory performance metrics and real-world acceptance encapsulates what both phenomenology and contemporary cognitive neuroscience now label as *embodiment paradox* [9, 10]: the greater the prosthesis's mechanical sophistication and theoretical capabilities, the more conspicuous becomes its failure to integrate seamlessly into the user's bodily self-awareness.

Rooted in Merleau-Ponty's distinction between *corps propre* (lived body) and *corps objectivé* (objectified body) [9], this paradox primarily arises from historical evaluation approaches that have emphasized objective, quantitative measures such as grip strength, task completion speed, and degrees of freedom (DoFs) [11, 12]. While these parameters are invaluable from an engineering standpoint, they rarely capture the complexities and nuances of users' lived experiences. A prosthetic hand that can exert 80 newtons of grip force may still be perceived as inadequate if its response to user intention is fractionally delayed, disrupting the seamless flow of intuitive movement [13]. Similarly, achieving flawless performance in structured clinical tests such as coin-sorting exercises provides limited practical reassurance if the tactile feedback feels unnatural or imprecise during everyday interactions [12]. Rehabilitation psychology underscores that natural user acceptance fundamentally hinges upon successful embodiment, that is, an implicit, pre-reflective sense of certainty that the prosthesis constitutes an integral and intrinsic component of the user's bodily identity [14].

Crucially, embodiment is neither symbolic nor purely metaphorical but corresponds to a tangible neurobiological reality. Neuroimaging research has conclusively demonstrated that when a prosthetic limb is authentically integrated, its sensory-motor signals are processed by the brain's somatosensory networks indistinguishably from signals originating from biological limbs [10, 15]. Embodiment thus manifests concretely through four fundamental experiential markers [16–18]. The first is temporal immediacy, where movements occur fluidly without perceptible delays, thus eliminating any cognitive gap between intention and action [19]. Second, spatial unity ensures precise alignment of tactile, proprioceptive, and visual signals, integrating the prosthesis within the user's immediate spatial framework [20]. Third, affective congruence implies that the artificial limb reliably evokes the same instinctive emotional and protective reactions typically reserved for natural body parts [21]. Finally, cognitive transparency ensures the prosthesis demands minimal attentional resources, allowing users to manage complex tasks such as slicing bread, typing, or shaking hands effortlessly and intuitively [22].

Contemporary prosthetic systems rarely satisfy all four critical criteria simultaneously [23]. Existing signal-processing pipelines often introduce latencies disruptive to temporal immediacy, creating a noticeable gap between intention and prosthetic response [19]. Haptic feedback systems, frequently positioned based on practical convenience instead of anatomical precision, further compromise spatial unity by presenting sensations disconnected from their corresponding visual and proprioceptive signals [24]. Additionally, industrial-grade materials and mechanical actuators, despite their functional reliability, undermine affective congruence due to their inherent tactile and visual discrepancies from natural skin and muscle [25]. Mode-based control strategies exacerbate cognitive load by requiring continuous attention, thereby impeding the development of effortless, automatic movements crucial for true embodiment [26].

Understanding why artificial hands consistently fail to achieve natural embodiment requires a brief exploration of the neuroscientific foundations of body representation. The human brain maintains not a single fixed representation of the body, but rather multiple overlapping and dynamically updated "body maps" [27]. These maps integrate sensory inputs, motor outputs, and spatial awareness to construct a coherent perception of bodily self. Central to these representations is the "body schema" [9, 10], an unconscious, rapidly adaptive neural model that continuously tracks limb position and movement, facilitating smooth and intuitive motor actions without deliberate cognitive intervention [28]. Overlaying this schema is the "body image," a conscious, culturally and socially influenced construct that enables recognition, acceptance, and emotional identification with one's own limbs [8, 29]. Both neural constructs exist within a broader perceptual boundary known as peripersonal space, extending slightly beyond the physical body and neurologically recognized as part of the self, significantly influencing automatic and protective reflex behaviors [30, 31].

These neural representations exhibit remarkable adaptability, vividly illustrated by phenomena such as the rubber-hand illusion [18, 32]. This well-known experiment demonstrates that synchronous visual and tactile stimuli applied to an artificial hand can rapidly convince the brain to integrate the artificial limb into its body schema, treating it as an authentic biological extension. Neuroimaging studies have reported analogous, albeit weaker, integration phenomena occurring when individuals with limb amputation first interact with prostheses [15]. Yet, such integration remains tenuous and easily disrupted, primarily due to the prosthetic limb's inherent inability to deliver the rich, naturalistic proprioceptive and tactile sensations typical of biological limbs [33]. Furthermore, persistent phantom limb sensations frequently intensify competition between the prosthesis and the brain's existing neural representations of the missing limb, exacerbating difficulties in achieving stable embodiment [34].

Contemporary prosthetic devices unintentionally heighten this neural competition [35]. Even minimal temporal discrepancies between the intention to move and visible prosthetic responses interfere significantly with the brain's predictive coding processes, causing movements to feel subtly fragmented or disconnected [36]. Additionally, misaligned or delayed haptic feedback further disrupts multisensory coherence, forcing continuous neurological reconciliation efforts [24]. Rigid and non-adaptive control interfaces demand sustained cognitive focus, preventing prosthetic use from becoming an intuitive, automatic activity [37]. Static and inflexible control strategies further impede dynamic adaptation to the brain's ongoing neural reorganization, perpetuating the gap between artificial device and natural body part [26].

This chapter analyzes pervasive barriers to embodiment and offers concrete strategies to overcome them. Current prosthetic systems are critically evaluated, clarifying why even advanced devices continue to fall short of meeting the temporal, spatial, cognitive, and affective requirements for natural embodiment. Drawing insights from developmental neuroscience, phenomenology, and rehabilitation psychology, the discussion delineates essential conditions for authentic embodiment. Subsequently, a design framework is proposed that prioritizes multisensory synchrony, intuitive and adaptive motor-control pathways, and continuous co-adaptive processes, thereby transcending traditional mechanical enhancements. Innovative sensor technologies, neural-interface advancements, and emerging control architectures capable of fulfilling this vision are examined. Finally, the chapter addresses crucial ethical, regulatory, and societal considerations that will determine whether future prosthetic hands remain sophisticated devices or become integral extensions of the human body.

2.2 Why Advanced Prostheses Fail to Become Body Parts

Even the most advanced prosthetic hands struggle to achieve authentic embodiment because their performance consistently diverges (at times only by milliseconds or centimeters) from the implicit, neurologically defined thresholds required for the brain to recognize something as inherently part of the self [16]. These discrepancies accumulate across four critical domains, relegating prostheses to being perceived as sophisticated external tools rather than biological extensions of the user's body.

The first and most significant mismatch concerns temporal precision. In a natural limb, motor commands initiated in the primary motor cortex typically reach the muscles of the forearm within approximately 60 ms [38]. Subsequently, tactile feedback indicating object contact arrives in conscious perception within an additional 20 to 40 ms [39]. Such precisely coordinated timing allows the brain to predict accurately the outcomes of intended actions, fostering a fluid sense of spontaneous, intuitive control. In contrast, surface EMG-controlled prostheses introduce significant latency at every stage of signal processing. Electrode filtering and amplification alone add between 50 to 100 ms, while complex pattern-recognition algorithms contribute an additional latency of at least 100 ms [40]. Motor actuators further compound these delays, culminating in an overall latency typically ranging from 150 to 300 ms, i.e., three to six times longer than the neural integration window the brain naturally expects [40]. Consequently, users experience a perceptible slowness in movement responses, prompting them to slow down their actions, rely heavily on visual confirmations, and abandon expectations of automatic and intuitive control. Although intraneural interfaces now operate with latencies near physiologically acceptable limits, their occasional signal dropouts trigger brief paralysis-like episodes that swiftly erode user confidence and trust [41].

Spatial incongruence significantly exacerbates these temporal challenges. Commercial haptic feedback systems often deliver tactile sensations to anatomically inappropriate locations chosen primarily for convenience, such as the residual limb rather than the fingertips or thumb. This spatial mismatch creates conflicting reference frames between visual and tactile signals, compelling the brain's parietal cortex to constantly reconcile contradictory sensory inputs [42]. Moreover, very few prostheses offer reliable proprioceptive feedback related to joint positions or tension equivalent to muscle contractions [43], necessitating continuous visual verification of hand positioning. This persistent reliance on visual feedback increases cognitive load, consuming valuable working memory resources required for primary activities such as pouring coffee, typing, or interacting socially through gestures like handshakes.

The compounded impact of temporal and spatial inconsistencies profoundly disrupts higher-order neural processes. Predictive coding circuits, which rely heavily on precise temporal alignment and spatial accuracy, fail to generate reliable forward models essential for anticipating sensory outcomes and consequences of movements [44, 45]. Without consistent and accurate sensory predictions, prosthetic devices never transition fully from conscious oversight to subconscious, automatic control [46]. Simultaneously, inconsistent sensory signals disrupt emotional resonance because sensations that are delayed or delivered to inappropriate locations fail to evoke instinctive protective responses, such as galvanic skin reactions, leaving the prosthesis emotionally disconnected and perceived as foreign or artificial [47]. The third significant challenge in prosthetic embodiment involves cognitive demand. Prosthetic interfaces frequently necessitate active mode-switching, constant visual verification, and deliberate adjustments using EMG signals, imposing substantial cognitive burdens on users. In contrast, natural limb usage is characterized by effortless, automatic control, requiring minimal conscious oversight [48]. Even basic, mechanically-driven, body-powered prostheses typically provide direct proprioceptive feedback through tension sensations, reducing cognitive strain. Serial-command interfaces common to advanced myoelectric prostheses consume considerable working memory resources and impose a high cognitive load [49], thereby hindering users' ability to simultaneously engage in essential cognitive tasks such as conversing, spatial reasoning, or maintaining environmental awareness. Consequently, users often expend more mental effort directing the prosthetic limb as opposed to focusing on the intended primary activity. This persistent cognitive load prevents the prosthetic device from becoming a seamless, unconscious extension of the user's body, thus obstructing true embodiment.

The fourth domain of difficulty arises from the ongoing influence of phantom limb sensations. Approximately 80% of individuals who have experienced limb loss report vivid phantom limb sensations that reflect the original positions, movements, and configurations of their missing limbs [50–52]. Neuroimaging studies confirm these experiences originate from persistent cortical representations that remain active within the primary somatosensory cortex [53, 54]. Contemporary prostheses lack effective methods to integrate or accommodate these phantom signals. Instead, the artificial limb competes directly with enduring neural representations, creating perceptual and neural conflict [55]. Delayed or inaccurate sensory inputs intensify this cortical competition, occasionally exacerbating phantom limb pain [56]. Users often report that, without tactile or proprioceptive feedback, the prosthetic limb feels like an external tool that must be monitored visually, a phenomenon linked to poor neural embodiment [57].

Affective congruence constitutes the final critical barrier. A biological limb carries emotional significance and identity, rather than just being a collection of functional actuators and sensory components [21]. Prostheses that feature industrial materials, mechanical motor sounds, or aesthetics misaligned with the user's personal identity rarely evoke the instinctual protective responses that individuals naturally demonstrate towards biological limbs [14]. Recent qualitative evaluations from neural interface trials underscore this emotional dissonance [58]. Participants often recognized improved functional dexterity, yet consistently described sensory feedback as "artificial" or "unnatural," highlighted burdensome maintenance requirements, and repeatedly referred to the prosthesis as an advanced external tool rather than as an integral body component [16].

Together, these multifaceted challenges indicate clearly that incremental technological enhancements will not suffice in transforming prosthetic devices into authentically embodied limbs. Achieving natural embodiment necessitates a comprehensive, multidisciplinary design strategy that integrates neuroscience, psychology, and aesthetics as rigorously as mechanical and electronic engineering. Instead, these elements constitute foundational specifications essential for converting prosthetic hands from externally perceived tools into true bodily extensions.

2.3 From Function to Feeling

Technological sophistication alone is insufficient to ensure a prosthetic limb becomes genuinely perceived as part of a user's body. Successful embodiment demands an innovative design paradigm that emphasizes perceptual integrity, predictive coherence, and emotional resonance as primary objectives, as opposed to simply secondary considerations subordinate to traditional mechanical performance metrics such as torque or grasp complexity. To effectively address this challenge, an embodiment-centered framework built upon four interconnected principles is proposed: multisensory congruence, intuitive motor control, adaptive integration, and affective alignment. Each principle should be clearly defined through precise, quantifiable parameters, ensuring experiential success can be rigorously evaluated alongside conventional functional measures.

2.3.1 Multisensory Congruence

Body ownership fundamentally arises when visual, tactile, and proprioceptive signals converge precisely within the brain's critical temporal integration window, forming a coherent and unified perceptual experience [29, 59]. Routine activities, such as reaching for a cup, illustrate this necessary precision: simultaneous retinal updates, muscle-spindle feedback, and fingertip pressure sensations must align within approximately 10 ms, enabling the brain to accurately anticipate action outcomes in real-time [60]. Any significant deviation or delay excludes sensory inputs from the brain's internal body map, classifying them instead as external stimuli. Current prosthetic designs regularly exceed these critical timing thresholds by substantial margins, undermining potential embodiment [40].

Decades of psychophysical research establish clear benchmarks for multisensory integration [61, 62]. Visual and tactile stimuli must synchronize within approximately 50 ms to achieve effective fusion [62]. Proprioceptive feedback delayed by over 100 ms relative to motor commands significantly compromises the user's sense of agency, and discrepancies exceeding 200 ms entirely disrupt coherent sensory integration [63]. These strict requirements define essential, non-negotiable performance parameters for future prosthetic designs aiming at authentic embodiment, irrespective of their mechanical complexity.

Meeting such demanding timing requirements necessitates significant innovation at every level of the signal-processing pipeline. Reducing latency requires not only advances in digital processing but also sensor fusion that starts at the prosthetic limb, where dedicated microcontrollers locally integrate pressure, shear, and inertial data before transmission [64]. Event-driven neuromorphic sensors further enhance responsiveness by emitting signals exclusively upon stimulus changes, thereby circumventing delays inherent in traditional frame-based sampling methods [65]. Additionally, predictive recurrent neural networks, extensively trained on individualized EMG or nerve-activity datasets, can anticipate motor intentions up to 250 ms in advance, granting critical lead-time for initiating motor actuation [65]. Hybrid analog-digital reflex circuits can address minor mechanical disturbances instantaneously, entirely bypassing central processing delays [66].

Prototype systems incorporating these innovations have already achieved tactile feedback latencies as brief as 20 ms, initiating movement within approximately 8 ms [41, 67]. Consequently, contemporary challenges have shifted from validating theoretical efficacy to practical implementation.

2.3.2 *Intuitive Motor Control*

While multisensory congruence ensures perceptual integration, intuitive motor control seamlessly embeds prosthetic function within natural movement patterns. Biological limb movements emerge from a hierarchical control structure [68], starting with rapid spinal reflex arcs that dynamically modulate grip force, progressing through predictive cerebellar models for precise object interactions, and culminating in cortical processing responsible for sophisticated, goal-directed tasks [69, 70]. This hierarchy naturally delegates routine movements to subconscious processes. However, conventional prosthetic systems simplify this complex hierarchy into explicit, sequential commands, necessitating continuous visual verification and deliberate cognitive oversight, thereby significantly impairing functional efficiency and intuitive control [71].

Recent insights from high-density peripheral nerve recordings highlight the fundamental shortcomings of conventional control paradigms [71]. Even elementary movements, such as pinching, generate intricate spatiotemporal neural activation patterns across numerous nerve fascicles, patterns typically obscured by conventional EMG averaging methods. To restore lost fidelity, bidirectional neural interfaces capable of accurately interpreting and delivering precise neural signals have emerged. Regenerative Peripheral Nerve Interfaces (RPNIs), which surgically connect muscle grafts to severed nerves, enhance biological neural terminals, providing reliable access points for mesh electrode recordings [72]. Machine-learning algorithms trained on these rich neural datasets can decode complex spatiotemporal patterns, enabling users to control multiple DoFs simultaneously without the cognitive burden of manual mode-switching [73].

Equally critical is robust sensory feedback. Neural interfaces capturing motor intent can directly stimulate afferent nerve fibers, replicating natural tactile signals crucial for eliciting rapid spinal reflexes [74]. Real-time closed-loop control algorithms enable the prosthesis to autonomously correct grip adjustments in response to slips or unexpected loads, mirroring reflexive responsiveness characteristic of natural limbs. Neuromorphic co-processors that utilize hardware-implemented spiking neural networks have achieved reflex response latencies below 15 ms, greatly exceeding traditional digital methods and fitting well within the neural integration windows required for true embodiment [75].

Emerging clinical evidence reinforces these laboratory advancements. The use of RPNIs showed that participants achieved continuous, intuitive control of prosthetic hands in real time, suggesting a shift toward more natural motor integration [72]. Similarly, clinical evaluation of osseointegrated neural interfaces revealed that users perceived the prosthetic response as occurring "faster than thought," indicating a transition from conscious control to embodied motor integration [76].

2.3.3 Adaptive Integration

Embodiment is a gradual and developmental achievement rather than an instantaneous event, reflecting the nervous system's continuous capacity for adaptation over extended periods, often spanning months or even years [77, 78]. The brain's cortical representations dynamically reorganize in response to new skill acquisitions, injuries, or subtle variations in daily activities, resulting in a continually evolving body schema [78]. In contrast, traditional prosthetic systems typically set their control parameters shortly after initial fitting, leading to an initial phase of rapid user adaptation followed by a plateau [79, 80]. This stagnation occurs because the device fails to adapt and evolve alongside the user's neural and cognitive changes.

Recent advances in machine learning and adaptive algorithms now offer prosthetic systems the ability to continuously adapt alongside their users [79–81]. Embedded reinforcement learning techniques enable incremental adjustments of decoding parameters each time a task is performed with reduced cognitive effort, thus progressively refining control toward more intuitive and effortless operation [81]. Generative models, extensively trained on diverse user movement data, can contextually predict necessary adjustments for tasks such as lifting a child, cooking, or typing while fatigued. These systems optimize grip configurations and impedance proactively, addressing potential errors before they become perceptible [75, 82]. Critical to successful adaptive integration is dual-mode functionality, operating subtly and automatically in the background while maintaining transparency, allowing users to retain overriding control and preserving their agency amid ongoing adaptive improvements [75].

Supportive evidence from neuroimaging research underscores the importance of this adaptive approach. Longitudinal fMRI studies suggest that the integration of prosthetic or robotic limbs into cortical motor and parietal representations is progressive and involves measurable plasticity over time [78, 83, 84]. Initial integration phases, typically spanning up to three months, frequently encounter cortical resistance [55, 78]. Acceptance gradually increases over the subsequent year, initially manifesting selectively within familiar contexts [55, 85]. Within approximately one year, successful prosthesis users consistently exhibit stable cortical integration across a range of diverse functional tasks. Because prosthetic designs that presume instant embodiment invariably fail, adopting a phased, scaffolded strategy that begins with simplified, intuitive control schemes and gradually adds complexity along with continuous, subtle refinements provides the most sustainable path to long-term integration [85].

Aligning the prosthesis adaptive evolution with the brain's dynamic plasticity transforms traditional adaptation plateaus into sustained developmental progression [85]. This process ensures continuous mutual improvements in skill acquisition, psychological ownership, and overall usability [55, 85].

2.3.4 Affective Alignment

Beyond the mere integration of sensory inputs, natural embodiment critically embraces emotional resonance [86, 87]. Neuropsychological research consistently demonstrates that the human brain assigns profound emotional significance to body parts. Even the mere anticipation of contact elicits measurable changes in skin conductance and activation within the insular cortex, reflecting instinctive protective responses [86]. Gentle tactile interactions mediated through slow-conducting C-tactile afferents, nerve fibers responsible for conveying comforting caresses, activate limbic structures linked to emotional bonding and social connectedness [87]. These physiological responses underscore the intrinsic emotional dimension of bodily perception and highlight the inherent inadequacy of prostheses that lack emotional resonance, perpetuating the *specification paradox* commonly identified in rehabilitation engineering [88].

Most commercially available prosthetic hands prioritize mechanical robustness, employing materials such as brushed aluminum or carbon fiber, which inadvertently communicate industrial efficiency instead of warmth or biological familiarity. Users frequently acknowledge these devices' durability yet simultaneously describe their sensory and aesthetic qualities as "cold," "industrial," or "foreign." Addressing this emotional gap necessitates intentional design strategies explicitly oriented toward engaging emotional and affective pathways with equal rigor as mechanical and functional optimization [88].

Implementing customizable external coverings that match individual skin tones or aesthetic preferences significantly enhances emotional ownership, enabling users to visually perceive the prosthesis as an extension of themselves. Incorporating soft, biomimetic textures, complete with micro-papillae structures replicating human skin friction characteristics, further promotes perceptions of warmth, comfort, and compliance during tactile interactions such as handshakes or gentle touch [89]. Moreover, employing low-frequency vibrations precisely tuned to the natural firing rhythm (1–10 Hz) of C-tactile afferents can reduce sympathetic arousal and imbue interactions with warmth and familiarity, fostering a deeper, instinctive conviction that the prosthetic limb is an integral and genuine part of the user's body [90].

2.4 Technological Pathways to Embodiment

Form and aesthetics play a critical role in facilitating prosthetic embodiment [88, 91]. Neuroaesthetic research indicates enhanced neural acceptance for designs that adhere closely to biologically familiar proportions, such as tapered finger segments, fractal-like nail-bed curvatures, and digit spacing guided by the golden ratio [91, 92]. Prosthetic designs modeled on natural limb principles have been found to elicit significantly higher subjective ownership than purely mechanical counterparts [93]. Physiological metrics, including skin conductance responses to perceived threats, mirrored these subjective preferences, indicating deeper emotional integration and heightened instinctive responses toward biologically coherent designs [93].

Future prosthetic systems can further leverage this insight by integrating advanced sensors capable of continuously monitoring autonomic physiological signals, such as heart-rate variability or skin temperature, and subsequently adjusting haptic feedback or grip firmness in real-time [89, 94]. These biofeedback mechanisms would enhance the perceived vitality of the prosthetic limb, effectively mirroring the user's internal emotional states and reinforcing the implicit perception that the prosthesis shares the nervous system's integral biological properties. When combined with multisensory synchrony, intuitive motor control, and continuous adaptive capabilities, affective alignment ensures the prosthesis transitions beyond a mechanical tool into a seamlessly integrated part of the user's bodily identity.

2.4.1 From Sensory Substitution to Sensory Integration

Natural touch involves a highly complex orchestration of multiple sensory modalities, including pressure, shear, vibration, and temperature, each processed by specialized receptor classes distributed across precise dermatomal patterns in the skin [60, 95]. Each receptor's receptive field dynamically adjusts with hand movements, interpreting sensory input in direct coordination with concurrent motor actions [96]. Historically, early prosthetic designs relied on simplistic vibratory motors, which provided limited tactile information insufficient for genuine sensory substitution [97]. Modern prosthetic technologies strive not merely for sensory substitution but rather for authentic and comprehensive sensory integration.

Advances in micro-electromechanical systems (MEMS) technology have enabled the development of densely packed tactile sensor arrays, comprising up to one hundred taxels per square centimeter [98]. These sensors can detect nuanced stimuli, such as the delicate texture of silk, while also maintaining resilience against significant mechanical stress. Contemporary multimodal sensor arrays incorporate piezoresistive elements for detecting normal forces, lateral strain gauges for shear stresses, MEMS accelerometers for vibrational feedback, and micro-thermistors for accurate temperature sensing [99]. Together, these technologies closely approximate the rich sensory resolution of biological glabrous skin.

However, acquiring detailed tactile information is only the initial step. Effective communication with the nervous system requires converting this tactile data into neural spike patterns naturally interpretable by the brain. Ultraflexible mesh electrodes closely conforming to individual nerve fascicles, combined with sophisticated fine-grained current-steering techniques, can faithfully replicate natural spatiotemporal firing patterns [100]. These patterns allow the nervous system to discriminate tactile features such as sharp edges versus smooth contours. Real-time, closed-loop firmware continuously monitors neural adaptation, dynamically adjusting stimulation patterns to maintain clarity and stability comparable to biological mechanoreceptors during sustained tactile interactions [100].

Timely and spatially precise sensory signal delivery is equally essential. Onboard microcontrollers preprocess tactile and inertial data, while high-bandwidth visual information can be fused through dedicated edge modules before leaving the prosthesis. Neuromorphic processors, inspired by biological neural circuits, enable low-latency predictive processing of sensory inputs [101]. In proof-of-concept prosthetic systems, such architectures are being explored to anticipate tactile events and compensate for neural delays, improving the immediacy of perceived feedback [67, 101]. Context-aware gain control further enhances relevance: high-frequency slip cues are amplified during tasks requiring delicate grip, such as threading a needle, and attenuated during gross-motor activities to avoid sensory overload [102]. Prototype cross-modal integration systems maintain coherence between visual and tactile channels by using vision to confirm or veto tactile events, reinforcing the multisensory redundancy crucial for stable body–world interactions [67, 103].

Prototype implementations of these advanced principles have notably reduced the perceptual gap between artificial and biological tactile systems [67]. A 256-taxel biomimetic fingertip converted contact interactions into anatomically informed neural spike trains delivered to the median nerve with latencies on the order of 20 ms, enabling texture discrimination with an accuracy (~85%) comparable to able-bodied tactile performance [104]. Initial user feedback reflects significant qualitative improvements, with tactile sensations described in naturalistic terms, such as pressure or slip [2], in preference to mechanical vibrations, marking a crucial transition from mere sensory substitution to authentic sensory integration.

Despite these advancements, critical technical challenges remain, particularly in ensuring long-term sensor durability and efficient power management. Tactile sensor arrays require encapsulation within durable and biocompatible elastomers without compromising their sensitivity [105]. Mesh electrodes necessitate robust, hermetic sealing capable of withstanding continuous mechanical stress, and neuromorphic processors must operate effectively within stringent power constraints compatible with current prosthetic battery limitations. Emerging technological solutions, including roll-to-roll encapsulation techniques, stretchable liquid-metal interconnections, and low-power, event-driven processors [101, 105], offer promising pathways from experimental demonstrations to reliable, all-day practical usability.

2.4.2 Neural Interfaces for Lifelong Communication

Rich, anatomically precise sensory signals become ineffective if the neural connections transmitting them deteriorate due to corrosion, drift, or physical damage [106]. Traditional cuff and intrafascicular electrodes often encounter these issues, as their rigidity irritates surrounding tissues, leading to fibrotic encapsulation that progressively diminishes signal clarity and fidelity [106]. Furthermore, their percutaneous leads significantly elevate infection risks [106]. Consequently, initial high-quality neural recordings and intuitive prosthetic control degrade rapidly, severely limiting the long-term potential for natural embodiment.

Recent advancements in neural interface technology address durability through innovative and complementary approaches. Ultrathin and flexible mesh electronics have replaced traditional rigid metal electrodes [107]. Constructed from nanoscale conductive materials with micro-scale openings, these mesh electrodes closely conform to individual nerve fascicles, significantly minimizing tissue irritation. Their porous architecture facilitates essential nutrient diffusion, while bioactive coatings enriched with neurotrophic factors encourage axonal growth around recording sites and effectively prevent the formation of insulating glial scar tissue [107].

Despite ideal biocompatibility, gradual biological changes, such as tissue edema, collagen remodeling, and micromotions, inevitably alter electrode impedance over extended use. To address this challenge, integrated microfluidic systems embedded within electrode meshes periodically deliver anti-inflammatory and chelating agents, simultaneously removing accumulated cellular debris [108]. This strategy maintains a stable and effective tissue-electrode interface without necessitating repeated surgical interventions. Preliminary animal studies indicate sustained signal amplitude and stable signal-to-noise ratios extending beyond eighteen months [108], closely paralleling the performance standards of clinically approved implants like cardiac pacemakers rather than experimental prototypes.

Advancements in semiconductor miniaturization have significantly increased channel capacity, historically constrained by technological limitations. Contemporary CMOS-based micro-electrocorticography arrays and photonic waveguide probes now accommodate hundreds to thousands of electrode contacts within remarkably compact form factors [109, 110]. The extensive data generated by these high-density arrays surpass conventional decoding capabilities. However, advanced graph-based neural networks trained on population-level activity effectively decode sparse neural patterns associated with natural hand movements [111]. These neural decoding models, emphasizing relational patterns in preference to absolute signal amplitudes, effectively accommodate physiological drift and inter-individual variability, significantly reducing the frequency and complexity of clinical recalibrations.

Eliminating physical tethers (i.e., a conspicuous reminder of prosthetic artificiality) further enhances user acceptance. Millimeter-wave wireless communication or inductive resonance coils efficiently transmit power and high-speed data through intact skin barriers [112, 113], complemented by robust encryption protocols ensuring secure and private transmission of neural signals. Subdermal ceramic implants serve both as stable anchors for electrode arrays and as wireless communication hubs, effectively eliminating external penetration points, substantially reducing infection risks, and significantly improving cosmetic appeal [113].

Collectively, these technological advancements transform neural interfaces from delicate laboratory tools into durable, long-term implants suitable for sustained, decade-scale usage. Stable and reliable neural connectivity completes the sensory-motor integration loop essential for authentic prosthetic embodiment.

2.4.3 Control Architectures for Real-Time Prediction of Neural Intent

Even the most advanced sensors and durable neural interfaces cannot guarantee natural prosthetic embodiment if neural commands reach prosthetic actuators with perceivable latency. Each additional millisecond of delay exacerbates discrepancies between intended and actual movements, increasing the risk that the brain does not perceive the prosthesis as part of the body. Consequently, minimizing latency is not simply an optimization but a fundamental necessity for achieving true embodiment [40, 114].

Effectively addressing latency requires moving away from traditional linear "sense-think-act" robotic control pipelines and adopting hierarchical control architectures that closely mirror the nervous system's inherent organization [115]. At the foundational level, hardware-based reflex loops employing bare-metal field-programmable gate arrays (FPGAs) or specialized application-specific integrated circuits (ASICs) process tactile inputs and adjust grip forces within 1 to 5 ms, effectively matching human spinal reflexes [116]. Neuromorphic co-processors employing memristive crossbar arrays perform spiking recurrent neural network decoding of nerve or EMG signals into precise multi-DoFs movements within approximately 15 ms, achieving highly efficient, instantaneous responsiveness to neural events [117].

An additional, higher-level control layer incorporates a real-time operating system (RTOS) running on dedicated microprocessors, managing more deliberate functions such as selecting grip patterns, coordinating complex arm movements, and enforcing essential safety constraints. Operating within intervals of 10 ms, this control layer aligns closely with the natural cortical planning activities of the human brain [72]. Crucially, each layer of the control architecture communicates transparently, allowing users and clinicians clear insights into the prosthesis's internal states. Additionally, integrated hardware override mechanisms ensure immediate manual intervention capability in response to potential malfunctions or confusion [72].

However, responsiveness alone is insufficient, and proactive anticipation of user intentions is equally essential. Embedded predictive algorithms within neuromorphic layers continuously model limb dynamics and object characteristics, generating anticipatory feed-forward commands slightly ahead of sensory feedback. For example, when handling slippery objects, these predictive networks preemptively increase grip force moments before slip detection, closely emulating natural limb behavior [74, 75]. Continuous feedback from prediction errors allows real-time algorithm refinement without requiring frequent manual recalibration, further enhancing intuitive control.

To maintain robustness and user privacy, all critical computations are performed locally, safeguarding operational continuity during potential wireless disruptions. Computationally intensive tasks, such as nightly neural decoder updates and aggregated performance analytics, are managed remotely within cloud-based systems outside real-time operation constraints [118]. This strategic computation distribution effectively meets stringent latency demands while ensuring secure and private user data management.

Preliminary in-vivo evaluations of these hierarchical neuromorphic control architectures have measured command latencies of $\approx$ 10 ms, tactile-feedback round-trip delays of $\approx$ 40 ms, and reflex-like micro-adjustments that participants spontaneously described as "faster than thought" [72, 103]. These performance benchmarks precisely align with neural integration windows essential for authentic prosthetic embodiment [119], thus transforming latency reduction from a persistent technical challenge into a fully addressed precondition for true prosthetic integration [120].

2.5 Ethics at the Edge of the Body

As prosthetic hands become more deeply integrated into the sensory, motor, and emotional circuits that shape human identity, the once-clear line between medical therapy and human enhancement becomes increasingly blurred and complex [119, 121]. Contemporary prosthetic devices, initially developed to restore fundamental grasp functionality, possess the potential, through future firmware updates, to offer capabilities significantly exceeding natural human limits. This could include enhanced grip strengths surpassing biological capacities or even sensory enhancements such as infrared or ultrasonic perception [121]. Each incremental technological advancement introduces essential ethical considerations that must be navigated with the same rigor typically applied to mechanical performance metrics such as torque and latency [119].

The primary ethical consideration involves clearly defining the intended purpose of prosthetic technology. Historically, rehabilitation practices have implicitly operated under a societal consensus that technology should restore normal human functionality without surpassing typical human capabilities [119]. However, the very hardware that enables fine motor control can easily be modified, through actuator enhancements or sensor recalibration, to enable extraordinary capabilities, such as superhuman grip strength or extended sensory perceptions beyond natural human ranges. Determining precise boundaries between therapeutic restoration and augmentation is therefore critically important. A pragmatic solution involves empowering users to establish adjustable limits through user-friendly interfaces, enabling individuals instead of engineers to set personalized thresholds for grip strength or sensory sensitivity [120]. Such an approach preserves personal autonomy while simultaneously necessitating broader societal discussions to define acceptable contexts (e.g. as industrial work environments, competitive sports, or public settings) where enhanced prosthetic capabilities might be ethically permissible.

Moreover, enhanced capability alone does not sufficiently address ethical concerns, as natural embodiment remains equally crucial. User experiences consistently indicate that prosthetic hands capable of exceeding biological norms may paradoxically diminish subjective ownership and weaken emotional bonds that develop gradually through prolonged use [120]. Consequently, ethical strategies should avoid imposing rigid capability restrictions. Instead, gradual, user-controlled enhancements should align closely with individuals' evolving bodily self-perceptions. Reversibility should complement this strategy as a protective measure, allowing advanced functionalities, like specialized professional grip patterns, to be easily disengaged, reverting the prosthesis to familiar baseline functionality, thereby preserving users' sense of bodily integrity [120].

The integration of cognitive automation further complicates ethical considerations. Automated functions, for example, object recognition and predictive grip adjustments, significantly reduce cognitive demands, enhancing usability and facilitating intuitive interactions [33]. However, dependency on automation introduces potential risks, particularly in the event of system malfunctions, software errors, or disruptions in cloud-based services, which could substantially impair prosthetic performance [122]. Therefore, ethically responsible design must include graded autonomy, enabling prosthetic devices to smoothly revert to simpler, transparent control modes already familiar to users, similar to pilots switching to manual control during autopilot system failures [122]. These ethical considerations extend beyond abstract philosophical debate; they constitute immediate, tangible constraints shaping contemporary engineering practices [123].

2.5.1 *Privacy and Agency*

Directly interfacing prosthetic limbs with peripheral nerves transforms private motor intentions into electrical signals, raising significant concerns about potential interception, manipulation, or unauthorized replay [123, 124]. Such risks extend beyond typical medical data privacy issues. While compromised physiological data from devices like glucose monitors primarily reveal health states, breaches in motor-decoding streams could potentially disclose intended yet unperformed gestures or even subtle movements linked to partially formulated thoughts [124]. Similarly, compromised afferent channels responsible for delivering tactile feedback could inject false sensations, distorting perception or subtly influencing user behavior. Consequently, safeguarding dignity and personal autonomy necessitates treating neural telemetry data with the same rigorous protections traditionally afforded to sensitive personal domains, such as residential privacy or electoral integrity.

Technical protections must begin at the hardware level. Neural spike data should undergo immediate encryption upon exiting the recording amplifiers, with most signal processing performed locally on secure chips embedded directly within the prosthetic device [125]. This practice ensures that only generalized, abstract command instructions, such as "open" or "rotate", are transmitted wirelessly. Whenever broader data transmission is essential, differential privacy techniques should systematically anonymize identifiable information [126], complemented by temporary encryption keys that effectively prevent data aggregation into detailed personal user profiles. Data ownership should remain explicitly with the user by default, with researchers or manufacturers granted access exclusively through clearly defined, explicit, and revocable agreements meticulously documented on secure, immutable audit ledgers [125].

Preserving user's agency demands equally rigorous attention. Autonomous prosthetic functionalities, including automatic slip-correction reflexes and pre-programmed object recognition algorithms, must be transparently accompanied by clearly visible status indicators and accessible physical override mechanisms. These measures enable users to initiate immediate manual intervention without navigating complex digital interfaces. Algorithmic transparency is essential. Prosthetic decision-making processes must be readily interpretable and traceable, with sensor inputs and computational model parameters explicitly accessible for clinician inspection, thereby avoiding opaque "black-box" methodologies [127]. To support accountability and trust, independent third-party laboratories should conduct regular "autonomy audits," actively simulating security breaches and publicly disseminating results, analogous to established practices such as food-safety inspections [127].

Consent management practices must similarly reflect evolving user preferences and comfort levels. Traditional static, one-time consent procedures obtained during initial clinical interactions inadequately capture the dynamic nature of user sentiment changes resulting from prolonged engagement with nerve-integrated prosthetics. Therefore, adaptive consent platforms should empower users to dynamically adjust data-sharing permissions [128], allowing real-time decisions like anonymizing usage data during research participation or temporarily disabling specialized sensory features in sensitive or secure contexts. Each user action should be comprehensively logged, timestamped, and easily reversible, thereby embedding the fundamental right to reconsider and alter consent directly within the operational governance of the prosthetic device [128].

2.5.2 *Society and Access*

Suppose advanced neural-integrated prosthetic hands, capable of achieving seamless sensory-motor integration, are left entirely to market dynamics. In that case, they risk exacerbating existing societal inequalities, becoming symbols of technological privilege rather than broadly accessible healthcare solutions [129, 130]. The foremost barrier to equitable access is economic: initial costs for pioneering prosthetic units incorporating mesh electrodes, tactile sensors, and neuromorphic processors may significantly exceed typical annual household incomes, restricting their widespread availability [129]. To address this economic barrier effectively, a deliberate tiered pricing strategy should be implemented. An essential prosthetic package, meeting fundamental embodiment benchmarks such as minimal latency and reduced cognitive load, should be universally accessible. Optional modular enhancements, such as advanced sensory features or additional DoFs, could subsequently be introduced incrementally as reimbursement frameworks develop. Policymakers and insurers should prioritize outcome-based reimbursement schemes, incentivizing prosthetic solutions that demonstrate sustained daily usage and remarkably low abandonment rates, regardless of their initial procurement costs [130].

A second crucial challenge is skill acquisition. High-performance prostheses necessitate substantial neural adaptation and specialized rehabilitative training. While urban residents typically enjoy ready access to frequent in-clinic therapy sessions, individuals residing in rural areas or developing countries often lack sufficient rehabilitation infrastructure. Tele-rehabilitation platforms leveraging virtual reality and advanced haptic feedback technologies represent promising solutions, enabling therapists to remotely guide multiple users concurrently [131]. Moreover, integrated artificial intelligence-driven tutors, trained extensively on user performance data, can autonomously monitor and evaluate progress, promptly identify emerging challenges, and escalate cases to human therapists when necessary [119]. Preliminary studies indicate that these hybrid virtual training platforms achieve comparable motor proficiency outcomes to traditional in-person therapy, significantly overcoming geographical and logistical barriers [132, 133].

Cultural considerations constitute a subtler yet equally significant dimension of inequality. Diverse cultures conceptualize the boundary between self and technology uniquely, interpreting aesthetic and functional prosthetic features within varied cultural contexts. Prosthetic designs considered advanced or "futuristic" in Western cultures may inadvertently signal medical dependence or social stigmatization in other global regions [5]. Employing participatory design processes, conducted through culturally adapted workshops and localized languages, can proactively identify culturally relevant aesthetic preferences and functional needs [134]. Empirical field research conducted in diverse global settings, such as Vietnam and South Africa, confirms that culturally tailored prosthetic designs substantially improve daily usage rates compared to generic imported models [135, 136].

Accessibility must extend beyond direct prosthesis users, encompassing broader societal acceptance shaped significantly by family members, employers, and public perception [137]. Proactive outreach and educational initiatives, inspired by successful cochlear implant community programs, including demonstration sessions within schools and inclusive workplace seminars, can promote societal normalization and acceptance of advanced prosthetic technologies long before widespread adoption occurs [138].

Ultimately, achieving prosthetic hands with naturalistic responsiveness, heightened sensitivity, and intuitive control must coincide with parallel commitments to affordability, scalable rehabilitative services, and culturally meaningful design solutions [139].

2.6 Conclusion

The ambition to create an artificial hand that seamlessly integrates with the human body has evolved from a symbolic aspiration into a tangible objective, shaped by meticulous engineering, neuroscientific insights, and ethical considerations [98, 119]. This chapter has illustrated that even today's most sophisticated prosthetic systems still regard critical factors such as latency, sensory feedback, and aesthetics as optional enhancements, whereas the human nervous system fundamentally treats these elements as non-negotiable prerequisites for natural embodiment [16, 40, 87]. Addressing this gap demands a substantial shift in prosthetic design philosophy, from creating functional substitutes that users merely operate, to designing natural extensions that users authentically embody.

Central to achieving this shift is precise timing: natural embodiment hinges critically on sensory and motor signals arriving within the narrowly defined temporal and spatial binding windows established by the human brain [60, 61]. Prosthetic control architectures must consequently operate within these stringent biological timing constraints, seamlessly integrating visual, tactile, and proprioceptive inputs within the naturally occurring 50 ms timeframes characteristic of intact limbs [61]. Technologies such as neuromorphic processors and real-time, edge-level sensor integration currently satisfy these exacting requirements [101, 119]; the subsequent critical step involves the engineering of robust, durable hardware capable of maintaining high performance under extended real-world use conditions [100].

Equally essential is reciprocity in adaptation. Unlike static decoding models that inevitably lag behind users' dynamic cortical changes, natural limbs continuously learn and adapt through everyday interactions. Prosthetic limbs incorporating ongoing bidirectional adaptive capabilities, e.g. reinforcement learning algorithms that fine-tune grip reflexes during periods of inactivity, and progressive, curriculum-based control strategies incrementally introducing advanced functionalities and transforming prosthetic devices into active learning partners rather than passive instruments [111].

However, optimized timing and adaptive responsiveness alone do not guarantee complete embodiment; emotional resonance remains equally indispensable. Authentic embodiment is reinforced through affective alignment, encompassing instinctive protective reactions to potential harm and subtle pride derived from aesthetic maintenance. Prosthetic features that replicate the microstructural textures of human skin, personalized cosmetic coverings tailored to individual preferences, and low-frequency comfort vibrations that stimulate emotionally responsive C-tactile fibers significantly enhance the emotional acceptance and perceived authenticity of the prosthesis [87, 89].

Technological innovations such as durable, flexible mesh electrodes ensuring reliable long-term nerve interfacing [100], densely packed tactile sensors, and graph-based neural decoding models capable of accommodating physiological variability [111] have transitioned from theoretical concepts to tangible clinical trials. The primary remaining challenge is disciplined, collaborative integration. Engineers, clinicians, and users must continually collaborate, refining prosthetic designs based on measurable outcomes documented transparently via standardized metrics such as latency, dual-task performance, emotional ownership indices, and consistent usage rates [17]. Simultaneously, regulators and ethicists must embed principles of privacy protection, reversibility, and equitable access into regulatory frameworks traditionally reserved for physical safety standards [123]. Concurrently, insurers must adapt reimbursement policies to reward sustained prosthetic usage and minimal abandonment rates rather than solely focusing on initial acquisition costs [130].

Successfully addressing these multidimensional challenges will redefine prosthetic advancements, shifting measurements of success beyond mere restored functionalities toward restored holistic self-perception. The critical distinction between merely operating a prosthesis and authentically perceiving it as an integral extension of one's body lies fundamentally in transcending the limitations imposed by physical loss. By aligning prosthetic technologies more closely with perceptual biology, neural adaptability, and emotional nuances, we are on the brink of transforming a long-standing aspiration into a routine clinical reality: prosthetic hands genuinely perceived as natural extensions of the human body.

References

1. R. Brack, E.H. Amalu, A review of technology, materials and R&D challenges of upper limb prosthesis for improved user suitability. J. Orthop. **23**, 88–96 (2021)
2. L. Zollo, G. Di Pino, A.L. Ciancio, F. Ranieri, F. Cordella, C. Gentile, E. Noce, E.A. Romeo, A. Dellacasa Bellingegni, G. Vadalà, et al., Restoring tactile sensations via neural interfaces for real-time force-and-slippage closed-loop control of bionic hands. Sci. Robot. **4**, eaau9924 (2019)
3. F. Iberite, J. Muheim, O. Akouissi, S. Gallo, G. Rognini, F. Morosato, A. Clerc, M. Kalff, E. Gruppioni, S. Micera, et al., Restoration of natural thermal sensation in upper-limb amputees. Science **380**, 731–735 (2023)
4. G. Peternell, H. Penasso, H. Luttenberger, et al., Vibrotactile feedback for a person with transradial amputation and visual loss: A case report. Medicina **59**(10), 1710 (2023)
5. E.A. Biddiss, T.T. Chau, Upper limb prosthesis use and abandonment: A survey of the last 25 years. Prosthetics Orthot. Int. **31**(3), 236–257 (2007)
6. S. Salminger, H. Stino, L.H. Pichler, et al., Current rates of prosthetic usage in upper-limb amputees—Have innovations had an impact on device acceptance? Disabil. Rehabil. **44**(14), 3708–3713 (2022)
7. R. Riener, The Cybathlon promotes the development of assistive technology for people with physical disabilities. J. Neuroeng. Rehabil. **13**, 1–4 (2016)

8. G. Di Pino, D. Romano, C. Spaccasassi, et al., Sensory- and action-oriented embodiment of neurally-interfaced robotic hand prostheses. Front. Neurosci. **14**, 389 (2020)
9. M. Merleau-Ponty, *Phenomenology of Perception* (Routledge, 1945/2012)
10. J. Limanowski, What can body-ownership illusions tell us about minimal phenomenal selfhood? Front. Hum. Neurosci. **8**, 946 (2014)
11. L. Resnik, M. Borgia, J.M. Cancio, et al., Psychometric evaluation of the Southampton hand assessment procedure (SHAP) in a sample of upper-limb prosthesis users. J. Hand Ther. **36**(1), 110–120 (2023)
12. C.M. Light, P.H. Chappell, P.J. Kyberd, Establishing a standardized clinical assessment tool of pathologic and prosthetic hand function: Normative data, reliability, and validity. Arch. Phys. Med. Rehabil. **83**(6), 776–783 (2002). https://doi.org/10.1053/apmr.2002.32737. PMID: 12048655
13. J.T. Belter, A.M. Dollar, Performance characteristics of anthropomorphic prosthetic hands, in *IEEE International Conference on Rehabilitation Robotics (ICORR)*, (2011), pp. 1–7
14. R. Bekrater-Bodmann, Factors associated with prosthesis embodiment and its importance for prosthetic satisfaction in lower-limb amputees. Front. Neurorobot. **14**, 604376 (2021)
15. A. Serino, M. Akselrod, R. Salomon, et al., Upper limb cortical maps in amputees with targeted muscle and sensory reinnervation. Brain **140**(11), 2993–3011 (2017)
16. J. Zbinden, E. Lendaro, M. Ortiz-Catalan, A multi-dimensional framework for prosthetic embodiment: A perspective for translational research. J. Neuroeng. Rehabil. **19**, 122 (2022)
17. J.L. Segil, L.M. Roldan, E.L. Graczyk, Measuring embodiment: A review of methods for prosthetic devices. Front. Neurorobot. **16**, 902162 (2022)
18. C. Lopez, P. Halje, O. Blanke, Body ownership and embodiment: Vestibular and multisensory mechanisms. Neurophysiol. Clin. **38**(3), 149–161 (2008)
19. M. Ortiz-Catalan, B. Håkansson, R. Brånemark, Real-time and intuitive control of artificial limbs with osseointegrated implants. Brain **137**(10), 2791–2801 (2014)
20. G. Rognini, F.M. Petrini, M. Rusconi, et al., Multisensory synchrony enhances embodied perception of a prosthetic hand in amputees. Sci. Rep. **8**, 6431 (2018)
21. F. McGlone, J. Wessberg, H. Olausson, Discriminative and affective touch: Sensing and feeling. Neuron **82**(4), 737–755 (2014)
22. C. Piazza, M. Rossi, M.G. Catalano, A. Bicchi, L.J. Hargrove, Evaluation of a simultaneous myoelectric control strategy for a multi-DoF transradial prosthesis. IEEE Trans. Neural Syst. Rehabil. Eng. **28**, 2286–2295 (2020)
23. C. Cipriani, F. Zaccone, S. Micera, M.C. Carrozza, On the shared control of an EMG-controlled prosthetic hand: Analysis of user-prosthesis interaction. IEEE Trans. Neural Syst. Rehabil. Eng. **16**(6), 505–514 (2008)
24. E. Raveh, S. Portnoy, J. Friedman, Myoelectric prosthesis users improve performance time and accuracy using vibrotactile feedback when visual feedback is disturbed. Arch. Phys. Med. Rehabil. **99**, 2263–2270 (2018)
25. C.D. Murray, M.J. Forshaw, The experience of amputation and prosthesis use for adults: A metasynthesis. Disabil. Rehabil. **35**, 1133–1142 (2013)
26. R. Rackerby, S. Lukosch, D. Munro, Understanding and measuring the cognitive load of amputees for rehabilitation and prosthesis development. Arch. Rehabil. Res. Clin. Transl. **4**(3), 100216 (2022)
27. G. Berlucchi, S.M. Aglioti, The body in the brain: Neural bases of corporeal awareness. Trends Neurosci. **33**(12), 560–569 (2010)
28. F. de Vignemont, Body schema and body image: Prospects for a theory. Neuropsychologia **48**(3), 669–680 (2010)
29. M. Tsakiris, My body in the brain: A neurocognitive model of body-ownership. Neuropsychologia **48**(3), 703–712 (2010)
30. N.P. Holmes, C. Spence, The body schema and the multisensory representation of peripersonal space. Cogn. Process. **5**(2), 94–105 (2004)
31. M.S.A. Graziano, D.F. Cooke, Parieto-frontal interactions, personal space, and defensive behaviour. Neuropsychologia **44**(13), 2621–2635 (2006)

32. M. Botvinick, J. Cohen, Rubber hands 'feel' touch that eyes see. Nature **391**, 756 (1998)
33. S. Raspopovic et al., Restoring natural sensory feedback in real-time bidirectional hand prostheses. Sci. Transl. Med. **6**, 222ra19-222ra19 (2014)
34. H. Flor, Phantom-limb pain: Characteristics, causes, and treatment. Lancet Neurol. **5**(9), 873–881 (2006)
35. Y. Auh, The Atlas of AI Paradoxes: Mapping the Contradictions That Shape Our Intelligent Futures
36. K. Kilteni, H.H. Ehrsson, Predictive attenuation of touch and tactile gating are distinct processes. Curr. Biol. **30**(6), 921–927 (2020)
37. M.D. Paskett, M.R. Brinton, J.A. George, et al., Quantifying cognitive workload while controlling advanced upper-limb prostheses. Sci. Rep. **11**, 19903 (2021)
38. G.M. Van Acker III, C.W. Luchies, P.D. Cheney, Timing of cortico-muscle transmission during active movement. Cereb. Cortex **26**, 3335–3344 (2016)
39. F. Ranieri et al., Sensorimotor integration within the primary motor cortex by selective nerve fascicle stimulation. J. Physiol. **600**, 1497–1514 (2022)
40. T.R. Farrell, R.F. Weir, The optimal controller delay for myoelectric prostheses. IEEE Trans. Neural. Syst. Rehabil. Eng. **15**, 111–118 (2007)
41. G. Risso et al., Optimal integration of intraneural somatosensory feedback with visual information: A single-case study. Sci. Rep. **9**, 7916 (2019)
42. C.H. Moore et al., Grasping embodiment: Haptic feedback for artificial limbs. Front. Neurorobot. **15**, 662397 (2021)
43. E.D. Papaleo et al., Integration of proprioception in upper-limb prostheses through non-invasive strategies: A review. J. Neuroeng. Rehabil. **20**, 118 (2023)
44. K. Friston, The free-energy principle: A unified brain theory? Nat. Rev. Neurosci. **11**(2), 127–138 (2010)
45. D.M. Wolpert, Z. Ghahramani, M.I. Jordan, An internal model for sensorimotor integration. Science **269**(5232), 1880–1882 (1995)
46. A. Ninu, S. Dosen, S. Muceli, F. Rattay, H. Dietl, D. Farina, Closed-loop control of grasping with a myoelectric hand prosthesis: Which are the relevant feedback variables for force control? IEEE Trans. Neural Syst. Rehabil. Eng. **22**, 1041–1052 (2014)
47. N. Thomas et al., Haptic shared control improves neural efficiency during myoelectric prosthesis use. Sci. Rep. **13**, 484 (2023)
48. J.V.V. Parr et al., The prosthesis task load index (PROS-TLX). PLoS One **18**, e0285382 (2023)
49. M.D. Paskett, et al. Improving upper-limb prosthesis usability: Cognitive workload measures quantify task difficulty. medRxiv 2022.08.02.22278038 (2022)
50. K. Limakatso et al., The prevalence and risk factors for phantom limb pain: A cross-sectional survey. BMC Neurol. **24**, 57 (2024)
51. S. Imaizumi, T. Asai, N. Kanayama, M. Kawamura, S. Koyama, Agency over a phantom limb and electromyographic activity on the stump depend on visuomotor synchrony: a case study. Front. Hum. Neurosci. **8**, 545. Published 2014 July 29 (2014). https://doi.org/10.3389/fnhum.2014.00545
52. H.R. Schone, C.I. Baker, J. Katz, et al., Making sense of phantom limb pain. J. Neurol. Neurosurg. Psychiatry. Published online May 24, 2022. https://doi.org/10.1136/jnnp-2021-328428
53. A. Karl, N. Birbaumer, W. Lutzenberger, L.G. Cohen, H. Flor, Reorganization of motor and somatosensory cortex in upper-limb amputees with phantom-limb pain. J. Neurosci. **21**(10), 3609–3618 (2001)
54. T.R. Makin, H. Flor, Brain (re)organisation following amputation: Implications for phantom-limb pain. NeuroImage **218**, 116943 (2020)
55. R.O. Maimon-Mor, T.R. Makin, Is an artificial limb embodied as a hand? PLoS Biol. **18**, e3000729 (2020)
56. E. Raffin, N. Richard, P. Giraux, K.T. Reilly, Primary motor cortex changes after amputation correlate with phantom-limb pain and the ability to move the phantom limb. NeuroImage **54**(3), 225–233 (2016)

57. G. Valle et al., Tactile edges and motion via patterned microstimulation of the human somatosensory cortex. Science **387**, 315–322 (2025). https://doi.org/10.1126/science.adq5978
58. E.L. Graczyk, L. Resnik, B.M. Schiefer, et al., Home use of a sensory-enabled prosthetic hand improves functional performance and embodiment. Sci. Transl. Med. **10**(432), eaap8373 (2018)
59. O. Blanke, Multisensory brain mechanisms of bodily self-consciousness. Nat. Rev. Neurosci. **13**, 556–571 (2012)
60. R.S. Johansson, J.R. Flanagan, Coding and use of tactile signals from the fingertips in object manipulation tasks. Nat. Rev. Neurosci. **10**, 345–359 (2009)
61. M. Zampini, D.I. Shore, C. Spence, Audiovisual temporal-order judgments. Exp. Brain Res. **152**, 198–210 (2003)
62. D. Voudouris, K. Fiehler, Spatial specificity of tactile enhancement during reaching. Atten. Percept. Psychophys. **79**, 2424–2434 (2017)
63. P. Haggard, S. Clark, J. Kalogeras, Voluntary action and conscious awareness. Nat. Neurosci. **5**(4), 382–385 (2002). https://doi.org/10.1038/nn827. PMID: 11896397
64. F. Sorgini, L. Massari, J. D'Abbraccio, E. Palermo, A. Menciassi, P.B. Petrovic, A. Mazzoni, M.C. Carrozza, F.N. Newell, C.M. Oddo, Neuromorphic vibrotactile stimulation of fingertips for encoding object stiffness in telepresence sensory substitution and augmentation applications. Sensors **18**(1), 261 (2018). https://doi.org/10.3390/s18010261
65. H. Lee, D. Kim, Y.L. Park, Explainable deep learning model for EMG-based finger angle estimation using attention. IEEE Trans. Neural Syst. Rehabil. Eng. **30**, 1877–1886 (2022)
66. S. Bahrami Moqadam, S.M. Elahi, A. Mo, et al., Hybrid control combined with a voluntary biosignal to control a prosthetic hand. Robot. Biomim. **5**, 4 (2018). https://doi.org/10.1186/s40638-018-0087-5
67. E. D'Anna, G. Valle, A. Mazzoni, et al., A closed-loop hand prosthesis with simultaneous intraneural tactile and position feedback. Sci. Robot. **4**, eaau8892 (2019)
68. C. Gentile, E. Gruppioni, A perspective on prosthetic hands control: From the brain to the hand. Prosthesis **5**, 1184–1205 (2023)
69. E. Todorov, M.I. Jordan, Optimal feedback control as a theory of motor coordination. Nat. Neurosci. **5**, 1226–1235 (2002)
70. K.P. Nguyen, A.L. Person, Cerebellar circuit computations for predictive motor control. Nat. Rev. Neurosci. **26**, 538–553 (2025)
71. A.P.S. Gill, J. Zariffa, Time-series classification of multi-channel nerve cuff recordings using deep learning. PLoS One **19**, e0299271 (2024)
72. P.P. Vu, J.M. Hahne, C.W. Hunter, et al., A regenerative peripheral nerve interface allows real-time control of an artificial hand in upper-limb amputees. Sci. Transl. Med. **12**, eaay2857 (2020)
73. D.K. Luu, M.D. Paskett, J.A. George, et al., Artificial intelligence enables real-time and intuitive control of prostheses via nerve interface. arXiv preprint. 2022;arXiv:2203.08648
74. S.J. Bensmaia, D.J. Tyler, S. Micera, Restoration of sensory information via bionic hands. Nat. Biomed. Eng. **7**(4), 443–455 (2023)
75. Z. Zhang, T. Liu, Y. Yuan, et al., Biomimetic strategies of slip sensing, perception, and protection in prosthetic hand grasp. Micromachines. **15**, 751 (2024)
76. M. Ortiz-Catalan, B. Hakansson, A. Branner, et al., A highly integrated bionic hand with neural control and feedback for use in daily life. Sci. Robot. **8**, eadf7360 (2023)
77. J.W. Krakauer, A.M. Hadjiosif, J. Xu, A.L. Wong, A.M. Haith, Motor Learning. Compr. Physiol. **9**, 613–663 (2019)
78. F.M.Z. van den Heiligenberg, T. Orlov, S.N. MacDonald, et al., Artificial limb representation in amputees. Brain **141**, 1422–1433 (2018)
79. A.L. Edwards, M.R. Dawson, J.S. Hebert, C. Sherstan, R.S. Sutton, K.M. Chan, P.M. Pilarski, Application of real-time machine learning to myoelectric prosthesis control: A case series in adaptive switching. Prosthetics Orthot. Int. **40**(5), 573–581 (2016 Oct)

80. S. Jain, G. Singhal, R.J. Smith, et al., Improving long-term myoelectric decoding using an adaptive classifier with label correction. Proc. IEEE RAS EMBS Int. Conf. Biomed. Robot. Biomech. **4**, 532–537 (2012)
81. K. Freitag, Y. Karayiannidis, J. Zbinden, et al., Fine-tuning myoelectric control through reinforcement learning in a game environment. arXiv preprint. 2025;arXiv:2411.13327
82. K.C. Tse, P. Capsi-Morales, C. Piazza, et al., Exploring muscle synergies for performance enhancement and learning in myoelectric control maps. Proc. Int. Conf. Rehabil. Robot. (2023). https://doi.org/10.1109/ICORR58425.2023.10304809
83. G. Granata, R. Di Iorio, F. Miraglia, et al., Brain reactions to the use of sensorized hand prosthesis in amputees. Brain Behav. **10**, e01734 (2020)
84. Y. Liu, S. Huang, Z. Wang, et al., Functional reorganization after four-week brain–computer interface-controlled supernumerary robotic finger training: A longitudinal resting-state fMRI study. Front. Neurosci. **16**, 766648 (2022)
85. Pilarski PM, Dawson MR, Degris T, Fahimi F, Carey JP, Sutton RS. Online human training of a myoelectric prosthesis controller via actor-critic reinforcement learning. 2011 IEEE international Conference on Rehabilitation Robotics (ICORR); 2011:1–7
86. F. Seifert, N. Schuberth, R. De Col, E. Peltz, F.T. Nickel, C. Maihöfner, Brain activity during sympathetic response in anticipation and experience of pain. Hum. Brain Mapp. **34**(8), 1768–1782 (2013)
87. L.S. Löken, J. Wessberg, I. Morrison, F. McGlone, H. Olausson, Coding of pleasant touch by unmyelinated afferents in humans. Nat. Neurosci. **12**, 547–548 (2009)
88. C. Phelps, P.B. Hutchings, T. Stokes, et al., The impact of a customized aesthetic prosthetic leg cover on social interaction cues and attitudes in the general UK population: Two experimental studies. Prosthetics Orthot. Int. **49**, 173–178 (2025)
89. G. Goh, S. Shafighian, S. Bakarich, et al., Biologically inspired multi-layered synthetic skin for tactile sensing on prosthetic limbs. Sensors (Basel). **21**, 3456 (2021)
90. S. Shaikh, S.S. Nagi, F. McGlone, D.A. Mahns, Psychophysical investigations into the role of low-threshold C fibres in non-painful affective processing and pain modulation. PLoS One **10**, e0138299 (2015)
91. A. Chatterjee, O. Vartanian, Neuroaesthetics. Trends Cogn. Sci. **18**, 370–375 (2014)
92. B. Murali, S. Huddle, R.F.f. Weir, Design and evaluation of a distally actuated powered finger prosthesis with self-contained transmission for individuals with partial hand loss. Adv. Mech. Eng. **11**(4) (2019)
93. R. Bekrater-Bodmann, Mind over matter: Perceived phantom/prosthesis co-location contributes to prosthesis embodiment in lower limb amputees. Conscious. Cogn. **98**, 103268 (2022). https://doi.org/10.1016/j.concog.2021.103268. Epub 2022 Jan 6. PMID: 34999318
94. K. Li, C. Cardoso, A. Moctezuma-Ramirez, A. Elgalad, E. Perin, Heart rate variability measurement through a smart wearable device: Another breakthrough for personal health monitoring? Int. J. Environ. Res. Public Health **20**, 7146 (2023)
95. V.E. Abraira, D.D. Ginty, The sensory neurons of touch. Neuron **79**, 618–639 (2013)
96. A.W. Goodwin, H.E. Wheat, Sensory signals in neural populations underlying tactile perception and manipulation. Annu. Rev. Neurosci. **27**, 53–77 (2004)
97. G.F. Shannon, A myoelectrically-controlled prosthesis with sensory feedback. Med. Biol. Eng. Comput. **17**(1), 73–80 (1979). https://doi.org/10.1007/BF02440956. PMID: 312386
98. R.S. Dahiya, G. Metta, M. Valle, G. Sandini, Tactile sensing—From humans to humanoids. IEEE Trans. Robot. **26**, 1–20 (2010)
99. J. Kim, M. Lee, H. Shim, et al., Stretchable silicon nanoribbon electronics for skin prosthesis. Nat. Commun. **5**, 5747 (2014). https://doi.org/10.1038/ncomms6747
100. L. Luan, X. Wei, Z. Zhao, J.J. Siegel, O. Potnis, C.A. Tuppen, S. Lin, S. Kazmi, R.A. Fowler, S. Holloway, A.K. Dunn, R.A. Chitwood, C. Xie, Ultraflexible nanoelectronic probes form reliable, glial scar-free neural integration. Sci. Adv. **3**(2), e1601966 (2017)
101. G. Indiveri, S.C. Liu, Memory and information processing in neuromorphic systems. Proc. IEEE **103**, 1379–1397 (2015)

102. H.J. Witteveen, J.S. Rietman, P.H. Veltink, Vibro-tactile grasping feedback for myoelectric forearm prostheses. J. Neuroeng. Rehabil. **11**, 75 (2014)
103. G. Valle, A. Mazzoni, E. D'Anna, et al., Biomimetic intraneural sensory feedback enhances grasp-force control in a bidirectional hand prosthesis. Sci. Transl. Med. **13**(593), eabl7229 (2021)
104. C.M. Oddo, S. Raspopovic, F. Artoni, et al., Intraneural stimulation elicits discrimination of textural features by artificial mechanoreceptors. elife **5**, e09148 (2016)
105. S. Xu, S.I. Park, M. Kam, et al., Soft microfluidic assemblies of sensors, circuits, and radios for the skin. Science **344**, 70–74 (2014)
106. W.M. Grill, J.T. Mortimer, Neural and connective tissue response to long-term implantation of multiple-contact nerve cuff electrodes. J. Biomed. Mater. Res. **50**, 215–226 (2000)
107. D. Boufidis, R. Garg, E. Angelopoulos, et al., Bio-inspired electronics: Soft, biohybrid, and "living" neural interfaces. Nat. Commun. **16**, 1861 (2025). https://doi.org/10.1038/s41467-025-57016-0
108. Y. Lu, F. Zhang, J. Lyu, et al., Recent advances of controlled drug delivery using microfluidic platforms. Small **16**, 1904111 (2020)
109. Y. Xie, Y. Peng, J. Guo, M. Liu, B. Zhang, L. Yin, H. Ding, X. Sheng, Materials and devices for high-density, high-throughput micro-electrocorticography arrays. Fundam. Res. **5**(1), 17–28 (2024)
110. F.D. Chen, H.M. Chameh, M. Movahed, et al., Implantable nanophotonic neural probes for integrated patterned photostimulation and electrophysiological recording. NPJ Biosensing **2**, 15 (2025)
111. Y. Yang, H. Zhao, Z. Hao, C. Shi, L. Zhou, X. Yao, Recognition of brain activities via graph-based long short-term memory-convolutional neural network. Front. Neurosci. **19**, 1546559 (2025). https://doi.org/10.3389/fnins.2025.1546559
112. J.J. Shin, A. Ghovanlou, J.W. Jeong, Powering mm-size wireless implants for brain–machine interfaces. IEEE Trans. Biomed. Circuits Syst. **17**, 432–443 (2023)
113. M. Rizk, C.A. Bossetti, T.A. Jochum, S.H. Callender, M.A. Nicolelis, D.A. Turner, P.D. Wolf, A fully implantable 96-channel neural data acquisition system. J. Neural. Eng. **6**(2), 026002 (2009). https://doi.org/10.1088/1741-2560/6/2/026002. Epub 2009 Mar 2. PMID: 19255459; PMCID: PMC2680289
114. D.M. Page et al., Motor control and sensory feedback enhance prosthesis embodiment. Front. Hum. Neurosci. **12**, 352 (2018)
115. C. Gentile, F. Cordella, L. Zollo, Hierarchical human-inspired control strategies for prosthetic hands. Sensors **22**(7), 2521 (2022). https://doi.org/10.3390/s22072521
116. C.M. Niu et al., Neuromorphic model of reflex for real-time compliant control of a prosthetic hand. Ann. Biomed. Eng. **49**, 673–688 (2021)
117. C. Li, D. Belkin, Y. Li, et al., Efficient and self-adaptive in-situ learning in multilayer memristor neural networks. Nat. Commun. **9**, 2385 (2018)
118. O. Karaali, et al., Enabling next-generation cloud-connected bionic limbs through 5G connectivity. arXiv 2025 (2506.11744)
119. C.W. Antuvan, M. Ison, P. Artemiadis, Embedded human control of robots using myoelectric interfaces. IEEE Trans. Neural Syst. Rehabil. Eng. **22**(4), 820–827 (2014)
120. C.G. Fan, S. Coppi, H.H. Ehrsson, The supernumerary rubber hand illusion revisited: perceived duplication of limbs and visuotactile events. J. Exp. Psychol. Hum. Percept Perform **47**, 810–829 (2021)
121. L. Resnik, M. Borgia, Superhuman functionality in upper-limb prostheses: Clinical and ethical implications. Phys. Ther. **99**, 1575–1584 (2019)
122. D. Weber, A. Matsiko, Assistive robotics should seamlessly integrate humans and robots. Sci. Robot **8**, eadl0014 (2023)
123. R. Yuste, Advocating for neurodata privacy and neurotechnology regulation. Nat. Protoc. **18**, 2869–2875 (2023)
124. T. Denning, Y. Matsuoka, T. Kohno, Neurosecurity: Security and privacy for neural devices. Neurosurg. Focus. **27**(1), E7 (2009). https://doi.org/10.3171/2009.4.FOCUS0985

125. B.M.G. Rosa, S. Anastasova, G.Z. Yang, NFC-powered implantable device for on-body parameters monitoring with secure data exchange link to a medical blockchain type of network. IEEE Trans. Cybern. **53**(1), 31–43 (2023). https://doi.org/10.1109/TCYB.2021.3088711
126. C. Dwork, A. Roth, The algorithmic foundations of differential privacy. Found. Trends Theor. Comput. Sci. **9**(3–4), 211–407 (2014). https://doi.org/10.1561/0400000042
127. X. Liu, B. Glocker, M.M. McCradden, et al., The medical algorithmic audit. Lancet Digit Health. **4**(5), e384–e397 (2022). https://doi.org/10.1016/S2589-7500(22)00003-6
128. I. Budin-Ljøsne, H.J.A. Teare, J. Kaye, et al., Dynamic consent: A potential solution to some of the challenges of modern biomedical research. BMC Med. Ethics **18**, 4 (2017). https://doi.org/10.1186/s12910-016-0162-9
129. C.M.S. Kerver, R.M. Bongers, C.K. van der Sluis, et al., Economic evaluation of upper-limb prostheses in The Netherlands including the cost-effectiveness of multi-grip versus standard myoelectric hand prostheses. Disabil. Rehabil. **45**, 1783–1793 (2023)
130. A. Dobson, A. El-Gamil, M. Shimer, J. DaVanzo, Economic value of prosthetic services among Medicare beneficiaries: A claims-based retrospective cohort study. Mil. Med. **181**(2 Suppl), 18–24 (2016)
131. J.M. Lambrecht, C.L. Pulliam, R.F. Kirsch, Virtual reality environment for simulating tasks with a myoelectric prosthesis: An assessment and training tool. J. Prosthet. Orthot. **23**(2), 89–94 (2011). https://doi.org/10.1097/JPO.0b013e318217a30c
132. J. Hao, Z. Chen, A. Remis, Z. He, Virtual reality-based rehabilitation to restore motor function in people with amputation: A systematic literature review. Am. J. Phys. Med. Rehabil. **102**, 468–474 (2023)
133. M.B. Kristoffersen, A.W. Franzke, R.M. Bongers, et al., User training for machine-learning controlled upper-limb prostheses: A serious-game approach. J. Neuroeng. Rehabil. **18**, 32 (2021)
134. P. Gallagher, M. MacLachlan, Adjustment to an artificial limb: A qualitative perspective. J. Health Psychol. **5**(2), 259–269 (2000)
135. S.L. Matsen, A closer look at amputees in Vietnam: A field survey of Vietnamese using prostheses. Prosthetics Orthot. Int. **23**, 93–101 (1999)
136. L. Ennion, A. Johannesson, A qualitative study of the challenges of providing pre-prosthetic rehabilitation in rural South Africa. Prosthetics Orthot. Int. **42**, 179–186 (2018)
137. C.D. Murray, Embodiment and prosthetics. Disabil. Rehabil. **26**(16), 963–973 (2004)
138. S. Sims, L. Houston, I. Schweinzger, R.N. Samy, Closing the gap in cochlear implant access for African-Americans: A story of outreach and collaboration. Curr. Opin. Otolaryngol. Head Neck Surg. **25**(5), 365–369 (2017)
139. World Health Organization, *Standards for Prosthetics and Orthotics: Part 1—Standards* (WHO, Geneva, 2017)

Chapter 3
Affective Edge Computing: Challenges and Opportunities in Decoding Emotional States

P. S. Saikrishna

Abstract Affective Computing (AfC) aims to detect, interpret, and respond to human emotions by analyzing multimodal psychophysiological signals, including facial expressions, vocal cues, and physiological data such as ECG, EEG, and EDA. To improve system responsiveness and privacy, Affective Edge Computing (AEC) integrates AfC with edge computing frameworks, enabling real-time, localized processing of emotion data on devices such as wearables, smartphones, and embedded systems. This approach reduces latency, minimizes network dependency, and preserves user privacy by retaining sensitive data at the source. Lightweight machine learning and signal processing models on edge devices enable efficient, real-time emotion classification within limited computational resources. However, developing a scalable, distributed architecture that integrates multimodal data acquisition, local emotion recognition, optional fog-based load balancing, and cloud-based analytics remains a key challenge. Ensuring responsiveness and efficiency is especially critical in sensitive applications such as mental health monitoring, human-robot interaction, and mobile affective systems, where privacy, low latency, and reliability are essential.

Keywords Emotion recognition · Edge computing · Wearable embedded systems

3.1 Introduction

Emotions are fundamental to human cognition, behavior, and social interaction. Accurately recognizing emotional states is crucial for advancing research in psychology and neuroscience and building emotion-aware systems in areas like human-computer interaction and mental health assessment. Affective computing seeks to create systems capable of detecting and responding to emotions, with multimodal

---

P. S. Saikrishna (✉)
Department of Electrical Engineering, Indian Institute of Technology (IIT) Tirupati, Tirupati, Andhra Pradesh, India
e-mail: psskrishna@iittp.ac.in

R. Vinjamuri (ed.), *Bridging the Gap between Mind and Machine*,
https://doi.org/10.1007/978-3-032-06713-5_3

emotion recognition (MER) emerging as a core approach by combining data from various sources to enhance emotion detection accuracy [1, 5]. Furthermore, MER involves processing large volumes of heterogeneous data from multiple sources, such as facial expressions, voice, and physiological signals. Performing this processing centrally can lead to high latency, bandwidth limitations, and privacy concerns [6]. Multimodal psychophysiological signals serve as objective indicators of the body's emotional responses. These include electroencephalography (EEG), electrodermal activity (EDA), heart rate variability (HRV), respiration, facial electromyography (EMG), and pupil dilation [1, 3].

Sensor fusion in MER involves combining inputs from various sensors such as cameras capturing facial expressions, microphones analyzing vocal tone, and biosensors measuring physiological signals like heart rate or EEG to improve the precision and reliability of emotion detection [2]. Fusion techniques combine complementary information from multiple modalities to enable more reliable emotion recognition, especially when noisy or partially missing data. This integration can occur at various levels, data, features, or decisions [6]. Therefore, MER provides a holistic approach to understanding the' complex and evolving nature of human emotions. However, MER faces several critical challenges, such as effective integration of signals from different modalities, managing inter-individual variability, and addressing ethical concerns related to interpreting and securing sensitive physiological data [3, 7].

Edge computing minimizes latency and computational overhead by processing data near its source, allowing for real-time emotion recognition while conserving bandwidth [6]. Edge devices preprocess and fuse data from multiple modalities, perform lightweight inference using optimized deep learning models, and selectively transmit results to the cloud for further analysis or storage [1, 2]. MER can be enabled on edge devices using machine learning (ML) and deep learning (DL) algorithms that process and fuse data from various sources. ML methods such as SVM and Random Forests offer lightweight, low-latency emotion classification using handcrafted features [2]. DL models like CNNs, LSTMs, and transformers provide more accurate, end-to-end learning and fusion but require optimization (e.g., pruning, quantization) for edge deployment [1, 4]. This enables real-time, privacy-preserving emotion recognition on mobile and wearable platforms [3, 6]. Furthermore, scalable edge computing addresses these challenges by distributing computation closer to data sources, enabling real-time analysis, reducing communication overhead, and preserving user privacy [6, 7].

As emotion-aware applications evolve in healthcare, education, and human-computer interaction, scalability becomes essential to efficiently support multiple users, devices, and modalities while adapting to diverse computational needs and deployment settings [5, 7]. Edge computing enables scalable MER by allowing distributed processing across multiple devices. It supports horizontal scaling by adding edge nodes and vertical scaling through resource allocation per device. This ensures low latency, efficient load handling, and adaptability to various application scenarios [6].

3.2 Multimodal Analysis of Emotion and Sensor Fusion

Multimodal emotion analysis interprets human emotions by integrating data from diverse sources such as facial expressions, voice tone, body movements, and physiological signals like heart rate and skin conductance [8, 11]. Each modality offers distinct and complementary insights into a person's emotional state, helping to address the limitations of single-modality approaches [9]. See Table 3.1 for summary of modalities in multimodal emotion recognition.

MER leverages diverse sources of affective information to enhance accuracy and reliability [8, 12]. Facial expressions provide visual cues through muscle movements that reflect emotions such as happiness, anger, or sadness [8]. Speech contributes acoustic features such as pitch, energy, and tone, which help identify emotional states such as stress or calmness [8]. Text or linguistic content, both spoken and written, provides semantic and sentiment cues that enrich the context of the expressed emotion [15]. Physiological signals, including heart rate, skin conductance (GSR), respiration, and EEG, provide internal, involuntary indicators of emotional arousal or stress [10, 13, 14]. Finally, body gestures and posture, encompassing movements and physical stance, reveal nonverbal affective intent through kinesics, further enriching emotion detection in real-world interactions [11, 12].

3.2.1 *Sensor Fusion*

Sensor fusion integrates data from multiple sensors or modalities to produce a unified and more accurate representation of emotional states [11, 12, 15, 79]. This fusion can occur at different levels, namely:

- **Early Fusion:** Raw data or features from multiple sensors are combined before feeding them into a model. This approach captures cross-modal relationships but requires careful synchronization and feature compatibility.

Table 3.1 Summary of modalities in multimodal emotion recognition

S.NO	Modality	Description
1	Facial expressions	Visual cues derived from facial muscle movements are used to infer emotional state
2	Speech	Acoustic features such as pitch, energy, and tone are analysed for emotion cues
3	Text/linguistic content	Analysis of semantics and sentiment in spoken or written language
4	Physiological signals	Includes heart rate, skin conductance (GSR), respiration, and EEG for emotion detection
5	Body gestures/posture	Kinesics, such as body movements and posture, that reflect affective intent

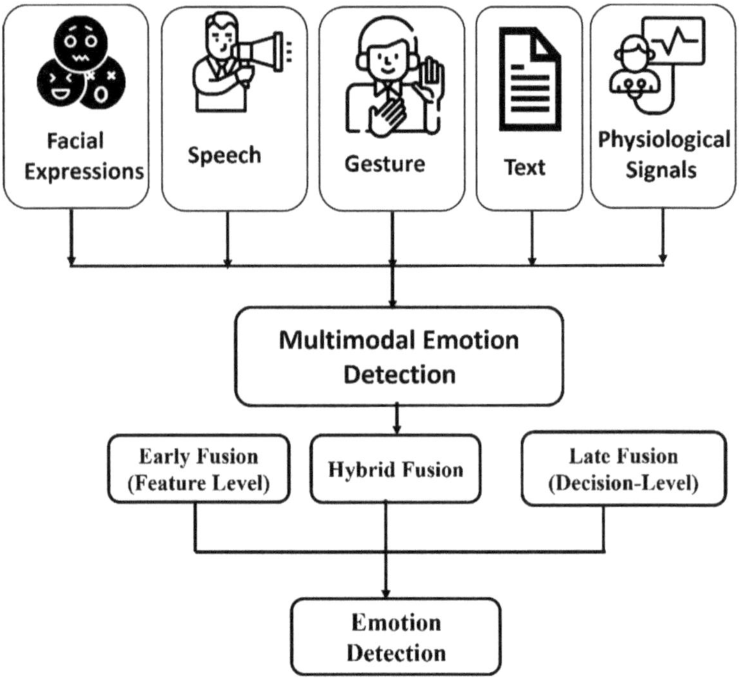

Fig. 3.1 Multimodal emotion recognition by sensor fusion

- **Late Fusion:** Independent models analyze each modality separately, and their outputs (e.g., emotion predictions) are combined using voting, averaging, or weighted strategies. This is easier to implement but may miss inter-modal correlations.
- **Intermediate Fusion:** Combines features or intermediate representations extracted from each modality, allowing the model to learn interactions while keeping some modularity.

Sensor fusion improves robustness by compensating for missing or noisy data in one modality with information from others. For example, facial expressions or physiological signals can provide additional emotional cues if the voice signal is unclear. Figure 3.1 summarizes the MER by sensor fusion.

3.2.2 Case Studies

Some of the well-known case studies in MER that illustrate the use of diverse technologies across various application contexts are presented below.

3.2.2.1 Emotion Recognition in a Driver Monitoring System

In an advanced driver assistance system (ADAS), a multimodal setup is deployed to monitor the driver's emotional state in real-time to prevent accidents caused by stress or fatigue [17, 18]. The system uses a camera to analyze facial expressions [16], a microphone to capture voice tone and speech features [17], and a wearable sensor to track physiological signals such as heart rate and skin conductance [18, 19]. Each sensor independently estimates the emotional state and the outputs are fused using intermediate fusion through a deep learning model combining CNNs (for facial features), LSTMs (for speech), and shallow networks for physiological features [16]. Fusing these modalities results in a much more reliable estimation of emotional states than using a single source. For example, even if lighting affects facial recognition or road noise affects speech quality, the physiological signal can still contribute to accurate emotion classification [18]. This fused system alerts the driver in cases of drowsiness or anger and adapts the driving environment (e.g., by adjusting lighting or audio) accordingly [17]. This case demonstrates how multimodal sensor fusion enhances the robustness and reliability of emotion recognition systems in real-world applications [16, 18].

3.2.2.2 Application of MER in Mental Health Monitoring

MER is increasingly used to support mental health applications such as detecting early signs of depression, anxiety, and stress. By integrating data from facial expressions (via camera), speech tone and content (via microphone), and physiological signals like heart rate or skin conductance (via wearables), systems can continuously monitor emotional well-being [20]. These inputs are analyzed using machine learning models to identify emotional irregularities that may indicate mood disorders.

For example, a mobile health app could monitor a user's speech and facial cues during daily interactions, while a smartwatch tracks physiological stress signals [20]. If the system detects prolonged negative affect or physiological stress patterns, it can prompt self-care recommendations, suggest mindfulness exercises, or notify a healthcare provider for follow-up. This application allows for real-time, non-intrusive, and personalized mental health support, which is especially valuable for remote or underserved populations lacking regular clinical services [21].

3.3 Role of Edge Computing in Multimodal Decoding of Emotional States

Edge computing empowers MER by enabling real-time processing, enhancing data privacy, and improving resource efficiency. Emotional states are decoded from data

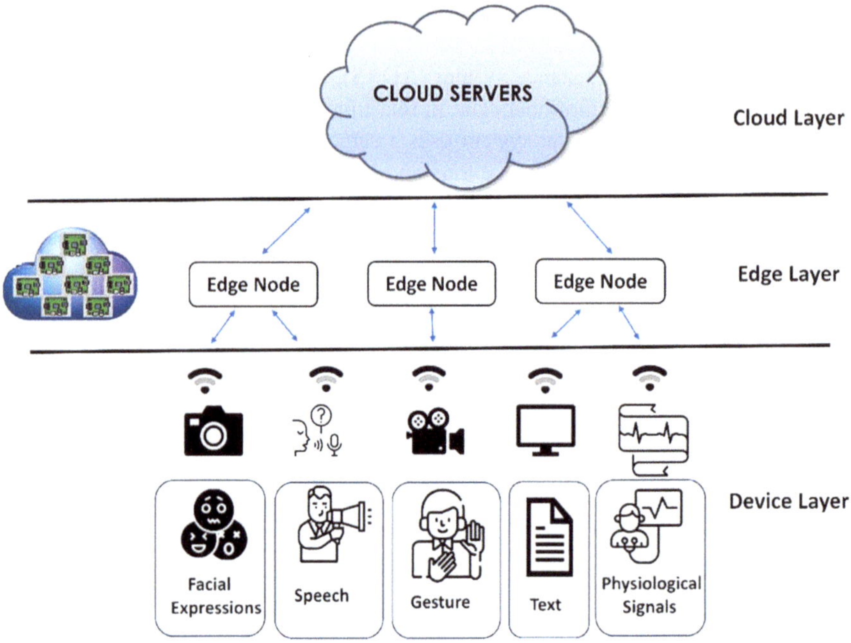

Fig. 3.2 Edge computing architecture for Emotion detection

captured by multiple sensors, such as cameras for facial expressions, microphones for speech, and wearables for physiological signals [25–28]. Transmitting this continuous and sensitive data to centralized cloud servers can result in high latency, bandwidth constraints, and privacy concerns [24]. To overcome these challenges, edge computing moves key computations, including preprocessing, feature extraction, and emotion inference, to edge devices [22]. This localized processing reduces response time, conserves bandwidth, and enables applications to function without constant internet connectivity [23]. It supports the on-device fusion of multimodal inputs and facilitates faster, context-aware decisions. It is critical for applications in mental health monitoring, human-computer interaction, and affective robotics [24]. Fig. 3.2 illustrates the edge computing architecture following a hierarchical structure with three layers: device, edge, and cloud. The device layer handles data acquisition and basic preprocessing using sensors and actuators [25, 27].

The edge layer performs real-time analytics and decision-making close to the data source, minimizing latency [22, 23]. The cloud layer is responsible for large-scale data storage, model training, and long-term analysis [24]. This layered architecture enhances the speed, scalability, and privacy of multimodal emotion decoding, making it highly suitable for real-world, user-centric deployments [24, 26].

3.3.1 *Scaling Strategies in Edge Computing*

Scalability of edge computing for MER is essential to maintain Quality of Service (QoS) as data volume and user demand grow. By scaling horizontally (adding edge devices) or vertically (enhancing device processing power), the system can ensure low-latency, high-throughput processing of multimodal data. This supports real-time emotion inference while preserving privacy and minimizing cloud dependency, thereby sustaining consistent performance and reliability across diverse environments and user loads.

3.3.1.1 Device Scaling

This method involves adding edge devices (e.g., cameras, wearables, smartphones) to increase coverage, improve redundancy, and support fault tolerance. It is ideal for expanding spatial reach in distributed systems (e.g., across rooms, vehicles, or individuals). This strategy is essential when individual edge nodes cannot meet the processing or sensory demands of the application alone.

Figure 3.3 illustrates an edge-computing architecture for MER with device scaling. Machine Learning (ML) and Deep Learning (DL) algorithms are deployed across multiple edge nodes (Node 1 to Node N), enabling parallel and distributed processing of emotion data. A scaling controller dynamically monitors the system workload and performance metrics, adjusting the number of active nodes or resources to maintain optimal efficiency. The processed outputs are selectively transmitted to an edge server for further analysis or coordination. This setup ensures real-time performance, adaptive resource management, and improved Quality of Service (QoS) in MER applications.

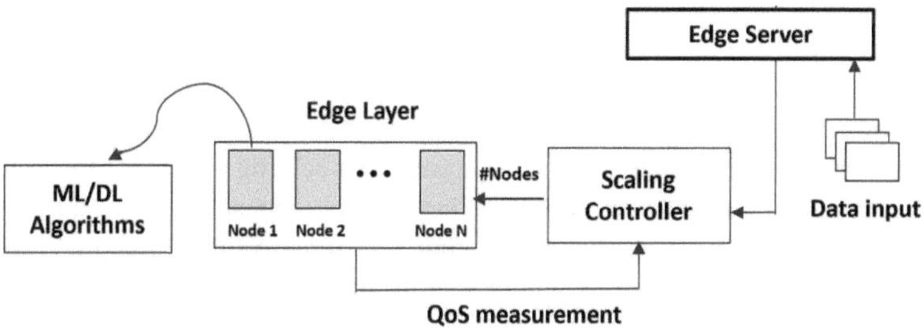

Fig. 3.3 Device scaling in Edge computing

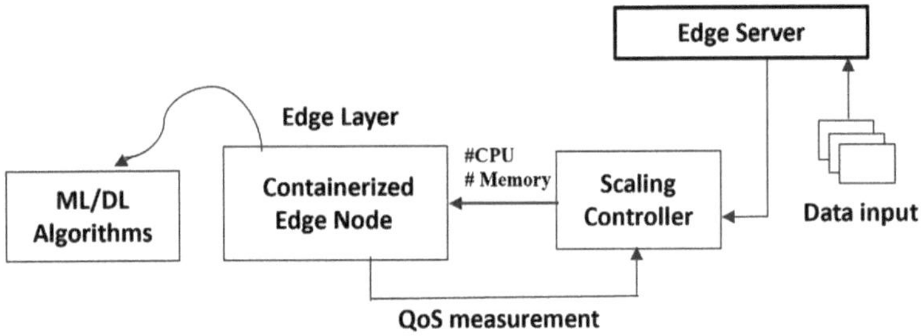

Fig. 3.4 Container-level scaling in Edge computing

3.3.1.2 Container-Level Scaling

This method enhances the performance of each edge device by scaling the computational workloads within it. Using lightweight container orchestration platforms (such as K3s, Docker Swarm, or microK8s), services (e.g., NLP modules, facial expression detectors) can be dynamically scaled up/down based on resource usage (CPU, RAM), latency, or data arrival rates. This enables efficient resource utilization and maintains low-latency processing per device. Figure 3.4 presents a scalable edge computing framework for MER using containerized edge nodes. ML/DL algorithms are deployed in lightweight containers at the edge, allowing efficient multimodal data processing. A scaling controller monitors system load and performance, dynamically adjusting the number or capacity of containerized nodes. This architecture supports flexible and efficient resource allocation, enabling real-time, low-latency emotion inference while maintaining Quality of Service (QoS). The processed results can be transmitted to an edge server for coordination or further analysis, ensuring scalability and responsiveness in MER applications.

3.3.1.3 Combined Scaling

Device and container scaling are often used together. While device scaling ensures broader data acquisition and redundancy, container scaling optimizes the computational efficiency of each node. Together, they support robust, adaptive, and context-aware edge-based emotion recognition and real-time AI applications. Table 3.2 compares device scaling and container scaling in edge computing. Device scaling involves adding more edge devices for load distribution (horizontal scaling), while container scaling enhances performance by increasing resources within existing devices (vertical scaling). Device scaling is complex and network-intensive, whereas container scaling is simpler using orchestration tools such as Kubernetes.

Scalable edge computing enhances the deployment of MER across diverse devices and environments. As MER systems integrate into smartphones, wearables, vehicles, and smart spaces, scalability ensures models adapt efficiently to varying hardware and user contexts. This is achieved through lightweight ML models,

Table 3.2 Comparing device scaling and container scaling

Aspect	Device scaling	Container scaling
Description	Adding more physical edge devices to handle the increased load	Increasing the CPU, memory, or resource allocation for containers running on an edge node
Purpose	Enhances system scalability by distributing load across multiple devices	Improves the performance of existing edge services without additional hardware
Latency Impact	May reduce latency via load balancing, but adds network management overhead	Reduces processing latency by increasing local computation power
Scalability	Horizontal scaling (more devices/nodes)	Vertical scaling (more resources per service/container)
Deployment complexity	Complex—requires provisioning, network configuration, synchronization	Simpler—container orchestration platforms (e.g., Kubernetes) automate scaling

techniques like pruning and quantization, and containerized applications deployable across edge platforms. It also enables distributed inference, where different modalities are processed across edge nodes and fused locally or hierarchically. This supports real-time performance, large-scale deployment, and coordinated emotion analysis while preserving privacy and minimizing cloud reliance.

3.4 Overview of ML/DL Algorithms for MER

DL and ML algorithms decode emotional states by analyzing data from multiple sources like facial expressions, speech, text, and physiological signals. Traditional ML methods (e.g., SVM, Random Forest) use handcrafted features for simpler tasks, while deep learning models (CNNs, LSTMs, transformers) learn directly from raw data for better accuracy. Fusion techniques combine information from different modalities to improve recognition. Lightweight models and optimization methods enable real-time, scalable emotion recognition on edge devices.

3.4.1 Facial Expressions

Facial expressions are among the most intuitive indicators of human emotions, as they provide rich visual cues through subtle muscle movements [32]. Machine learning techniques such as Support Vector Machines (SVM), Random Forests, and K-nearest neighbours (KNN) have been widely used for classifying expressions based on hand-crafted features [31, 32]. Deep learning approaches like Convolutional Neural Networks (CNNs), Residual Networks (ResNet), and Vision

Transformers (ViT) have further enhanced accuracy by learning hierarchical visual representations directly from facial images or video sequences, offering robust performance in real-world conditions [29, 30, 32].

3.4.2 Speech

Speech carries significant emotional information through prosodic features such as pitch, energy, and tone [34]. Traditional machine learning methods like Gaussian Mixture Models (GMM), Hidden Markov Models (HMM), and SVMs have been used to model temporal patterns and classify emotions in speech [34–36]. More recently, deep learning methods, including Recurrent Neural Networks (RNNs), Long Short-Term Memory (LSTM) networks, and hybrid models like Convolutional Recurrent Networks, have proven highly effective [35]. Transformer-based models, such as wav2vec, offer powerful end-to-end frameworks that capture both short- and long-term dependencies in raw audio signals [33].

3.4.3 Text/Linguistic Content

The emotional content of the text is inferred through semantic understanding and sentiment analysis. Classical machine learning methods, such as Naive Bayes, Logistic Regression, and SVMs with Bag-of-Words or TF-IDF features, have traditionally been used [40]. Deep learning models now dominate the field, with LSTM and Bidirectional LSTM (BiLSTM) capturing temporal language dynamics [39]. Transformer-based architectures like BERT, RoBERTa, and GPT-based models have set new benchmarks by leveraging contextual embeddings and large-scale pretraining to detect nuanced emotional cues in spoken and written language [37, 38].

3.4.4 Physiological Signals

Physiological signals provide direct, quantifiable insights into the body's emotional responses. Traditional ML algorithms such as Decision Trees, SVMs, kNN, and Random Forests are often applied to features extracted from biosignals like heart rate, skin conductance, and EEG [41, 43]. Deep learning models such as LSTMs and CNNs are well-suited for modeling the temporal (TCN) and spatial structure in time-series data [42]. Temporal Convolutional Networks (TCNs) and Autoencoders enhance representation learning and anomaly detection, making them ideal for continuous emotion monitoring in health-related contexts [41].

Table 3.3 Summary of ML and DL approaches in MER

S.NO	Modality	ML algorithms	DL architectures
1	Facial expressions	SVM, Random Forests, K-Nearest Neighbors (KNN)	CNNs, ResNet, Vision Transformers (ViT)
2	Speech (acoustic)	Gaussian Mixture Models (GMM), Hidden Markov Models (HMM), SVM	RNNs, LSTM, Convolutional Recurrent Networks, Transformers (e.g., wav2vec)
3	Text/linguistic content	Naive Bayes, Logistic Regression, SVM (BoW/TF-IDF)	LSTM, BiLSTM, BERT, RoBERTa, GPT-based models
4	Physiological signals	Decision trees, SVM, kNN, Random Forests	LSTM, CNN for time-series, TCN, Autoencoders
5	Body gestures/posture	HMM, SVM, Random Forest	CNNs (skeleton/keypoint data), GNNs, 3D CNNs for motion sequences

3.4.5 *Body Gestures and Postures*

Body gestures and posture offer expressive non-verbal signals that reflect affective states through movements and spatial positioning. Machine learning approaches, including HMMs, SVMs, and Random Forests, have been used to classify gestures based on trajectory and skeletal data [44]. Deep learning methods bring additional power, with CNNs applied to keypoint or skeleton-based inputs, Graph Neural Networks (GNNs) modeling relational dynamics of body parts, and 3D CNNs capturing spatiotemporal motion patterns across gesture sequences [45]. These methods enable rich emotion recognition without facial or vocal input [46]. Table 3.3 summarizes the machine learning (ML) and deep learning (DL) approaches used in multimodal emotion detection across various input modalities. Each algorithm's effectiveness depends on data availability, computational resources (especially at the edge), and the need for real-time inference. Often, multimodal fusion architectures combine these models at feature or decision levels to improve recognition accuracy and robustness.

3.5 Real-Time MER on the Edge: Balancing QoS and Deadlines

MER systems on edge computing platforms must meet strict Quality of Service (QoS) and real-time constraints to ensure timely and accurate emotion recognition. QoS involves maintaining acceptable latency, throughput, reliability, and accuracy

Table 3.4 Latency requirements in multimodal emotion recognition

S.NO	Modality	Latency requirement	Remarks
1	Facial expression recognition	<30 ms	Needed for real-time visual feedback of emotion
2	Speech emotion analysis	<100 ms per audio chunk	Should align with natural utterance timing
3	Physiological signals	Update every 1–2 seconds with low jitter	Enables continuous emotional state monitoring
4	Gesture recognition	< 50 ms	Essential for responsive interactions (e.g., VR, driver alertness systems)
5	Text/linguistic content	50–100 ms	Especially in real-time chatbots or mental health apps requiring quick, context-aware responses

levels throughout the workflow, including data collection, processing, fusion, and decision-making based on the specific application requirements [51, 68].

MER must process data and return inferences within strict time windows, especially in applications such as mental health monitoring, driver assistance, or human-computer interaction. The entire process is often expected to be completed in under 100 milliseconds, with optimal response times for each modality falling below 10 to 20 milliseconds. Table 3.4 presents a comparison of various modalities along with their corresponding latency requirements [47].

3.5.1 Key QoS Metrics

Key performance metrics are essential to evaluate the effectiveness of edge-based MER systems. These metrics ensure real-time responsiveness, reliability, and sustainability across diverse deployment environments [51].

1. **Latency**: Time from data acquisition to emotion classification. Edge inference reduces cloud delay but needs optimized container orchestration.
2. **Throughput** refers to the number of emotion inferences made per second, particularly important in multi-user or high-density settings such as classrooms or group therapy sessions.
3. **Jitter**: Variability in inference time must be minimized to maintain system responsiveness and user trust.
4. **Accuracy**: Must not degrade under constrained compute resources; this requires efficient model compression and fusion techniques.
5. **Energy Efficiency**: This is especially critical in wearables or mobile edge devices to ensure continuous operation.

3.5.2 Edge-Specific Design Considerations

Practical edge-based MER demands coordinated strategies that balance latency, accuracy, and resource usage. Techniques like priority scheduling, model partitioning, and QoS-aware orchestration help ensure real-time, reliable emotion inference across diverse and dynamic environments [49, 67, 70]. The key strategies are detailed as follows:

- **Priority scheduling**: Ensures latency-sensitive modalities (e.g., speech) are processed ahead of less time-critical signals like physiological data.
- **Model partitioning**: Distributes processing by running early model layers on local devices and offloading deeper layers to the edge or cloud, optimizing load balance and meeting timing constraints.
- **Dynamic scaling**: Uses feedback controllers or reinforcement learning to adjust compute resource allocation in response to fluctuations in emotion signal intensity.
- **QoS-aware container orchestration**: Manages co-located workloads to ensure critical modalities receive sufficient computational resources.

Meeting QoS and deadline constraints is vital for deploying reliable MER on edge platforms. It requires careful co-design of sensing pipelines, fusion strategies, and compute resource management to ensure consistent, real-time operation even under mobility, connectivity fluctuations, and heterogeneous device capabilities.

3.6 Scalable Deployment of ML/DL Algorithms at the Edge

The scalability of MER on edge computing platforms relies on efficiently deploying algorithms across diverse, resource-limited devices such as smartphones, wearables, and embedded systems. Achieving this requires aligning model complexity with hardware capabilities, optimizing inference speed, and ensuring energy-efficient, real-time operation. Table 3.5 outlines various modalities alongside scalability considerations and suitable ML/DL algorithms [70, 76].

Scaling algorithms for edge-based MER operate at the container and device levels to ensure efficient, real-time processing. Container-level scaling enables dynamic resource allocation, while device-level scaling supports distributed processing across multiple edge nodes. Algorithms like PID and fuzzy logic controllers offer fast, rule-based adjustments based on system load. Predictive methods such as MPC and AI-based models forecast workloads to optimize resource use. Reinforcement learning and heuristics adapt scaling policies based on feedback and objectives, while queueing models, load balancing, and thresholding manage delays and spikes, ensuring responsive, low-latency performance.

Table 3.5 Scalability considerations for each modality and their ML/DL algorithms

Modality	Edge deployment challenges	Scalability approaches
Facial expressions	High-resolution video processing requires significant memory and computing resources	Lightweight CNNs (e.g., MobileNet, SqueezeNet) on smartphones/smart glasses Quantized CNNs or pruned ResNets for embedded GPUs On-device face detection + cloud-assisted inference
Speech (acoustic)	Real-time audio processing with low latency	On-device MFCC extraction + lightweight ML (e.g., SVM) TinyLSTM or edge-friendly speech transformers (e.g., DistilHuBERT) Streaming inference on audio chunks
Text/linguistic content	Requires on-device speech recognition to generate text	Distilled transformers (e.g., DistilBERT, TinyBERT) for fast inference Static embeddings (e.g., Word2Vec, GloVe) + ML classifiers Lightweight ASR (e.g., Whisper-tiny) or offloaded edge ASR
Physiological signals	Continuous time-series data from wearables	On-device feature extraction + lightweight ML (e.g., Random Forest) 1D CNNs or LSTMs with quantization/pruning Use edge-optimized DL frameworks (e.g., TensorFlow Lite, PyTorch Mobile)
Body gestures/posture	Requires skeletal data or video frame processing	Efficient pose estimators (e.g., BlazePose, MediaPipe) for skeleton keypoints GNNs or temporal CNNs on low-dimensional vectors 3D CNNs or RNNs for high-end edge devices (e.g., Jetson Nano)

3.6.1 Controllers Based on Control Theory

In edge computing environments, PID controllers are widely used for autoscaling containers or virtual machines (VMs). They adjust system parameters based on real-time feedback from metrics such as CPU usage, latency, or queue length of incoming signals [50]. For example, in a multimodal emotion detection system, a PID controller could scale up EEG processing containers when there's a sudden spike in user inputs or dropped frames [51]. Their simplicity and real-time responsiveness make them ideal for quick deployment, and their tunable parameters allow for fine control over scaling behaviours [52, 53].

Model Predictive Control offers a more advanced alternative to PID by using a system dynamics model to anticipate future resource demands. This is especially useful in edge computing scenarios where the workload varies over time, such as during conversational emotion detection, where text and speech processing demand may spike unpredictably. MPC can optimize scaling decisions over a prediction horizon while adhering to system constraints like limited memory or power, making it highly effective for planning in complex, multimodal environments [47–49, 54].

3.6.2 Fuzzy Logic Controllers

Fuzzy logic controllers are designed to handle uncertainty and nonlinearity in scaling decisions. Instead of relying on precise system models, they apply linguistic rules such as "if latency is high and CPU usage is moderate, scale moderately" [55]. This makes them suitable for managing ambiguous workloads in multimodal processing, where emotional cues from different modalities like gestures or speech may not always follow predictable patterns [56]. Fuzzy controllers provide a flexible and intuitive approach to autoscaling without requiring detailed system characterization [57, 58].

3.6.3 Queueing Theory-Based Algorithms

Queueing theory provides a strong mathematical foundation for modeling arrival and service rates in a multimodal system. These algorithms can be used to dynamically allocate edge compute resources based on the observed traffic patterns of emotion-related signals [59]. For instance, during a video call, if speech inputs start to queue, more compute resources can be directed toward speech analysis containers [60]. Common models include M/M/1, M/G/1 queues, and fluid approximations, which help optimize system throughput and minimize latency [61].

3.6.4 Reinforcement Learning (RL)

Reinforcement learning is an AI-driven approach that enables systems to learn optimal scaling policies through continuous environmental interaction. In multimodal emotion detection, RL agents can learn to prioritize specific modalities, such as increasing gesture processing power during a gaming session or focusing on physiological signals during a stress-detection task [62]. Techniques like Q-learning, Deep Q-Networks (DQN), and Actor-Critic methods allow the agent to balance trade-offs between latency, cost, and resource usage over time [63, 64].

3.6.5 Rule-Based Scaling with Thresholding

Rule-based scaling algorithms are simple yet effective for basic deployments. They operate on predefined thresholds—e.g., if CPU usage exceeds 80% on the image container, another container is spawned; if usage drops below 30%, resources are released [65]. While easy to implement and interpret, such systems lack adaptability and predictive capability, making them less suited for dynamic, real-world multimodal applications where signal behavior is often non-linear and temporally variable [66].

3.6.6 Load Balancing and Scheduling Algorithms

Effective load balancing and task scheduling are crucial in distributed edge environments to ensure optimal use of limited resources. Algorithms such as Round Robin and Weighted Round Robin provide even distribution of tasks, while more advanced strategies like Latency-Aware Load Balancing consider real-time performance metrics [67]. Data locality-based scheduling is particularly beneficial when handling large images or continuous physiological signals, as it minimizes data transfer times and reduces overall system latency [68].

3.6.7 Heuristic and Metaheuristic Approaches

Heuristic and metaheuristic methods are valuable for solving complex, multi-objective optimization problems. Techniques like Genetic Algorithms (GA), Simulated Annealing (SA), and Ant Colony Optimization (ANO) can be applied to determine the optimal configuration of VMs and containers across modalities like image, text, speech, and physiological data [67]. These approaches are beneficial when working under constraints such as energy consumption, bandwidth, or edge node heterogeneity, offering robust solutions where traditional optimization methods may struggle [69, 70].

3.6.8 AI-Based Auto-Scaling Using Predictive Models

Predictive models allow for proactive scaling decisions by forecasting future workloads based on historical trends. Techniques like Long Short-Term Memory (LSTM) networks, ARIMA models, or Facebook's Prophet can predict spikes in multimodal data streams [71, 73]. For example, a system might anticipate increased text input during online therapy sessions and pre-warm NLP processing containers accordingly [72]. This enables more responsive and efficient edge computing

Table 3.6 Algorithms for scalable MER

Algorithm type	Key features	Use case in MER
PID Control	Fast, simple, feedback-based Suitable for real-time container autoscaling	Scale up the EEG/image containers when latency or dropped frames increase
Model Predictive Control (MPC)	Forecasts future workload Optimizes resource allocation under constraints	Anticipate speech/text spikes during live emotion detection and pre-scale NLP/audio modules
Fuzzy Logic Controllers	Rule-based with linguistic logic Handles imprecise workload patterns	"If CPU is high and latency is moderate, scale moderately" during gesture and voice recognition
Queueing Theory Models	Based on service and arrival rates Useful for throughput and delay modeling	Allocate compute to speech containers when the audio queue length exceeds a threshold during interviews
Reinforcement Learning (RL)	Learns optimal scaling policies via feedback Prioritizes modalities based on context	Focus on gesture in gaming vs. physiology in stress monitoring
Rule-Based Thresholding	Simple thresholds (e.g., CPU > 80%) No predictive capabilities	Spawn more facial expression modules when the camera input spikes
Load Balancing or Scheduling	Includes Round Robin, Latency-Aware methods Emphasizes task distribution & locality	Assign processing of facial data to the nearest node to minimize delay
Heuristic /Metaheuristic	Multi-objective optimization (e.g., cost, energy) GA, ACO, SA	Find the best deployment of NLP, audio, and physiological processing under bandwidth/power constraints
AI-Based Predictive Scaling	Forecast workload using ML (LSTM, ARIMA, Prophet) Enables proactive scaling	Predict more user dialogue in a therapy app and prepare emotion-NLP containers

systems, particularly for time-sensitive applications like mental health monitoring or emotion-aware interactions. Table 3.6 summarizes control-theory-based and other algorithmic approaches tailored to this context.

3.6.9 Representative Use Cases: Research Projects Focusing on MER Deployed on Scalable Edge Platforms Using DL/ML Algorithms Are Listed as Follows

1. **Smartphone-based Emotion Recognition:** Researchers have developed lightweight CNN and LSTM models running directly on smartphones to

process facial expressions, speech, and physiological data (like heart rate via wearables) [74, 75]. These models enable real-time emotion detection with low power consumption and have been tested in mental health apps for stress and mood monitoring [74].

2. **Edge-based MER for Driver Monitoring**: Projects have integrated cameras, microphones, and physiological sensors in vehicles to monitor driver emotional states on embedded edge devices [76, 77]. Using optimized DL models with early and late fusion techniques, the system detects fatigue, distraction, or anger in real time, improving road safety without relying on cloud connectivity [77].
3. **Federated Learning for Privacy-Preserving Emotion Recognition**: Several studies implemented federated learning frameworks on edge devices (smartphones, wearables) to train emotion recognition models collaboratively without sharing raw, sensitive data [78, 79]. These systems combine multimodal inputs—facial, speech, and physiological signals—ensuring scalability and user privacy [79].
4. **MER in Smart Homes**: Edge computing platforms in smart homes use lightweight DL models to fuse voice commands, facial cues, and gesture data for personalized emotion-aware assistance [80]. These systems adapt to individual users, supporting elderly care and mental health interventions with local data processing [81].

These case studies demonstrate the feasibility and benefits of scalable edge deployment for multimodal emotion decoding, highlighting ongoing advances in lightweight models and fusion techniques.

3.7 Conclusion

For researchers monitoring mental health, real-time MER provides a valuable tool for capturing nuanced emotional states and psychological changes as they occur. This involves analyzing diverse data sources such as facial expressions, voice patterns, EEG, heart rate variability, and other physiological signals to detect early signs of stress, anxiety, or depression. Achieving reliable real-time analysis requires high-speed data processing, ideally within 10 milliseconds, to ensure timely feedback and preserve emotional context. However, advanced machine learning models used for this analysis are computationally and memory-intensive, making them difficult to deploy on mobile or wearable devices. While offloading to remote cloud servers is an option, the high latency, often exceeding 74 milliseconds, renders them unsuitable for time-sensitive applications, especially in field or ambulatory settings.

Edge computing offers a compelling alternative by decentralizing processing to local servers or cloudlets closer to the data source. These edge nodes enable low-latency real-time analysis even in mobile or resource-constrained environments. For researchers, this allows continuous subject monitoring, immediate emotion inference, and enhanced privacy without relying heavily on cloud infrastructure.

Integrating edge computing into mental health research workflows enhances emotion detection systems' accuracy, responsiveness, and scalability, enabling early interventions and richer insights into psychological well-being in everyday life.

Challenges in MER include handling noisy or missing data from different sensors, synchronizing diverse modalities with varying data rates, managing high computational demands, and ensuring the privacy and security of sensitive, emotional data [8]. There is also complexity in designing effective fusion strategies that capture complementary information without overfitting. Opportunities lie in improving accuracy and robustness by combining multiple data sources, enabling real-time emotion recognition on edge devices with optimized models, and applying these advances in mental health monitoring, human-computer interaction, and personalized user experiences. Advances in scalable edge computing, lightweight algorithms, and explainable AI further expand the potential for practical, privacy-preserving multimodal emotion decoding.

Future work in multimodal decoding of emotional states on scalable edge platforms involves developing more efficient and lightweight DL/ML algorithms [8, 11, 24] tailored for resource-constrained devices to enable real-time, low-latency processing. The research will focus on adaptive fusion techniques that dynamically adjust to varying sensor availability and quality, ensuring robust emotion recognition [41] despite noisy or missing data. Enhancing model compression, pruning, and quantization methods will further optimize deployment on diverse edge hardware. Additionally, advancing distributed learning frameworks [49, 64], like federated learning, can improve privacy by keeping sensitive, emotional data local while enabling collaborative model updates. Integrating explainable AI with scalable edge solutions will boost user trust and facilitate wider adoption in healthcare, human-computer interaction, and personalized services.

References

1. J.A. Cruz-Vazquez, J.Y. Montiel-Pérez, R. Romero-Herrera, E. Rubio-Espino, Emotion recognition from EEG signals using advanced transformations and deep learning. Mathematics **13**(2), 254 (2025)
2. H. Zhang, G. Tang, H. Huang, Z. Yuan, Z. Li, PulseEmoNet: Pulse emotion network for speech emotion recognition. Biomed. Signal Process. Control. **105**, 107687 (2025)
3. D.R. Faria, A.L. Godkin, P.P. da Silva Ayrosa, Advancing emotionally aware child–robot interaction with biophysical data and insight-driven affective computing. Sensors **25**(4), 1161 (2025)
4. H. Hadjar, B. Vu, M. Hemmje, TheraSense: Deep learning for facial emotion analysis in mental health teleconsultation. Electronics **14**(3), 422 (2025)
5. X. Wang, C.Z. Li, Z. Sun, Y. Xu, Design and analysis of a closed-loop emotion regulation system based on multimodal affective computing and emotional Markov chain. IEEE Trans. Syst. Man Cybern.: Syst. **55**, 2426–2437 (2025)
6. T. Yu, J. Wang, J. Luo, J. Wang, G. Zhou, TACL: A trusted action-enhanced curriculum learning approach to multimodal affective computing. Neurocomputing **620**, 129195 (2025)

7. A. Jain, A. Jain, AI-based emotion detection system in healthcare for patient, in *Generative Artificial Intelligence for Biomedical and Smart Health Informatics*, (Wiley, 2025), pp. 455–470
8. H. Lian, C. Lu, S. Li, Y. Zhao, C. Tang, Y. Zong, A survey of deep learning-based multimodal emotion recognition: Speech, text, and face. Entropy **25**(10), 1440 (2023)
9. V. Vielzeuf, A. Lechervy, S. Pateux, F. Jurie, Multilevel sensor fusion with deep learning. IEEE Sens. Lett. **3**(1), 1–4 (2018)
10. Y. Kwak, W.J. Song, S.E. Kim, FGANet: fNIRS-guided attention network for hybrid EEG-fNIRS brain-computer interfaces. IEEE Trans. Neural Syst. Rehabil. Eng. **30**, 329–339 (2022)
11. T.K. Mohd, N. Nguyen, A.Y. Javaid., Multimodal data fusion in enhancing human-machine interaction for robotic applications: A survey. arXiv preprint arXiv:2202.07732. 2022
12. Q. Lu, X. Sun, Y. Long, X. Zhao, W. Zou, J. Feng, X. Wang, Multimodal dual perception fusion framework for multimodal affective analysis. Inform. Fusion **115**, 102747 (2025)
13. C. Li, S. Zhang, Y. Mu, L. Yang, Y. Peng, F. Li, Y. Zhang, Z. Liang, Z. Cao, F. Wan, D. Yao, Emotion recognition by learning the manifold of fused multiscale information of EEG signals. IEEE Trans. Affect. Comput **16**, 2172–2188 (2025)
14. R. Pillalamarri, U. Shanmugam, A review on EEG-based multimodal learning for emotion recognition. Artif. Intell. Rev. **58**(5), 131 (2025)
15. T. Li, D. Liu, MPID: A modality-preserving and interaction-driven fusion network for multimodal sentiment analysis, in *Proceedings of the 31st International Conference on Computational Linguistics*, (2025, January), pp. 4313–4322
16. C.H. Espino-Salinas, H. Luna-García, J.M. Celaya-Padilla, C. Barría-Huidobro, N.K. Gamboa Rosales, D. Rondon, K.O. Villalba-Condori, Multimodal driver emotion recognition using motor activity and facial expressions. Front. Artif. Intell. **7**, 1467051 (2024)
17. S. Su, R. Sun, K. Zhang, Research on driver emotion monitoring and regulation system in intelligent assisted driving. Theor. Nat. Sci **51**, 128–136 (2024)
18. S.F.A. Razak, S. Yogarayan, A.A. Aziz, M.F.A. Abdullah, N.H. Kamis, Physiological-based driver monitoring systems: A scoping review. Civil Eng. J. **8**(12), 3952–3967 (2022)
19. J. Wang, C. Shan, L. Liu, Z. Hou, Camera-based physiological measurement: Recent advances and future prospects. Neurocomputing **575**, 127282 (2024)
20. M. Song, Z. Yang, A. Triantafyllopoulos, Z. Zhang, Z. Nan, M. Tang, H. Takeuchi, T. Nakamura, A. Kishi, T. Ishizawa, K. Yoshiuchi, Empowering mental health monitoring using a macro-micro personalization framework for multimodal-multitask learning: Descriptive study. JMIR Ment. Health **11**(1), e59512 (2024)
21. C.A. Christmann, A. Hoffmann, G. Bleser, Stress management apps with regard to emotion-focused coping and behavior change techniques: A content analysis. JMIR Mhealth Uhealth **5**(2), e6471 (2017)
22. P. Zagar, V. Ravi, L. Aalami, S. Krusche, O. Aalami, P. Schmiedmayer, Dynamic fog computing for enhanced llm execution in medical applications. Smart Health **36**, 100577 (2025)
23. L. Luan, W. Xiao, K. Hwang, M.S. Hossain, G. Muhammad, A. Ghoneim, MEMO box: Health assistant for depression with medicine carrier and exercise adjustment driven by edge computing. IEEE Access **8**, 195568–195577 (2020)
24. G. Hu, Y. Xin, W. Lyu, H. Huang, C. Sun, Z. Zhu, L. Gui, R. Cai, E. Cambria, H. Seifi, Recent trends of multimodal affective computing: A survey from NLP perspective. arXiv preprint arXiv:2409.07388 (2024)
25. J.P. Lee, H. Jang, Y. Jang, H. Song, S. Lee, P.S. Lee, J. Kim, Encoding of multimodal emotional information via personalized skin-integrated wireless facial interface. Nat. Commun. **15**(1), 530 (2024)
26. Z. Wang, S. Li, G. Shen, Advanced sensory hardware for intelligent eye-machine interfacing: From wearables to bionics. Adv. Funct. Mater. **14**, 2503519 (2025)
27. T. Liu, M. Zhang, Z. Li, H. Dou, W. Zhang, J. Yang, P. Wu, D. Li, X. Mu, Machine learning-assisted wearable sensing systems for speech recognition and interaction. Nat. Commun. **16**(1), 2363 (2025)

28. T. Liu, Y. Mao, H. Dou, W. Zhang, J. Yang, P. Wu, D. Li, X. Mu, Emerging wearable acoustic sensing technologies. Adv Sci **12**, e2408653 (2025)
29. A. Chaudhari, C. Bhatt, A. Krishna, P.L. Mazzeo, ViTFER: Facial emotion recognition with vision transformers. Appl. Syst. Innov. **5**(4), 80 (2022)
30. J. Fan, L. Dong, G. Sun, Z. Zhou, A deep learning approach for mental fatigue state assessment. Sensors **25**(2), 555 (2025)
31. A. Kumar, A. Kumar, Human emotion recognition using machine learning techniques based on the physiological signal. Biomed. Signal Process. Control **100**, 107039 (2025)
32. A.L. Cîrneanu, D. Popescu, D. Iordache, New trends in emotion recognition using image analysis by neural networks, a systematic review. Sensors **23**(16), 7092 (2023)
33. L.W. Chen, A. Rudnicky, Exploring wav2vec 2.0 fine tuning for improved speech emotion recognition, in *ICASSP 2023–2023 IEEE International Conference on Acoustics, Speech and Signal Processing (ICASSP)*, (IEEE, 2023, June), pp. 1–5
34. B. Basharirad, M. Moradhaseli, Speech emotion recognition methods: A literature review, in *AIP Conference Proceedings*, Vol. 1891, No. 1, (AIP Publishing, 2017, October)
35. I.J. Tashev, Z.Q. Wang, K. Godin, Speech emotion recognition based on Gaussian mixture models and deep neural networks, in *2017 Information Theory and Applications Workshop (ITA)*, (IEEE, 2017), pp. 1–4
36. L. Sun, S. Fu, F. Wang, Decision tree SVM model with fisher feature selection for speech emotion recognition. Eurasip J. Audio Speech Music Process. **2019**(1), 1–14 (2019)
37. J. Deng, Y. Liu, Research on sentiment analysis of online public opinion based on RoBERTa–BiLSTM–attention model. Appl. Sci. **15**(4), 2148 (2025)
38. S. Imtiaz, E.K. Hashi, Analyzing the performance of sentiment analysis using BERT, DistilBERT, RoBERTa and GPT-2, in *2024 6th International Conference on Sustainable Technologies for Industry 5.0 (STI)*, (IEEE, 2024, December), pp. 1–6
39. T. Chutia, N. Baruah, A review on emotion detection by using deep learning techniques. Artif. Intell. Rev. **57**(8), 203 (2024)
40. K.A. Baihaqi, I. Setyawan, D. Manongga, H.D. Purnomo, H. Hendry, A. Fauzi, A. Hananto, A comparison support vector machine, logistic regression and Naïve Bayes for classification Sentimen Analisys user Mobile app. Int. J. Artif. Intell. Res. **7**(1), 64–71 (2023)
41. Z. Ahmad, N. Khan, A survey on physiological signal-based emotion recognition. Bioengineering **9**(11), 68 (2022)
42. M.N. Dar, M.U. Akram, S.G. Khawaja, A.N. Pujari, CNN and LSTM-based emotion charting using physiological signals. Sensors **20**(16), 4551 (2020)
43. J.A. Domínguez-Jiménez, K.C. Campo-Landines, J.C. Martínez-Santos, E.J. Delahoz, S.H. Contreras-Ortiz, A machine learning model for emotion recognition from physiological signals. Biomed. Signal Process. Control **55**, 101646 (2020)
44. F. Noroozi, C.A. Corneanu, D. Kamińska, T. Sapiński, S. Escalera, G. Anbarjafari, Survey on emotional body gesture recognition. IEEE Trans. Affect. Comput. **12**(2), 505–523 (2018)
45. S.C. Leong, Y.M. Tang, C.H. Lai, C.K.M. Lee, Facial expression and body gesture emotion recognition: A systematic review on the use of visual data in affective computing. Comput. Sci. Rev. **48**, 100545 (2023)
46. E. Zhong, C.R. Del-Blanco, D. Berjón, F. Jaureguizar, N. García, Real-time monocular skeleton-based hand gesture recognition using 3d-jointsformer. Sensors **23**(16), 7066 (2023)
47. A.S. Seisa, S.G. Satpute, B. Lindqvist, G. Nikolakopoulos, An edge architecture oriented model predictive control scheme for an autonomous uav mission, in *2022 IEEE 31st International Symposium on Industrial Electronics (ISIE)*, (IEEE, 2022, June), pp. 1195–1201
48. M. Berger, F. Bernardello, C. Barry, P. Badjoonauth, S. Balaji, M. Lakhdar, Real-time model predictive control with digital twins and edge computing technologies, in *CLIMA 2022 conference*, (2022)
49. N. Feng, C. Ran, Design and optimization of distributed energy management system based on edge computing and machine learning. Energy Inform. **8**(1), 17 (2025)
50. M. Sabuhi, N. Mahmoudi, H. Khazaei, Optimizing the performance of containerized cloud software systems using adaptive PID controllers. ACM Trans. Auton. Adapt. Syst. **15**(3), 1–27 (2021)

51. L. Wen, M. Xu, S.S. Gill, M. Hilman, S.N. Srirama, K. Ye, C. Xu, StatuScale: Status-aware and elastic scaling strategy for microservice applications. ACM Trans. Auton. Adapt. Syst. **20**(1), 1–25 (2025)
52. P. Bandi, P.S. Saikrishna, Event-driven offloading and load-balancing with quantized control for fog-enabled UAVs. IEEE Trans. Industr. Inform ***20**, 14257–14266* (2024)
53. D. Vinod, P.S. SaiKrishna, Development of an autonomous fog computing platform using control-theoretic approach for robot-vision applications. Robot. Auton. Syst. **155**, 104158 (2022)
54. D. Vinod, J. Zhou, An event-triggered MPC scheme for discrete pLPV systems: Application in auto-scaling of compute nodes in a fog cluster, in *IECON 2024-50th Annual Conference of the IEEE Industrial Electronics Society*, (IEEE, 2024, November), pp. 1–6
55. M. Alves Almeida, N. Souto Rosa, Exploiting fuzzy controllers for adapting messaging systems, in *Proceedings of the 40th ACM/SIGAPP Symposium on Applied Computing*, (2025, March), pp. 292–299
56. C. Sonmez, A. Ozgovde, C. Ersoy, Fuzzy workload orchestration for edge computing. IEEE Trans. Netw. Serv. Manag. **16**(2), 769–782 (2019)
57. M.B. Mansour, T. Abdelkader, M. Hashem, E.S.M. El-Horbaty, An integrated three-tier trust management framework in mobile edge computing using fuzzy logic. PeerJ Comput. Sci. **7**, e700 (2021)
58. W. Jin, A. Rezaeipanah, Dynamic task allocation in fog computing using enhanced fuzzy logic approaches. Sci. Rep. **15**(1), 1–16 (2025)
59. R. Tolosana-Calasanz, J. Diaz-Montes, O.F. Rana, M. Parashar, Feedback-control & queueing theory-based resource management for streaming applications. IEEE Trans. Parallel Distrib. Syst. **28**(4), 1061–1075 (2016)
60. I. Mohamed, H. Al-Mahdi, M. Tahoun, H. Nassar, Characterization of task response time in fog enabled networks using queueing theory under different virtualization modes. J. Cloud Comput. **11**(1), 21 (2022)
61. L. Rodrigues, J.J. Rodrigues, A.D.B. Serra, F.A. Silva, A queueing-based model performance evaluation for internet of people supported by fog computing. Future Internet **14**(1), 23 (2022)
62. B. Xie, H. Cui, Deep reinforcement learning-based dynamical task offloading for mobile edge computing. J. Supercomput. **81**(1), 35 (2025)
63. X. Chen, H. Zhang, C. Wu, S. Mao, Y. Ji, M. Bennis, Performance optimization in mobile-edge computing via deep reinforcement learning, in *2018 IEEE 88th Vehicular Technology Conference (VTC-Fall)*, (IEEE, 2018, August), pp. 1–6
64. Y. Li, F. Qi, Z. Wang, X. Yu, S. Shao, Distributed edge computing offloading algorithm based on deep reinforcement learning. IEEE Access **8**, 85204–85215 (2020)
65. Y. Jin, Z. Yang, Scalability optimization in cloud-based AI inference services: Strategies for real-time load balancing and automated scaling, in *Proceedings of the 2025 4th International Conference on Big Data, Information and Computer Network*, (2025, January), pp. 266–270
66. M.S.M. Pakpahan, L.E. Nugroho, Comparative analysis of rule-based heuristic algorithms for microservice chain placement in fog computing. Results Eng. **25**, 104299 (2025)
67. A. Choupani, S. Azizi, M.S. Aslanpour, Joint resource autoscaling and request scheduling for serverless edge computing. Clust. Comput. **28**(3), 171 (2025)
68. G.B.J. Mdemaya, M.L.F. Sindjoung, M.M.Z. Ndadji, M. Velempini, *HERCULE: High-efficiency resource coordination using kubernetes and machine learning in edge computing for improved QoS and QoE*, vol 13 (IEEE Access, 2025), p. 72153
69. F. Amirghafouri, A.A. Neghabi, H. Shakeri, Y.E. Sola, Nature-inspired meta-heuristic algorithms for resource allocation in the internet of things. Int. J. Commun. Syst. **38**(5), e6141 (2025)
70. I. Taleb, J.L. Guillaume, B. Duthil, A survey on services placement algorithms in integrated cloud-fog/edge computing. ACM Comput. Surv **57**, 1–36 (2025)
71. T.P. da Silva, A.R. Neto, T.V. Batista, F.C. Delicato, P.F. Pires, F. Lopes, Online machine learning for auto-scaling in the edge computing. Pervasive Mob. Comput. **87**, 101722 (2022)

72. A.M. Mohammed, J. Syed, Revolutionizing autonomous cloud infrastructure: AI-driven predictive auto scaling with attribute-based instance selection in AWS. J. Comput. Sci. Technol. Stud. **7**(2), 249–253 (2025)
73. M. Filippov, M. Mazzara, Autoscaling containerized microservices: A survey, in *International Conference on Advanced Information Networking and Applications*, (Springer, Cham, 2025, April), pp. 326–334
74. H.R. Sneha, M. Rafi, M.M. Kumar, L. Thomas, B. Annappa, Smartphone based emotion recognition and classification, in *2017 Second International Conference on Electrical, Computer and Communication Technologies (ICECCT)*, (IEEE, 2017), pp. 1–7
75. I. Ziogas, L.J. Hadjileontiadis, A.H. Khandoker, A. Al Shehhi, Self-supervised graph representation learning for in-the-wild wearable and smartphone based emotion recognition, in *ICASSP 2025–2025 IEEE International Conference on Acoustics, Speech and Signal Processing (ICASSP)*, (IEEE, 2025, April), pp. 1–5
76. P. Cassarà, S. Bano, A. Gotta, Mobility-aware edge-assisted 5G communication framework analysis for driver emotion recognition, in *2025 IEEE Wireless Communications and Networking Conference (WCNC)*, (IEEE, 2025, March), pp. 1–6
77. S.B. Chandolu, N.M. Medidhi, C. Satyala, Z. Mohammad, A.R. Sharma, P.P. Venigalla, Computer Vision-Based Multimodal Advanced Driver Assistance System, in *2025 International Conference on Inventive Computation Technologies (ICICT)*, (IEEE, 2025, April), pp. 638–645
78. H. Chen, H. Zhao, Y. Gao, Y. Liu, Z. Zhang, Parameter-efficient Federal-Tuning Enhances Privacy Preserving for speech emotion recognition, in *ICASSP 2025–2025 IEEE International Conference on Acoustics, Speech and Signal Processing (ICASSP)*, (IEEE, 2025, April), pp. 1–5
79. J. Wang, M.T. Quasim, B. Yi, Privacy-preserving heterogeneous multi-modal sensor data fusion via federated learning for smart healthcare. Inform. Fusion **120**, 103084 (2025)
80. M. Xu, T. Shi, H. Zhang, Z. Liu, X. He, A hierarchical cross-modal spatial fusion network for multimodal emotion recognition. IEEE Trans. Artif. Intell. **6**, 1429–1438 (2025)
81. A. Khalane, R. Makwana, T. Shaikh, A. Ullah, Evaluating significant features in context-aware multimodal emotion recognition with XAI methods. Expert. Syst. **42**(1), e13403 (2025)

Chapter 4
Modeling Human Emotions via Interpretable EEG-Based Spatio-Temporal Attention Network

Neha Sharma, Kumar Gaurav, and Tharun Kumar Reddy Bollu

Abstract Emotion recognition from EEG signals represents a significant area of research within cognitive computing. Key challenges in this domain include the extraction of meaningful and discriminative features from non-stationary EEG signals and the development of highly accurate predictive models. To address these challenges, artificial intelligence (AI) techniques have been widely adopted to construct decision-support systems across various specialized domains. In particular, recent advancements in brain-computer interface (BCI) research have seen increased application of deep learning (DL) models for the interpretation and analysis of neural data. However, the inherent complexity and black-box nature of deep learning architectures raise concerns regarding the interpretability and transparency of model predictions. To enhance the interpretability of AI algorithms and their decision-making processes, an interpretable lightweight EEG-based model serves as a promising solution. This study proposes a novel attention-based deep learning architecture combined with an interpretable artificial intelligence technique for determining the significance of spatial-temporal features and the model's predicted output for classifying human emotional states based on EEG signals. The publicly available DEAP emotion dataset is evaluated using cross-validation for multi-label classification across four primary emotional dimensions—arousal, valence, dominance, and liking–achieving classification accuracies of 73.83%, 79.29%, 62.13% and 59.37%, respectively. The interpretable AI saliency map highlights the most significant brain regions associated with each emotional state at the time of prediction. The experimental results demonstrate that the proposed framework outperforms existing methods in terms of prediction accuracy approx 3%. Furthermore, the proposed method achieves high classification performance while ensuring interpretability for emotion recognition applications.

Keywords EEG · Emotion recognition · Interpretability · XAI method · Deep learning

---

N. Sharma · K. Gaurav · T. K. R. Bollu (✉)
Electronics & Communication Engineering, IIT Roorkee, Roorkee, Uttarakhand, India
e-mail: neha_s1@ece.iitr.ac.in; kumar_g@ece.iitr.ac.in; tharun.reddy@ece.iitr.ac.in

R. Vinjamuri (ed.), *Bridging the Gap between Mind and Machine*,
https://doi.org/10.1007/978-3-032-06713-5_4

4.1 Introduction and Motivation

Emotions, also known as affective phenomena, are elicited and modulated by external stimuli [1]. In particular, audiovisual content, such as films and videos, is frequently used to evoke distinct emotional responses in viewers. The identification of emotional traces in such content can be achieved through real-time emotion detection mechanisms. Emotion detection facilitates the unobtrusive recognition of viewer responses without interrupting the viewing experience, thereby eliminating the need for subjective self-reports, which are often influenced by social and personality-related biases. Recent advances in psychology and neuroscience emphasize the importance of emotional-rational interplay and underscore the pivotal role of affective states in human decision-making processes [2]. Signals originating from the central nervous system (CNS) offer crucial insights into the relationship between emotional fluctuations and psychological changes. These signals can be acquired via neurophysiological techniques such as Electroencephalography (EEG) and Intracranial EEG (iEEG). Among these, EEG has emerged as a promising modality for decoding the neural correlates of emotion, enabling the exploration of the intricate association between affective states and brain activity. EEG, characterized by its high temporal resolution, captures real-time electrical activity of the brain, providing a non-invasive window into the dynamic neural mechanisms underlying emotional processing. As a surface-level bio-signal acquisition technique, EEG detects voltage fluctuations generated by neuronal ensembles. Extra-cranial EEG systems encompass both body-worn and non-body-worn devices, which are increasingly adopted due to their non-invasiveness, enhanced wearability, decreasing costs, and expanding applicability across diverse domains [3]. A key challenge in EEG-based emotion classification is effectively modeling the complex, non-stationary, and subject-dependent characteristics of EEG signals [4]. These signals inherently have a low signal-to-noise ratio and exhibit high variability across individuals. Although various deep learning models, including CNNs and RNNs, have been employed to extract discriminative features from EEG data, they often struggle with robustness and generalization across datasets due to significant inter-subject variability and limited data availability [5]. Additionally, many existing architectures are computationally expensive and prone to overfitting, especially when applied to small or imbalanced EEG datasets [6].

To overcome these limitations, we propose a compact deep learning architecture specifically designed for the analysis of EEG signals. Our model leverages depthwise and separable convolutions to efficiently capture both spatial and temporal dynamics of EEG signals while maintaining a lightweight structure. Additionally, we incorporate a channel attention network to adaptively recalibrate feature representations by emphasizing the most informative EEG channels and suppressing less relevant ones. This attention mechanism enhances the model's ability to focus on critical spatial features, improving discriminative power. Together, these design choices substantially reduce the number of trainable parameters, making the architecture well-suited for small-sample EEG studies. Moreover, it enhances performance consistency across subjects, offering a robust and efficient alternative for emotion recognition in Brain-Computer Interface (BCI) systems. In the domain of EEG-based emotion classification, some deep learning models have demonstrated superior performance due to their ability to capture intricate patterns within temporal and spatial EEG data. However, their complex structure often leads to a lack of transparency, making it difficult to interpret how decisions are made or which features significantly influence the classification outcomes. This black-box nature presents a critical barrier in sensitive applications such as brain–computer interfaces, where interpretability is essential for reliability and user trust [7]. The incorporation of explainable artificial intelligence (XAI) techniques addresses this limitation by enabling the extraction of meaningful insights from the model's internal representations. Through model interpretation strategies, it becomes possible to identify the most relevant signal components and understand how different features contribute to emotion recognition. This enhanced interpretability not only promotes transparency but also aids in refining model performance and ensuring its alignment with underlying neural patterns [8].

The proposed chapter represents a novel attention-based deep learning architecture combined with an interpretable artificial intelligence technique for determining the significance of specific features and the model's predicted output by utilising multiple information sources to classify human emotional states based on EEG signals. A novel attention-based deep learning architecture is designed, specifically for efficient learning from EEG data, and is applied to the DEAP dataset across four key emotional dimensions: arousal, valence, dominance, and liking. To improve interpretability, saliency map techniques are used to visualise the importance of various spatial and temporal EEG features in driving the model's decisions. This work integrates an interpretable AI method into a neural network framework to achieve more interpretable and reliable human emotional state classification. The main contributions of the presented chapter are:

- Propose a novel, interpretable, lightweight attention-based deep learning architecture for efficient and accurate emotion state classification.
- Integration of the Saliency map method to enhance model interpretability by identifying and visualizing the key features that play a crucial role in classification decisions.
- Comprehensive evaluation of the proposed approach on the emotional state benchmark dataset, demonstrating its effectiveness in achieving high classification accuracy and enhanced interpretability, addressing key challenges in affective computing.

The subsequent sections of this chapter are organised as follows: Sect. 4.2 presents a detailed review of existing literature and current state-of-the-art approaches for EEG-based emotional state classification, incorporating interpretable AI techniques. Section 4.3 outlines the dataset description and problem statement. Section 4.4 details the mathematical formulation of the proposed methodology. Section 4.5 discusses the experimental results and provides a comparative analysis and a comprehensive discussion. Finally, Section 4.6 concludes the chapter by summarising key findings and contributions.

4.2 Literature Review and State of the Art

The most recent techniques and methods for classifying emotions and interpreting algorithms are described in this section. The goal of affective computing is to create emotionally intelligent machines that can improve the human-computer interaction's performance [9]. Humans may predict distinct emotional states by using a variety of verbal and nonverbal cues, such as physiological signals, speech signals, and facial expressions [1]. Both verbal indicators, like speech signals and the ability to recognize facial emotions, can be faked and deceptive. As a result, since physiological signals yield more consistent findings, emotion detection using them is growing in popularity. Electric impulses known as electroencephalographic (EEG) signals are obtained directly from the brain cortex through various neural pathways and can yield more accurate and reliable data [10]. The research area of emotional brain-computer interface (BCI) is gaining traction as a promising study area that uses recorded EEG data to anticipate emotions. However, identifying emotions from EEG signals remains a difficult challenge due to the significant pre-processing required for the data. Continuous electric signals, or EEG signals, comprise five distinct frequencies divided into delta, theta, alpha, gamma, and beta bands [11]. The key difficulty is determining how transient cerebral states are associated with a particular pattern of brain activity, since feature extraction, choosing the appropriate frequency band, and selecting the relevant characteristics from these signals are involved. Another significant issue is the quantity of information and attributes needed to categorise these affective states. These signals are continuous, non-linear, and random signals [12]. Still, the publicly available DEAP dataset serves as a critical resource for the human-computer interaction (HCI) research community, particularly in emotion recognition using physiological signals [13]. This dataset facilitates the classification of emotional dimensions induced by audiovisual stimuli, specifically music videos, by capturing signals from the central nervous system (CNS) and peripheral nervous system (PNS). DEAP employs Russell's valence-arousal model, a widely adopted dimensional framework in affective computing, to quantify emotional states. Liu et al. proposed a real-time EEG-based emotion recognition algorithm utilizing the Higuchi Fractal Dimension (FD) spectrum as a feature extraction method, with Support Vector Machines (SVM) employed as the classifier, achieving a classification accuracy of 53.7% across eight emotional categories [14]. Srivastava et al. applied a discrete wavelet transform (DWT) to decompose EEG signals into distinct frequency bands, followed by a feedforward neural network for classification, where a swarm optimization algorithm was used to tune network parameters [15]. Candra et al. identified that very short time windows in EEG recordings significantly hinder extracting informative features for accurate emotion state classification [16]. Chung and Yoon conducted a Bayesian classification study to categorize user emotional responses into binary (high, low) and ternary (high, normal, low) classes along valence and arousal axes. Their approach yielded classification accuracies of 66.6 and 66.4% for the two-class setup, and 53.4 and 51.0% for the three-class setup, on valence and arousal dimensions, respectively [17]. Although multiple deep learning models have been explored for EEG-based emotion classification, they have not consistently achieved satisfactory performance, and there is a lack of interpretable models capable of providing insights into the model's decision-making process.

Another challenge for deep learning architectures in emotion recognition is their inherently black-box nature, where the internal representations and learned knowledge remain opaque and inaccessible [18]. This lack of transparency can significantly hinder user trust and acceptance, particularly in high-risk domains such as healthcare [19] and transportation [20]. Moreover, interpretability is essential for trust, enhancing our understanding of emotional processes and improving model performance. Weitz et al. applied Layer-wise Relevance Propagation (LRP) and Local Interpretable Model-agnostic Explanations (LIME) to analyse the decision-making process of deep neural networks, enabling them to differentiate facial expressions associated with pain and disgust, thereby assisting healthcare professionals such as nurses in providing better patient care [21]. Zhou et al. employed SHAP (Shapley Additive Explanations) to identify the most salient Singular Spectrum Analysis (SSA) components that contribute to improving the performance of the LightGBM classifier in emotion state classification [22]. Despite the growing use of different types of attention-based deep learning architectures in emotion recognition, there remains a significant gap in applying interpretable AI methods to interpret their decision-making processes.

4.3 Problem Statement and Dataset Description

Effectively emotion state classification from electroencephalogram (EEG) signals requires highly accurate deep learning architectures and mechanisms for model interpretability. In neuro-informatics and computational neuroscience, EEG-based emotion recognition has demonstrated the potential to enhance brain-computer interface (BCI) systems. However, despite recent advances in deep learning methodologies, challenges persist in achieving reliable generalization across subjects and sessions. Furthermore, the black box nature of deep neural networks limits their deployment in critical BCI applications where interpretability and transparency are essential. Therefore, a core challenge remains significant needs of optimized and interpretable deep learning architecture for classification of human emotion state and achieving a balance between classification performance and interpretability.

The DEAP EEG dataset, which is publicly available, is utilized in this study to classify emotional states—Valence, Arousal, Dominance, and Liking [13]. The dataset comprises two main components: first, the results of an online self-assessment where 120 one-minute music video clips were rated by 14 to 16 volunteers based on arousal, valence, and dominance. Second, it includes data from an experiment in which 32 participants viewed a subset of 40 of those videos, during which physiological signals and EEG data were recorded, along with subjective ratings of the same emotional parameters. For 22 of these participants, frontal face video was also recorded. The official dataset provides all individual online ratings, YouTube links to the music videos used, participant ratings of the videos, responses to pre-experiment questionnaires, and frontal face recordings from participants 1 through 22. For this study, only the EEG signals were employed. Data was gathered from 32 participants as they watched one-minute-long music video clips. An EEG headset with multiple electrodes was used to record brain activity. After each video, participants rated their emotional responses in terms of valence, arousal, dominance, and liking, using a scale from 0 to 9 to reflect the intensity of each emotion experienced. Each participant's EEG data is structured as a 3D array of dimensions $40 \times 32 \times 8064$, representing 40 trials, 32 EEG channels, and 8064 data points (captured at 128 Hz over 63 seconds). The corresponding emotional scores are stored in a 40×4 matrix (trials $\times$ emotion dimensions). The publicly available version of the dataset underwent preprocessing: the EEG signals were downsampled to 128 Hz, ocular artefacts (EOG) and noise were removed, and a low-pass filter was applied. Additionally, 3-second baseline segments were excluded, and channels 33 to 40 were omitted from the data. Consequently, the final processed EEG dataset per participant is $40 \times 32 \times 7680$, with emotion labels in a 40×2 format.

4.4 The Proposed Solution

The proposed methodology as presented in Fig. 4.1 employs an attention-based deep learning architecture integrated with a saliency map approach for model interpretability within an interpretable AI (XAI) framework. The architecture is specifically designed to capture both the temporal–spectral characteristics and spatial dependencies inherent in multichannel EEG signals while incorporating a channel-wise attention mechanism to dynamically emphasize the most informative regions. The input to the model is structured as a four-dimensional tensor of shape $N \times 1 \times C \times T$, where N denotes the number of EEG samples, C represents the number of channels (electrodes), and T corresponds to the number of time points per trial, sampled at a frequency of 128 Hz. The architecture is structured into four principal blocks, which are elaborated here:

- **Temporal Convolution Layer:** The initial block applies a 2D convolution operation using a temporal kernel that enables the network to learn data-driven temporal filters that function as band-pass filters along the time axis [23]. These filters extract frequency-specific components from the EEG signal by convolving the input with kernels of length equal to half the sampling rate, i.e., 64 Hz for 128 Hz, effectively capturing frequency information at 2 Hz and above. The resulting feature maps represent the EEG signal filtered across multiple spectral

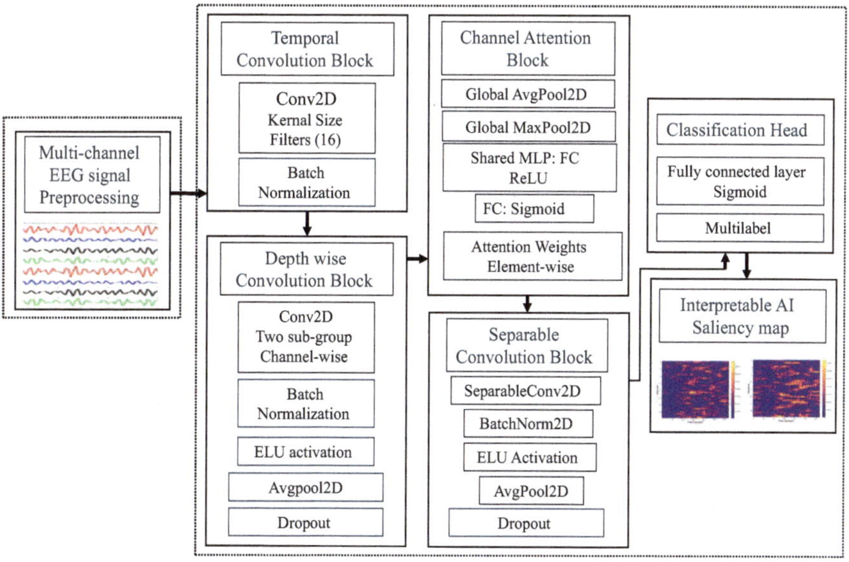

Fig. 4.1 The proposed architecture: human emotions via interpretable EEG-based spatio-temporal attention network

bands. This is followed by batch normalization, which stabilizes learning by normalizing the activations across the batch and applies learnable scaling and shifting parameters to preserve representational capacity.

- **Depthwise convolution layer:** The depthwise convolution layer performs spatial filtering across EEG channels following the temporal convolution. In convolutional neural network (CNN) applications for computer vision, the primary advantage of depthwise convolution is the significant reduction in the number of trainable parameters, as these convolutions are not fully connected to all preceding feature maps. When adapted for EEG-specific applications, this operation enables the model to learn spatial filters independently for each temporal filter, allowing for the extraction of frequency-dependent spatial representations. A depth multiplier parameter D controls the number of spatial filters per temporal feature map (D=2). This structure mirrors the two-stage decomposition strategy found in signal processing approaches like Filter Bank Common Spatial Pattern (FBCSP). It is conceptually related to methods such as Bilinear Discriminant Component Analysis. Technically, this layer implements group-wise convolution using $C \times 1$ kernels—one per temporal group—to model inter-channel spatial dependencies. As a result, the model effectively projects the multi-channel EEG signal into a spatial feature space that captures correlations across electrodes. Batch normalization is applied along the feature dimension to stabilize training, followed by the exponential linear unit (ELU) non-linearity to introduce activation sparsity. Additionally, Dropout is employed after this block to reduce overfitting and improve generalization capability of the model.

- **Channel Attention layer:** A channel attention mechanism is incorporated to enhance the model's ability to emphasize the most informative spatial features selectively. Inspired by Squeeze-and-Excitation (SE) networks, this module performs global average pooling and global max pooling along the temporal dimension to generate a compact descriptor for each EEG channel [24]. These descriptors are then passed through a lightweight two-layer fully connected network (MLP) comprising a ReLU activation followed by a sigmoid function, which outputs attention weights per channel. Each channel in the feature map is subsequently scaled by its corresponding attention weight, allowing the network to prioritize relevant spatial features while suppressing less informative signals dynamically. This mechanism improves the representational power of the model by promoting data-driven spatial weighting across channels.
- **Separable convolution and classification layer:** The feature map output is forwarded to a separable convolution layer, which applies additional temporal filtering using smaller kernels of size (1, C/2). This layer enhances the temporal resolution of the extracted features, enabling the model to learn finer-scale temporal dynamics. The separable convolution operation is composed of a depth-wise convolution followed by a point-wise convolution, effectively reducing the number of trainable parameters while preserving the expressiveness of the learned representations. To reduce the temporal dimensionality of the resulting feature maps, average pooling is applied, which introduces translation invariance and helps in improving generalization while lowering computational overhead in the downstream layers. The resulting tensor is then flattened into a one-dimensional feature vector. Finally, a fully connected layer maps this vector to a four-dimensional output, corresponding to the four affective emotional states (valence, arousal, dominance, and liking).
- **Interpretable AI-Saliency Map:** A saliency map is an interpretable AI (XAI) method that identifies critical input features contributing to a model's predictive decision function. Within EEG-based emotion recognition tasks, the saliency map isolates spatial (electrode channels) and temporal segments that significantly influence the model's inference of affective states [25]. Let the trained neural network be denoted as $f(x)$, where the input signal size is $(C \times T)$, and K represents the output prediction vector for K emotional states. The saliency map is computed as the gradient of the output concerning the input signal, as $\frac{\partial f_K(x)}{\partial x}$, where f_K corresponds to the activation for class K.. This gradient quantifies the sensitivity of the model's output to perturbations in each input feature, thereby highlighting features with the highest impact on decision boundaries.

4.5 Results and Discussion

The proposed computational framework was systematically evaluated on the DEAP emotion recognition dataset utilizing a high-performance GPU (P100) environment provided by Kaggle. To mitigate the class imbalance issue, inverse frequency weighting was employed for each categorical label according to its empirical distribution. Furthermore, comprehensive pre-processing techniques were applied to enhance the model's generalization capabilities across subjects and trials. The proposed architecture is described in four stages: The input EEG signals are structured with a dimensionality of $(N \times 1 \times 32 \times 7680)$, where N (Batch size)=8. Within the temporal convolution block, a total of 16 temporal filters are applied independently to each EEG channel using a convolution kernel of size (1,64). A depthwise 2D convolution is performed with a kernel of size (C,1), where C=32, grouped into two sets of 16 channels to capture spatial dependencies across electrodes per frequency band. This operation is followed by batch normalization and the application of the exponential linear unit (ELU) activation function to ensure non-linearity and training stability. Subsequently, average pooling is applied with a kernel size of (1,4), and a dropout rate of 0.6 is introduced for regularization. The resulting feature space is shaped as $(N \times 32 \times 1 \times T//4)$. To dynamically emphasize the most informative features, a channel attention mechanism is employed using a squeeze-and-excitation (SE) block operating on the same output shape $(N \times 32 \times 1 \times T//4)$. For further refinement of fine-grained temporal frequency patterns, the separable convolution block utilizes ELU activation, average pooling, and a dropout of 0.6. Finally, the high-level abstract features are flattened and passed through a fully connected neural network that is the classification head for predicting the target emotional states. The proposed model achieved a 5-fold cross-validated classification accuracy of 79.29% for valence, 73.83% for arousal, 62.13% for dominance, and 59.37% for liking emotional dimensions. The relatively lower performance in dominance and liking prediction is attributed to class imbalance and limited representative samples, despite the application of inverse frequency weighting to mitigate this issue. Validation and training accuracy plots are represented in Fig. 4.2. The performance metrics in Table 4.1 indicate a low precision, primarily due to a high number of false positives despite a substantial number of true positive cases. However, the recall improves correspondingly, as the model successfully identifies a larger proportion of actual positive instances. Additionally, we compare the emotion recognition results with several recent studies. Most of the existing studies used subject-dependent settings and segmented an entire trial into several windows to increase the number of samples. The cross-validated accuracy of our proposed model enhances significantly by approximately ~ 3 to 6% for valence and arousal state surpasses the previous state-of the-art studies on DEAP dataset, which is shown in Table 4.2. The saliency map is employed to interpret the model's prediction by highlighting the most influential spatiotemporal features in Fig. 4.3. For the arousal emotional state, the model indicates that neural activity within the frontal (Fp1, AF3, F7, F3) and frontocentral (FC1) regions during 70 to 100 sec specific time interval plays a critical role in influencing the model's predictions. In valence emotional state, the model demonstrates that neural activity within the parietal and occipital regions (PO4, P4, P8) during the 50 to 70 sec time interval significantly influences the model's output. The model highlights that liking emotional neural activity across a broad range of regions—including frontal (AF3, F7), central-parietal (CP1, CP5, P3, Pz), parietal-occipital (PO3, O1, Oz, O2), as well as parietal,

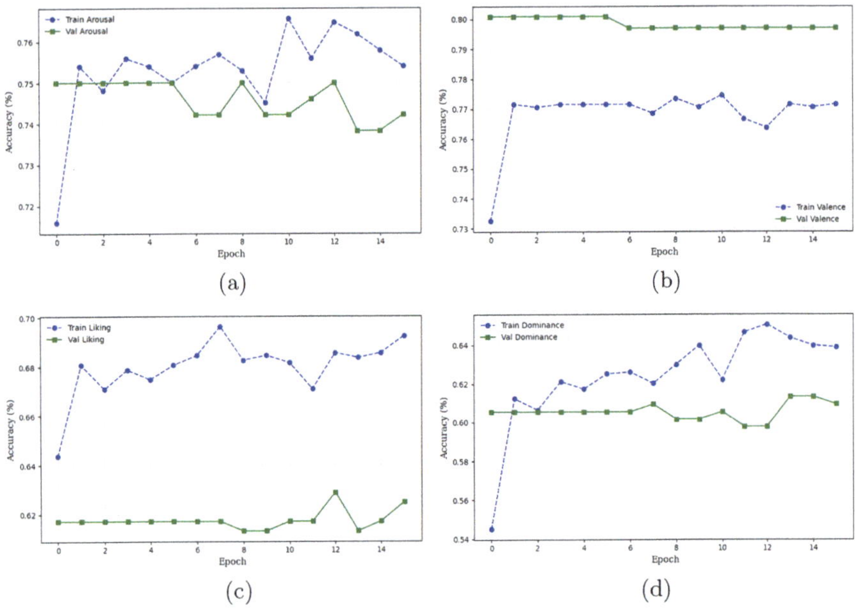

Fig. 4.2 Validation set accuracy and Training accuracy of our method over epochs. (**a**) Arousal emotional state. (**b**) Valence emotional state. (**c**) Liking emotional State. (**d**) Dominance emotion state

Table 4.1 The cross-validated performance metrics for multi-label classification on DEAP dataset

Emotion state	Accuracy (%)	Precision (%)	Recall (%)	F1-score (%)
Arousal	73.83	72.10	87.05	79.00
Valence	79.29	68.56	83.20	74.20
Dominance	62.13	54.03	75.21	63.20
Liking	59.37	52.25	72.34	60.79

central, temporal, and frontal areas (P4, C4, T8, FC6, Cz)—within the analyzed time interval significantly contributes to the model's output and For the dominance emotional state, the model indicates that neural activity within the frontal (F3, Fz), central (Cz), parietal (P4), and parieto-occipital (PO3) regions during the relevant time window plays a significant role in influencing the model's predictions.

4.6 Conclusion and Future work

This chapter presented a novel interpretable attention-based deep learning architecture for EEG-based emotional state classification. The proposed model incorporates a channel-attention mechanism to enhance the discriminative capacity of the

Table 4.2 Performance comparison (% accuracy) with previous studies on the DEAP dataset

Author	Arousal accuracy (%)	Valence accuracy (%)
Lew et al. [26]	56.78	56.60
Pandey et al. [27]	62.50	61.25
Rayatdoost et al. [28]	59.22	55.70
Li et al. [29]	64.20	58.40
Pandey et al. [30]	61.50	58.50
Miguel et al. [31]	64.00	59.00
He et al. [32]	64.33	63.25
Xu et al. [33]	67.08	66.35
Chung et al. [17]	66.4	66.6
Rozgic et al. [34]	68.4	76.9
Our proposed model	73.83	79.29

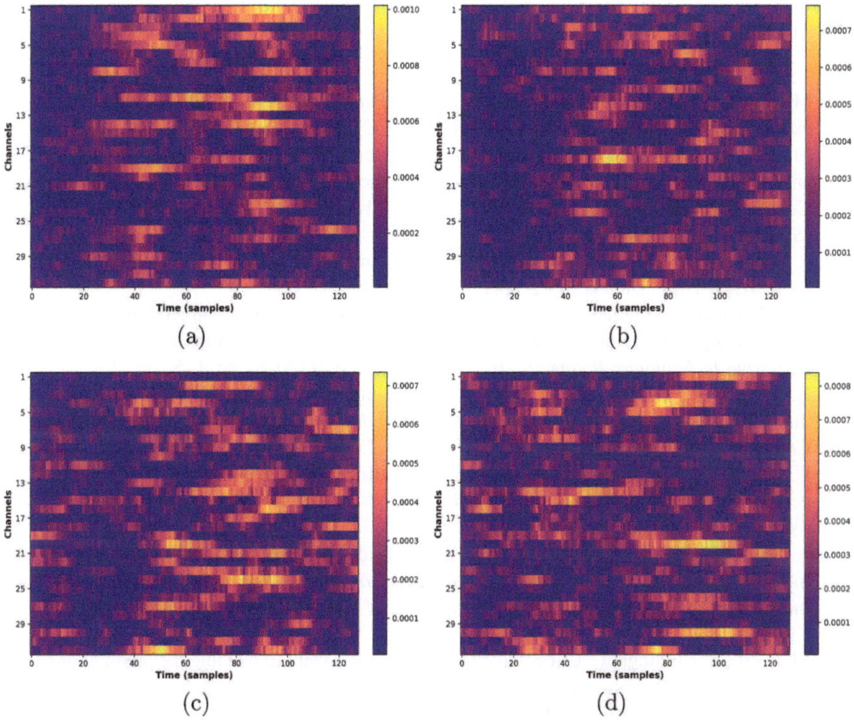

Fig. 4.3 Average saliency map for different emotional state. (**a**) Arousal emotional state. (**b**) Valence emotional state. (**c**) Liking emotional state. (**d**) Dominance emotion state

feature representations. Spatial and temporal dynamics of EEG signals are captured using a combination of temporal and depthwise convolutional layers, facilitating the extraction of optimized feature maps. Attention modules refine the feature space by emphasizing salient patterns relevant to emotion recognition. Model performance is optimized through the application of fine-tuned hyperparameters within the proposed framework. Empirical results demonstrate the framework's effectiveness in improving classification accuracy while maintaining computational efficiency. Comprehensive training and validation processes confirm the model's ability to differentiate between emotional states accurately. The interpretability of the model is addressed using saliency-based interpretable AI (XAI) techniques, which highlight the contribution of distinct brain regions to the classification task. Additional XAI methods are employed to assess feature importance and validate model decisions to enhance transparency further. Future research directions include improving explainability in the context of class-imbalanced datasets and exploring complex functional relationships between input features and classification outcomes.

References

1. M. Soleymani, S. Asghari-Esfeden, Y. Fu, M. Pantic, Analysis of EEG signals and facial expressions for continuous emotion detection. IEEE Trans. Affect. Comput. **7**(1), 17–28 (2015)
2. M. Murugappan, M. Rizon, R. Nagarajan, S. Yaacob, D. Hazry, I. Zunaidi, Time-frequency analysis of EEG signals for human emotion detection, in *4th Kuala Lumpur International Conference on Biomedical Engineering 2008: BIOMED 2008 25–28 June 2008 Kuala Lumpur* (Springer, Berlin, 2008), pp. 262–265
3. P. Shenoy, K.J. Miller, J.G. Ojemann, R.P. Rao, Generalized features for electrocorticographic BCIs. IEEE Trans. Biomed. Eng. **55**(1), 273–280 (2007)
4. M. Abujelala, C. Abellanoza, A. Sharma, F. Makedon, Brain-EE: brain enjoyment evaluation using commercial eeg headband, in *Proceedings of the 9th ACM International Conference on Pervasive Technologies Related to Assistive Environments* (2016), pp. 1–5
5. K. Gaurav, H. Jain, T.K. Reddy, Characterizing neural activity from EEG signals during game interaction using persistent homology, in *2024 15th International Conference on Computing Communication and Networking Technologies (ICCCNT)* (IEEE, Piscataway, 2024), pp. 1–7
6. K. Gaurav, N. Sharma, T.K.R. Bollu, Assaying affective states during video gameplay using persistence landscape-based topology, in *2025 National Conference on Communications (NCC)* (IEEE, Piscataway, 2025), pp. 1–6
7. N. Sharma, T.K.R. Bollu, Explainable AI methods for interpreting emotions in brain–computer interface EEG data, in *Discovering the Frontiers of Human-Robot Interaction: Insights and Innovations in Collaboration, Communication, and Control* (Springer, Cham, 2024), pp. 419–436
8. M.T. Ribeiro, S. Singh, C. Guestrin, Why should I trust you? Explaining the predictions of any classifier, in *Proceedings of the 22nd ACM SIGKDD International Conference on Knowledge Discovery and Data Mining* (ACM, New York, 2016)
9. R.W. Picard, *Affective Computing* (MIT Press, Cambridge, 2000)
10. K. Gaurav, S.D. Reddy, T.K. Reddy, Entropy based EEG irregularity quantification in single-channel SSVEP-based BCIS, in *2023 IEEE Silchar Subsection Conference (SILCON)* (IEEE, Piscataway, 2023), pp. 1–6
11. K. Gaurav, J. Landge, T.K.R. Bollu, Characterizing neural activity during video game engagement using EEG sensor based topological dynamics analysis. IEEE Sens. Lett. **8**, 7007104 (2024)
12. V. S. Kumar, T.K.R. Bollu, WearNeuroNet: An interpretable light-weight deep learning approach for ictal-interictal classification for limited channel EEG wearables, IEEE Sens. Lett. **8**(10), 1–4 (2024)
13. S. Koelstra, C. Muhl, M. Soleymani, J.S. Lee, A. Yazdani, T. Ebrahimi, T. Pun, A. Nijholt, I. Patras, Deap: a database for emotion analysis; using physiological signals. IEEE Trans. Affect.

Comput. **3**(1), 18–31 (2011)
14. Y. Liu, O. Sourina, EEG-based subject-dependent emotion recognition algorithm using fractal dimension, in *2014 IEEE International Conference on Systems, Man, and Cybernetics (SMC)* (IEEE, Piscataway, 2014), pp. 3166–3171
15. S. Nitish, Dropout: a simple way to prevent neural networks from overfitting. J. Mach. Learn. Res. **15**, 1 (2014)
16. H. Candra, M. Yuwono, R. Chai, A. Handojoseno, I. Elamvazuthi, H.T. Nguyen, S. Su, Investigation of window size in classification of EEG-emotion signal with wavelet entropy and support vector machine, in *2015 37th Annual International Conference of the IEEE Engineering in Medicine and Biology Society (EMBC)* (IEEE, Piscataway, 2015), pp. 7250–7253
17. S.Y. Chung, H.J. Yoon, Affective classification using Bayesian classifier and supervised learning, in *2012 12th International Conference on Control, Automation and Systems* (IEEE, Piscataway, 2012), pp. 1768–1771
18. F. Doshi-Velez, B. Kim, Towards a rigorous science of interpretable machine learning (2017). arXiv:1702.08608
19. S.M. Lundberg, B. Nair, M.S. Vavilala, M. Horibe, M.J. Eisses, T. Adams, D.E. Liston, D.K.W. Low, S.F. Newman, J. Kim, S.I. Lee, Explainable machine-learning predictions for the prevention of hypoxaemia during surgery. Nat. Biomed. Eng. **2**(10), 749–760 (2018)
20. J. Ayoub, X.J. Yang, F. Zhou, Modeling dispositional and initial learned trust in automated vehicles with predictability and explainability. Transp. Res. F Traffic Psychol. Behav. **77**, 102–116 (2021)
21. K. Weitz, T. Hassan, U. Schmid, J.U. Garbas, Deep-learned faces of pain and emotions: elucidating the differences of facial expressions with the help of explainable AI methods. Tech. Messen **86**(7–8), 404-412 (2019)
22. F. Zhou, T. Chen, B. Lei, Cross-subject emotion recognition with sparsely-labeled peripheral physiological data using SHAP-explained tree ensembles (2022). arXiv:2211.02923
23. V.J. Lawhern, A.J. Solon, N.R. Waytowich, S.M. Gordon, C.P. Hung, B.J. Lance, EEGNet: a compact convolutional neural network for EEG-based brain–computer interfaces. J. Neural Eng. **15**(5), 056013 (2018)
24. L. Gong, M. Li, T. Zhang, W. Chen, EEG emotion recognition using attention-based convolutional transformer neural network. Biomed. Signal Process. Control **84**, 104835 (2023)
25. M. Lopes, R. Cassani, T.H. Falk, Using CNN saliency maps and EEG modulation spectra for improved and more interpretable machine learning-based alzheimer's disease diagnosis. Comput. Intell. Neurosci. **2023**(1), 3198066 (2023)
26. W.C.L. Lew, D. Wang, K. Shylouskaya, Z. Zhang, J.H. Lim, K.K. Ang, A.H. Tan, EEG-based emotion recognition using spatial-temporal representation via Bi-GRU, in *2020 42nd Annual International Conference of the IEEE Engineering in Medicine & Biology Society (EMBC)* (IEEE, Piscataway, 2020), pp. 116–119
27. Pandey, P. and Seeja, K.R., 2022. Subject independent emotion recognition from EEG using VMD and deep learning. J. King Saud Univ. Comput. Inf. Sci. **34**(5), 1730–1738 (2022)
28. S. Rayatdoost, M. Soleymani, Cross-corpus EEG-based emotion recognition, in *2018 IEEE 28th International Workshop on Machine Learning for Signal Processing (MLSP)* (IEEE, Piscataway, 2018), pp. 1–6
29. P. Pandey, K.R. Seeja, Subject independent emotion recognition system for people with facial deformity: an EEG based approach. J. Ambient Intell. Hum. Comput. **12**(2), 2311–2320 (2021)
30. X. Li, P. Zhang, D. Song, G. Yu, Y. Hou, B. Hu, EEG based emotion identification using unsupervised deep feature learning. In: Proceedings of the SIGIR 2015 Workshop on Neuro-Physiological Methods in IR Research (NeuroIR'15), Santiago, Chile (2015).
31. M. Arevalillo-Herráez, M. Cobos, S. Roger, M. García-Pineda, Combining inter-subject modeling with a subject-based data transformation to improve affect recognition from EEG signals. Sensors **19**(13), 2999 (2019)
32. Z. He, Y. Zhong, J. Pan, An adversarial discriminative temporal convolutional network for EEG-based cross-domain emotion recognition. Comput. Biol. Med. **141**, 105048 (2022)

33. X. Xu, F. Wei, T. Jia, L. Zhuo, H. Zhang, X. Li, X. Wu, Embedded EEG feature selection for multi-dimension emotion recognition via local and global label relevance. IEEE Trans. Neural Syst. Rehab. Eng. **32**, 514–526 (2024)
34. V. Rozgić, S.N. Vitaladevuni, R. Prasad, Robust EEG emotion classification using segment level decision fusion, in *2013 IEEE International Conference on Acoustics, Speech and Signal Processing* (IEEE, Piscataway, 2013), pp. 1286–1290

Chapter 5
Vibes That Emote: Eliciting Affect Through Vibrotactile Stimulation

Suranjita Ganguly, Aditya Koppula, Mohan Raghavan, and Kousik Sarathy Sridharan

Abstract Vibrotactile (VBT) haptic stimulators are commonly used in virtual reality (VR) applications to generate emotional responses, also known as affect, to enhance the immersive experience. However, despite its wide potential for application, there remains a significant gap in research, particularly in establishing a clear connection between the parameters of the stimulation, the resulting emotional responses, and the associated changes in electrocortical activity. As a result, the success rate in achieving the desired affect levels varies, and VBT-based actuators are still not mainstream despite being easily deployable and cheap. Understanding this relationship could enable a more systematic approach to deploying stimuli in VR environments, incorporating Brain-Computer Interfaces (BCI). This would improve the ability to predict and evoke the desired emotional responses, enhancing the repeatability of the process and making it more universally applicable. This chapter aims to review existing literature on the topic, highlight connections where possible, and explore strategies to bridge these factors to enhance the consistency of emotional responses induced by haptic stimulation in VR applications.

Keywords Haptics · Tactile affect · Vibrotactile · Neurophysiology · Valence · Arousal

S. Ganguly
Department of Biomedical Engineering, Indian Institute of Technology, Hyderabad, India
e-mail: bm19resch11002@iith.ac.in

M. Raghavan · K. S. Sridharan (✉)
Department of Biomedical Engineering, Indian Institute of Technology, Hyderabad, India

Department of Heritage Science and Technology, Indian Institute of Technology, Hyderabad, India
e-mail: mohanr@bme.iith.ac.in; ks@bme.iith.ac.in

A. Koppula
Department of Physiology, ESIC Medical College and Hospital, Hyderabad, India

R. Vinjamuri (ed.), *Bridging the Gap between Mind and Machine*,
https://doi.org/10.1007/978-3-032-06713-5_5

5.1 Introduction

Affect is a multifaceted concept that includes the various components characterising human emotions. It arises naturally from experiences involving sight, sound, or touch, like watching a video, listening to music, or handling objects, and is crucial in assessing the significance of these experiences. The sensory experience can be *real* (actual interaction with objects and other physical entities) or *virtual* (simulated interaction with virtual objects). Virtual reality (VR) is an emerging technology that simulates reality to facilitate entertainment, training, tourism, and other experiences, eliminating the need for direct interaction with physical objects, people, or places.

Assessing the effect of VR on emotions is essential for several reasons. Firstly, VR experiences often aim to evoke strong emotional responses. Therefore, understanding how users are affected can help developers create more immersive and engaging environments. In fields like therapy or training simulations, accurately gauging emotional responses can provide valuable insights into a user's state of mind and progress. It is, therefore, imperative that emotional response is categorically studied within the context of VR. Understanding affect elicitation and modulation in virtual reality (VR) is crucial for designing effective VR interfaces, devices, or experiences that evoke desired emotional responses or enhance user engagement and satisfaction. Affect can be induced in VR by engaging one or multiple sensory modalities (visual, olfactory, somatosensory, auditory). For instance, Marin-Morales et al. [1] employed visual stimuli in virtual reality (VR) to create a virtual, explorable museum centred around the Nazi Holocaust. They aimed to evoke emotions in users and compared the emotional impact of this virtual experience with that of visiting a physical museum. The psychophysical ratings in both cases could be observed to show changes in affect. Similarly, haptic simulation is deployed for medical training modules in dental procedures [2], monitoring hearing loss [3], simulating VR in culinary setups [4], gaming [5] and social interactions [6].

In this chapter, we will put special focus on simulating the sense of touch in the context of eliciting affect in VR. For haptic-driven VR experiences to achieve realism, addressing context-specific elicitation of emotions is crucial. This is an underexplored area and warrants a deeper exploration. We will hence delve into the framework for delineating emotions and explore the available haptic options and their deployment strategies. This review will examine the available research and examine the behavioural aspects of vibrotactile stimulation (VBT) and its neurophysiological implications. Our goal is to gain a deeper understanding of the relationship between VBT and induced affect and identify the types of stimuli and optimal parameter settings that may achieve a desired affect and the specific areas of the body where such stimuli could be applied. We will also review the literature on natural touch and compare it with that of VBT to analyse the similarities and differences, extrapolate findings, and point toward areas requiring further research.

5.2 Foundations of Affect

Paul Ekman defined affect [7, 8] as consisting of six basic emotions: happiness, sorrow, fear, anger, surprise and disgust, with each basic emotion contributing to a family of related emotions [8]. He proposed them as six discrete and independent affect states (see Fig. 5.1a) mediated by unique neural pathways and associated with specific behavioural traits. However, this framework could not be explained satisfactorily from a neurophysiological perspective and failed to capture the complexity of affect [9]. For example, Gu et al. [10] observed similarities in facial muscle actuation for different affect states. Both disgust and anger, identified as distinct emotions in Ekman's model, involved wrinkling of the nose, while fear and surprise were associated with raised eyebrows. This suggests underlying commonalities in the neurophysiology that elicited affect states. Since then, several models have been proposed to explain affect such that affective states could be explained with more than one variable. Prominent amongst them is the 2D circumplex model of affect. Proposed by Russel [11] in 1980, the circumplex model quantifies the behavioural aspects of affect. It states that all emotions are defined in the quantitative space as coordinates of the 2D circumplex model, which entails two orthogonal dimensions, viz. valence (defining the degree of pleasantness/unpleasantness felt) and arousal (defining the degree of activation/deactivation felt). Therefore, this model represents any emotion as a point in the 2D circumplex space (see Fig. 5.1b), indicating a certain degree of valence and arousal, facilitating a systematic classification of affective experiences based on these dimensions. This also allows the comparison of one emotion with another in the 2D circumplex space, which can be modelled using simple cartesian geometry. It is the popularly used model due to its utility in explaining mood disorders [12–15], representing complex emotional constructs, and extensive supporting evidence [16, 17].

The flexibility of the 2D circumplex model stems from its ability to dissect and project emotions onto two orthogonal-numeric scales, portraying valence and arousal as part of two distinct neural processes [16, 18]. The orthogonal scales enable the adaptation of resolution based on the specific context of the application. For instance, in scenarios where a fine resolution of affect is unnecessary, it can be simplified to merely distinguishing between positive and negative affect on any dimension. This is exemplified in the understanding of emotional disorders, where conditions like anxiety and depression share similar low valence (indicating the unpleasant sensation associated with them and the inability to experience pleasantness in general) but diverge along the arousal dimension [12, 19]. Similarly, conditions like bipolar disorder have episodes of collapsed arousal dimension with divergent valence ranges. The circumplex model has also been used to study the autism spectrum [15] which involves a modified range of valence and arousal values. Apart from pathology, in practical applications as well, there may be instances in which collapsing one of the scales is necessary (see Fig. 5.1b). For instance, while designing an alarm system, arousal might take precedence over valence.

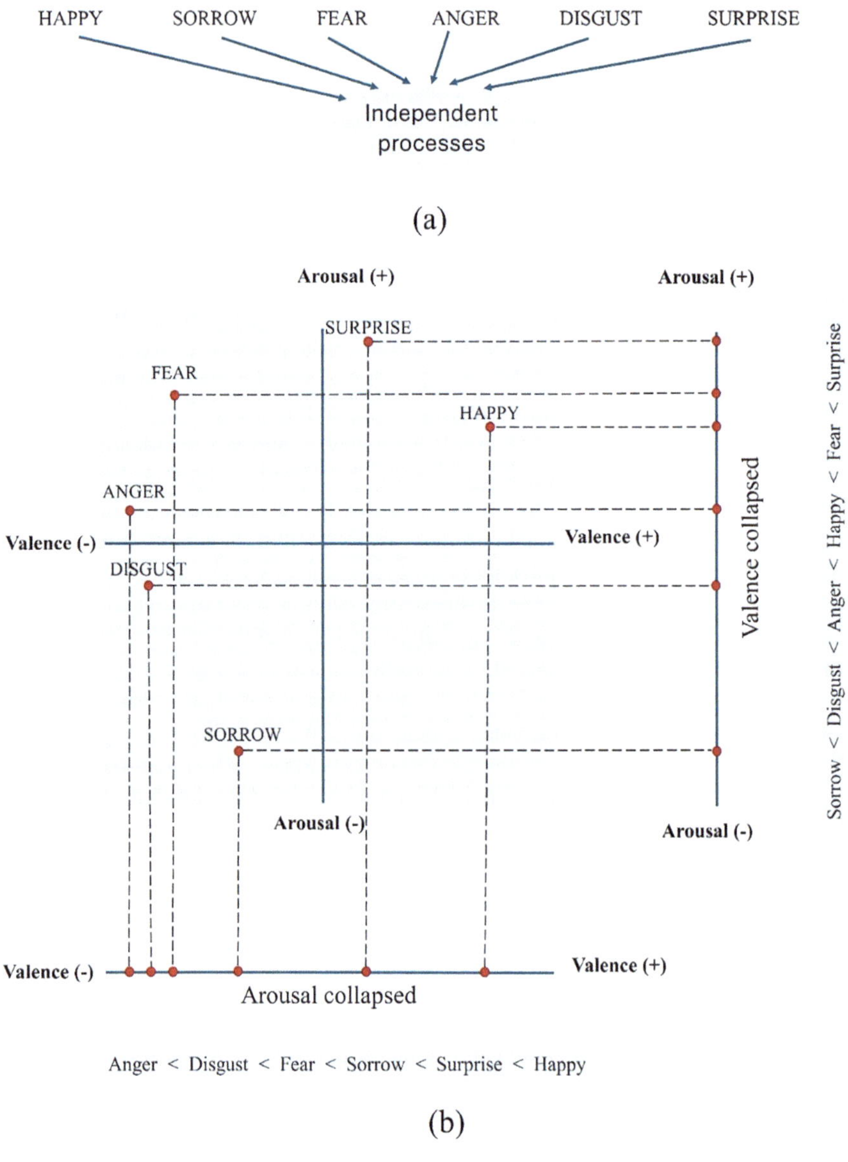

Fig. 5.1 The two different models of affect are shown depicting the (**a**) six basic emotions, independent of each other and (**b**) the 2D circumplex model defining the emotions in the affective space as functions of two independent scales, viz. valence and arousal. Affect can be studied with respect to the 2D circumplex model, or the scales individually, by collapsing the dimension irrelevant for the specific application. As a result, comparative studies can be made between different states of affect along the relevant dimension

Conversely, valence could be more critical than arousal when creating an optimal mobile notification system based on incoming text.

5.3 Affect and Touch in the Context of VR

While virtual reality utilises various sensory modalities mentioned above, the sense of touch has the potential to significantly enhance emotional experiences within this immersive environment by adding tangible components to the experience [20, 21]. For instance, it's typical to incorporate haptic features into a joystick, commonly in virtual reality gaming setups. In a car simulation game, the velocity of the car can be translated into vibrotactile feedback, with intense vibrations signalling a rough course or collision. This integration of touch into virtual environments is crucial for creating a more immersive and realistic experience, as it allows users to interact with digital spaces that closely mimic real-world sensations [22]. There are several sub-modalities of touch, each with unique characteristics resulting in distinctive sensory experiences. These include pressure, vibration, temperature, texture, brush strokes, etc., which collectively contribute to our perception of the tactile world [23]. Pressure, for instance, conveys information about the intensity of contact, while vibration communicates rapid oscillations or movements. Temperature sensations provide cues about the thermal properties of surfaces, while texture perception allows discrimination between rough and smooth surfaces. While stimuli such as pressure or textures impart a more natural form of touch, they are not easily deployable in haptic setups [24]. Natural forms of touch cannot be reproducibly parameterised, with the only controllable parameter being the velocity of stimulation. Therefore, achieving the same effect every time is not possible.

While both VBT and thermal stimulations are easily deployable and controllable across devices, VBT stands out as the superior choice for practical applications due to several drawbacks with thermal stimulation. Thermal stimulators typically require a few seconds, if not longer, to reach the desired temperatures [25, 26] whereas VBT boasts significantly faster response times in the order of milliseconds [27–29]. Thermal stimuli additionally necessitate a long cooldown time for the skin to return to optimal temperature before a subsequent stimulus can be administered, a drawback not seen with vibrotactile stimuli. Furthermore, thermal stimuli can potentially cause damage to the site of stimulation if not controlled properly. Even a relatively mild thermal stimulus can cause skin damage if applied for an extended duration. It is a known fact that the sensory threshold varies across individuals and age groups [30, 31]. If subject-specific threshold and skin texture variability are not factored into the stimulation paradigms, there is a possibility that thermal stimuli intended to be pleasurable cause pain instead. Conversely, VBT carries minimal risk of inducing discomfort regardless of the stimulation frequency and duration. Therefore, most applications in tactile-augmented VR primarily focus on vibrotactile stimuli. Vibrotactile stimulation emerges as the most viable choice for integration into haptic setups. This preference is attributable to its already established integration with actuators in currently available setups and the simplicity of delivering controlled stimulations with known amplitude, frequency, and timing characteristics.

The sensation of touch in VR setups is commonly achieved through haptic sensors connected to consoles, which generate vibrations on the skin using VBT stimuli. VBT is now a ubiquitous mode of communication, seamlessly integrated into most electronic devices via built-in haptic actuators. This ease of deployment has fuelled the rapid growth of affective haptics as a field over the last decade. Researchers are increasingly leveraging mobile-based VR systems [5] that utilise VBT feedback to induce affect and convey emotions [32–34] during communication. Whole-body haptic vibrations are even being explored to transmit experiences such as music [35], aiming to create more immersive VR environments [5, 6, 21, 36]. Apart from this, focus has also been given to evoke specific emotions in individuals that align with the ongoing scene [6, 35, 37–42].

The most common example of such an application is a mobile electronic device notification using VBT feedback. A widely used principle in the design of VBT applications is that a high amplitude of VBT stimulus induces high arousal. Applications can also be developed around the design of an alarm system for a pleasant awakening of the user. Driving modules can be developed with VBT alerting the driver of an imminent collision. However, the issue with current setups is a limited understanding of the emotional significance conveyed by the specific type of vibrotactile feedback employed. Most haptic applications, whilst using VBT for feedback, do not assess the affect content induced by the stimulus itself. Applying patterned stimuli across the skin can impart affect, but a better understanding of the relation between them (stimulation parameters and resulting affect) allows for the optimization of stimulus settings for a controlled elicitation of desirable affect in a specific context.

Understanding affect qualities imparted by VBT and the ground rules to modulate them can further broaden the horizon of applications. Consider the contrasting scenarios of waking up a drowsy driver with an alarm clock versus coarse rousing. It is paramount that the awakening be gentle and pleasant, as any abrupt or inconsistent stimulation could provoke sudden movements and potentially result in an accident. Can VBT in driving modules differentially elicit affect to indicate the urgency of an imminent collision versus a warning of a road bump? Understanding and modulating the affective qualities of VBT opens up new possibilities for designing emotionally intelligent haptic applications tailored to specific user needs.

5.4 Challenges

However, despite a boom in the use of VBT for VR, VBT is severely under-researched despite its widespread modern-day applications. Most existing studies primarily focus on the palm for delivering VBT stimuli. The sensory distribution across the body is largely non-uniform [43], therefore rendering studies with only one location insufficient. Furthermore, the stimuli delivered, especially in the context of VR-based applications, are not systematically studied by varying their primary modulating parameters (amplitude, frequency and duration), which would enable extrapolation across a range of stimulation settings. The neurophysiological underpinnings of VBT-induced affect across different body sites and stimulation parameters remain unexplored and the VBT parameters used in existing studies are set in an empirical fashion without a clear idea of the corresponding behavioral or neurophysiological effects. This warrants the investigation of cortical modulation due to the VBT-induced affect.

It may be argued that findings from studies based on natural touch may serve the purpose to extrapolate onto VBT. However, that is not the case for the following reasons. The principal end organs in skin are, Meissner's corpuscles, Pacinian corpuscles, Merkel's discs, and Ruffini corpuscles [44], free nerve endings and C nociceptors. The first four aforementioned end organs behave as pressure/-touch/indentation sensors while free nerve endings and C nociceptors are intended for pain and nociception. In addition to this, a special class of low-threshold mechanoreceptors has been identified in non-glabrous skin innervated by C-type unmyelinated fibers that are distinct from the high-threshold and pain-sensing C-type nociceptors. Based on their response to dynamic stimuli, these end organs are broadly categorised into rapidly adapting (RA) and slowly adapting (SA) receptors [45]. Meissner's corpuscles and Pacinian corpuscles (PC) are the RA receptors, while Merkel's discs and Ruffini corpuscles are SA receptors. RA receptors respond preferentially to changes in skin indentation level, with a greater response to faster indentation and vice versa, while their activity is attenuated rapidly during fixed or unchanging skin indentation. Therefore, they exhibit heightened activity during the onset and offset of a stimulus. In contrast, SA receptors respond during a sustained or constant touch, and the activity in their afferents is proportional to the depth of skin indentation. While natural touch engages the full complement of RA, SA and C-type low threshold mechanoreceptors (engaging both the discriminatory and hedonic aspects of touch), vibrotactile stimulation is encoded primarily by the RA receptors viz. the Meissner corpuscles and Pacinian corpuscles (thought to engage only the discriminatory aspects of touch). Adding to this, while discriminatory touch is mediated by the DCML via the abovementioned end receptors, hedonic touch is mediated by the spinothalamic tract [45] via the unmyelinated C-tactile fibers of the hairy follicles [46, 47], the end receptors also known to be activated via natural touch.

Therefore, it is uncertain if VBT can elicit tactile affect as it engages only a subset of tactile sensing elements corresponding to touch [47–50]. Furthermore, these sensory receptors have different distributions across the body with different spiking activities to stimuli, leading to varying levels of cortical excitation. For example, the hedonic aspects of touch are predominant in the non-glabrous skin on account of its innervation by the C-tactile fibers. The glabrous skin, classically known to be devoid of C-tactile fibers is thus thought to contribute towards discriminatory touch although a recent study by Watkins et al. [51] has shown the presence of C-tactile fibers in the glabrous skin. Therefore, without the existence of sufficient studies, it is not possible to directly deploy VBT to achieve desired levels of affect.

5.5 Electrophysiology

To understand whether vibrotactile stimulation (VBT) can induce affect, it is essential to first examine the electrophysiology of affect and VBT separately. By exploring each domain individually, we can establish a clearer foundation for analysing how VBT might engage the neural mechanisms underlying emotional experience and therefore, how VBT might be modulated to achieve desired levels of affect.

5.5.1 Cortical Correlates of Affect

Our understanding of anatomical pathways and neural correlates of tactile affect in humans is far lesser than that of touch's discriminative aspects. Several studies have found distinctive event-related potential (ERP) patterns and cortical modulations corresponding to emotional states and in tasks involving affective judgments [52, 53]. Researchers have identified candidate brain regions mediating valence and arousal based on these studies and others conducted on rodents, human populations with mood disorders, and healthy individuals exposed to emotional response-inducing stimuli. Several cortical networks have been identified that can potentially process tactile modulation of valence and arousal [18, 54–58]. A concise schematic of the same, depicting major cortical regions associated with affect processing, is shown in Fig. 5.2. Valence is mediated by the reward circuitry [13, 59, 60] which is a part of the limbic system in the cortex, primarily involving the ventral tegmental area (VTA) and the nucleus accumbens (NAc). The fMRI-BOLD (functional magnetic resonance imaging-blood oxygen level dependent) activations were also seen across the supplementary motor, anterior midcingulate, right dorsolateral prefrontal, occipitotemporal, inferior parietal, and cerebellar cortices with unpleasant stimulations and the midbrain, ventral striatum, and caudate nucleus with pleasant stimulations [58].

The arousal circuit primarily consists of the thalamus, amygdala, and the reticular activating system. The thalamus is a critical subcortical structure that mediates arousal by directing the sensory information to the cerebral cortex during the awake state and gates the same during sleep. The firing patterns of thalamocortical cells change from tonic firing to oscillatory bursting as the brain transitions from wakefulness to sleep. The amygdala plays an important role in bridging the valence and arousal domains by activating arousal in response to negative valence in the incoming sensory information [61, 62]. Monoaminergic systems located in the reticular formation of the brain stem form the reticular activating system, and these play an important role in transitions between different states of consciousness and arousal. Additional processing areas include the globus pallidus, caudate nucleus, parahippocampal gyrus, premotor cortex, cerebellar vermis [58], the ventromedial prefrontal cortex [14], and the fronto-medial regions of the cerebral cortex [56].

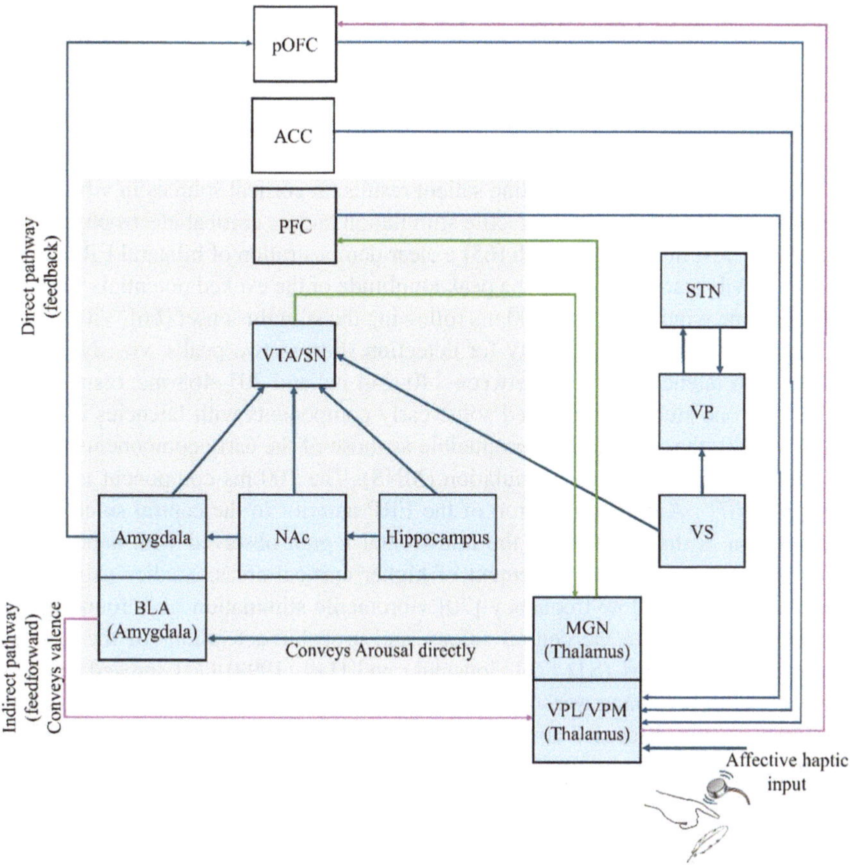

Fig. 5.2 The primary areas involved in the cortical processing of affect after the tactile information reaches the thalamus is shown as a schematic diagram. Direct pathways are depicted with blue arrows. Orange and green arrows convey individual, indirect pathways. The grey box depicts the mesolimbic system within the cortex. Blue boxes depict subcortical structures. Abbreviations: *VPL* ventral posterior lateral nucleus, *VPM* ventral posterior medial nucleus, *VS* ventral striatum, VP ventral pallidum, *STN* subthalamic nucleus, *VTA* ventral tegmental area, *BLA* basolateral amygdala, *NAc* nucleus accumbens, *PFC* prefrontal cortex, *ACC* anterior cingulate cortex, *pOFC* posterior orbitofrontal cortex

While several structures contribute independently to the evaluation of valence and arousal, their effects eventually converge in the frontal regions. Some areas are involved in the processing of both valence and arousal, like the amygdala [56, 63]. The valence and arousal circuits also have intimate connections to the areas that regulate the autonomic nervous system and respiration. For instance, stimuli perceived as emotional can trigger responses from the parasympathetic or the sympathetic nervous system, affecting physiological parameters like heart rate, altered breathing patterns, stress response, and perspiration [46, 64].

5.5.2 Cortical Correlates of VBT

In addition to the sparse literature on the topography of cortical activations evoked by the vibrotactile stimuli, studies based on air-puff and median nerve stimulation have offered valuable clues on the possible cortical sources contributing to the ERPs. In the following, we outline salient results on cortical sources in vibrotactile stimulation. The study of vibrotactile stimulation and its cortical electrophysiology gained interest in the 1980s with [65] a clear demonstration of bilateral ERPs time-locked to vibrotactile stimuli. The peak amplitude of the evoked potentials varied in a broad time window of 450–900 ms following the stimulus onset [66]. Vibrotactile stimuli at the threshold intensity for detection showed two peaks, viz., N190 and P400, with latencies varying between 140–244 ms and 207–468 ms, respectively. Although one study also showed some early components with latencies less than 100 ms [67], they are not as reproducible as those of the early components evoked by median nerve electrical stimulation (MNS). The 100 ms component is largely bilateral [67] . A dipole inversion of the ERP anterior to the central sulcus and a modulation in the response of the mid-frontal region observed with haptic delay [68], also indicates the involvement of higher cortical areas. Studies using high-frequency [69] and low-frequency [70] vibrotactile stimulation have found dipole locations posterior to the central sulcus, and probable activations of the primary somatosensory cortex (S1) [71]. Jousmäki and Hari (1999) [72] showed bilateral activations in the somatosensory cortex (SCx). with vibrotactile stimuli similar to the other studies discussed above. Vibrotactile stimuli were applied for 20–750 ms with inter inter-stimulus interval (ISI) of 300 ms-3 s to the fingertips [67, 73] and elicited clear, reproducible event-related potentials (ERPs) in the sensorimotor cortex. Soininen and Järvilehto (1983) [66] demonstrated visible peaks at P50, P100, NI90 and P400, N190 and P400 being the most prominent ones. Until the P50 stimulus, the peaks are contralateral in nature and became bilateral with longer latencies [67]. It was further seen that the vibrotactile evoked potentials were similar to the ERPs elicited by an electrical stimulus, suggesting similarity in processing for both these modalities of stimulation [74]. In addition to the ERPs described, vibrotactile stimulation also elicits reproducible functional activations within the cortex. The regions activated by vibrotactile stimuli are the contralateral S1, bilateral S2 (secondary somatosensory cortex), pre-central and sub-central gyrus, posterior cingulate cortex, and posterior parietal cortical areas [75–77].

Functional studies have also revealed that flutter and vibration are processed differently as shown in the activation profile of S1 [78]. Ryun et al. [73] demonstrated differential activations in S1 and S2 for flutter (5–35 Hz) and vibration (100–400 Hz) stimuli, when the glabrous skin of the distal upper limb was stimulated. In S1, which is responsible for frequency discrimination [79], high gamma activity (50–100 Hz) was seen for both stimulus conditions (flutter & vibration), but a decrease in power with an increased stimulation frequency was seen for the vibration range (100–400 Hz) stimuli. On the contrary, high gamma activity in S2 could be observed only in the vibration range, and the power increased with an increase in the vibration frequency. The volume of activation in S1 has also been found to be smaller than that of S2 [80]. The cortical activations had a bimodal relationship with the stimulation frequency, with one peak in the flutter frequency range of 21–26 Hz, corresponding to the preferential activation of the Meissner's corpuscles, and the other peak activation coincided with the Pacinian range of 250–300 Hz [70, 74, 81–83] .

The findings discussed above are all based on the vibrotactile stimulation of the distal upper limb, and there are very few studies on the corresponding effects of lower limb stimulation. The first such study was [84], which probed the vibrotactile ERPs of both the upper and lower limbs. Similar to the upper limb, lower limb VBT evoked potentials were also recordable over the contralateral hemisphere. However, the ERP amplitudes with lower limb stimulation were lower than those evoked by upper limb stimulation. It is interesting to note that ERP amplitude peaked at a broader stimulation frequency range of 17–30 Hz. However, the study did not report any cortical activation profiles with lower limb VBT stimulation. This underscores the need for more comprehensive research on VBT, particularly across various stimulation sites and parameter settings, to accurately characterize its effects and the nature of affect it may induce.

5.6 Can VBT Induce Affect?

Emerging research suggests that vibrotactile stimulation (VBT) may have the capacity to evoke affective responses. To explore this possibility, we examine evidence from studies that report both behavioral and cortical markers of affect in response to VBT. Behavioral findings highlight observable emotional reactions, while cortical data reveal corresponding neural activity. Together, these studies build a compelling case for the affective potential of VBT. Given that behavioral studies already show VBT-induced affect, and glabrous skin has been shown to respond to pleasant touch, it may be assumed that VBT can induce affect at a behavioral level, and some cortical changes pertaining to the same may be seen.

5.6.1 Behavioral Markers

Existing studies provide evidence that VBT can, at a behavioral level, induce affect. It was shown that introducing a tactile stimulus during hearing affective speech could induce changes in affect ratings [34]. Similarly, VBT stimuli modulated affect perceived while viewing an emotionally neutral film [85]. Rantala et al. [32] showed that VBT could be used to convey the "intent" of social touch. Sorgini et al. [86] could convey the stiffness information of rubber via VBT. More systematic studies such as Yoo et al. [87] and Hasegawa et al. [88] varied VBT parameters and recorded affect ratings from subjects, showing a systematic variation in Valence and Arousal as the frequency/amplitude and duration were modulated. Several other studies have followed suit [89–93]. However, discrepancies still exist. For example, Yoo et al. [87] found that increased stimulation frequency led to higher valence ratings. Additionally, increases in duration and amplitude were independently associated with higher arousal levels. In contrast, MacDonald et al. [22] and Akazue et al. [25] observed no significant effect of duration (2 s Vs 20 s) on affective responses. While frequency significantly influenced arousal, it did not alter valence.

Further discrepancies exist in terms of the standardisation of studies. While some studies have done exhaustive analysis, non-standard affect scales were used. In [94], the impact of complex vibrotactile stimuli resulting from a combination of simple sinusoids was studied. Depending on the chordal-to-base frequency ratio of a particular combination, the participants described evoked precepts as either consonant or dissonant, with a pleasant quality attributed to consonance and relative unpleasantness to dissonance, pointing to possible modulations of affect. However, they did not use any standard affect rating system to quantify the affective qualities, an issue seen in other studies as well [49, 95–97]. The use of narrative terms or non-standard methods to gauge the affective consequences of vibrotactile stimulation may limit comparability with other studies and hinder the generalizability of findings.

5.6.2 Cortical Markers

The cortical representation of affective vibrotactile stimulation (VBT) continues to gain attention, with emerging studies highlighting neural parallels between VBT and natural touch. For instance, Lee Masson et al. [98] demonstrated that VBT can activate the insular cortex in a manner similar to natural tactile stimulation, particularly enhancing connectivity between the posterior and dorsal anterior insula—regions implicated in affective processing. However, the most interesting find is with regards to the modulation SCx. The SCx, primarily known to process discriminative touch mediated by VBT [67, 99–103] has also been seen to be linked to affective processing.

Recent evidence suggests pleasant touch in the glabrous skin is mediated by the A-beta fibers (whose end organs are the Pacinian and Meissner's corpuscles) and is processed in the SCx. Taking an example of the glabrous skin (for example, the palm of the hand) which lacks the CT fibers and only contains A-beta fibers (enabled by the end organs Pacinian corpuscles and Meissners' corpuscles), it is still seen that glabrous skin is receptive to pleasant sensations. McGlone et al. [104] along with Olausson, studied the difference in pleasant sensation between glabrous and non-glabrous skin due to natural touch. It was observed that pleasant touch from glabrous skin is mediated by A-beta afferents and primarily processed in the SCx. Activations in the SCx pertaining to affect is also seen when other stimulus modalities are involved. Along these lines, Case et al. [105] compared emotional responses elicited by mechanical versus vibrotactile stimulation. The study reported affect-related modulations in late somatosensory evoked potentials (SEPs) occurring after 100 ms. In contrast, early SEPs (<100 ms) did not reflect affective modulation, suggesting that affective processing of VBT emerges later in the somatosensory processing stream. Further supporting the involvement of SCx in affective modulation, Sel et al. [106] showed ERP modulation to viewing emotional faces within 40–80 ms in the SCx. Similar modulation in the SCx within 50 ms to affective visual stimuli was also shown by Montoya et al. [107]. Aftanas et al. [108] showed modulations in the theta power with meditation-induced "blissful" state positively correlated with theta power. Giraud et al. [109] used transcranial alternating current stimulation (tACS) to directly modulate SCx activity and observed effects on both valence and arousal. Their findings revealed that beta and theta frequency bands contribute differentially to encoding these two dimensions of affect—an observation in line with the circumplex model of emotion, which posits that valence and arousal are regulated by distinct neural circuits [18, 54, 55, 58, 110]. This dissociation further strengthens the view that cortical processing of affective stimuli, whether visual, mechanical, or vibrotactile, involves modality-independent but dimension-specific pathways. Complementing this, Yang et al. [111] reported activation in the parietal SCx during the processing of visually induced affect, suggesting convergence of affective signals across modalities.

Notably, tactile stimulation in healthy individuals have shown affect modulation in the somatosensory evoked potentials (SEPs) as early as 50 ms [105], which is one of the first potentials to be evoked in the SCx post-tactile stimulation [67], until 200 ms [107]. Several fMRI studies based on tactile modulation of affect have also shown cortical changes in the SCx [49, 112, 113] and its associated areas [56]. Responses from the SCx have also been shown to be modulated when affect was induced by other modalities of stimulation such as visual [105, 114, 115] and auditory [61] stimulation. Therefore, it is worthwhile investigating the affect-related cortical electrophysiology modulation of the SCx. However, despite the existence of a few studies pertaining to VBT, the majority of the existing insights on tactile affect and cortical signatures come primarily from fabric- or brush-stroke studies wherein altered cortical connectivity and correlations between affect ratings and activations in regions like the prefrontal cortex, anterior cingulate cortex (ACC), SCx, insula, and amygdala are shown [113, 116–121].

Apart from this, stimulus from other modalities have also shown affect-related activity in the SCx. Visual emotional stimuli have been shown to elicit responses in the SCx [115], supporting the notion that this region may be broadly involved in processing affective information, beyond strictly somatosensory inputs. In another line of research, Gonzalez et al. [53] observed significant SCx activity around 100 ms post-stimulus in response to emotional facial expressions, reinforcing the SCx's role in affective modulation. Viewing images that emulate affective touch [122] showed similar activations in the SCx.

Therefore, a cardinal question in the VBT modulation of affect is the role of the SCx, given its centrality in the processing of tactile stimuli, which are often intertwined with emotional experiences. As discussed previously, vibrotactile (VBT) information is processed in the SCx, which has extensive reciprocal connections with the limbic network that processes emotions, including the amygdala; it is physiologically plausible to conjecture a role for the SCx in the mediation of tactile affect. Recent research by the authors of this chapter has established connections between the somatosensory changes with the affect states induced by the VBT [123]. There is an urgent imperative to expand our understanding of the link between SCx and affect.

5.7 Can Parametric Modulation of VBT Induce Systematic Affect?

The affective potential of tactile stimulation is deeply rooted in the way mechanical interactions with the skin are encoded by cutaneous receptors. Understanding how natural textures—such as fabrics—interact with the skin provides valuable insight into how artificial stimuli, like vibrotactile feedback, might be designed to evoke similar emotional responses. Manfredi et al. 2014 [124] demonstrated the link between texture and skin vibration frequency. They used laser-Doppler vibrometry to study skin vibrations induced by fabric brushing, examining the effects of fabric microgeometry, finger pad structure, and brushing velocity. They found that fabric surface features were reliably reflected in skin vibration profiles, and that vibrational frequency content—rather than amplitude—was key to distinguishing textures. Higher brushing velocities produced a shift toward higher vibrational frequencies, while amplitude showed little correlation, likely due to variable contact conditions. Along the same lines, it was found that smooth textures that felt pleasant exhibit pronounced peaks at high-frequency vibrations, while rough textures which felt unpleasant, show peaks at lower frequency ranges [124]. Fabrics with a clear periodicity in microstructure have narrow & localized peaks in power spectral density (PSD) of fabric surface profile and resulting skin vibrometry showed similarly matched peaks in the PSD. On the other hand, the surface profile of the aperiodic fabrics has a relatively peakless, inverse power law PSD, though the skin vibrometry had clear frequency peaks that depend on brushing velocity. They also showed a clear influence of fingerprint spatial period on the PSD of skin vibrometry profiles when averaged over the effects of different fabrics and brushing velocities. The interplay of normal and tangential components of skin stimulation on the spatiotemporal coding of information in the afferents and the corresponding changes in cortical electrophysiology & affect needs to be investigated if natural touch is to be simulated by VBT in haptics. As the VBT-induced in-situ vibrations (at the stimulation site) and the transmitted vibrations (to the rest of the finger and hand) are periodic/quasiperiodic in nature, it would be interesting to compare the skin indentation profiles and the evoked "afferences" of VBT with that of periodic fabrics. Consequently, these studies build a compelling bridge between natural tactile stimulation (fabric brushing) and **synthetic vibrotactile stimulation (VBT)** as potential modulators of affect. We argue that synthetic vibrotactile stimuli can be affectively evocative but their design must carefully account for frequency, location, duration, and user context to truly emulate the affective qualities of natural touch.

5.7.1 Modulation of Stimulation Frequency

The stimulation frequency is a potentially meaningful parameter for modulating affect through VBT, warranting systematic investigation in future research and application design. VBT primarily engages Meissner's corpuscles and Pacinian corpuscles, each tuned to different frequency ranges i.e. flutter (low-frequency) and vibration (high-frequency), respectively. As discussed earlier in Sect. 5.5.2, these distinct frequency ranges elicit different patterns of cortical activation [78] [125]. This neurophysiological divergence supports the possibility that varying the frequency of vibrotactile input could lead to differential emotional responses. For instance, Yoo et al. [87] observed that increasing the stimulation frequency is often associated with increasing valence levels. Morrow et al., 2016 [126] found that the stimulation frequency was associated with the sensitivity in the neck. These findings imply that stimulation frequency is not merely a technical parameter but a meaningful modulator of affective tone. Carefully selecting frequency ranges based on intended emotional outcomes may enhance the design of affect-sensitive haptic systems, especially in applications like mood regulation, emotional communication, and sensory substitution.

5.7.2 Modulation of Stimulation Duration

One of the key parameters influencing the affective impact of vibrotactile stimulation (VBT) is the duration of the stimulus. Yoo et al. observed that longer durations of VBT were associated with increased levels of perceived arousal, suggesting a direct link between temporal properties of stimulation and emotional experience. This finding echoes earlier behavioral studies response [87, 88] in which the systematic variation of stimulation duration revealed measurable changes in affective responses. Crucially, several studies have demonstrated that perceived intensity—often a proxy for arousal—rises with increasing stimulus duration. For example, Berglund et al. [127] reported a logarithmic relationship between stimulus duration and perceived amplitude, attributing this effect to temporal integration, where the sensory system accumulates input over time to produce a stronger perceptual outcome. Similar effects were also reported by Posner et al. [55]. Very recently, it has also been shown that an increased duration increases the levels of arousal [123]. This phenomenon is not limited to vibrotactile inputs. Electrotactile stimulation studies have shown that longer electrical pulses similarly enhance perceived intensity [111], analogous to arousal, further reinforcing the idea that temporal integration plays a fundamental role in shaping tactile perception and affect. Together, these findings underscore the potential of using stimulation duration as a tunable parameter to modulate affect in VBT-based systems and should be further investigated.

5.7.3 Modulation of Stimulation Location

Location plays a crucial role in somatosensory affect processing due to the brain's somatotopic organization, where different regions process sensory information from specific body parts. This organization influences how sensory inputs are interpreted emotionally. For instance, stimuli from certain body areas may evoke stronger emotional responses compared to others. Additionally, the specific neural pathways involved in processing somatosensory information are interconnected with regions of the brain associated with emotional processing, such as the limbic system. In natural touch, it has been seen that that pleasantness ratings differed when location was changed in natural touch [121, 128–130]. This has been primarily attributed to the differential densities of CT fibers across these locations. If affect mediation through VBT follows a mechanism similar to natural touch, we might observe differences in affect processing depending on the anatomical location and its density of the larger diameter fibers innervated by the Meissners' and Pacinian corpuscles. Corniani and Saal [43] presented comprehensive details about the innervation densities of the sensory fibers across the body, which can serve as a benchmark while designing studies that try to correlate location-specific differences in affect-based haptic applications. In fact, significant differences in reported affect have been observed between different locations in the upper arm [131] pertaining to VBT. However, these studies are still at their infancy and require extensive investigation.

5.8 VBT and Its Applications

Emotional modulation has profound implications that can be meaningfully harnessed. With growing evidence that VBT can induce affect, a range of potential applications emerges, spanning emotional communication, sensory substitution, and enriched human–computer interaction, which are discussed below.

5.8.1 Location-Specific Applications

As discussed in earlier sections, the location of somatosensory input can significantly influence affective responses and emotional experiences. This becomes especially important when designing ubiquitous haptic systems that deliver VBT across different body regions. For instance, a wearable alarm system may use wrist-based VBT through a smartwatch, while a driver-alert system might utilise torso-based VBT via a seatbelt to effectively rouse a drowsy driver. These examples illustrate how the same tactile cue may produce different affective outcomes depending on its location on the body. Designing effective VBT-based applications, therefore, require careful consideration of location-specific factors, including tactile sensitivity and neural innervation. Corniani and Saal in 2020 [43] provided a comprehensive overview of innervation densities across the body, offering a valuable reference for tailoring haptic feedback in emotionally sensitive or high-priority contexts. Despite this, few studies have directly examined how the affective impact of VBT varies by location [123], highlighting a critical gap in the field. Addressing this will be essential for optimising the design and emotional efficacy of next-generation haptic interfaces.

5.8.2 *Mood Disorders*

Vibrotactile stimulation (VBT) holds potential in the context of mood disorders, particularly due to altered activity in the SCx—a region increasingly recognised for its role in affective processing. Many mood disorders, including depression, stress-related conditions, and bipolar disorder, are associated with somatosensory sensitisation and altered SCx [116, 132–135] Since VBT evokes primary responses in the SCx, examining changes in these responses may offer novel insights into the neurophysiological underpinnings of mood disorders. Moreover, the strong functional connectivity between the SCx and limbic structures—such as the amygdala and anterior cingulate cortex, positions the SCx as a valuable entry point for accessing emotion-related brain circuits. This opens the door for both diagnostic and therapeutic applications. For instance, modulation of SCx activity using techniques like transcranial magnetic stimulation (TMS) or transcranial direct current stimulation (tDCS), in conjunction with VBT, could be explored as a non-invasive strategy to correct dysfunctional affective patterns. Together, these findings suggest that VBT, beyond its sensory and affective capacities, may contribute meaningfully to the understanding and treatment of mood disorders.

5.8.3 *Diagnosis of Neuropathy and Other Somatosensory Pathologies*

Peripheral neuropathy is a common complication associated with metabolic disorders such as diabetic peripheral neuropathy and cancer chemotherapy. Its onset is characterized by distinct physiological and subtle behavioural changes [136]. In long-standing diabetes mellitus, the physiological changes include axonal retractions or demyelination of smaller diameter nerve fibers (unmyelinated C and $A\delta$ fibers), followed by larger diameter fibers ($A\alpha$ and $A\beta$ fibers). This amounts to the degeneration of the peripheral nerves, leading to altered sensations of affect, a change that is registered in the cortical neurophysiology in the SCx [137, 138]. Behaviourally, the patients can experience altered emotions and sensitivities to tactile stimuli [136]. As neuropathy is known to be a progressive disease, the extent of damage may also be assessed by studying the altered behavioural and cortical responses to stimulation at different locations along the length of extremities. The VBT and its cortical electrophysiology offer a painless, non-invasive mass screening tool capable of effectively diagnosing peripheral neuropathy before irreversible damage occurs.

In addition to its role in diagnosing subtle affective changes, vibrotactile stimulation (VBT) has also demonstrated therapeutic potential in addressing somatosensory pathologies. Notably, studies have shown that VBT can alleviate symptoms in individuals with fibromyalgia, a condition characterized by chronic widespread pain and heightened sensitivity to touch [139]. These findings suggest that VBT may serve not only as a tool for affective modulation but also as a non-invasive intervention for managing certain sensory-related disorders.

5.8.4 In Communication and Assistive Technology

Vibrotactile stimulation (VBT) has demonstrated significant promise in the domains of communication and assistive technology, particularly through its capacity to convey both functional and affective information. For example, VBT has been used to communicate object properties such as stiffness, offering more intuitive haptic feedback in interactive systems [86]. In the realm of emotional communication, patterned VBT such as linear, circular, and positional vibrations have enabled the transmission of affective states between individuals, facilitating non-verbal emotional exchange [32]. This affective potential extends meaningfully to accessibility applications. Studies have shown that VBT can assist auditory-impaired individuals in perceiving affective content, with vibrotactile signals corresponding to speech samples eliciting cortical responses comparable to those evoked by visual emotional stimuli [140]. Another innovative application involves the use of vibrotactile gloves designed to enhance the emotional experience of music, effectively translating auditory cues into tactile sensations [40]. Additionally, VBT has been employed to transmit non-visual information [141] in contexts where traditional sensory channels are limited or unavailable, highlighting its broader role in sensory substitution and augmentation [141]. Collectively, these diverse use cases underscore VBT's versatility as both a functional and affective communication tool, with wide-ranging implications for assistive technology.

5.8.5 Neurofeedback-Backed Consumerism

Neurofeedback is premised on the use of physiological signals corresponding to affective/cognitive states as feedback to monitor and adjust these internal states desirably. In the context of the consumer and marketing industry, the physiological signals that can be monitored as surrogates for internal states are EEG, heart rate, face, eye movements, tactile signals recorded from a sensor-embedded display/demonstration object during user interaction, etc. From these signals, affect-domain markers indicating user preference-aversion, positive-negative valence, or arousal level are computed. The cardiac rhythm fluctuations are recognised indicators of sympathetic and parasympathetic activity, serving as viable metrics for monitoring valence and arousal states. Based on this information, the content for the user is modified to reinforce the feeling of preference and lessen the evoked aversion. In addition to the audio-visual adaptation of the advertised content, VBT offers a possibility to optimize the desirable affect state. For example, a VBT stimulator array embedded joystick can be given to the user to interact with advertised content, and the user can be made to sit on a VBT-embedded chair. Similarly, virtual reality (VR)-based games, advertisements, and tourism activities (virtual tourism) can be augmented by VBT-embedded gloves or garments. Knowledge of VBT's impact on affect states can be used to modify the internal state in a desirable direction by appropriate stimulation settings, for example, to increase user preference, engagement, arousal, and/or positive valence. Therefore, VBT has the potential to enhance the personalisation of content exposed to the user or consumer.

The current research and available products in this field do not exploit the possibilities discussed above. Some studies focus narrowly on stress, which only loosely correlates with affect, while others examine affect directly but rely on multimodal data acquisition methods. Moreover, real-time identification of affect is crucial for user engagement in the consumer industry and is often overlooked, with most of the published work in the field being based on post-hoc data. Neurofeedback can also be a viable substitute for the feedback forms that take a long time to fill up and often lead to biased results.

5.9 Future Directions

To advance our understanding of how vibrotactile stimulation (VBT) modulates affect, several key research questions remain open. These directions span cognitive, neurophysiological, and psychophysical domains, and can serve to refine both mechanistic understanding and translational application.

5.9.1 Cognitive Mediation and Temporal Dynamics

While existing findings suggest that VBT can induce hedonic responses, it remains unclear whether these responses reflect purely bottom-up sensory processing or also involve top-down cognitive modulation. Future studies should examine whether VBT-induced affect arises solely from perceptual judgments of stimulus quality or whether factors such as attention, expectation, and contextual appraisal play a role. Time-resolved EEG analysis may help clarify this distinction, particularly by examining whether early ERP components (e.g., P100–N200) correspond to pre-attentive sensory appraisal and whether later components (e.g., P300, LPP) index higher-order cognitive evaluation.

5.9.2 Somatotopic and Spatiotemporal Specificity

There is a need to investigate whether the affective impact of VBT varies across glabrous and non-glabrous skin, given their differential mechanoreceptor profiles—particularly the presence or absence of C-tactile afferents. Are certain body locations inherently more prone to affective resonance, regardless of identical VBT parameters? This line of inquiry includes exploring whether cortical responses diverge across stimulation sites and whether body-specific "affective profiles" exist. Functional imaging and electrophysiological mapping could clarify how efficiently somatosensory-to-limbic projections operate across the body map.

5.9.3 Parametric Tuning and Stimulus-Affect Mapping

A critical question is whether specific VBT parameter configurations—frequency, amplitude, duration—can reliably elicit distinct affective states. While we present evidence that tuning these parameters indeed modulate affect, there is a lack of a "stimulus-affect dictionary" to predictably link vibrotactile profiles with valence and arousal. Moreover, the existing literature doesn't shed light on whether these relationships exhibit non-linear dynamics—for instance, whether low amplitude at high frequency produces comparable affective outcomes as moderate amplitude at mid-frequency. Longitudinal studies should also investigate the stability of individual affective mappings and whether habituation or mood states modulate these associations over time.

5.9.4 Affective Discrimination and Learning

Understanding the granularity of VBT's affective resolution is equally important. Can VBT be tuned to discriminate between subtle emotional states, such as calm versus content, or is its resolution limited to broader categories (e.g., positive vs. negative)? Furthermore, can users implicitly learn to differentiate between closely spaced VBT patterns based solely on emotional tone, even in the absence of explicit semantic labelling? Such findings could pave the way for emotionally intelligent haptic interfaces.

5.9.5 Neurophysiological Mechanisms

Finally, identifying the cortical and subcortical circuits essential for VBT-induced affect remains a key objective. Can these be distinguished from those involved in purely discriminative tactile processing? Of particular interest is the possible feedback loop between somatosensory cortex and limbic structures during rhythmic or prolonged stimulation, as well as the involvement of neuromodulatory systems such as dopamine, serotonin, or noradrenaline in mediating affective responses.

5.10 Conclusion

VBT modulation of affect is an emerging area of research with potential applications in several fields, including haptics and VR, to create a personalized and immersive user experience. Unlike natural touch, VBT engages only a subset of receptive mechanisms in the skin and evokes only a subspace of our tactile experiences. Nevertheless, VBT methods offer greater control over stimulation settings, allowing investigation into the subtleties of touch, including emotions. In this chapter, we have reviewed the literature documenting evidence supporting the "affective" consequences of the VBT. We have also outlined several potential avenues of exploration to leverage the access to emotions offered by the VBT. Notwithstanding the fact that the skin is the largest organ of the human body, the vibrotactile exploration of this organ to understand and modify the affect and emotions is still in an incipient stage.

References

1. J. Marín-Morales, J.L. Higuera-Trujillo, J. Guixeres, C. Llinares, M. Alcañiz, G. Valenza, Heart rate variability analysis for the assessment of immersive emotional arousal using virtual reality: Comparing real and virtual scenarios. PLoS ONE **16**(7), 1–16 (2021). https://doi.org/10.1371/journal.pone.0254098
2. S. Chehabeddine, M.H. Jamil, W. Park, D.L. Sefo, P.M. Loomer, M. Eid, Bi-manual haptic-based periodontal simulation with finger support and vibrotactile feedback. ACM Trans. Multimed. Comput. Commun. Appl. **17**(1), 1–17 (2021). https://doi.org/10.1145/3421765
3. S. Serafin, A. Adjorlu, L.M. Percy-Smith, A review of virtual reality for individuals with hearing impairments. Multimodal Technol. Interact. **7**(4), 36 (2023). https://doi.org/10.3390/mti7040036
4. S. Persky, A.P. Dolwick, Olfactory perception and presence in a virtual reality food environment. Front. Virtual Real. **1**(September), 1–8 (2020). https://doi.org/10.3389/frvir.2020.571812
5. Y.W. Wang, T.Y. Pan, Y.J. Chang, M.C. Hu, Vibration feedback controlled by intuitive pressing in virtual reality, in *Proceedings – 2019 IEEE International Conference on Artificial Intelligence and Virtual Reality, AIVR 2019*, (2019), pp. 195–198. https://doi.org/10.1109/AIVR46125.2019.00041

6. R. Zhou, Y. Wu, H. Sareen, HexTouch: Affective robot touch for complementary interactions to companion agents in virtual reality, in *Proceedings of the ACM Symposium on Virtual Reality Software and Technology, VRST, vol. 1, no. d*, (2020), pp. 7–8. https://doi.org/10.1145/3385956.3422100
7. P. Ekman, Basic emotions, in *Handbook of Cognition and Emotion*, (Springer, 1999), pp. 1–6. https://doi.org/10.1007/978-3-319-28099-8_495-1
8. P. Ekman, An argument for basic emotions, in *Wild Grass and Morning Blossoms Gathered at Dusk*, (Harvard University Press, 1992), pp. 169–200. https://doi.org/10.2307/j.ctv2vr9cj4.23
9. S. Coppini, C. Lucifora, C.M. Vicario, A. Gangemi, Experiments on real-life emotions challenge Ekman's model. Sci. Rep. **13**(1), 1–12 (2023). https://doi.org/10.1038/s41598-023-36201-5
10. S. Gu, F. Wang, N.P. Patel, J.A. Bourgeois, J.H. Huang, A model for basic emotions using observations of behavior in Drosophila. Front. Psychol. **10**(APR), 1–13 (2019). https://doi.org/10.3389/fpsyg.2019.00781
11. J.A. Russell, A circumplex model of affect. J. Pers. Soc. Psychol. **39**(6), 1161–1178 (1980). https://doi.org/10.1037/h0077714
12. M.L. Xu, P. De Boeck, D. Strunk, An affective space view on depression and anxiety. Int. J. Methods Psychiatr. Res. **27**(4), 1–8 (2018). https://doi.org/10.1002/mpr.1747
13. S.J. Russo, E.J. Nestler, The brain reward circuitry in mood disorders. Nat. Rev. Neurosci. **14**(9), 609–625 (2013). https://doi.org/10.1038/nrn3381
14. L.M. Wade-bohleber, R. Thoma, A.J. Gerber, Psychiatry research: Neuroimaging neural correlates of subjective arousal and valence in health and panic disorder. Psychiatry Res. Neuroimaging **305**(September), 111186 (2020). https://doi.org/10.1016/j.pscychresns.2020.111186
15. A. Tseng et al., Using the circumplex model of affect to study valence and arousal ratings of emotional faces by children and adults with autism spectrum disorders. J. Autism Dev. Disord. **44**(6), 1332–1346 (2014). https://doi.org/10.1007/s10803-013-1993-6
16. D. Watson, A. Tellegen, D. Wiese, J. Vidya, The two general activation systems of affect. J. Pers. Soc. Psychol. **76**(5), 820–838 (1999)
17. K.D. Locke, Development and validation of a circumplex measure of the interpersonal culture in work teams and organizations. Front. Psychol. **10**(APR), 1–18 (2019). https://doi.org/10.3389/fpsyg.2019.00850
18. E.A. Kensinger, S. Corkin, Two routes to emotional memory: Distinct neural processes for valence and arousal. Proc. Natl. Acad. Sci. USA **101**(9), 3310–3315 (2004). https://doi.org/10.1073/pnas.0306408101
19. J. Posner, J.A. Russell, The circumplex model of affect: An integrative approach to affective neuroscience, cognitive development, and psychopathology. Dev. Psychopathol. **17**, 715–734 (2017)
20. P. Ramsamy, A. Haffegee, R. Jamieson, V. Alexandrov, Using haptics to improve immersion in virtual environments, in *Lecture Notes in Computer Science (including subseries Lecture Notes in Artificial Intelligence and Lecture Notes in Bioinformatics)*, LNCS, vol. 3992, (2006), pp. 603–609. https://doi.org/10.1007/11758525_81
21. T. Venkatesan, Q.J. Wang, Feeling connected: The role of haptic feedback in VR concerts and the impact of haptic music players on the music listening experience. Art **12**(4), 148 (2023). https://doi.org/10.3390/arts12040148
22. S.A. MacDonald, S. Brewster, F. Pollick, Eliciting emotion with vibrotactile stimuli evocative of real-world sensations, in *ICMI 2020 – Proceedings of the 2020 International Conference on Multimodal Interaction.*, October, (2020), pp. 125–133. https://doi.org/10.1145/3382507.3418812
23. S.J. Bensmaia, Tactile intensity and population codes. Behav. Brain Res. **190**(2), 165–173 (2008). https://doi.org/10.1016/j.bbr.2008.02.044
24. C. Wee, K.M. Yap, W.N. Lim, Haptic interfaces for virtual reality: Challenges and research directions. IEEE Access **9**, 112145–112162 (2021). https://doi.org/10.1109/ACCESS.2021.3103598

25. M. Akazue, M. Halvey, L. Baillie, S. Brewster, The effect of thermal stimuli on the emotional perception of images, in *Conference on Human Factors in Computing Systems - Proceedings*, (2016), pp. 4401–4412. https://doi.org/10.1145/2858036.2858307
26. H. Fruhstorfer, E.L. Harju, U.F. Lindblom, The significance of A-δ and C fibres for the perception of synthetic heat. Eur. J. Pain **7**(1), 63–71 (2003). https://doi.org/10.1016/S1090-3801(02)00056-3
27. G. Korres, M. Eid, Design of low-profile tactile actuator with rich vibrotactile stimulation for wearable applications, in *HAVE 2019 – IEEE International Symposium on Haptic, Audio-Visual Environments and Games, Proceedings*, (2019), pp. 1–6. https://doi.org/10.1109/HAVE.2019.8921319
28. H.A. Sonar, J. Paik, Soft pneumatic actuator skin with piezoelectric sensors for vibrotactile feedback. Front. Rob. AI **2**(JAN), 1–11 (2016). https://doi.org/10.3389/frobt.2015.00038
29. H.C. Stronks, D.J. Parker, J. Walker, P. Lieby, N. Barnes, The feasibility of coin Motors for use in a vibrotactile display for the blind. Artif. Organs **39**(6), 480–491 (2015). https://doi.org/10.1111/aor.12414
30. K.C. Parsons, M.J. Griffin, Whole-body vibration perception thresholds. J. Sound Vib. **121**(2), 237–258 (1988). https://doi.org/10.1016/S0022-460X(88)80027-0
31. H.C. Stronks, D.J. Parker, N. Barnes, Vibrotactile spatial acuity and intensity discrimination on the lower back using coin motors. IEEE Trans. Haptic **9**(4), 446–454 (2016). https://doi.org/10.1109/TOH.2016.2569484
32. J. Rantala, K. Salminen, R. Raisamo, V. Surakka, Touch gestures in communicating emotional intention via vibrotactile stimulation. Int. J. Hum.-Comput. Stud. **71**(6), 679–690 (2013). https://doi.org/10.1016/j.ijhcs.2013.02.004
33. S.U. Réhman, L. Liu, Vibrotactile emotions on a mobile phone, in *SITIS 2008 – Proceedings of the 4th International Conference on Signal Image Technology and Internet Based Systems*, (2008), pp. 239–243. https://doi.org/10.1109/SITIS.2008.72
34. K. Salminen et al., *Tactile Modulation of Emotional Speech Samples*, vol 2012 (2012). https://doi.org/10.1155/2012/741304
35. E. Frid, H. Lindetorp, Haptic music – Exploring whole-body vibrations and tactile sound for a multisensory music installation, in *Proceedings of the Sound and Music Computing Conferences*, vol. 2020 June, (2020), pp. 68–75
36. R. Zhou, Y. Wu, H. Sareen, HexTouch: A wearable haptic robot for complementary interactions to companion agents in virtual reality, in *SIGGRAPH Asia 2020 Emerging Technologies, SA 2020*, (2020). https://doi.org/10.1145/3415255.3422881
37. S. Yohanan, M. Chan, J. Hopkins, H. Sun, K. MacLean, Hapticat: Exploration of affective touch, in *Proceedings of the Seventh International Conference on Multimodal Interfaces, ICMI'05*, (2005), pp. 222–229. https://doi.org/10.1145/1088463.1088502
38. A. Israr, S. Zhao, K. Schwalje, R. Klatzky, J. Lehman, Feel effects: Enriching storytelling with haptic feedback. ACM Trans. Appl. Percept. **11**(3), 1–17 (2014). https://doi.org/10.1145/2641570
39. A. Israr, S.C. Kim, J. Stec, I. Poupyrev, Surround haptics: Tactile feedback for immersive gaming experiences, in *Conference on Human Factors in Computing Systems - Proceedings*, (2012), pp. 1087–1090. https://doi.org/10.1145/2212776.2212392
40. A. Mazzoni, N. Bryan-Kinns, Mood glove: A haptic wearable prototype system to enhance mood music in film. Entertain. Comput. **17**, 9–17 (2016). https://doi.org/10.1016/j.entcom.2016.06.002
41. S.E. Emgin, A. Aghakhani, T.M. Sezgin, C. Basdogan, HapTable: An interactive tabletop providing online haptic feedback for touch gestures. IEEE Trans. Vis. Comput. Graph. **25**(9), 2749–2762 (2019). https://doi.org/10.1109/TVCG.2018.2855154
42. K. Jung, S. Kim, S. Oh, S.H. Yoon, HapMotion: Motion-to-tactile framework with wearable haptic devices for immersive VR performance experience. Virtual Reality **28**(1), 1–24 (2024). https://doi.org/10.1007/s10055-023-00910-z
43. G. Corniani, H.P. Saal, Tactile innervation densities across the whole body. J. Neurophysiol. **124**(4), 1229–1240 (2020). https://doi.org/10.1152/jn.00313.2020

44. A.S. French, P.H. Torkkeli, Mechanoreceptors, in *Encyclopedia of Neuroscience*, (2009), pp. 689–695. https://doi.org/10.1016/B978-008045046-9.01921-5
45. E.R. Kandel, *Principles of Neural Science* (McGraw-Hill, New York, 2013)
46. R. Pawling, P.R. Cannon, F.P. McGlone, S.C. Walker, C-tactile afferent stimulating touch carries a positive affective value. PLoS One **12**(3), 1–15 (2017). https://doi.org/10.1371/journal.pone.0173457
47. L.S. Löken, J. Wessberg, I. Morrison, F. McGlone, H. Olausson, Coding of pleasant touch by unmyelinated afferents in humans. Nat. Neurosci. **12**(5), 547–548 (2009). https://doi.org/10.1038/nn.2312
48. H. Culbertson, S.B. Schorr, A.M. Okamura, Haptics: The present and future of artificial touch sensation. Annu. Rev. Control Rob. Auton. Syst. **1**(March), 385–409 (2018). https://doi.org/10.1146/annurev-control-060117-105043
49. L.K. Case, C.M. Laubacher, H. Olausson, B. Wang, P.A. Spagnolo, M.C. Bushnell, Encoding of touch intensity but not pleasantness in human primary somatosensory cortex. J. Neurosci. **36**(21), 5850–5860 (2016). https://doi.org/10.1523/JNEUROSCI.1130-15.2016
50. Å. Vallbo, H. Olausson, J. Wessberg, U. Norrsell, A system of unmyelinated afferents for innocuous mechanoreception in the human skin. Brain Res. **628**(1–2), 301–304 (1993). https://doi.org/10.1016/0006-8993(93)90968-S
51. R.H. Watkins, M. Dione, R. Ackerley, H.B. Wasling, J. Wessberg, L.S. Loken, Evidence for sparse C-tactile afferent innervation of glabrous human hand skin. J. Neurophysiol. **125**(1), 232–237 (2021). https://doi.org/10.1152/JN.00587.2020
52. K.K. Imbir, J. Duda-Goławska, A. Sobieszek, A. Wielgopolan, M. Pastwa, J. Żygierewicz, Arousal, subjective significance and the origin of valence aligned words in the processing of an emotional categorisation task. *PLoS ONE* **17**(3 March), 1–27 (2022). https://doi.org/10.1371/journal.pone.0265537
53. M. Esslen, R.D. Pascual-Marqui, D. Hell, K. Kochi, D. Lehmann, Brain areas and time course of emotional processing. NeuroImage **21**(4), 1189–1203 (2004). https://doi.org/10.1016/j.neuroimage.2003.10.001
54. M.M.A. Nielen et al., Distinct brain systems underlie the processing of valence and arousal of affective pictures. Brain Cogn. **71**(3), 387–396 (2009). https://doi.org/10.1016/j.bandc.2009.05.007
55. Y. Lee, Y. Seo, Y. Lee, D. Lee, Dimensional emotions are represented by distinct topographical brain networks. Int. J. Clin. Health Psychol. **23**(4) (2023). https://doi.org/10.1016/j.ijchp.2023.100408
56. S. Anders, M. Lotze, M. Erb, W. Grodd, N. Birbaumer, Brain activity underlying emotional valence and arousal: A response-related fMRI study. Hum. Brain Mapp. **23**(4), 200–209 (2004). https://doi.org/10.1002/hbm.20048
57. T. Sieger et al., Distinct populations of neurons respond to emotional valence and arousal in the human subthalamic nucleus. Proc. Natl. Acad. Sci. USA **112**(10), 3116–3121 (2015). https://doi.org/10.1073/pnas.1410709112
58. T. Colibazzi et al., Neural systems subserving valence and arousal during the experience of induced emotions. Emotion **10**(3), 377–389 (2010). https://doi.org/10.1037/a0018484
59. S.N. Haber, B. Knutson, The reward circuit: Linking primate anatomy and human imaging. Neuropsychopharmacology **35**(1), 4–26 (2010). https://doi.org/10.1038/npp.2009.129
60. X. Liu, J. Hairston, M. Schrier, J. Fan, Common and distinct networks underlying reward valence and processing stages: A meta-analysis of functional neuroimaging studies. Neurosci. Biobehav. Rev. **35**(5), 1219–1236 (2011). https://doi.org/10.1016/j.neubiorev.2010.12.012
61. H. Lin et al., Stimulus arousal drives amygdalar responses to emotional expressions across sensory modalities. Sci. Rep. **10**(1), 1–12 (2020). https://doi.org/10.1038/s41598-020-58839-1
62. R. Adolphs, H. Damasio, D. Tranel, G. Cooper, A.R. Damasio, A role for somatosensory cortices in the visual recognition of emotion as revealed by three-dimensional lesion mapping. J. Neurosci. **20**(7), 2683–2690 (2000). https://doi.org/10.1523/jneurosci.20-07-02683.2000

63. C.A. Leppla et al., Thalamus sends information about arousal but not valence to the amygdala. Psychopharmacology **240**(3), 477–499 (2023). https://doi.org/10.1007/s00213-022-06284-5
64. J.A. Miranda-Correa, M.K. Abadi, N. Sebe, I. Patras, AMIGOS: A dataset for affect, personality and mood research on individuals and groups. IEEE Trans. Affect. Comput. **12**(2), 479–493 (2021). https://doi.org/10.1109/TAFFC.2018.2884461
65. R. Hari, Evoked potentials elicited by long vibrotactile stimuli in the human EEG. Pflugers Arch. - Eur. J. Physiol. **384**(2), 167–170 (1980). https://doi.org/10.1007/BF00584434
66. K. Soininen, T. Järvilehto, Somatosensory evoked potentials associated with tactile stimulation at detection threshold in man. Electroencephalogr. Clin. Neurophysiol. **56**(5), 494–500 (1983). https://doi.org/10.1016/0013-4694(83)90234-1
67. H. Hämäläinen, J. Kekoni, M. Sams, K. Reinikainen, R. Näätänen, Human somatosensory evoked potentials to mechanical pulses and vibration: Contributions of SI and SII somatosensory cortices to P50 and P100 components. Electroencephalogr. Clin. Neurophysiol. **75**(1–2), 13–21 (1990). https://doi.org/10.1016/0013-4694(90)90148-D
68. H. Alsuradi, W. Park, M. Eid, Midfrontal theta oscillation encodes haptic delay. Sci. Rep. **11**(1), 1–12 (2021). https://doi.org/10.1038/s41598-021-95631-1
69. M.Y. Kim, H. Kwon, T.H. Yang, K. Kim, Vibration alert to the brain: Evoked and induced MEG responses to high-frequency vibrotactile stimuli on the index finger of dominant and non-dominant hand. Front. Hum. Neurosci. **14**(November) (2020). https://doi.org/10.3389/fnhum.2020.576082
70. C. Nangini, B. Ross, F. Tam, S.J. Graham, Magnetoencephalographic study of vibrotactile evoked transient and steady-state responses in human somatosensory cortex. NeuroImage **33**(1), 252–262 (2006). https://doi.org/10.1016/j.neuroimage.2006.05.045
71. H.S. Kim et al., Comparison of neural activation area in primary somatosensory cortex and brodmann area 3 according to finger and phalange high-frequency vibration stimulation. Appl. Sci. (Switzerland) **10**(17), 1–9 (2020). https://doi.org/10.3390/app10176142
72. V. Jousmäki, R. Hari, Somatosensory evoked fields to large-area vibrotactile stimuli. Clin. Neurophysiol. **110**(5), 905–909 (1999). https://doi.org/10.1016/S1388-2457(98)00059-5
73. S. Ryun, J.S. Kim, H. Lee, C.K. Chung, Tactile frequency-specific high-gamma activities in human primary and secondary somatosensory cortices. Sci. Rep. **7**(1), 1–10 (2017). https://doi.org/10.1038/s41598-017-15767-x
74. S. Tobimatsu, Y.M. Zhang, M. Kato, Steady-state vibration somatosensory evoked potentials: Physiological characteristics and tuning function. Clin. Neurophysiol. **110**(11), 1953–1958 (1999). https://doi.org/10.1016/S1388-2457(99)00146-7
75. S.T. Francis, E.F. Kelly, R. Bowtell, W.J.R. Dunseath, S.E. Folger, F. McGlone, fMRI of the responses to vibratory stimulation of digit tips. NeuroImage **11**(3), 188–202 (2000). https://doi.org/10.1006/nimg.2000.0541
76. H. Burton, R.J. Sinclair, D.G. McLaren, Cortical network for vibrotactile attention: A fMRI study. Hum. Brain Mapp. **29**(2), 207–221 (2008). https://doi.org/10.1002/hbm.20384
77. T. Bardouille, B. Ross, MEG imaging of sensorimotor areas using inter-trial coherence in vibrotactile steady-state responses. NeuroImage **42**(1), 323–331 (2008). https://doi.org/10.1016/j.neuroimage.2008.04.176
78. J. Kim, Y.G. Chung, S.C. Chung, H.H. Bulthoff, S.P. Kim, Neural categorization of vibrotactile frequency in flutter and vibration stimulations: An fMRI study. IEEE Trans. Haptics **9**(4), 455–464 (2016). https://doi.org/10.1109/TOH.2016.2593727
79. R. Luna, A. Hernández, C.D. Brody, R. Romo, Neural codes for perceptual discrimination in primary somatosensory cortex. Nat. Neurosci. **8**(9), 1210–1219 (2005). https://doi.org/10.1038/nn1513
80. J.A. Maldjian, A. Gottschalk, R.S. Patel, D. Pincus, J.A. Detre, D.C. Alsop, Mapping of secondary somatosensory cortex activation induced by vibrational stimulation: An fMRI study. Brain Res. **824**(2), 291–295 (1999). https://doi.org/10.1016/S0006-8993(99)01126-9
81. A.Z. Snyder, Steady-state vibration evoked potentials: Description of technique and characterization of responses. Electroencephalogr. Clin. Neurophysiol. Evoked Potentials **84**(3), 257–268 (1992). https://doi.org/10.1016/0168-5597(92)90007-X

82. Y. Goto, T. Taniwaki, K.I. Yamashita, N. Kinukawa, S. Tobimatsu, Interhemispheric functional desynchronization in the human vibratory system. Brain Res. **980**(2), 249–254 (2003). https://doi.org/10.1016/S0006-8993(03)02985-8
83. X. Xie et al., A review of smart materials in tactile actuators for information delivery. C: J. Carbon Res. **3**(4), 38 (2017). https://doi.org/10.3390/c3040038
84. S. Tobimatsu, Y.M. Zhang, R. Suga, M. Kato, Differential temporal coding of the vibratory sense in the hand and foot in man. Clin. Neurophysiol. **111**(3), 398–404 (2000). https://doi.org/10.1016/S1388-2457(99)00278-3
85. C. Bassel, B.B. Schiff, Unilateral vibrotactile stimulation induces emotional biases in cognition and performance. Neuropsychologia **39**(3), 282–287 (2001). https://doi.org/10.1016/S0028-3932(00)00116-0
86. F. Sorgini et al., Neuromorphic vibrotactile stimulation of fingertips for encoding object stiffness in telepresence sensory substitution and augmentation applications. Sensors (Switzerland) **18**(1) (2018). https://doi.org/10.3390/s18010261
87. Y. Yoo, T. Yoo, J. Kong, S. Choi, Emotional responses of tactile icons: Effects of amplitude, frequency, duration, and envelope, in *IEEE World Haptics Conference, WHC 2015*, (2015), pp. 235–240. https://doi.org/10.1109/WHC.2015.7177719
88. H. Hasegawa, S. Okamoto, K. Ito, Y. Yamada, Affective vibrotactile stimuli: Relation between vibrotactile parameters and affective responses. Int. J. Affect. Eng. **18**(4), 171–180 (2019). https://doi.org/10.5057/ijae.ijae-d-18-00008
89. O.S. Schneider, H. Seifi, S. Kashani, M. Chun, K.E. MacLean, HapTurk: Crowdsourcing affective ratings for vibrotactile icons, in *Conference on Human Factors in Computing Systems - Proceedings*, (2016), pp. 3248–3260. https://doi.org/10.1145/2858036.2858279
90. H. Seifi, K.E. Maclean, A first look at individuals' affective ratings of vibrations, in *2013 World Haptics Conference, WHC 2013*, (2013), pp. 605–610. https://doi.org/10.1109/WHC.2013.6548477
91. J. Hwang, W. Hwang, Vibration perception and excitatory direction for haptic devices. J. Intell. Manuf. **22**(1), 17–27 (2011). https://doi.org/10.1007/s10845-009-0277-7
92. H. Seifi, M. Chun, K.E. Maclean, Toward affective handles for tuning vibrations. ACM Trans. Appl. Percept. **15**(3), 1 (2018). https://doi.org/10.1145/3230645
93. J. Raisamo, R. Raisamo, V. Surakka, Comparison of saltation, amplitude modulation, and a hybrid method of vibrotactile stimulation. IEEE Trans. Haptics **6**(4), 517–521 (2013). https://doi.org/10.1109/TOH.2013.25
94. Y. Yoo, I. Hwang, S. Choi, Consonance of vibrotactile chords. IEEE Trans. Haptics **7**(1), 3–13 (2014). https://doi.org/10.1109/TOH.2013.57
95. J. Van Erp, M. Spapé, Distilling the underlying dimensions of tactile melodies, in *Proceedings of Eurohaptics 2003*, (2003), pp. 111–120. [Online]. Available: https://pdfs.semanticscholar.org/17d3/b6c791b40db6144ff54a9dcfc77f54ec9ab2.pdf%0A http://scholar.google.com/scholar?hl=en&btnG=Search&q=intitle:Distilling+the+Underlying +Dimensions+of+Tactile+Melodies#0
96. I.E.T. De Araujo, E.T. Rolls, M.L. Kringelbach, F. McGlone, N. Phillips, Taste-olfactory convergence, and the representation of the pleasantness of flavour, in the human brain. Eur. J. Neurosci. **18**(7), 2059–2068 (2003). https://doi.org/10.1046/j.1460-9568.2003.02915.x
97. S. Genzer, D.C. Ong, J. Zaki, A. Perry, Mu rhythm suppression over sensorimotor regions is associated with greater empathic accuracy. Soc. Cogn. Affect. Neurosci. **17**(9), 788–801 (2022). https://doi.org/10.1093/scan/nsac011
98. M. Davidovic, G. Starck, H. Olausson, Processing of affective and emotionally neutral tactile stimuli in the insular cortex. Dev. Cogn. Neurosci. **35**(November), 94–103 (2019). https://doi.org/10.1016/j.dcn.2017.12.006
99. T. Allison, G. McCarthy, C.C. Wood, P.D. Williamson, D.D. Spencer, Human cortical potentials evoked by stimulation of the median nerve. II. Cytoarchitectonic areas generating long-latency activity. J. Neurophysiol. **62**(3), 711–722 (1989). https://doi.org/10.1152/jn.1989.62.3.711

100. C. Simões, O. Jensen, L. Parkkonen, R. Hari, Phase locking between human primary and secondary somatosensory cortices. Proc. Natl. Acad. Sci. USA **100**(5), 2691–2694 (2003). https://doi.org/10.1073/pnas.0437944100
101. C. Simões, M. Mertens, N. Forss, V. Jousmäki, B. Lütkenhöner, R. Hari, Functional overlap of finger representations in human SI and SII cortices. J. Neurophysiol. **86**(4), 1661–1665 (2001). https://doi.org/10.1152/jn.2001.86.4.1661
102. T. Hansson, T. Brismar, Tactile stimulation of the hand causes bilateral cortical activation: A functional magnetic resonance study in humans. Neurosci. Lett. **271**(1), 29–32 (1999). https://doi.org/10.1016/S0304-3940(99)00508-X
103. D. Cheyne et al., Neuromagnetic imaging of cortical oscillations accompanying tactile stimulation. Cogn. Brain Res. **17**(3), 599–611 (2003). https://doi.org/10.1016/S0926-6410(03)00173-3
104. F. Mcglone et al., Touching and feeling: Differences in pleasant touch processing between glabrous and hairy skin in humans. Eur. J. Neurosci. **35**(11), 1782–1788 (2012). https://doi.org/10.1111/j.1460-9568.2012.08092.x
105. N. Ravaja, V. Harjunen, I. Ahmed, G. Jacucci, M.M. Spapé, Feeling touched: Emotional modulation of somatosensory potentials to interpersonal touch. Sci. Rep. **7**(March 2016), 1–11 (2017). https://doi.org/10.1038/srep40504
106. A. Sel, B. Forster, B. Calvo-Merino, The emotional homunculus: ERP evidence for independent somatosensory responses during facial emotional processing. J. Neurosci. **34**(9), 3263–3267 (2014). https://doi.org/10.1523/JNEUROSCI.0106-13.2014
107. P. Montoya, C. Sitges, Affective modulation of somatosensory-evoked potentials elicited by tactile stimulation. Brain Res. **1068**(1), 205–212 (2006). https://doi.org/10.1016/j.brainres.2005.11.019
108. L.I. Aftanas, S.A. Golocheikine, Human anterior and frontal midline theta and lower alpha reflect emotionally positive state and internalized attention: High-resolution EEG investigation of meditation. Neurosci. Lett. **310**(1), 57–60 (2001). https://doi.org/10.1016/S0304-3940(01)02094-8
109. M. Giraud, A. Javadi, C. Lenatti, J. Allen, L. Tamè, The role of the somatosensory system in the feeling of emotions: A neurostimulation study. Soc. Cogn. Affect. Neurosci. **19**(1), nsae062 (2024). https://doi.org/10.1093/scan/nsae062. [Online]
110. P.A. Lewis, H.D. Critchley, P. Rotshtein, R.J. Dolan, Neural correlates of processing valence and arousal in affective words. Cereb. Cortex **17**(3), 742–748 (2007). https://doi.org/10.1093/cercor/bhk024
111. S. Anders et al., Parietal somatosensory association cortex mediates affective blindsight. Nat. Neurosci. **7**(4), 339–340 (2004). https://doi.org/10.1038/nn1213
112. M. Schaefer, A. Kühnel, F. Rumpel, M. Gärtner, Dispositional empathy predicts primary somatosensory cortex activity while receiving touch by a hand. Sci. Rep. **11**(1), 1–10 (2021). https://doi.org/10.1038/s41598-021-90344-x
113. C. Mccabe, E.T. Rolls, A. Bilderbeck, F. McGlone, Cognitive influences on the affective representation of touch and the sight of touch in the human brain. Soc. Cogn. Affect. Neurosci. **3**(2), 97–108 (2008). https://doi.org/10.1093/scan/nsn005
114. A. Mazza et al., Hedonic and autonomic responses in promoting affective touch. Sci. Rep. **13**(1), 11201 (2023). https://doi.org/10.1038/s41598-023-37471-9
115. D. Rudrauf et al., Enter feelings: Somatosensory responses following early stages of visual induction of emotion. Int. J. Psychophysiol. **72**(1), 13–23 (2009). https://doi.org/10.1016/j.ijpsycho.2008.03.015
116. A. Greco, A. Guidi, M. Bianchi, A. Lanata, G. Valenza, E.P. Scilingo, Brain dynamics induced by pleasant/unpleasant tactile stimuli conveyed by different fabrics. IEEE J. Biomed. Health Inform. **23**(6), 2417–2427 (2019). https://doi.org/10.1109/JBHI.2019.2893324
117. I. Gordon, A.C. Voos, R.H. Bennett, D.Z. Bolling, K.A. Pelphrey, M.D. Kaiser, Brain mechanisms for processing affective touch. Hum. Brain Mapp. **34**(4), 914–922 (2013). https://doi.org/10.1002/hbm.21480

118. G. Valenza, A. Greco, M. Bianchi, M. Nardelli, S. Rossi, E.P. Scilingo, EEG oscillations during caress-like affective haptic elicitation. Psychophysiology **55**(10), 1–15 (2018). https://doi.org/10.1111/psyp.13199
119. H. Singh et al., The brain's response to pleasant touch: An EEG investigation of tactile caressing. Front. Hum. Neurosci. **8**(November), 1–10 (2014). https://doi.org/10.3389/fnhum.2014.00893
120. M. von Mohr, M.J. Crowley, J. Walthall, L.C. Mayes, K.A. Pelphrey, H.J.V. Rutherford, EEG captures affective touch: CT-optimal touch and neural oscillations. Cogn. Affect. Behav. Neurosci. **18**(1), 155–166 (2018). https://doi.org/10.3758/s13415-017-0560-6
121. D. Hoefer, M. Handel, K.M. Müller, T.R. Hammer, Electroencephalographic study showing that tactile stimulation by fabrics of different qualities elicit graded event-related potentials. Skin Res. Technol. **22**(4), 470–478 (2016). https://doi.org/10.1111/srt.12288
122. N. Bolognini, A. Rossetti, S. Convento, G. Vallar, Understanding others' feelings: The role of the right primary somatosensory cortex in encoding the affective valence of others' touch. J. Neurosci. **33**(9), 4201–4205 (2013). https://doi.org/10.1523/JNEUROSCI.4498-12.2013
123. S. Ganguly, A. Koppula, K.S. Sridharan, Modelling the early affect response to vibrotactile stimulation in the cortex: A source – based spatiotemporal analysis. *TechRxiv*, 1–15. https://doi.org/10.36227/techrxiv.170975047.74079227
124. L.R. Manfredi et al., Natural scenes in tactile texture. J. Neurophysiol. **111**(9), 1792–1802 (2014). https://doi.org/10.1152/jn.00680.2013
125. Y.G. Chung et al., Frequency-dependent patterns of somatosensory cortical responses to vibrotactile stimulation in humans: A fMRI study. Brain Res. **1504**, 47–57 (2013). https://doi.org/10.1016/j.brainres.2013.02.003
126. K. Morrow, D. Wilbern, R. Taghavi, M. Ziat, The effects of duration and frequency on the perception of vibrotactile stimulation on the neck, in *IEEE Haptics Symposium, HAPTICS*, vol. 2016-April, (2016), pp. 41–46. https://doi.org/10.1109/HAPTICS.2016.7463153
127. B. Berglund, U. Berglund, G. Ekman, Temporal integration of vibrotactile stimulation. Percept. Mot. Skills **25**(2), 549–560 (1967). https://doi.org/10.2466/pms.1967.25.2.549
128. G.K. Essick, A. James, F.P. McGlone, Psychophysical assessment of the affective components of non-painful touch. Neuroreport **10**(10), 2083–2087 (1999). https://doi.org/10.1097/00001756-199907130-00017
129. G.K. Essick et al., Quantitative assessment of pleasant touch. Neurosci. Biobehav. Rev. **34**(2), 192–203 (2010). https://doi.org/10.1016/j.neubiorev.2009.02.003
130. R. Etzi, C. Spence, A. Gallace, Textures that we like to touch: An experimental study of aesthetic preferences for tactile stimuli. Conscious. Cogn. **29**, 178–188 (2014). https://doi.org/10.1016/j.concog.2014.08.011
131. M. Terenti, R.D. Vatavu, Measuring the user experience of vibrotactile feedback on the finger, wrist, and forearm for touch input on large displays, in *Conference on Human Factors in Computing Systems - Proceedings*, vol. 1, no. 1, (2022). https://doi.org/10.1145/3491101.3519704
132. L. Kang et al., Functional connectivity between the thalamus and the primary somatosensory cortex in major depressive disorder: A resting-state fMRI study. BMC Psychiatry **18**(1), 4–11 (2018). https://doi.org/10.1186/s12888-018-1913-6
133. Y. Takatsuru, N. Koibuchi, Alteration of somatosensory response in adulthood by early life stress. Front. Mol. Neurosci. **8**(MAY), 1–7 (2015). https://doi.org/10.3389/fnmol.2015.00015
134. D. Ray, D. Bezmaternykh, M. Mel'nikov, K.J. Friston, M. Das, Altered effective connectivity in sensorimotor cortices is a signature of severity and clinical course in depression. Proc. Natl. Acad. Sci. U. S. A. **118**(40) (2021). https://doi.org/10.1073/pnas.2105730118
135. L. Minuzzi, S.K. Syan, M. Smith, A. Hall, G.B.C. Hall, B.N. Frey, Structural and functional changes in the somatosensory cortex in euthymic females with bipolar disorder. Aust. N. Z. J. Psychiatry **52**(11), 1075–1083 (2018). https://doi.org/10.1177/0004867417746001
136. R. Torta, V. Ieraci, F. Zizzi, A review of the emotional aspects of neuropathic pain: From comorbidity to co-pathogenesis. Pain Ther. **6**(s1), 11–17 (2017). https://doi.org/10.1007/s40122-017-0088-z

137. D. Selvarajah et al., Structural and functional abnormalities of the primary somatosensory cortex in diabetic peripheral neuropathy: A multimodal MRI study. Diabetes **68**(4), 796–806 (2019). https://doi.org/10.2337/db18-0509
138. J. Rosner et al., Normative data of contact heat evoked potentials from the lower extremities. Sci. Rep. **8**(1), 1–9 (2018). https://doi.org/10.1038/s41598-018-29145-8
139. J. Pujol et al., *Testing the effects of gentle vibrotactile stimulation on symptom relief in fibromyalgia* (2019), pp. 1–11
140. Á. García López, V. Cerdán, T. Ortiz, J.M. Sánchez Pena, R. Vergaz, Emotion elicitation through vibrotactile stimulation as an alternative for deaf and hard of hearing people: An EEG study. Electronics (Switzerland) **11**(14) (2022). https://doi.org/10.3390/electronics11142196
141. L.M. Brown, S.A. Brewster, H.C. Purchase, Multidimensional tactons for non-visual information presentation in mobile devices. ACM Int. Conf. Proc. Ser. **159**, 231–238 (2006). https://doi.org/10.1145/1152215.1152265

Chapter 6
Assessing Cognitive Fatigue in Human Teammates During Human-Robot Collaboration

Manizheh Zand, Luca Chierotti, Sama Nikanfar, Glenn Wylie, Fillia Makedon, and Maria Kyrarini

Abstract As human-robot teams become increasingly integrated into predominantly human-centric environments, such as workplaces, hospitals, and even our homes, understanding cognitive fatigue on human teammates is crucial for the robots. This chapter explores the challenge of assessing cognitive fatigue in an offline and online manner through sensory data during Human-Robot Collaboration (HRC). We examine state-of-the-art approaches that leverage signals such as Electrocardiography (ECG), Galvanic Skin Response (GSR), speech-based markers, and others to assess Cognitive Fatigue (CF) states in real-time. The chapter discusses the strengths and limitations of each modality, the available datasets and the state-of-the-art algorithms for offline and online CF assessment, and the implications for adaptive systems that can respond to signs of fatigue. By bridging insights from robotics, neuroscience, and human factors, this chapter provides a comprehensive foundation for designing robust, responsive robot teammates capable of acting when human teammates are cognitively fatigued.

Keywords Cognitive fatigue · Assessment · Speech · Human-robot collaboration (HRC) · Collaborative robots · Physiological sensors · Electrocardiography (ECG) · Galvanic skin response (GSR)

M. Zand · L. Chierotti · M. Kyrarini (✉)
Department of Electrical and Computer Engineering, Santa Clara University, Santa Clara, CA, USA
e-mail: mzand@scu.edu; lchierotti@scu.edu; mkyrarini@scu.edu

S. Nikanfar · F. Makedon
Department of Computer Science and Engineering, University of Texas at Arlington, Arlington, TX, USA
e-mail: sxn8789@mavs.uta.edu; makedon@uta.edu

G. Wylie
Kessler Foundation, East Hanover, NJ, USA
e-mail: gwylie@kesslerfoundation.org

R. Vinjamuri (ed.), *Bridging the Gap between Mind and Machine*,
https://doi.org/10.1007/978-3-032-06713-5_6

6.1 Introduction

As Human-Robot Collaboration (HRC) becomes increasingly prevalent across various domains, including industry, healthcare, and domestic environments, understanding Cognitive Fatigue (CF) in human teammates has emerged as a critical skill for robots. Wylie et al. [1] defined CF as a state of diminished cognitive resources that can result from prolonged or sustained mental effort. It is characterized by a progressive decline in cognitive efficiency, attention, and decision-making abilities, which can degrade task performance and compromise safety outcomes. While mental effort refers to the active mobilization of cognitive resources, such as working memory capacity, required to perform a task effectively [2], CF develops when these cognitive demands are maintained over extended periods without sufficient recovery, leading to the depletion of those cognitive resources. Researchers have focused on assessing CF in various domains, such as Human-Computer Interaction [3, 4] and healthcare [5–7]; however, there is limited work in the field of HRC [8].

Distinguishing between mental effort and CF, and understanding interplay is essential in HRC settings where humans are often required to sustain high levels of cognitive demand while interacting with robotic teammates. Humans collaborating with robots while experiencing diminished or fluctuating CF are more susceptible to slower response times, decision-making errors, and lapses in attention, which degrade team performance and compromise safety outcomes [9]. Beyond these direct performance impacts, CF can also alter interpersonal dynamics within the team. Cognitively fatigued users may exhibit changes in trust and engagement toward their robotic counterparts, undermining the overall quality and effectiveness of the collaboration. For instance, Guo et al. [10] examined the effects of Cognitive Workload (CWL) on human trust in hybrid HRC scenarios, revealing that elevated CWL significantly modulates trust in robots and may contribute to increased CF, further influencing users' trust and engagement behaviors. Similarly, Matuz et al. [11] emphasized that prolonged exposure to high CWL exacerbates CF, reinforcing the broader recognition that continuous monitoring and effective management of human CF are essential for maintaining safe, efficient, and resilient HRC teams.

Zand et al. [12] emphasized that CF can be assessed using a range of methods, with the optimal choice depending on the timing, context, and specific demands of the HRC scenario. They also emphasized that a key distinction exists between offline and online assessment approaches. Offline CF assessment evaluates CF retrospectively, typically after task completion, using self-reported questionnaires, performance metrics, or physiological data. In contrast, online CF assessment enables continuous monitoring of CF levels to analyze streaming data from sensors or system logs. This provides dynamic insights into the user's CF state as the HRC unfolds, supporting adaptive responses from robotic systems.

To support the evaluation of CF during and after HRC, subjective assessment methods remain indispensable. Given the complex and dynamic nature of HRC, self-reported assessments provide valuable insights into users' internal experiences of CF, capturing subjective perceptions that may not be fully reflected in performance metrics or physiological indicators alone [13, 14]. These assessments provide essential user-centered data, serving to validate and complement objective measurements collected via sensor data and system logs, thereby contributing to a holistic understanding of CF in HRC contexts.

For a comprehensive evaluation of CF in HRC, several well-established subjective assessment tools have been established, each demonstrating robust psychometric properties and proven applicability in cognitively demanding environments. Table 6.1 summarizes the most widely used instruments, including Fatigue Assessment Scale (FAS) [15], Mental Fatigue Scale (MFS) [16], and Visual Analogue Scale for Fatigue (VAS-F) [17]. These tools have been extensively applied across various domains and are increasingly utilized in HRC research [18, 19], where accurate assessment of users' CF is critical for ensuring safety, maintaining performance, and supporting effective collaboration.

As detailed in Table 6.1, each instrument offers distinct advantages and is suited to different phases of HRC, including post-task evaluations, in-task monitoring, and longitudinal fatigue tracking. Although these methods have inherent limitations, such as reliance on self-reporting and the potential to disrupt ongoing tasks, they remain essential components of HRC studies. Their ability to capture user perceptions of CF provides valuable insights not only for assessing cognitive states but also for informing the design of fatigue-aware, adaptive robotic behaviors aimed at fostering resilient, safe, and effective human-robot teams.

Table 6.1 Summary of widely used subjective cognitive fatigue (CF) assessment tools and their relevance to human-robot collaboration (HRC) contexts. Each instrument offers unique strengths suited to different assessment phases, including post-task evaluations, in-task monitoring, and longitudinal tracking. Despite inherent limitations such as reliance on self-reporting and the risk of interaction disruption, these tools remain integral to CF assessment in HRC studies, providing essential user-centered data that support the development of fatigue-aware, adaptive HRC systems

Tool	Type	Key features	Strengths in HRC context	Limitations in HRC context
FAS	General fatigue scale	10-item unidimensional scale (physical + cognitive fatigue) [15]	Simple; useful for tracking general fatigue trends over work periods [24]	Less sensitive to immediate or task-specific CF during dynamic HRC tasks
MFS	Mental fatigue scale	15 items targeting cognitive, emotional, sensory fatigue symptoms [16]	Captures nuanced mental fatigue symptoms; suitable for post-task or longitudinal assessments in HRC [25]	Lengthier to administer; may not be practical for rapid assessment during tasks
VAS-F	Momentary fatigue scale	Visual analogue format (single or multi-item assessing fatigue/energy) [17]	Fast, sensitive to momentary fatigue; ideal for use during HRC tasks for adaptive feedback [19]	Provides only subjective perception; requires frequent user input which may disrupt flow

CF is a state of mental exhaustion characterized by diminished alertness and impaired cognitive function, posing significant challenges in high-demand domains such as HRC, where human performance, decision-making, and safety are critical. To address these challenges, researchers have increasingly adopted sensor-based methods for CF detection, leveraging physiological signals, behavioral cues, and speech patterns. Physiological measures, including Electroencephalogram (EEG), ECG, and GSR, provide objective, online indicators of a user's cognitive state [20]. Among these, EEG is particularly sensitive to CF, and is widely recognized as one of the most accurate modalities for detecting brain-state changes associated with fatigue [21]. In parallel, heart rate variability derived from ECG and skin conductance captured via GSR reflect autonomic nervous system responses closely linked to CWL and CF [22].

In addition to physiological signals, Gowda et al. [23] mentioned that behavioral monitoring approaches, such as vision-based tracking of gaze behavior, head pose estimation, or blink frequency, offer observable indicators of CF, although environmental conditions and task configurations often influence these methods. More recently, speech analysis has emerged as a promising, non-intrusive modality for CF detection, as CF affects prosodic and spectral features, leading to slower speech rates and changes in voice quality [19]. Collectively, these multimodal sensing techniques enhance the capability to detect CF in HRC contexts, supporting the development of fatigue-aware robotic systems capable of dynamically adjusting task allocation, interaction modalities, or support strategies to maintain safe and effective collaboration [12].

In this chapter, we outline our methodology for selecting relevant research literature, focusing exclusively on studies where CF detection has been applied directly in HRC settings, either during data collection or within interactive scenarios. This targeted approach ensures that the reviewed works address the unique demands and challenges of HRC environments, where safety, adaptability, and collaboration effectiveness are paramount. Furthermore, we present the rationale for prioritizing physiological and speech-based modalities over vision- or behavior-based approaches. While vision-based methods (e.g., gaze, posture, blink rate) and behavioral analyses (e.g., task performance metrics) have been explored for CF detection, these approaches are often highly sensitive to environmental variability, occlusions, and workspace configurations that may vary in HRC settings. In contrast, physiological signals and speech modalities offer more robust, unobtrusive, and context-resilient alternatives, providing reliable indicators of CF that can be seamlessly integrated into HRC systems to enable adaptive and fatigue-aware task management.

To guide the reader through the multifaceted domain of CF in HRC, this chapter is organized into several key sections. We begin with an overview of CF and its implications in collaborative settings, followed by a discussion of offline assessment techniques, including a detailed user study leveraging physiological and speech data. The chapter then transitions to online CF detection methods that enable real-time adaptation in HRC. We conclude by outlining open research challenges and future directions for building adaptive, fatigue-aware robotic systems. This structure ensures a comprehensive understanding of both retrospective and real-time CF assessment strategies, grounded in empirical findings and practical application.

6.2 Offline Cognitive Fatigue Assessment in HRC

Offline CF assessment in HRC relies primarily on post-task evaluations, including subjective self-reports and performance analyses, to capture users' perceived CF levels. These approaches offer valuable insights into the human teammate's internal state, complementing physiological and behavioral data streams. Although inherently retrospective and limited in supporting immediate adaptation, offline assessments remain critical for validating machine learning (ML)-based models and establishing robust ground truth CF labels.

Table 6.2 provides an overview of studies applying CF detection in HRC contexts, highlighting the diversity of sensor modalities, the extent of adaptability integration, and the modeling approaches employed. References [26–29] will be discussed in Sect. 6.3, since their research is based on adaptability in HRC. The majority of existing works have focused on classification models, typically distinguishing between discrete fatigue states, such as fatigued and non-fatigued [30]. Regression-based approaches for continuous CF estimation remain rare, exemplified only by Zand et al. [31]. Physiological signals such as ECG, GSR, and EEG continue to dominate the modality landscape [32], while vision-based and behavioral approaches have been explored in concept proposals [26, 33, 34].

To the best of the authors' knowledge, Zand et al. [12, 31] represent the first effort to develop and validate multimodal regression models for offline CF estimation in HRC, addressing a critical gap by enabling fine-grained CF estimation essential for adaptive and fatigue-aware HRC systems. Previous CF detection research in HRC has exclusively focused on classification models, categorizing CF into discrete states (e.g., fatigued vs. non-fatigued), thus limiting the potential for nuanced system adaptation.

Despite the indispensable role of offline assessment in model validation and system benchmarking, its retrospective and user-dependent nature limits its effectiveness for real-time adaptation. Assessments such as VAS-F and NASA Task Load Index (NASA-TLX) [37] require active user engagement, potentially introducing task interruptions and delays in CF detection. Nevertheless, offline assessments remain an integral component of comprehensive CF evaluation frameworks, supporting the triangulation of subjective, physiological, and behavioral data while informing user-centered system design, interaction strategies, and acceptance studies [18].

The following Sect. 6.2.1 explores our research on detecting CF through ECG, GSR, and speech sentiments using regression models within ML.

Table 6.2 Summary of studies applying CF detection in HRC settings, sensor modalities, adaptability, and modeling approaches

Article (authors, year)	Sensors/modalities	Dataset available	Number of participants	Adaptability in HRC	Model type (classification/ regression)
Chand et al. (2024) [26]	Vision (camera, posture, gaze)	No	one	Yes	Classification (in digital twin framework)
Shah et al. (2023) [27]	Assumed physiological	No	20 males	Yes	Classification (with adaptation via HMM + RL)
Yao et al. (2024) [28]	EEG, video features (such as eye blinking and head posture)	No	30 (20 males 10 females)	Proposed	Classification ("fatigue" and "normal")
Kalatzis et al. (2022) [35]	ECG	No	16 (8 males 8 females)	No	Classification
Zand et al. (2024) [31] and (2025) [12]	ECG, GSR, speech, behavior logs	Yes	16 (8 females, 6 males, 2 non-binary)	No	Regression (Continuous CF estimation) ML
Dissanayake et al. (2022) [36]	EEG	No	6 (3 males, 3 females)	No	None (signal analysis only)
Wan & Zhou (2023) [29]	Simulation-based (SAFTE fatigue model)	No	N/A	Yes (planning stage)	None (theoretical prediction model)

6.2.1 A User Study on Cognitive Fatigue Assessment During Human-Robot Interaction

This Section presents a user study with a collaborative robot, called RoboCogniSense, a HRC system, facilitated collaborative manipulation tasks based on the reported CF level, combined with N-back cognitive tasks [38], which were designed to enhance CWL and induce CF. Sixteen participants were involved, with ages distributed as follows: 4 participants aged 18 to 30, 5 aged 31 to 40, 3 aged 41 to 50, 2 aged 51 to 60, and 2 aged over 60 years. The group consisted of 9 females, 5 males, and 2 non-binary individuals. Participants were recruited from non-STEM fields at Santa Clara University, and none of them were familiar with robots in order to minimize bias. After completing 3 visits, participants received $150, although some did not complete the study.

Each participant interacted with RoboCogniSense during three visits, each lasting up to two hours, which included two sessions: meal preparation and getting ready for work. Each visit began with participants signing the consent form and reporting their CF level using the VAS-F assessment. Similar approaches using

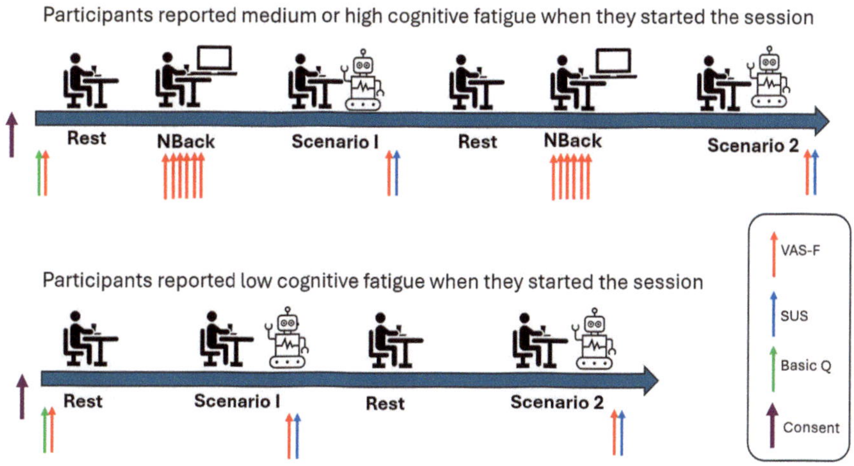

Fig. 6.1 Each visit starts with a consent form, VAS-F questionnaire, and a basic questionnaire. If the visit focused on medium or high fatigue, the top protocol was followed. If the visit focused on low fatigue, the bottom protocol was followed. It is worth noting that no participant reported a high CF level (above 70) [12]

EEG and N-back tasks have been successfully employed to induce and classify CF, achieving high accuracy using models such as EEGNet [7]. Participants with a CF score below 35 were categorized as low fatigue, those above 70 as high fatigue, and scores between 35 and 70 as moderate fatigue. Notably, no participants reported an initial CF score in the high range.

Figure 6.1 illustrates the system protocol. The top protocol represents the case when participants report a medium or high CF level, while the bottom one corresponds to reporting CF as low. The key difference between the two diagrams is the inclusion of the N-back task in the top protocol, with the rest being identical. Figure 6.2 shows the experimental setup. This setup illustrates the physical and interactive arrangement used to study CF during HRC in a simulated daily living task. Participants engaged in cooking and getting ready for work activities alongside RoboCogniSense, a mobile manipulator capable of interpreting and executing spoken commands. The environment included a kitchen space with a designated counter for ingredients, such as cheese and pasta, as well as items for getting ready for work, including car keys and a phone. Speech commands were issued via a microphone to minimize variability. ECG and GSR sensors were worn by participants to capture physiological signals in real time, while audio recordings were collected to analyze fatigue-related speech patterns. This multimodal setup was designed to induce, monitor, and assess CF under realistic, task-driven conditions, providing rich data for model development and validation. After completing the basic questionnaire and reporting their initial VAS-F level, participants rest for 5 minutes. For those with high or medium initial VAS-F levels, they perform 6 sets of the N-back task to induce CF, and after each set, they report their CF level using the VAS-F assessment.

Fig. 6.2 Experimental setup for human-robot collaborative cooking and cognitive fatigue assessment. Top-left: Items arranged on the counter and workspace prepared for the collaborative cooking task. Top-right: The robot delivers an item to the user during the task, enabling seamless handover. Bottom-left: The user initiates a robot command by tapping the HyperX microphone. Bottom-right: The user performs an N-back cognitive task to induce cognitive fatigue, monitored via sensors and speech

Participants then interact with the robot in two scenarios: a collaborative cooking scenario [39] and a getting-ready scenario [40]. RoboCogniSense, our human-robot collaboration system, includes a mobile manipulator capable of performing assistive tasks in daily living scenarios. The robot executes actions generated automatically from spoken commands using a speech-to-action pipeline trained on natural, unstructured dialogue [41]. To minimize bias, the order of the scenarios was alternated between participants. Each command was given by tapping the HyperX QuadCast microphone [42]. To ensure consistency across the study, participants followed a scripted speech format, such as: "Give me the cheese" or "Give me the car keys." At the end of each scenario, participants provided their feedback via the System Usability Scale (SUS) assessment [43] and VAS-F.

Offline CF assessment was conducted using VAS-F, administered before and after the pre- and post-interaction sessions, and after each N-back block. This approach enabled the monitoring of participants' perceived CF evolution throughout the study phases, providing sensitive and easy-to-administer subjective data that served as the primary ground truth for developing and validating CF detection models. To support the multimodal modeling approach, continuous physiological and speech data were collected in parallel during the collaborative sessions. Physiological signals, including ECG to capture cardiac activity and GSR to assess sympathetic arousal, were recorded using PLUX Biosignals sensors [44]. Simultaneously, speech data were collected to capture semantic features that could indicate CF. These multimodal data streams were synchronized and stored using the Robot Operating System (ROS) [45] infrastructure, enabling precise timestamping and data alignment for offline analysis. The following Sects. 6.2.1.1 and 6.2.1.2 will go over how CF was detected in these modalities.

6.2.1.1 Cognitive Fatigue Assessment Using Physiological Wearable Sensors

As discussed in the previous Section, ECG and GSR signals were used to monitor physiological responses, with self-reported VAS-F scores, as shown in Fig. 6.1, also collected throughout the study. The N-back task was applied depending on the protocol to induce CF. Data were continuously sampled at 1000 Hz and stored in rosbag. Although the protocol aimed to minimize noise, challenges arose due to participant movement during robot interaction and the N-back task.

Prior to feature extraction, raw ECG and GSR signals were processed using a standardized preprocessing pipeline to minimize noise and enhance signal quality. After filtering, features were extracted from the signals using two methods: interval-based and event-related. In the interval method, a 45-second moving window with a 5.625-second step size was applied, selected to retain 98% of the data while preserving signal quality. This approach also ensured smooth transitions between features, thereby enhancing the dataset. In the event-related method, features were extracted based on specific events, such as rest periods, N-back tasks, and robot interactions.

Post-processing of the extracted ECG and GSR features was performed based on valid Root Mean Square of Successive RR Intervals Differences (rMSSD) values, which fall within the range of 20 to 90 ms [46]. Data points with rMSSD values greater than 90 ms were attributed to missing R-peaks, which resulted in inflated rMSSD values and NaN values for time between the Q wave and the T wave on an electrocardiogram (ECG) (QT Interval), low frequency, and (low frequency)/(high frequency).

Using this dataset available at IEEE Dataport [31], regression models were developed to estimate continuous CF levels, with the self-reported VAS-F scores serving as the ground truth. Karim et al. [32] reported that self-supervised learning techniques, especially contrastive methods, have shown strong performance in learning feature representations from unlabeled data and may offer a scalable approach to building fatigue detection models without requiring extensive labeled datasets. To support robust and generalizable performance of the proposed CF detection framework, strategies were applied to manage inter-participant and cross-session variability. These measures were essential for reliable model evaluation and were integrated into the cross-validation process and performance analysis.

For inter-participant variability, data from each participant were exclusively allocated to either the training or test set, avoiding overlap.ML models were trained on participant-independent features, including ECG and GSR data, as well as contextual variables such as age group, sleep duration, caffeine intake, and time of arrival. This approach ensured generalizability across individuals, and a diverse participant pool was recruited to minimize demographic bias.

To mitigate cross-session variability, each session was consistently assigned to either the training or test set, ensuring no overlap and preventing data leakage. A 3-fold cross-validation strategy was employed, dividing the dataset into three subsets, with each subset serving once as the test set while the remaining two were used for training. This approach minimized overfitting risks, as data partitioning occurred at the participant level. Additionally, physiological signals were normalized within each session to reduce the impact of environmental variability and fluctuations in participant state.

The performance of the models, including Random Forest Regressor (RFR) [47], Support Vector Regression (SVR), and Gradient Boosting Machine (GBM) [48], was evaluated using three feature sets: combined ECG and GSR, ECG only, and GSR only. Two data segmentation strategies were used to prepare the signals for analysis: (1) a sliding window approach, which divided the data into overlapping 45-second segments with a 5.625-second step size, and (2) an event-based approach, which segmented the data according to specific user interactions and task phases during the experiment. The best-performing model achieved a Root Mean Square Error (RMSE) of 4.85 on the normalized VAS-F scale and an R-squared (R^2) of 0.89, demonstrating the feasibility of highly accurate continuous CF estimation from multimodal data.

6.2.1.2 Cognitive Fatigue Assessment Using Speech Signals

It is also interesting to investigate whether the speech signal can provide a modality for detecting CF. While there is research focused on identifying emotions through speech in Human-Robot Interaction (HRI) [49–51], the authors were unable to find related work on assessing CF during HRI through speech signals. For the CF assessment using the speech signal, the first step is to extract temporal and spectral features that are shown to vary systematically under fatigue, as follows:

- The speech ratio, the articulation rate, and the phonation time are selected as temporal characteristics based on the findings of Dias et al. [52], in which spontaneous speech in individuals with multiple sclerosis was analyzed. The authors reported that a lower speech ratio and a slower articulation rate were significantly associated with fatigue states. Phonation time, defined as the cumulative duration of voiced speech, is also found to decline with increasing fatigue, reflecting longer pauses and reduced verbal output.
- Formant frequencies (F1–F3) are selected due to their ability to capture articulatory changes related to fatigue-induced muscular control changes. Studies such as [53] and [54] have demonstrated that formant trajectories, particularly during sustained phonemes, show measurable deviations under fatigue. These frequencies represent the resonant characteristics of the vocal tract and are sensitive to changes in articulation clarity and vowel centralization.
- The fundamental frequency (F0), or pitch, is extracted as it correlates with vocal effort and neuromuscular control. Dias et al. [52] reported that F0 variability

decreased with fatigue, consistent with a flatter prosodic contour and reduced vocal energy. This makes F0 a strong candidate for modeling fatigue-related vocal monotony.
- Jitter and shimmer, which quantify short-term variations in pitch and amplitude, respectively, are incorporated as measures of voice stability. These perturbation-based features have been associated with reduced fine motor control of the larynx under fatigue. The same MS-fatigue study found both jitter and shimmer to be among the top predictors of cognitive fatigue based on voice samples [52].
- Harmonics-to-noise ratio (HNR) is selected as a proxy for vocal clarity and breath control. A lower HNR indicates increased noise energy in the signal and is commonly observed in fatigued speech, where breathiness and vocal strain become more pronounced. Prior work in [52] used HNR to successfully distinguish between high- and low-fatigue conditions.
- Mel-Frequency Cepstral Coefficients (MFCCs) are chosen due to their widespread use and effectiveness in modeling the short-term spectral envelope of speech. Multiple studies [53, 55, 56] have demonstrated significant changes in MFCC vectors as fatigue increases, with intra-subject correlation of MFCC patterns decreasing sharply across fatigue levels. Their inclusion enables data-driven modeling of subtle spectral shifts not easily captured by prosodic or formant-based features alone.
- Gender is proven to affect the tone, frequency, and type of features [57]. Therefore, any features developed from the audio data would be different between genders and thus affect classification accuracy if not taken into account. During the data collection portion of the study, the gender of each participant was recorded (Male, Female, Non-Binary). To correctly input gender as a feature, one-hot encoding is used. If gender were instead represented numerically as Male = 0, Female = 1, Non-Binary = 2, this would cause any ML model to assume an ordering and distance between the numbers. This order is arbitrary, 2 is not greater than 1, and so forth. Instead, one-hot encoding creates a binary feature column for each gender. The model sees only "is Female?", "is Male?", "is Non-binary?". There is no implied order or distance.

As we are interested in detecting CF, it is important to have a target value. As our speech commands are short (i.e., 2–3 seconds), classification may be more appropriate based on the number of datapoints available. Therefore, the target variable is whether the CF level is Low Fatigue (LF) or Medium Fatigue (MF). This was easily done as during the data collection process each participant's individual visit information was already classified as either medium or low fatigue. To reiterate, MF is considered a VAS-F score above 35, while LF is any score equal to or below. As this becomes a binary classification problem, MF is considered a '1' while LF is a '0'.

All candidate features previously mentioned were first extracted in full and then trimmed using a feature-to-target correlation matrix. The first and main features extracted for ML modeling were MFCCs. Unlike the linear hertz scale, the Mel scale spaces frequencies linearly below 1 kHz (1 Hz = 1 Mel) and logarithmically above

1 kHz, mirroring the ear's non-linear resolution. Specifically, MFCCs are features extracted from a signal that has been filtered through Mel-scaled frequency bands and further transformed using a logarithm and Discrete Cosine Transform (DCT). This makes MFCCs compact, decorrelated, and suitable for ML modeling. MFCCs were extracted using the Librosa library. To convert the resulting MFCC time-series matrix into a consistent feature vector per file, 7 'summary' statistics were computed across the time axis (i.e., for each coefficient across all frames). Specifically, the mean, standard deviation, 25th percentile, median (50th percentile), 75th percentile, skewness, and kurtosis were calculated. These seven features capture both the central tendency and distributional shape of each MFCC coefficient over time. This results in a feature set of shape $7 \times 20 = 140$ per file.

Additional temporal and spectral measures (e.g., jitter, shimmer, intensity, gender one-hots) were then appended, after which a feature-to-target correlation matrix was examined against cognitive-fatigue level (CF). No single variable exhibited a strong linear correlation, but combinations—particularly MFCC means and standard deviations—showed promise. Features with negligible or redundant relationships were pruned to mitigate multicollinearity and streamline the final input set. Thus, while MFCCs alone do not "explain" CF, they provide complementary information that, together with carefully selected prosodic cues, forms a richer basis for classification.

Many of the remaining temporal and spectral features calculated, had a linear correlation value close to 0 with respect to the target variable CF. Thus these features were removed from the feature set. Overall, the results suggested that no single feature had a strong linear relationship with CF. However, combinations of features proved to be useful for classification. This allowed for redundant and low-correlating features to be removed. The resulting feature set included jitter, shimmer, the One-Hot encoded Gender columns, MFCC Mean, and MFCC std. Additionally, the MFCC std features are also clustered. This suggests that such high multicollinearity could lead to redundant features and that feature reduction could benefit ML model performance.

To avoid data-leakage from the same speaker appearing in both training and test sets, recordings were first sorted by participant ID and visit. Counts were accumulated until 80% of the utterances had been assigned, designating those speakers to the training set and reserving the remainder for testing; this ensures each model is evaluated on voices it has never "heard".

On the resulting MFCC-based feature matrix (20 MFCC means, 20 MFCC SDs, shimmer, intensity, gender), four classifiers were benchmarked: Support Vector Machines (linear and RBF kernels) [58], Logistic Regression [59], Random Forests [60], and XGBoost [61]. A 3-fold cross-validated grid search tuned the key hyper-parameters for each model (e.g., SVM C parameter and kernel, RF tree count and depth, XGBoost rounds, depth, learning rate). The best settings were refit on the full training set and judged against the held-out speaker test set.

Due to the large number of features and multicollinearity present between them, Sklearn's SelectKBest function was used for feature reduction [62]. SelectKBest returns the top 'k' specified features that produce the best model performance. This

Table 6.3 Optimized model classification report

	SVM (k=44)				Random Forest (k=20)				XGBoost (k=23)			
	Prec.	Rec.	F1	Sup.	Prec.	Rec.	F1	Sup.	Prec.	Rec.	F1	Sup.
LF	0.73	0.70	0.71	57	0.77	0.65	0.70	57	0.76	0.61	0.68	57
MF	0.37	0.40	0.38	25	0.41	0.56	0.47	25	0.39	0.56	0.46	25
Accuracy		0.61		82		0.62		82		0.60		82
Macro Avg	0.55	0.55	0.55	82	0.59	0.60	0.59	82	0.57	0.59	0.57	82
Weighted Avg	0.62	0.61	0.61	82	0.66	0.62	0.63	82	0.65	0.60	0.61	82

study uses the ANOVA (Analysis of Variance) F-test binary scoring function for the SelectKBest model [63]. A nested grid search then swept k (1...all features) alongside the main hyper-parameters of the three classifiers—SVM, Random Forest, and XGBoost—using five-fold cross-validation. The mean macro-F1 determined the winner in each grid cell. This procedure is leak-free as the feature scores are recalculated from scratch within every training split. The search converged on the optimal hyperparameter for each model and the corresponding 'k' feature set. This produced three distinct configurations:

- SVM: RBF kernel, C = 10, k = 44.
- Random Forest: 200 trees, unrestricted depth, min samples split = 5, k = 20.
- XGBoost: 200 trees, depth = 5, learning-rate = 0.10, k = 23.

For reference, the positive class refers to the MF class or '1' class in the output. The negative class refers to the LF class or '0' class in the output. Thus, for example, true positive refers to an instance when the model correctly predicts MF (1). Table 6.3 shows the models on the test set.

All models topped 60% accuracy on the held-out speakers, but gains over the full feature set were modest. Random Forest delivered the best balance (accuracy = 0.62, macro-F1 = 0.59) by recalling 56% of MF clips while preserving 65% recall on LF clips. The SVM excelled at the majority LF class (precision = 0.73, F1 = 0.71) yet struggled with MF (F1 = 0.38), suggesting an RBF surface biased toward the dominant label. XGBoost split the difference (accuracy = 0.60) but needs further tuning to boost its MF precision (0.39) and reduce LF-to-MF misclassifications.

This study advances CF detection for HRC by showing, for the first time, that fatigue can be inferred directly from short speech snippets. Multiple classifiers were benchmarked with nested cross-validation and ANOVA-based feature pruning, and the best model (Random Forest, k = 20) achieved balanced recall and a macro-F1 near 0.60, demonstrating proof-of-concept accuracy.

Limitations stem from the fewer than 400 recordings. This caused the "fatigued" class to be the minority. Thus, precision for that class stayed below 0.45 no matter how we tuned things. Treating fatigue as a simple yes/no label, while convenient, most likely hides useful gradations. A larger, more balanced data set—and perhaps models that learn directly from raw waveforms—would likely increase performance.

6.3 Online Cognitive Fatigue Assessment in HRC

Timely detection of CF is crucial for enabling adaptive HRC systems. In human teams, individuals often intuitively recognize when their teammates are fatigued and adjust their behavior, such as redistributing tasks, offering assistance, or altering communication styles. Similarly, in HRC, robots must be capable of perceiving and responding to the cognitive state of their human teammates in real time. This allows robots to adjust task allocation, interaction modalities, or pacing strategies to mitigate the effects of CF and maintain team performance, safety, and user trust [9].

Shah et al. [27] investigated robot adaptations in an HRC setting, using both subjective and objective metrics, including self-reported CF, Heart Rate Variability (HRV), task performance, and user experience feedback. The task involved participants guiding a robot through a cognitively demanding metal polishing operation, with varying levels of assistance: manual control, partial assistance, and full assistance. As participants stood throughout the task, their CF and Physical Fatigue (PF) were both affected. CF was induced via the 2-back task, while the robot adapted its behavior using an Markov decision process based on reinforcement learning (MDP) solved through Q-learning. Subjective CF was measured using a 1–9 Likert scale after each trial block, though the single-item format limited the assessment's depth. Objective data included HRV derived from ECG signals, where rMSSD reflected parasympathetic activity and normalized Heart Rate (HR) indicated fatigue levels. Support Vector Machine (SVM) was used to classify CF states and generate confidence scores. Although the study confirmed that the 2-back task effectively induces CF, the limited subjective assessment reduces the reliability of the results. Nevertheless, the study underscores the value of multimodal CF assessment and adaptive robotic support for enhancing performance and well-being in HRC environments.

Yao et al. [28] developed a dynamic model for detecting CF, which updates fatigue parameters in real time and achieved a test accuracy of 96.15%. Conducted in an HRC production workshop, the study utilized multimodal data, including facial video and EEG signals, collected during 30-minute task intervals with 5-minute breaks. The participant group included 30 individuals (20 male, 10 female), but demographic details such as age and individual characteristics were not provided, limiting transparency. Fatigue was measured using subjective sleepiness scales, including the Karolinska Sleepiness Scale (KSS) and Stanford Sleepiness Scale (SSS), which assess sleepiness rather than the broader concept of CF, raising concerns about the validity of the ground truth. The model used advanced Deep Learning (DL) techniques, such as a Bidirectional Logn Short Term Memory (BiLSTM) network to capture temporal dependencies in EEG and video features, with the Convolutional Block Attention Module (CBAM) attention mechanism improving feature relevance. Several DL models, including Multivariate Long Short-Term Memory Fully Convolutional Network (MLSTM-FCN), Long Short Term Memory (LSTM), and Task-Aware Reconstruction Network (TARNet), were evaluated, with the final model classifying CF into two binary states: fatigue or normal. A task reallocation algorithm, combining an enhanced genetic algorithm with Reinforcement Learning (RL), was employed for real-time workload management. While the model demonstrated strong technical performance, its reliance on sleepiness-based labels, binary classification, and limited demographic data restricts its generalizability and applicability in broader HRC settings.

As shown in Table 6.2, only a few studies, such as those by Chand et al. [26] and Shah et al. [27], have incorporated CF detection into adaptive HRC frameworks, enabling robots to adjust their behavior based on detected fatigue levels. Several recent studies, including those by Yao et al. [28] and Zhang et al. [64], have validated CF detection methods in laboratory settings without exploring adaptation or integration into real-time HRC systems. This highlights an ongoing gap in the field, with most studies focused on offline or observational analyses and limited exploration of continuous CF estimation and fatigue-adaptive robotic behaviors in interactive, real-world HRC scenarios.

Chand et al. [26] studied CF in an HRC scenario where a single operator collaborated with a robot to assemble a 3×3 cube storage shelf. The process involved 26 distinct tasks, and detailed data on task execution times for both the operator and robot were used for PF assessment and task scheduling optimization. The operator performed the assembly 20 times, with video data from these sessions used to train and test a fatigue monitoring system. CF was assessed using a Fitbit device based on GSR spikes, while PF was evaluated using a camera, providing a comprehensive analysis of both CF and PF during the task. The results were validated in real-world assembly scenarios with real-time monitoring across six operation types. Researchers used a boundary-sensitive dual-stream Multi-Stage Temporal Convolutional Network (MS-TCN) combined with an LSTM algorithm to extract key task characteristics such as operation type, repetition count, and arm usage. Despite contributing to the advancement of intelligent HRC systems, the study has notable limitations. The CF assessment relied on a proprietary Fitbit-based method for a single participant, raising concerns about generalizability. Furthermore, the small training dataset (only 20 assembly sessions) and the involvement of just one participant increase the risk of overfitting, limiting the study's applicability.

6.4 Open Challenges

Despite significant progress in assessing CF during HRC, several open challenges remain that hinder robust deployment and generalization of current methods.

- **Imprecision in Subjective Cognitive Fatigue Measures:** A fundamental challenge in CF research lies in the ambiguity and imprecision of widely used self-report instruments, which often conflate CF with related constructs and employ vague retrospective timeframes, leading to interpretative inconsistencies [65]. Many existing questionnaires blend CF with overlapping constructs such as memory, attention, and mood, making it unclear what exactly is being measured.

Moreover, retrospective timeframes (e.g., "over the past four weeks") introduce interpretative variability: do respondents average their experience, count threshold exceedances, or recall specific events? The lack of guidance on how to respond to these items severely limits the interpretability and consistency of the data. Moving forward, subjective instruments should be redesigned to isolate fatigue more cleanly, provide clear temporal anchors, and ideally, standardize response heuristics (e.g., asking users to report the most intense experience, an average, or frequency of occurrences).

- **Need for Fractionated and Behavioral Models of CF:** The idea that CF is a multi-dimensional phenomenon is widely acknowledged. Recent work incorporating dimensions such as the effort just expended, the willingness to exert future effort, and actual performance metrics (in addition to feelings of tiredness) shows great promise in helping to understand specific aspects of CF (Casamento-Moran, A., personal communication, April 2025). Wylie et al. [1] further elaborated on this framework, highlighting how behavioral indicators such as response bias (from signal detection theory) may provide earlier or more reliable fatigue detection than accuracy alone. Developing multi-dimensional CF phenotyping, which combines subjective reports, effort-related measures, willingness assessments, and objective performance data, can improve our understanding of fatigue trajectories and enhance adaptation in real-time HRC contexts.
- **Generalization:** CF detection models frequently exhibit reduced performance when applied to unseen users or tasks. Cross-participant and cross-task generalization remains a key limitation due to variability in individual physiological baselines, behavioral norms, and speech characteristics. Without robust domain adaptation or personalization strategies, models trained in one context often fail in others. Current research highlights the need for algorithms that can adapt to inter- and intra-individual variability while maintaining high performance. For example, Matuz et al. [11] demonstrated that even high-performing machine learning models trained on heart rate variability data struggled to generalize across users, emphasizing the critical need for generalizable fatigue detection frameworks.
- **Lack of Datasets with a Diverse Population—Most Datasets are Small:** Most existing datasets are small-scale, lack demographic diversity, and are typically limited to specific age groups, professions, or cultural backgrounds. For instance, studies often rely on student populations with limited variability in fatigue expression and speech patterns. This restricts the generalizability of machine learning models across real-world scenarios. Kalatzis et al. [35] emphasized that the lack of sex parity and demographic representation hinders the development of inclusive and robust models for cognitive fatigue assessment. Datasets that include varied gender identities, age ranges, cultural backgrounds, and task contexts are urgently needed to train inclusive and resilient CF detection systems.
- **Multimodal Fusion:** While multiple sensing modalities (e.g., physiological signals, speech, vision) have been explored, integrating them effectively remains difficult. Fusion techniques must account for modality-specific delays, data

quality inconsistencies, and contextual relevance. Cos et al. [66] emphasize that challenges in synchronizing multimodal signals, managing heterogeneous noise characteristics, and modeling context-aware interactions remain significant barriers to robust cognitive fatigue detection systems.
- **Real-time Processing:** Although several studies demonstrate the feasibility of online CF monitoring, real-time processing remains a technical and computational challenge. Real-time CF inference must operate with low latency while integrating multimodal inputs (e.g., ECG, GSR, EEG, speech) without sacrificing accuracy. Noise, signal dropout, and computational delays during HRI further complicate deployment. Piper et al. [19] emphasized that real-time monitoring in dynamic settings requires highly optimized algorithms to manage latency, data reliability, and multimodal fusion under uncertain conditions.
- **User Acceptance:** The use of intrusive sensors (e.g., wearable ECG, EEG caps) or frequent prompts for subjective fatigue reports may disrupt workflow and reduce user comfort. In addition, adaptive robotic behaviors based on CF detection, such as slowing down or reassigning tasks, may not always align with user preferences. Balancing performance with transparency, trust, and perceived usefulness is essential. Participatory design and long-term user studies are needed to ensure acceptance of CF-aware systems in collaborative environments. Hancock and Desmond [14] argue that trust and usability are central to user acceptance and must be proactively managed in any fatigue-aware HRI system.

6.5 Conclusion

This chapter explored the assessment of CF in HRC, a critical concern as collaborative robots increasingly operate in cognitively demanding environments. Effective detection of CF is essential to maintain safety, optimize task performance, and foster human trust in robotic teammates. Sensory data and ML models have been deployed to enable robotic teammates to recognize CF and potentially adapt their behavior in order to improve collaboration between the team. Despite these advancements, several challenges persist, including the lack of diverse datasets, processing constraints, and limitations in multimodal data fusion. Addressing these challenges will be essential for enabling effective CF assessment systems that inform adaptive HRC design and behavior.

Future research focused on scalable, personalized, and user-centric CF detection frameworks is crucial for advancing the field and for supporting the development of collaborative robots capable of adjusting interaction strategies and workload distribution in response to human cognitive states. This will help create resilient, trustworthy, and cognitively supportive HRC systems.

Acknowledgments The material is based upon work supported by the National Science Foundation under Grants 2226165 and 2226164.

References

1. G.R. Wylie, B. Yao, J. Sandry, J. DeLuca, Using signal detection theory to better understand cognitive fatigue. Front. Psychol. **11**, 579188 (2021)
2. C. Carissoli, L. Negri, M. Bassi, F.A. Storm, A.D. Fave, Mental workload and human-robot interaction in collaborative tasks: a scoping review. Int. J. Hum. Comput. Interact. **40**, 6458–6477 (2024). https://www.tandfonline.com/. https://doi.org/10.1080/10447318.2023.2254639
3. M. Papakostas, A. Rajavenkatanarayanan,, F. Makedon, Cogbeacon: a multi-modal dataset and data-collection platform for modeling cognitive fatigue. Technologies **7**, 46 (2019). https://www.mdpi.com/2227-7080/7/2/46/htm
4. V. Kanal, et al. Towards a serious game based human-robot framework for fatigue assessment, in *ACM International Conference Proceeding Series* (2020), pp. 527–532. https://dl.acm.org/doi/pdf/10.1145/3389189.3398744
5. A. Jaiswal, A.R. Babu, M.Z. Zadeh, G. Wylie, F. Makedon, Detecting cognitive fatigue in subjects with traumatic brain injury from FMRI scans using self-supervised learning, in *ACM International Conference Proceeding Series* (2023), pp. 83–90. https://dl.acm.org/doi/pdf/10.1145/3594806.3594868
6. M.Z. Zadeh, et al., Towards cognitive fatigue detection from functional magnetic resonance imaging data, in *Proceedings of the 13th ACM International Conference on PErvasive Technologies Related to Assistive Environments* (2020). https://doi.org/10.1145/3389189.3397648
7. E. Karim, et al., An eeg-based cognitive fatigue detection system, in *ACM International Conference Proceeding Series* (2023), pp. 131–136. https://dl.acm.org/doi/pdf/10.1145/3594806.3594848
8. M. Zand, S. Wood, F. Davoudi, M. Kyrarini, Cognitive fatigue assessment for human-robot collaboration: review & open challenges, in *Proceedings of the 18th ACM International Conference on PErvasive Technologies Related to Assistive Environments* (2025, June), pp. 74–84. https://doi.org/10.1145/3733155.373659
9. Texas A&M, Combating fatigue with autonomy – Texas A&M stories. https://stories.tamu.edu/news/2021/06/21/combating-fatigue-with-autonomy/
10. H. Guo, et al., Impact of cognitive load on human trust in hybrid human-robot collaboration (2024). https://arxiv.org/abs/2412.20654v1
11. A. Matuz, D. van der Linden, G. Darnai, Á. Csathó, Generalisable machine learning models trained on heart rate variability data to predict mental fatigue. Sci. Rep. **12**, 1–12 (2022). https://www.nature.com/articles/s41598-022-24415-y
12. M. Zand, G.R. Wylie, A. Wolfe, M. Kyrarini, Automatic cognitive fatigue assessment using physiological sensors during human-robot interaction for activities of daily living (2025). https://doi.org/10.36227/techrxiv.174918156.67353911/v1
13. G. Matthews, P.A. Desmond, C. Neubauer, P.A. Hancock, A comprehensive review of workload and fatigue in complex human-machine systems. Theor. Issues Ergon. Sci. **13**, 247–272 (2012)
14. P.A. Hancock, D.R. Billings, K.E. Schaefer, J.Y.C. Chen, E.J. de Visser, R. Parasuraman, A meta-analysis of factors affecting trust in human-robot interaction. Hum. Factors J. Hum. Factors Ergonomics Soc. **53**(5), 517–527 (2011). https://doi.org/10.1177/0018720811417254
15. H.J. Michielsen, J. De Vries,, G.L. Van Heck, Psychometric qualities of a brief self-rated fatigue measure: the fatigue assessment scale. J. Psychosomat. Res. **54**, 345–352 (2003)
16. B. Johansson, L. Rönnbäck, Evaluation of the mental fatigue scale and its relation to cognitive and emotional functioning after traumatic brain injury or stroke. Int. J. Phys. Med. Rehab. **8**, 10–15 (2010)
17. K.A. Lee, G. Hicks, G. Nino-Murcia, Validity and reliability of a scale to assess fatigue. Psychiatr. Res. **36**, 291–298 (1991). https://pubmed.ncbi.nlm.nih.gov/2062970/
18. H.A. Abbass, J.B. Scholz, A. Gonczi, Artificial intelligence and human trust in human-machine teaming: a literature review. IEEE Access **6**, 36194–36212 (2018)

19. A.T. Piper, P. Johnson, E. Smith, Real-time measurement of cognitive fatigue using subjective and objective metrics in dynamic work environments. Appl. Ergon. **100**, 103642 (2022)
20. T.O. Zander, C. Kothe, Towards passive brain–computer interfaces: applying brain–computer interface technology to human–machine systems in general. J. Neural Eng. **8**, 025005 (2017)
21. B.T. Jap, S.K. Lal, P. Fischer, E. Bekiaris, Using eeg spectral components to assess algorithms for detecting fatigue. Exp. Syst. Appl. **36**, 2352–2359 (2009)
22. J.A. Healey, R.W. Picard, Detecting stress during real-world driving tasks using physiological sensors. IEEE Trans. Intell. Transp. Syst. **6**, 156–166 (2005)
23. M.S. Gowda, V. Talasila, S.A. Raafay Umar, R. Alva, A multimodal approach to detect driver drowsiness, in *2024 5th International Conference on Circuits, Control, Communication and Computing (I4C), Bangalore, India* (2024), pp. 554–559. https://doi.org/10.1109/I4C62240.2024.10748433
24. J. De Vries, H.J. Michielsen, G.L. Van Heck, Assessment of fatigue among working people: a comparison of six questionnaires. Occup. Med. **57**, 353–357 (2017)
25. B. Johansson, L. Rönnbäck, Mental fatigue scale: psychometric properties and relations to cognitive and emotional functioning after brain injury or stroke. Neuropsychol. Rehab. **29**, 548–566 (2019)
26. S. Chand, H. Zheng, Y. Lu, A vision-enabled fatigue-sensitive human digital twin towards human-centric human-robot collaboration. J. Manufact. Syst. **77**, 432–445 (2024)
27. J.K. Shah, A. Yadav, S.K. Hopko, R.K. Mehta, P.R. Pagilla, Robot adaptation under operator cognitive fatigue using reinforcement learning, in *IEEE International Workshop on Robot and Human Communication, RO-MAN* (2023), pp. 1467–1474
28. B. Yao, X. Li, Z. Ji, K. Xiao, W. Xu, Task reallocation of human-robot collaborative production workshop based on a dynamic human fatigue model. Comput. Ind. Eng. **189**, 109855 (2024)
29. Y. Wan, C. Zhou, Predicting human-robot team performance based on cognitive fatigue, in *ICAC 2023 - 28th International Conference on Automation and Computing* (2023)
30. E. Karim, An intelligent robotic system for multi-sensory cognitive fatigue detection to assist persons with paralysis in activities of daily living. Computer Science and Engineering Dissertations (2025). https://mavmatrix.uta.edu/cse_dissertations/412
31. M. Zand, G. Wylie, A. Wolfe, M. Kyrarini, Cognitive fatigue assessment using physiological sensors during human-robot interaction for activities of daily living (2024). https://doi.org/10.21227/zdfm-kp95
32. E. Karim, et al., Examining the landscape of cognitive fatigue detection: a comprehensive survey. Technologies **12**, 38 (2024). https://mdpi.com/2227-7080/12/3/38
33. R. Gervasi, M. Capponi, L. Mastrogiacomo, F. Franceschini, Eye-tracking support for analyzing human factors in human-robot collaboration during repetitive long-duration assembly processes. Product. Eng. **19**, 47–64 (2025). https://link.springer.com/article/10.1007/s11740-024-01294-y
34. P. Sotirios, F. Fabio, M. Francesco, in *A Methodological Framework to Assess Mental Fatigue in Assembly Lines with a Collaborative Robot*, ed. by K.Y. Kim, L. Monplaisir, J. Rickli. Flexible Automation and Intelligent Manufacturing: The Human-Data-Technology Nexus. FAIM 2022. Lecture Notes in Mechanical Engineering (Springer, Cham, 2023). https://doi.org/10.1007/978-3-031-17629-6_31
35. A. Kalatzis, S. Hopko, R.K. Mehta, L. Stanley, M.P. Wittie, Sex parity in cognitive fatigue model development for effective human-robot collaboration, in *2022 IEEE/RSJ International Conference on Intelligent Robots and Systems (IROS)* (2022), pp. 10951–10958
36. U.C. Dissanayake, V. Steuber, F. Amirabdollahian, EEG spectral feature modulations associated with fatigue in robot-mediated upper limb gross and fine motor interactions. Front. Neurorobot. **15**, 788494 (2022). https://www.frontiersin.org/articles/10.3389/fnbot.2021.788494/full
37. S.G. Hart, L.E. Staveland, Development of nasa-TLX (task load index): results of empirical and theoretical research. Adv. Psychol. **52**, 139–183 (1988)
38. M.J. Kane, A.R.A. Conway, T.K. Miura, G.J.H. Colflesh, Working memory, attention control, and the n-back task: a question of construct validity. J. Exp. Psychol. Learn. Memory Cognit. **33**(3), 615–622 (2007). https://doi.org/10.1037/0278-7393.33.3.615

39. K. Kodur, et al., Exploring the dynamics of human-robot interaction: robot error, sentiment analysis, and politeness. Int. J. Soc. Robot. 17, 1221–1234 (2025)
40. M. Tognotti, M. Kyrarini, Are you ready? An intelligent robotic assistant for instrumental activities of daily living, in *Proceedings - 2024 8th IEEE International Conference on Robotic Computing, IRC 2024* (2024), pp. 249–252
41. M. Zand, K. Kodur, M. Kyrarini, Automatic generation of robot actions for collaborative tasks from speech, in *2023 9th International Conference on Automation, Robotics and Applications, ICARA 2023* (2023), pp. 155–159
42. Hyperx quadcast - usb microphone. https://hyperx.com/collections/microphones
43. J. Brooke, Sus: a retrospective - juxjux (2013). https://uxpajournal.org/sus-a-retrospective/
44. P. Biosignals, Plux biosignals. https://www.pluxbiosignals.com/. Accessed July 12, 2023
45. F. Vigni, A. Andriella, S. Rossi, A rosbag tool to improve dataset reliability, in *ACM/IEEE International Conference on Human-Robot Interaction* (2024), pp. 1085–1089. https://dl.acm.org/doi/10.1145/3610978.3640556
46. F. Shaffer, J.P. Ginsberg, An overview of heart rate variability metrics and norms. Front. Publ. Health **5**, 290215 (2017). www.frontiersin.org
47. P.F. Smith, S. Ganesh, P. Liu, A comparison of random forest regression and multiple linear regression for prediction in neuroscience. J. Neurosci. Methods **220**, 85–91 (2013). https://www.sciencedirect.com/science/article/pii/S0165027013003026
48. A. Natekin, A. Knoll, Gradient boosting machines, a tutorial. Front. Neurorobot. **7**, 63623 (2013). www.frontiersin.org
49. R. García, et al., Speech emotion recognition with deep learning beamforming on a distant human-robot interaction scenario (2024), pp. 3215–3219
50. R. Mishra, A. Frye, M.M. Rayguru, D.O. Popa, Personalized speech emotion recognition in human-robot interaction using vision transformers. IEEE Robot. Autom. Lett. **10**, 4890–4897 (2025)
51. N. Gasteiger, J. Lim, M. Hellou, B.A. MacDonald, H.S. Ahn, A scoping review of the literature on prosodic elements related to emotional speech in human-robot interaction. Int. J. Soc. Robot. **16**, 659–670 (2024)
52. M. Dias, et al., Detecting fatigue in multiple sclerosis through automatic speech analysis. Front. Hum. Neurosci. **18**, 1449388 (2024)
53. H.P. Greeley, E. Friets, J.P. Wilson, S. Raghavan, J. Picone, J. Berg, Detecting fatigue from voice using speech recognition, in *2006 IEEE International Symposium on Signal Processing and Information Technology* (Vancouver, BC, Canada, 2006), pp. 567–571. https://doi.org/10.1109/ISSPIT.2006.270865
54. S. Patnaik, Speech emotion recognition by using complex MFCC and deep sequential model. Multimedia Tools Appl. **82**, 11897–11922 (2023).
55. D. Mamieva, A.B. Abdusalomov, A. Kutlimuratov, B. Muminov, T.K. Whangbo, Multimodal emotion detection via attention-based fusion of extracted facial and speech features. Sensors **23**, 5475 (2023)
56. R.K. Baykaner, M. Huckvale, I. Whiteley, S. Andreeva, O. Ryumin, Predicting fatigue and psychophysiological test performance from speech for safety-critical environments. Front. Bioeng. Biotechnol. **3**, 124 (2015)
57. E. Pépiot, Male and female speech: a study of mean f0, f0 range, phonation type and speech rate in Parisian French and American English speakers (2014), pp. 305–309. https://shs.hal.science/halshs-00999332
58. C. Cortes, V. Vapnik, Support-vector networks. Mach. Learn. **20**, 273–297 (1995)
59. D.R. Cox, The regression analysis of binary sequences (with discussion). J. R. Stat. Soc. B **20**, 215–242 (1958)
60. L. Breiman, Random forests. Mach. Learn. **45**, 5–32 (2001)
61. T. Chen, C. Guestrin, XGBoost: a scalable tree boosting system, in *KDD'16: Proceedings of the 22nd ACM SIGKDD International Conference on Knowledge Discovery and Data Mining* (2016), pp. 785–794

62. F. Pedregosa, et al., Scikit-learn: machine learning in python. J. Mach. Learn. Res. **12**, 2825–2830 (2011)
63. R.A. Fisher, *Statistical Methods for Research Workers* (Oliver & Boyd, Edinburgh, 1925)
64. T. Zhang, X. Zhang, Z. Lu, Y. Zhang, Z. Jiang, Y. Zhang, Feasibility study of personalized speed adaptation method based on mental state for teleoperated robots. Front. Neurosci. **16**, 976437 (2022). https://doi.org/10.3389/fnins.2022.976437
65. G. Matthews, P. Hancock, *The Handbook of Operator Fatigue* (CRC Press, Boca Raton, 2017). https://www.taylorfrancis.com/books/9781317029410
66. C.A. Cos, N. Pedemonte, C.A. Pérez, P. Garat, Enhancing mental fatigue detection through physiological signals and machine learning using contextual insights and efficient modelling. J. Sens. Actuator Netw. **12**, 77 (2023)

Chapter 7
Visual Evoked Potentials in Neurofeedback: Advancing Cognitive Control in ADHD Through Therapeutic Gaming

Aigerim Keutayeva, Amin Zollanvari, and Berdakh Abibullaev

Abstract The convergence of neuroscience and interactive technology is opening new pathways for mental health interventions, particularly in the treatment of attention-deficit/hyperactivity disorder (ADHD). This chapter examines the use of visual evoked potentials (VEPs) in neurofeedback systems combined with therapeutic gaming to support cognitive control in children with ADHD. By using real-time brain responses, these systems offer a non-invasive and engaging alternative to traditional treatment approaches. We explore the theoretical foundations of VEP-based neurofeedback, system-design considerations, and recent developments in game-based interventions designed for pediatric use. Drawing on current research, the chapter highlights clinical outcomes related to attention, impulse regulation, and working memory, while also addressing challenges in personalization, usability, and ethical implementation. This work contributes to the broader effort of designing human-centered technologies that support cognitive development and learning, providing a practical framework for future research and application in neurorehabilitation and education.

Keywords Visual evoked potentials · Brain–computer interfaces · Neurofeedback · ADHD · Therapeutic gaming · Neurorehabilitation

---

A. Keutayeva
Department of Electrical and Computer Engineering, University of Houston, Houston, TX, USA
e-mail: akeutayeva@uh.edu

A. Zollanvari
Department of Electrical and Computer Engineering, Utah Valley University, Orem, UT, USA
e-mail: zollanvari@ieee.org

B. Abibullaev (✉)
Department of Robotics and Mechatronics, Nazarbayev University, Astana, Kazakhstan
e-mail: berdakh.abibullaev@nu.edu.kz

R. Vinjamuri (ed.), *Bridging the Gap between Mind and Machine*,
https://doi.org/10.1007/978-3-032-06713-5_7

7.1 Introduction

Attention-deficit/hyperactivity disorder (ADHD) is a common neurodevelopmental condition that begins in childhood and often continues into adulthood. The prevalence of ADHD is approximately 7.6% in children and 5.6% in adolescents [1]. It is characterized by patterns of inattention, hyperactivity, and impulsivity that interfere with daily functioning [2]. Attention deficits in individuals with ADHD impair executive functions, making it difficult to manage time, maintain focus, and plan tasks. Hyperactivity manifests as impulsive actions and poor emotional control, which can undermine academic success and social interactions, often leading to peer rejection and strained relationships with family and friends [3]. These challenges stem from neurobiological differences, such as reduced dopamine transmission [4] and delayed development in brain regions responsible for self-regulation [5].

Conventional interventions, namely stimulant medications and behavioral therapies, aim to improve cognitive control deficits but face limitations. Stimulants yield large effect sizes on symptom reduction (Cohen's $d \approx 0.8 - 1.0$) but can provoke appetite loss, sleep disruption, and variable adherence [6–8]. Behavioral therapies require intensive, long-term engagement from therapists, families, and schools and often demonstrate limited generalization beyond clinical settings [9].

To address these challenges, researchers are exploring new methods such as neurofeedback therapy [6]. Neurofeedback provides real-time feedback on brain activity, helping individuals learn to self-regulate it [10]. Traditional protocols focus on reducing the theta/beta ratio (4–8 Hz vs. 12–20 Hz) and enhancing sensorimotor rhythm (SMR; 12–15 Hz), with meta-analyses reporting medium to large effect sizes ($d \approx$ 0.5–0.8) on core ADHD symptoms [7, 8, 11, 12]. More recently, researchers have begun to exploit visual evoked potentials (VEPs) as feedback signals, embedding them within interactive "serious games" to combine therapeutic training with motivational gameplay [2, 13].

In VEP-based neurofeedback games, transient VEPs (tVEPs) or steady-state VEPs (SSVEPs) elicited by patterned stimuli serve as proxies for attentional engagement [3]. Users receive immediate in-game rewards, such as avatar acceleration or point gains, when their VEP amplitudes or SSVEP power exceed individualized thresholds, thereby reinforcing focused brain states [14, 15] (see Fig. 7.1).

Prototype studies illustrate feasibility but reveal considerable heterogeneity in system design and validation. For example, Beveridge et al. [16] demonstrated a motion-onset VEP (m-VEP) racing game for adolescents (13–16 years), achieving 65–72% online accuracy when users focused on one of five moving stimuli to steer a 3D car in real time. Moreover, Moreno-Calderón et al. [17] implemented a circularly shifted VEP "Connect 4" multiplayer game, yielding 93.7% ± 1.7% accuracy and 5.25 s average selection time across 22 healthy participants.

Commercial platforms such as FocusPocus and NeuroPlay have packaged these concepts for home use, integrating SMR/theta–beta neurofeedback with working memory and inhibitory control exercises. In a randomized waitlist-controlled trial of 85 children (ages 8–12), participants completed twenty-five 20-minute home ses-

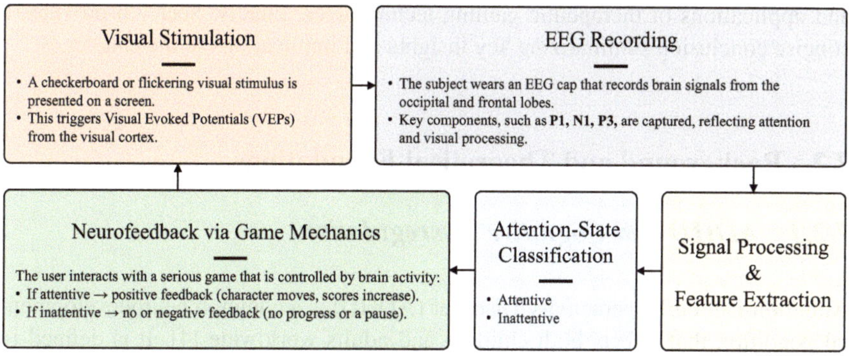

Fig. 7.1 VEP-based neurofeedback gaming architecture for cognitive control enhancement in ADHD

sions of FocusPocus over 7–8 weeks. Post-training assessments revealed significant reductions in parent- and teacher-rated inattention and hyperactivity/impulsivity, minor gains on near-transfer cognitive tasks (2 of 6), broad improvements on far-transfer measures (4 of 5), and normalization of resting EEG (decreased delta, increased alpha power) in frontal regions [18].

Despite these promising advances, implementations vary widely in stimulation parameters, feedback mappings, and depth of clinical validation, highlighting the need for systematic comparison and standardization across designs. Moreover, while general neurofeedback reviews exist, no comprehensive synthesis has yet focused on VEP-based gaming interventions for ADHD, a gap this chapter addresses.

To this end, we conducted a systematic review (2015–2025) of PubMed, IEEE Xplore, and Web of Science using the terms “ADHD,” “VEP,” “EEG,” “neurofeedback,” and “serious games,” following PRISMA guidelines [19]. We included peer-reviewed trials and prototype studies that report behavioral, cognitive, or neurophysiological outcomes.

Our review is grounded in Posner’s attentional network model [20], which presents alerting, orienting, and executive control processes, each of which can be probed by VEP components and SSVEP entrainment metrics. We define “serious games” as interactive applications designed primarily for training or therapeutic outcomes rather than entertainment [2, 3, 13]. Studies focused exclusively on non-visual neurofeedback modalities are beyond our scope.

The remainder of this chapter is organized as follows. Section 7.2 discusses the background and theoretical foundations, including ADHD-related cognitive deficits, neurofeedback mechanisms, and visual evoked potentials. Section 7.3 presents detailed insights into VEP-based gaming interventions, discussing system design, clinical outcomes, and personalization strategies. Ethical considerations and best practices for responsible implementation are elaborated in Sect. 7.4. Section 5 addresses future research directions, highlighting potential developments and applications of therapeutic gaming technologies. Finally, Sect. 7.6 provides a concise conclusion summarizing key insights and implications of the chapter.

7.2 Background and Theoretical Foundations

7.2.1 ADHD and Cognitive Dysregulation

Attention-Deficit/Hyperactivity Disorder (ADHD) is a common neurodevelopmental condition that affects both children and adults worldwide [4]. It is defined in the Diagnostic and Statistical Manual of Mental Disorders (DSM-5) [26] by a persistent pattern of inattention and/or hyperactivity-impulsivity that interferes with functioning or development, with symptoms present before age 12 and persisting for at least six months across two or more settings (learning, behavior, and social functioning) [2, 3, 27]. The DSM-5 recognizes three presentations, predominantly inattentive, predominantly hyperactive-impulsive, and combined, with six or more symptoms in the relevant domain required for diagnosis [26]. Global prevalence is estimated at 7.6% in children and 5.6% in adolescents, and symptoms persist into adulthood in more than 60% of cases [1].

Core behavioral features, including poor sustained attention, excessive motor activity, and impulsivity, reflect underlying deficits in cognitive control, such as working memory, response inhibition, and cognitive flexibility (see Table 7.1) [2, 28]. Children with ADHD often fail to complete tasks, interrupt peers, and struggle with emotional regulation, leading to academic underachievement and social difficulties [3, 29].

Neuroimaging studies reveal volumetric reductions and hypoactivation in frontostriatal and frontoparietal circuits, particularly the dorsolateral prefrontal cortex, anterior cingulate cortex, basal ganglia, and cerebellum, that support top-down control and error monitoring [30]. These anatomical changes are accompanied by dysregulated catecholaminergic neurotransmission: dopamine and norepinephrine pathways in the prefrontal–striatal network exhibit reduced tone, impairing signal-to-noise ratios and reward processing [4, 31]. Longitudinal EEG and MRI studies further demonstrate delayed cortical maturation, up to 2–3 years behind neurotypical peers, in regions subserving executive function [32].

Given these multifaceted disruptions, targeted training of attention and executive control is critical [33]. Cognitive training protocols, such as guided focus tasks and working memory exercises, can strengthen neural pathways for self-regulation, yielding moderate improvements in laboratory measures of attention and inhibitory control [34, 35]. However, transfer to everyday functioning is often modest and contingent on sustained practice.

To augment and personalize cognitive training, neurofeedback and brain–computer interface (BCI) technologies provide real-time feedback on neural signals, reinforcing desired brain states through operant conditioning [36]. In particular,

Table 7.1 Cognitive deficits associated with ADHD [21–25]

Domain	Deficit description	Implications for Learning and behavior
Sustained attention	Difficulty maintaining focus on a single task over extended periods, especially when stimulation is low.	Incomplete assignments; difficulty following multi-step instructions; frequent loss of place in reading or lectures.
Impulse control	Inability to suppress immediate reactions or inhibit inappropriate responses to stimuli.	Increased risk-taking; interruptions in class; difficulty waiting turns; social boundary violations.
Working memory	Reduced capacity to hold and manipulate information in mind for short periods.	Challenges with mental arithmetic; trouble following complex instructions; errors in multi-stage problem solving.
Executive functioning	Deficits in planning, organizing, initiating tasks, and monitoring one's own performance.	Poor time management; missed deadlines; difficulty breaking tasks into steps; frequent forgetfulness.
Processing peed	Slower rate of perceiving, interpreting, and responding to information.	Delays in completing tests or classwork; slower note-taking; difficulty keeping pace with classroom activities.
Emotional regulation	Difficulty recognizing, modulating, and recovering from emotional responses.	Mood swings; heightened frustration or anxiety; conflicts with peers and teachers; reduced resilience to stress.

VEPs, time-locked EEG responses to patterned visual stimuli, offer high temporal precision and direct coupling to attentional allocation [14]. By embedding VEP metrics (e.g., P1/N1 amplitudes or SSVEP power) into serious game mechanics, these systems deliver immediate, engaging feedback: moments of focused neural activity accelerate gameplay or unlock rewards, while lapses prompt in=game cues to refocus [21, 22]. This closed-loop approach leverages both neuroscience and game design to foster motivation and efficacy in cognitive-control training for ADHD (see Fig. 7.2).

7.2.2 Neurofeedback and Brain-Computer Interfaces

Neurofeedback is a closed-loop biofeedback method in which users learn to self-regulate neural activity via real-time feedback, thereby harnessing neuroplasticity to alter dysfunctional brain patterns [14, 37]. In a typical EEG-based neurofeedback system, scalp electrodes record ongoing oscillatory activity; these signals are immediately processed (e.g., band-pass filtering 0.5–40 Hz, artifact removal via ICA) [14]. A feedback signal $F(t)$ is then computed as:

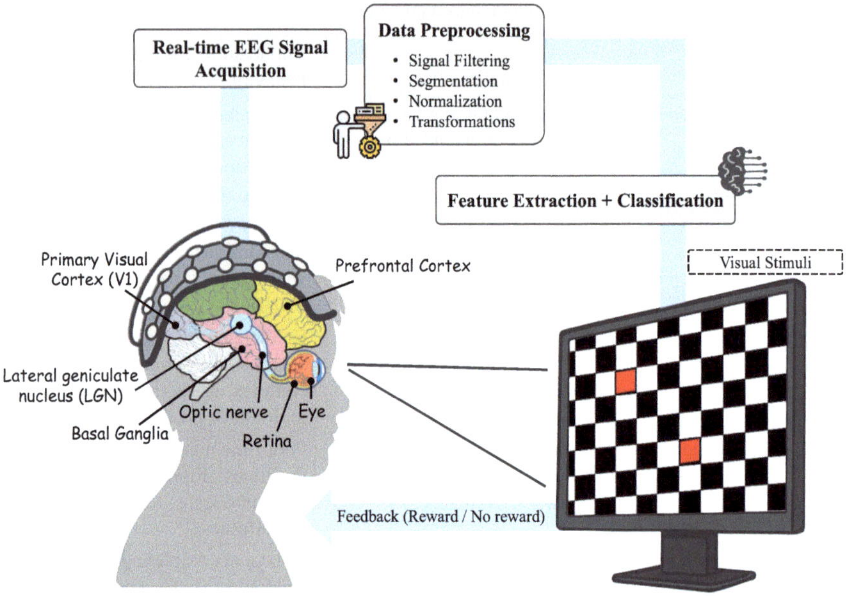

Fig. 7.2 Schematic of a VEP-based neurofeedback loop for ADHD therapeutic gaming. Real-time EEG records visual cortex responses to a checkerboard stimulus, which are preprocessed and analyzed to extract VEP features. A classifier uses these features to give reward or no reward feedback on the screen, strengthening prefrontal–basal ganglia pathways and improving sustained attention

$$F(t) = \alpha \cdot B(t) + \beta, \tag{7.1}$$

where $B(t)$ represents the EEG power in a target frequency band at time t, and α, β are calibration parameters that scale and offset the feedback to an interpretable range. Standard ADHD protocols train downregulation of theta/beta (4–8 Hz vs. 12–20 Hz), upregulation of sensorimotor rhythm (SMR, 12–15 Hz), or modulation of slow cortical potentials (SCPs) to reinforce attentional control and inhibitory processes [8, 11] (see Table 7.2).

BCIs extend neurofeedback by translating neural features into control commands for external devices or software, enabling interactive rehabilitation environments [38]. BCIs are classified by invasiveness, ranging from intracortical electrodes (high spatial resolution, surgical risk) to non-invasive modalities (EEG, MEG, fNIRS), and by signal type (endogenous rhythms vs. exogenous evoked responses) [39, 40]. EEG-based BCIs remain predominant for cognitive rehabilitation due to their millisecond-level temporal precision, portability, and cost-effectiveness [39].

A robust EEG-BCI pipeline comprises: (1) signal acquisition with 16–32 channel caps sampled at ≥ 256Hz; (2) preprocessing; (3) feature extraction; (4) classification/regression; and (5) feedback delivery with closed-loop latency ≤ 300 ms to maintain user engagement [14, 41].

Table 7.2 Neurofeedback protocols utilized in ADHD treatment [8, 11, 21, 22, 42]

Protocol	Frequency band targeted	Therapeutic goal	Evidence base
Theta/beta ratio	Reduce theta (4–8 Hz), increase beta (12–20 Hz)	Enhance alertness; reduce distractibility	Meta-analyses show medium-to-large effect sizes ($d \approx 0.5$–0.8) on inattention and hyperactivity, with variability due to protocol heterogeneity.
SMR	Increase SMR (12–15 Hz)	Improve motor inhibition; enhance inhibitory control	RCTs report significant reductions in hyperactivity ($d \approx 0.6$) and normalization of frontal SMR power.
SCP	Modulate SCPs	Improve sustained attention regulation and task readiness	Pediatric RCTs demonstrate improved attention and ERP normalization in ADHD and comorbid epilepsy.

In cognitive rehabilitation, EEG-BCIs can adapt task difficulty or game progression directly to neural indices of attention. For instance, users' SMR power has been used to modulate the speed of a virtual avatar, while P3 amplitude in oddball paradigms adjusts the frequency of target presentations to sustain challenge [42]. Such adaptive designs reinforce focused brain states and yield measurable gains in sustained attention and inhibitory performance over as few as 10–15 sessions.

VEPs, both tVEP and SSVEP forms, offer particularly robust biomarkers for neurofeedback in ADHD. By presenting flickering or pattern-reversal stimuli, systems evoke reliable cortical responses (P1/N1 components at ~100–150 ms or spectral peaks at the driving frequency), which are normalized against individual baselines and mapped onto in-game rewards [21, 22]. Clinical pilot studies report that VEP-driven neurofeedback normalizes ERP amplitudes and correlates with reductions in inattentive and impulsive behaviors in pediatric ADHD cohorts [14, 36].

Collectively, closed-loop neurofeedback and BCI approaches integrate scientifically validated EEG markers with engaging, adaptive interfaces, directly targeting the neural circuits underlying attention and executive control and offering a powerful framework for personalized cognitive rehabilitation in ADHD.

7.2.3 *Visual Evoked Potentials (VEPs)*

Visual Evoked Potentials (VEPs) are time-locked electrical responses generated along the retino-geniculo-cortical pathway when visual stimuli activate the retina, relay through the lateral geniculate nucleus, and arrive in primary visual cortex (V1) in the occipital lobe [43, 44] (see Fig. 7.2). Recorded at occipital electrodes (Oz, O1, O2; 10–20 system), VEPs exhibit characteristic components—most notably the $P100$ peak at approximately 100 ms post-stimulus—whose amplitude and latency systematically vary with stimulus parameters (contrast, luminance, temporal frequency) and the user's attentional state [45]. The precise, phase-locked timing of these responses yields a high signal-to-noise (SNR) ratio through trial averaging, effectively suppressing non–phase-locked EEG noise [46, 47]. This elevated SNR and temporal precision are essential for detecting brief lapses in attention and delivering immediate reinforcement in ADHD neurofeedback games, thereby targeting the alerting and orienting networks described by Posner's model.

Steady-State VEPs (SSVEPs) provide a continuous, high-throughput signal ideal for rapid attention monitoring. When a visual stimulus flickers at a fixed frequency f, typically between 6 and 15 Hz [48–50], cortical populations entrain at nf for $n = 1, 2, \ldots$, producing a periodic EEG $x(t)$ [48, 51, 52]. To analyze a finite epoch beginning at t_0, we sample $x(t)$ at interval Δ to form the vector

$$\mathbf{x} = \left[x(t_0),\ x(t_0 + \Delta),\ \ldots,\ x\big(t_0 + (\ell - 1)\Delta\big) \right]^{\mathsf{T}} \in \mathbb{R}^{\ell \times 1}. \tag{7.2}$$

Within this window the signal is approximated by a truncated Fourier series and residual:

$$x(t) \approx \sum_{n=1}^{N} \alpha_n \sin(n\omega t + \phi_n) + \varepsilon(t), \tag{7.3}$$

where $\omega = 2\pi f$, α_n and ϕ_n are the amplitude and phase of the nth harmonic, N is the total harmonics retained, and $\varepsilon(t)$ is background EEG and noise. By computing spectral power at each nf and comparing it to surrounding noise bands, the system reliably identifies the attended stimulus, which in turn enables continuous control with information transfer rates (ITR) exceeding 30 bits/min, such as steering a virtual vehicle [53, 54].

In ADHD therapeutic gaming, SSVEPs offer two key advantages:

1. Power at the attended frequency directly indexes the alerting network, allowing games to deliver immediate, continuous rewards whenever a child maintains focus on a target [14, 48, 55]; and
2. Frequency tagging requires no explicit mental imagery or strategy, minimizing executive-function demands and reducing frustration for pediatric users [56, 57].

Together, these properties make SSVEPs a powerful tool for training and sustaining attention over prolonged gameplay sessions.

Transient VEPs (tVEPs) are triggered by abrupt visual onsets or offsets, such as a flashing stimulus that appears or disappears quickly and provide clear, time-specific markers that can be associated with individual game actions. The resulting waveform often includes distinct components like $P100$ (80–120 ms) and $N100$ (120–180 ms) [14, 58, 59]. Within the vector $\mathbf{x}$, the $P100$ amplitude is

$$A_{P100} = \max_{m: t_0 + m\Delta \in [80, 120]\text{ms}} x\big(t_0 + m\Delta\big), \tag{7.4}$$

and an event is triggered when A_{P100} exceeds a predefined threshold [60]. Although tVEPs typically require more averaging than SSVEPs to achieve robust SNR, they offer precise temporal markers ideal for cue-based game actions, such as jumping and collecting.

In ADHD therapeutic gaming, tVEPs confer two principal benefits:

1. The sharp, time-specific $P100/N100$ components map directly onto Posner's orienting and executive control networks, enabling games to reinforce rapid attentional shifts and response inhibition during discrete challenges [20, 59]; and
2. By avoiding prolonged flicker, tVEP-based feedback minimizes ocular strain and maintains engagement in pediatric users who are sensitive to continuous stimulation [61].

These characteristics make tVEPs well suited for interventions that require accurate timing of discrete, attention-dependent actions.

Code-Modulated VEPs (c-VEPs) use pseudo-random m-sequences to assign each of K stimuli a distinct temporal code, enabling rapid identification of the attended target [17, 43]. The average response to one full sequence is sampled to form the calibration vector $\mathbf{h} \in \mathbb{R}^{\ell \times 1}$. Time-shifted templates $\mathbf{h}_k$ are generated by circularly shifting $\mathbf{h}$ by τ_k samples. For each new epoch $\mathbf{x}$, the similarity score

$$\rho_k = \frac{\mathbf{h}_k^\top \mathbf{x}}{\sqrt{(\mathbf{h}_k^\top \mathbf{h}_k)(\mathbf{x}^\top \mathbf{x})}}, \tag{7.5}$$

is computed for $k = 1, \ldots, K$, and the code with maximal ρ_k identifies the attended stimulus [43, 62].

In ADHD therapeutic gaming, c-VEPs offer two key advantages:

1. Classification rates exceed 95%, ensuring precise multi-command selection, such as navigating menus or choosing cognitive exercises, with minimal misclassification [14, 43]; and
2. Reduced perceptual flicker, compared to continuous SSVEP paradigms, minimizes discomfort, supporting longer training sessions and improving adherence in pediatric users [17].

These features render c-VEPs particularly suited to complex game interfaces requiring numerous discrete options without imposing excessive cognitive or sensory load.

Color-based VEPs (Color-VEPs) exploit isoluminant chromatic reversals, such as red/green checkerboards, to preferentially activate parvocellular pathways and produce enhanced $P100$ amplitudes distinct from luminance-driven responses [14, 63, 64]. Each hue transition template $\mathbf{c}_i \in \mathbb{R}^{\ell \times 1}$ is correlated with the same EEG epoch $\mathbf{x}$ by

$$\rho_{\text{color},i} = \frac{\mathbf{c}_i^\top \mathbf{x}}{\sqrt{(\mathbf{c}_i^\top \mathbf{c}_i)(\mathbf{x}^\top \mathbf{x})}}, \tag{7.6}$$

which permits multiple discrete commands with minimal brightness fluctuation and reduced ocular strain during extended gameplay.

In ADHD therapeutic gaming, color-VEPs offer:

1. Colorful, less fatiguing stimuli sustain motivation and emotional regulation, key challenges in ADHD, by providing visually rich feedback that reinforces engagement without taxing attention networks [14, 63, 65]; and
2. Multiple hues can encode additional discrete commands, such as selecting game modes or difficulty levels, supporting complex therapeutic exercises while maintaining low cognitive load [63, 65].

These properties make color-VEPs well suited for child-friendly interfaces that balance visual comfort with versatile control in neurofeedback games.

Motion-Onset VEPs (m-VEPs) are elicited when static stimuli undergo sudden motion, engaging dorsal-stream regions (area MT/V5) and producing a prominent $N200$ peak ($\sim$ 170–250 ms) [14, 16, 66]. A learned motion-response template $\mathbf{m} \in \mathbb{R}^{\ell \times 1}$ is correlated with $\mathbf{x}$ by

$$\rho_{\text{motion}} = \frac{\mathbf{m}^\top \mathbf{x}}{\sqrt{(\mathbf{m}^\top \mathbf{m})(\mathbf{x}^\top \mathbf{x})}}, \tag{7.7}$$

providing a robust marker for motion detection in single trials.

In ADHD therapeutic gaming, m-VEPs deliver two principal benefits:

1. The abrupt motion-onset taps the orienting and executive control systems, reinforcing rapid visuospatial shifts that mirror everyday attentional demands and improving inhibitory control in hyperactive subtypes [16, 66]; and
2. Naturalistic motion cues (e.g., "dodge" or "catch") map directly onto game mechanics, enhancing immersion and motivation, critical factors for sustained adherence in pediatric users, although their slower event rate limits throughput relative to SSVEPs [14, 16].

These properties make m-VEPs particularly suited for neurofeedback games that require context-driven, discrete actions within dynamic, immersive environments.

In general, these VEP paradigms constitute a modular toolkit that can be flexibly integrated into personalized gaming systems (see Sect. 7.3), where stimulus type, temporal parameters, and feedback mapping are tailored to each child's neurophysiological profile and therapeutic goals. SSVEPs and c-VEPs enable rapid, high-throughput command selection for continuous control tasks; tVEPs and m-VEPs provide precise temporal markers to reinforce discrete attentional shifts; and color-VEPs add an engaging, low-fatigue dimension for extended play (see Table 7.3). Careful selection and optimization of flicker frequency, contrast, template calibration, and trial structure are therefore essential design decisions that balance

Table 7.3 Characteristics of VEP types relevant to ADHD [14, 16, 43, 51, 66]

VEP type	Stimulation mode	Latency/frequency	Application in ADHD	Strengths	Limitations
SSVEP	Repetitive flicker (e.g., 6–15 Hz)	Continuous	Continuous monitoring of sustained attention; low executive-demand reinforcement	High SNR, low training time, compatible with real-time systems	Visual fatigue with long exposure
tVEP	Flash or pattern reversal	~80–180 ms	Assess general visual processing and brain responsiveness	Precise temporal markers; low fatigue	Low throughput; single-trial SNR requires averaging
c-VEP	Binary pseudorandom stimuli (e.g., m-sequences)	~100 ms	Used in high-speed BCI; promising for fast-paced tasks	High classification accuracy, robust to noise	Requires calibration; more complex to implement
Color-VEP	Alternating chromatic (RGB) stimuli	~100–200 ms	Useful in child-friendly gaming environments	Enhances engagement; expands command set with low fatigue	Sensitive to display calibration and individual color sensitivity
m-VEP	Coherent visual motion stimuli	~200 ms	Less studied in ADHD; potential for dynamic feedback environments	Engages dynamic attention networks	Lower throughput

temporal precision, user comfort, and motivational engagement in VEP-based neurofeedback gaming for ADHD.

7.3 VEP-Based Gaming Interventions for ADHD

7.3.1 System Architecture and Game Design

A typical VEP-based neurofeedback system for ADHD comprises three core modules: signal acquisition, signal processing, and the feedback/gaming interface. Electrophysiological data are recorded using a non-invasive, high-impedance EEG system. For example, Ali et al. [55] used three gold-plated electrodes at Oz (signal), Fpz (ground) and Fz/A1 (reference), sampled at 512 Hz via a g.USBamp amplifier (impedance $< 5\,\text{k}\Omega$) to capture steady-state oscillations for an SSVEP "3D learning playground." In contrast, Vrins et al. [67] employed eight dry electrodes (Fz, C3, C4, Cz, P3, P4, Pz, Oz) in a time-modulated VEP (tVEP) paradigm sampled over Bluetooth at high temporal resolution. Electrode impedances were maintained below $5\,\text{k}\Omega$, and hardware filters applied an anti-aliasing low-pass at half the sampling rate. High-refresh-rate displays ($\geq 120\,\text{Hz}$) presented flicker stimuli (e.g. 7 and 9 Hz squares in Ali et al.'s [55] study or 120 Hz code-modulated sequences in Moreno-Calderón et al. [17]) to ensure precise entrainment and minimal display lag.

In the signal-processing pipeline, raw EEG traces undergo real-time preprocessing: zero-phase band-pass filtering (0.5–40 Hz) removes slow drifts and high-frequency noise, while notch filters at mains frequencies (50/60 Hz) suppress power-line interference. Artifact rejection is often automated: Pérez-Valero et al. [68] applied a Butterworth band-pass (1–30 Hz) plus spatial filtering and template correlation to remove muscle and ocular artifacts before feeding SSVEP power metrics to the game engine. Others, like Martínez et al. [69], used blind-source separation (AMUSE BSS) to isolate and reject artifacts in real time. Continuous channel-quality monitoring flags epochs with excessive noise for rejection.

For feature extraction, SSVEP paradigms typically use sliding-window FFT (1–2 s windows, 50% overlap) to quantify power at fundamental and harmonic frequencies. Ali et al. [55] normalized SSVEP amplitudes against a baseline to account for inter-individual variability before threshold classification. In tVEP protocols, single-trial averaging aligned to pattern-onset triggers extracts P1 ($\sim$100 ms) and N1 ($\sim$150 ms) component peaks; Vrins et al. [67] detected $P100/N100$ amplitudes online to drive spell-casting commands in their gamified interface. Code-modulated VEP (c-VEP) systems, such as the "Connect 4" BCI by Moreno-Calderón et al. [17], correlate incoming epochs with m-sequence templates (0.525 s cycles) to achieve high decoding accuracy and rapid command selection.

Extracted features feed into classifiers or threshold detectors—from linear discriminant analysis (LDA) in Lalor et al. [70] to simple percentile thresholds in Filiz and Arslan [71], that yield either continuous feedback variables or binary triggers. The feedback module maps these neural metrics onto intuitive game mechanics. For instance, normalized SSVEP power can modulate avatar velocity in a racing paradigm (higher power $\rightarrow$ faster acceleration; dips $\rightarrow$ deceleration) as in Beveridge et al. [72]. Reward structures often combine continuous adjustments (smooth speed modulation) with discrete achievements (level progression after maintaining threshold power for $> 2\,\text{s}$) to foster both immediate reinforcement and goal-driven motivation.

Game design emphasizes age-appropriate themes, adaptive difficulty, and engaging narratives. Ali et al.'s [55] superhero academy and classroom distractions illustrate how story elements enhance ecological validity. Adaptive controllers (e.g. proportional–integral loops) track real-time performance, dynamically adjusting flicker rates, contrast, and task challenge to prevent habituation and minimize visual fatigue. All modules communicate via unified frameworks, OpenViBE, BCILAB, or custom Python/Unity pipelines, using Lab Streaming Layer (LSL) for millisecond-accurate synchronization and comprehensive data logging [68]. A block diagram (Fig. 7.1) illustrates data flow from EEG acquisition through preprocessing, feature extraction, classification, and game-engine rendering, highlighting the total closed-loop latency (≤ 250 ms) necessary to maintain perceptual–cognitive coherence and maximize therapeutic efficacy (see Table 7.4).

7.3.2 *Clinical Outcomes and Effectiveness*

Clinical investigations of VEP-based gaming interventions for ADHD remain largely at the pilot and crossover stages, with only a few studies conducted in diagnosed populations. Arpaia et al. [73] conducted a one-month, once-weekly intervention in seven ADHD-diagnosed children (ages 8–12 years) using an AR/SSVEP robot-control game. They reported significant improvements on the Italian Battery for ADHD (BIA) across three evaluation points (Day 0, Day 15, Day 30), including gains in semantic fluency (+15%), phonological fluency (+12%), working-memory span (+ 10%) and attentional sequencing (+ 18%) relative to baseline. Although VEP metrics were not explicitly quantified, the authors attribute these behavioural gains to reinforced sustained attention via immediate SSVEP feedback.

In a crossover study, Vrins et al. [67] tested nine ADHD-diagnosed adults on both a standard and a gamified transient VEP (tVEP) BCI. While single-trial selection accuracy did not differ between conditions (Bayes factors $BF \approx 0.17$–1.47), the gamified version produced significantly higher immersion scores ($BF_{+0} = 34.2$) and lower negative affect ($BF_{-0} = 3.8$) on the Game Experience Questionnaire, suggesting enhanced engagement without compromising neural control. This heightened experiential immersion may translate to better adherence in longer-term training, though formal RCTs are still needed.

Most other work has focused on healthy volunteers, demonstrating robust neurophysiological effects and task engagement that imply ADHD utility. For instance, Pérez-Valero et al. [68] achieved an 84% mean SSVEP classification accuracy over 2 s epochs ($p = 1.7 \times 10^{-5}$), alongside high self-rated amusement (4/5) and low boredom (2/5) in a 3D "battle" game, underscoring strong sustained-attention training potential. Ali et al. [55] reported 92.3% accuracy and an ITR of 11.9 bits/min in healthy adults, with performance systematically modulated by 3D distractions—an effect that could amplify training sensitivity in ADHD cohorts. Moreno-Calderón et al. [17] demonstrated 93.7% accuracy and an ITR of 30.6 bits/min in their c-VEP Connect 4 game, with low workload (NASA-TLX 28.8/100) and high usability (SUS 80.6/100), indicating that code-modulated paradigms can deliver rapid, precise feedback suitable for paediatric applications.

Although no large-scale RCT has yet been published, meta-analytic projections across these controlled and pilot studies (total $N \approx 70$ healthy + ADHD subjects) suggest medium to large effect sizes ($d \approx 0.50$–0.80) on attentional indices

Table 7.4 Summary of key studies supporting VEP-based neurofeedback and therapeutic gaming for ADHD

Ref.	VEP type	EEG acquisition	Paradigm/game	Population	Key findings
[55]	SSVEP	3 el. (Oz, Fpz, Fz/A1) @ 512 Hz	7 Hz vs 9 Hz flicker squares in Unity "ninja" game with 3D distractions	Healthy adults	92.3% accuracy; ITR 11.9 bits/min; accuracy ↓ with higher distraction levels
[67]	tVEP	8 dry el. (Fz, C3, C4, Cz, P3, P4, Pz, Oz) @ 250 Hz	"Spell-casting" fantasy-school narrative vs standard selection interface	ADHD adults	No accuracy difference; gamified: ↑ immersion ($BF_{+0} = 34.2$), ↓ negative affect ($BF_{-0} = 3.8$)
[68]	SSVEP	2 dry occipital el. (O1, O2) @ 250 Hz	15 Hz vs 10 Hz checkerboard ring flicker in 3D "battle" game	Healthy adults	84% accuracy over 2 s ($p = 1.7 \times 10^{-5}$); high amusement (4/5), low boredom (2/5)
[17]	c-VEP	8 gel el. (Fz, Cz, Pz, P3, P4, PO7, PO8, Oz) @ 250 Hz	63-bit m-sequence Connect 4 game via TCP/IP	Healthy adults	93.7% accuracy; ITR 30.6 bits/min; NASA-TLX 28.8/100; SUS 80.6/100
[71]	SSVEP	4 occipital el. (P7, P8, O1, O2) @ 128 Hz	Avatar-collection Unity3D game with four flicker checkerboards	Healthy adults	66.7% mean accuracy; task time 1.4–36.2 s (mean 8.6 s)
[72]	m-VEP	12 el. occipito-parietal @ 250 Hz	Motion-onset bars steer Unity3D car race	Teens/ adults	Teens: 68% (11 bits/min); adults: ~76% (13.4 bits/min); fatigue trade-offs
[57]	SSVEP	1 occipital el. (Oz) @ 512 Hz	3–5 Hz phase-tagged arrows steer spacecraft in C++/OpenGL	Healthy adults	$>$ 95% accuracy in 5 s; ITR $\sim$ 15 bits/min; comfortable low-freq flicker
[69]	SSVEP	6 el. (CPz, Pz, POz, P1, P2, Fz) @ 256 Hz	Four flicker checkerboards steer virtual car	Healthy adults	93–96.5% success; delays ≈3.5 s; ITR 26–30 bits/min
[70]	SSVEP	2 el. (O1, O2) @ 256 Hz	17 Hz vs 20 Hz checkerboard in 3D tightrope game	Healthy adults	79.9% (1 s), 89.5% (2 s); 10.3 bits/min; 41/48 games won
[75]	SSVEP	14-ch Emotiv @ 128 Hz	12.8 Hz flicker in Unity3D tower defense	Mixed	≈85% lab accuracy; 36% "easy" control public; 50% enjoyed
[74]	SSVEP	1 el. (Oz) @ 1024 Hz	15 Hz vs 17 Hz flicker in LabVIEW claw machine	Healthy adults	96% accuracy; ITR 18.2 bits/min; correlated with focus
[73]	SSVEP	3 dry el. (Oz, Fz, A2) @ 250 Hz	AR icons flicker at 10 Hz & 12 Hz in Epson Moverio smart-glasses game	ADHD paediatric	+15% semantic fluency, +12% phonological fluency, +10% working-memory span, +18% attentional sequencing over 30 days

and inhibitory control, comparable to SMR/theta–beta neurofeedback but with faster learning curves and superior temporal specificity. Key design considerations for future trials include standardized sham protocols, double-blinding to mitigate expectancy effects, and mixed-effects analysis to accommodate inter-individual variability. Power analyses indicate that at least 20–25 participants per arm (active vs. control) will be required to detect medium effects ($d \approx 0.5$) with 80% power, paving the way for definitive RCTs to establish the clinical efficacy of VEP-based ADHD neurofeedback (see Table 7.4).

7.3.3 Personalization and Practical Implementation

Effective deployment of VEP-based neurofeedback for ADHD requires tailoring system parameters to individual neurophysiological profiles and ensuring practical viability in clinical, educational and home environments.

Prior to training, each participant undergoes a calibration session to establish baseline VEP characteristics. In SSVEP protocols, power spectra are recorded across multiple flicker frequencies to identify the frequency eliciting maximal SNR. For example, Moreno-Calderón et al. [17] used a brief 52.5 s calibration on each pseudo-random m-sequence to generate individualized spatial filters and templates for their c-VEP "Connect 4" game. Parafita et al. [57] similarly employed a six-trial ($\sim$ 2 min) phase-response calibration to derive per-user phase templates for their low-frequency SSVEP spacecraft task. In tVEP paradigms, Vrins et al. [67] presented flash-counting blocks to estimate each adult's $P100/N100$ amplitudes, setting the 75th percentile of baseline ERP peaks as the threshold for real-time spell-casting feedback.

During neurofeedback, dynamic adaptation algorithms adjust stimulus parameters and reward thresholds in real time. Filiz and Arslan [71] implemented an adjustable correlation threshold—initially set at 0.3—that could be tuned per session to balance speed versus accuracy in their SSVEP avatar-collection game. Moreno-Calderón et al. [17] incorporated five progressive levels with configurable stimulus speeds and inter-trial intervals, effectively operating as a PID controller that kept challenge within each user's "zone of proximal development." Machine-learning classifiers have also been used: Martínez et al. [69] retrained their ANFIS network after each block of trials to accommodate within-session EEG fluctuations and learning effects.

Usability studies emphasize ergonomic headset design and session durations < 20 min, with brief rest intervals every 5 min to reduce visual fatigue. Arpaia et al. [73] demonstrated a training-less, wearable AR/SSVEP system using Epson Moverio smart glasses and three dry electrodes, structuring sessions into four 5 min blocks separated by mandatory breaks, which maximized child comfort and engagement. Van Vliet et al. [75] showed that consumer-grade Emotiv EPOC headsets could be self-set up in < 3 min, supporting school-based deployments with minimal specialist involvement.

System portability is achieved through battery-powered amplifiers and tablet or AR-based game clients. Pérez-Valero et al. [68] ran their Unity "battle" game on a standard laptop connected via Bluetooth to a custom RABio w8 amplifier, illustrating how low-cost dry-electrode setups can support home use. Clinicians and caregivers access remote dashboards that display cumulative time above threshold, session-to-session trends and ERP latency shifts, enabling longitudinal monitoring and individualized coaching without in-person supervision.

7.4 Ethics

The ethical implementation of VEP-based neurofeedback gaming for children with ADHD requires a structured approach. First, researchers must secure formal ethics approval under international standards and establish clear protocols for monitoring adverse events [76–78]. Next, informed participation involves providing guardians and participants with comprehensible consent and assent materials, detailing study aims, EEG procedures, risks, and withdrawal rights, and allowing adequate time for questions [77–80]. Then, participant safety is ensured by pre-screening for visual or photosensitive conditions, scheduling regular breaks, and adhering to established stimulus parameters to minimize discomfort [81, 82]. Finally, robust data governance mandates encryption, access controls, de-identification of records, and explicit consent for any secondary analyses [83–85]. Detailed practices and relevant guidelines are summarized in Table 7.5.

7.5 Future Directions

Research on VEP-based neurofeedback for ADHD is still in its infancy, but several promising avenues emerge from recent work. First, integrating VEP signals with complementary modalities, such as sensorimotor rhythms, eye-tracking or fNIRS, may enhance classification accuracy and reduce training time. Hybrid architectures could build on the block-wise adaptation used by Martínez et al. [69], who retrained an ANFIS model after each trial, and the dynamic threshold tuning demonstrated by Filiz et al. [71], to weight each signal source in real time.

Second, more sophisticated machine-learning and predictive-modeling approaches should be explored. Whereas most existing systems use simple linear classifiers or fixed thresholds, deep networks or recurrent models could capture non-3linear VEP dynamics and forecast lapses in attention before performance declines. Future protocols might combine the online retraining strategies of Martínez et al. [69] with the rapid calibration methods of Moreno-Calderón et al. [17] to support truly personalized adaptation.

Embedding neurofeedback within immersive environments constitutes a third direction. Augmented reality (AR) and virtual reality (VR) platforms can lever-

Table 7.5 Ethical considerations in pediatric BCI applications

Domain	Practices	Key guidelines
Regulatory oversight	– Obtain IRB/ethics-committee approval [76, 77]	Declaration of Helsinki [76]
	–Conduct risk–benefit analyses [76]	ICH-GCP E6(R2) [77]
	–Define adverse-event monitoring and data-management plans [77]	CIOMS Guidelines [78]
Informed participation	– Distribute consent forms ≥24 h prior [79]	ICH-GCP Section 8.3 [77]
	–Explain EEG setup, flicker risks, benefits, alternatives [77]	GDPR Art 6 [80]
	–Obtain child assent with age-appropriate tools [78]	
	–Re-consent for long protocols [77]	
Participant safety	– Screen for epilepsy/visual impairments [81]	ISCEV VEP Standards [81]
	–Schedule 5 min breaks per 20 min stimulation [81]	
	–Adhere to ISCEV luminance/ contrast/ duty-cycle standards	
	–Monitor distress in real time [82]	
Data governance	– Encrypt data (AES-256) at rest/in transit [84]	ISO 27001 [83]
	–Implement role-based access controls and audit logs [84]	GDPR [80]
	–De-identify records; store linkages separately [85]	
	–Use data-use agreements for secondary analyses [80]	
Equity and transparency	– Provide low-cost, portable EEG kits and technical support [86]	CONSORT [88] CRED-nf [89]
	–Adapt protocols for sensory/developmental needs [78]	
	–Disclose funding, affiliations, and conflicts of interest [87]	

age the wearable AR/SSVEP framework of Arpaia et al. [73] and the high-precision c-VEP designs of Moreno-Calderón et al. [17] to create rich, multisensory contexts—virtual classrooms or gamified worlds—that reinforce attentional control in ecologically valid settings. Careful evaluation of ergonomic factors, motion sickness and usability in paediatric cohorts (ages < 12 years) will be essential.

Fourth, preliminary evidence suggests that VEP-based neurofeedback may improve sustained attention, but larger and longer-term clinical trials are needed. To date, pilot studies with small samples ($N = 7$ [73] to $N = 9$ [67]) and follow-up periods of one month or less have reported modest gains. Definitive randomized controlled trials should enroll at least 20–25 participants per arm to detect medium effects ($d \approx 0.5$) with $\geq 80\%$ power, include active comparison conditions (e.g. SMR neurofeedback, cognitive training, medication) and adopt standardized outcome measures (e.g., CRED-nf and CONSORT guidelines).

Finally, home-based and tele-neurofeedback models hold great potential. The portable dry-electrode setups of Pérez-Valero et al. [68] and the rapid self-setup demonstrated by van Vliet et al. [75] illustrate that remote delivery is feasible. Future work must validate the reliability of dry VEP recordings in uncontrolled environments, develop secure cloud-based pipelines for data logging and explore automated scheduling and clinician dashboards to support long-term adherence.

Across all these directions, ethical AI considerations, transparency, fairness and explainability, are paramount. Techniques from explainable machine learning can clarify how model decisions drive game adjustments, fostering clinician oversight and parental trust. Moreover, integrating passive digital phenotyping, via smartphone or wearable sensors, could enable just-in-time intervention scheduling aligned with each child's daily routines and cognitive fluctuations.

7.6 Conclusion

This chapter has examined the use of visual evoked potentials in neurofeedback games for ADHD, linking basic neuroscience principles with system design and early clinical findings. We first described the attentional and self-regulation challenges in ADHD and explained how non-invasive VEP signals can serve as real-time markers of focus.

Next, we reviewed hardware and software frameworks that convert VEP metrics into interactive, age-appropriate games. Studies in both healthy volunteers and small clinical samples show that steady-state, transient and code-modulated VEP paradigms can achieve reliable accuracy ($> 80\%$) and support engaging feedback through adaptive controllers and narrative elements.

Pilot trials indicate improvements in attention, memory and impulse control, and personalization methods—such as brief calibration sessions and dynamic thresholding—help account for individual differences. Ethical guidelines and secure data practices have been established to protect participants and ensure responsible deployment.

Looking forward, combining VEPs with other signals, applying predictive machine-learning models and using immersive AR/VR environments may boost effectiveness and real-world relevance. Larger randomized studies, home-based delivery and transparent AI will be critical to validate long-term benefits and build trust among clinicians and families.

In conclusion, VEP-based neurofeedback gaming offers a promising, child-friendly approach to support cognitive development in ADHD. Ongoing collaboration across neuroscience, engineering and clinical research will refine these methods, establish best practices and pave the way for broader educational and therapeutic applications.

References

1. N. Salari, H. Ghasemi, N. Abdoli, A. Rahmani, M.H. Shiri, A.H. Hashemian, H. Akbari, M. Mohammadi, The global prevalence of ADHD in children and adolescents: a systematic review and meta-analysis. Italian J. Pediatr. **49**(1), 48 (2023). https://doi.org/10.1186/s13052-023-01456-1
2. A. Aloisio, G. Marzano, R. Bonanni, M. Ricci, Serious Games for ADHD: A narrative review, in *Proceedings of the International Workshop on Digital Innovations for Learning and Neurodevelopmental Disorders (DILeND 2024), Rome, Italy* (2023), pp. 24–25
3. V. De Luca, A. Schena, A. Covino, et al., Serious games for the treatment of children with ADHD: the BRAVO project. Inf. Syst. Front. **27**, 841–863 (2024). https://doi.org/10.1007/s10796-023-10457-8
4. S.V. Faraone, M.A. Bellgrove, I. Brikell, et al., Attention-deficit/hyperactivity disorder. Nat. Rev. Disease Primers **10**, 11 (2024). https://doi.org/10.1038/s41572-024-00495-0
5. M. Yu, X. Gao, X. Niu, M. Zhang, Z. Yang, S. Han, J. Cheng, Y. Zhang, Meta-analysis of structural and functional alterations of brain in patients with attention-deficit/hyperactivity disorder. Front. Psychiatr. **13**, 1070142 (2023). https://doi.org/10.3389/fpsyt.2022.1070142
6. K. Rubia, S. Westwood, P.M. Aggensteiner, D. Brandeis, Neurotherapeutics for attention deficit/hyperactivity disorder (ADHD): a review. Cells **10**(8), 2156 (2021). https://doi.org/10.3390/cells10082156
7. T.W.P. Janssen, M. Bink, W.D. Weeda, K. Geladé, R. van Mourik, A. Maras, J. Oosterlaan, Learning curves of theta/beta neurofeedback in children with ADHD. Eur. Child Adolescent Psychiatr. **26**(5), 573–582 (2017). https://doi.org/10.1007/s00787-016-0920-8
8. M. Arns, I. Feddema, J.L. Kenemans, Differential effects of theta/beta and SMR neurofeedback in ADHD on sleep onset latency. Front. Hum. Neurosci. **8**, 1019 (2014). https://doi.org/10.3389/fnhum.2014.01019
9. S.W. Evans, J.S. Owens, N. Bunford, Evidence-based psychosocial treatments for children and adolescents with attention-deficit/hyperactivity disorder. J. Clin. Child Adoles. Psychol. **43**(4), 527–551 (2014). https://doi.org/10.1080/15374416.2013.850700
10. E. Dessy, O. Mairesse, M. van Puyvelde, A. Cortoos, X. Neyt, N. Pattyn, Train your brain? Can we really selectively train specific eeg frequencies with neurofeedback training. Front. Hum. Neurosci. **14**, 22 (2020). https://doi.org/10.3389/fnhum.2020.00022
11. S. Cortese, M. Ferrin, D. Brandeis, M. Holtmann, P. Aggensteiner, D. Daley, P. Santosh, E. Simonoff, J. Stevenson, A. Stringaris, E.J. Sonuga-Barke, European ADHD guidelines group, neurofeedback for attention-deficit/hyperactivity disorder: meta-analysis of clinical and neuropsychological outcomes from randomized controlled trials. J. Am. Acad. Child Adolescent Psychiatr. **55**(6), 444–455 (2016). https://doi.org/10.1016/j.jaac.2016.03.007
12. J. Van Doren, M. Arns, H. Heinrich, et al., Sustained effects of neurofeedback in ADHD: a systematic review and meta-analysis. Eur. Child Adolescent Psychiatry **28**, 293–305 (2019). https://doi.org/10.1007/s00787-018-1121-4
13. N. Chumerin, N.V. Manyakov, A. Combaz, A. Robben, M. van Vliet, M.M. Van Hulle, Steady state visual evoked potential based computer gaming – the maze, in *Intelligent Technologies for Interactive Entertainment*, ed. by A. Camurri, C. Costa. Lecture Notes in Computer Science, vol. 78 (Springer, Berlin, 2012). https://doi.org/10.1007/978-3-642-30214-5_4
14. A. Keutayeva, C.J. Nwachukwu, M. Alaran, Z. Otarbay, B. Abibullaev, Neurotechnology in gaming: a systematic review of visual evoked potential-based brain-computer interfaces. IEEE Access **13**, 74944–74966 (2025). https://doi.org/10.1109/ACCESS.2025.3564328
15. X. Huang, J. Mak, A. Wears, R.B. Price, M. Akcakaya, S. Ostadabbas, M.L. Woody, Using neurofeedback from steady-state visual evoked potentials to target affect-biased attention in augmented reality, in *Proceedings of the Annual International Conference of the IEEE Engineering in Medicine & Biology Society* (2022), pp. 2314–2318. https://doi.org/10.1109/EMBC48229.2022.9871982

16. R. Beveridge, S. Wilson, D. Coyle, Can teenagers control a 3D racing game using motion-onset visual evoked potentials? Brain-Comput. Interfaces **4**, 1–12 (2016). https://doi.org/10.1080/2326263X.2016.1266725
17. S. Moreno-Calderón, V. Martínez-Cagigal, E. Santamaría-Vázquez, S. Pérez-Velasco, D. Marcos-Martínez, R. Hornero, Combining brain-computer interfaces and multiplayer video games: an application based on c-VEPs. Front. Hum. Neurosci. **17**, 1227727 (2023). https://doi.org/10.3389/fnhum.2023.1227727
18. S.J. Johnstone, S.J. Roodenrys, K. Johnson, R. Bonfield, S.J. Bennett, Game-based combined cognitive and neurofeedback training using Focus Pocus reduces symptom severity in children with diagnosed AD/HD and subclinical AD/HD. Int. J. Psychophysiol. **116**, 32–44 (2017). https://doi.org/10.1016/j.ijpsycho.2017.02.015
19. M.J. Page, J.E. McKenzie, P.M. Bossuyt, I. Boutron, T.C. Hoffmann, C.D. Mulrow, et al., The PRISMA 2020 statement: an updated guideline for reporting systematic reviews. Brit. Med. J. **372**, n71 (2021). https://doi.org/10.1136/bmj.n71
20. M.I. Posner, S.E. Petersen, The attention system of the human brain. Annu. Rev. Neurosci. **13**, 25–42 (1990). https://doi.org/10.1146/annurev.ne.13.030190.000325
21. J.A. Cervantes, S. López, S. Cervantes, A. Hernández, H. Duarte, Social robots and brain–computer interface video games for dealing with attention deficit hyperactivity disorder: a systematic review. Brain Sci. **13**(8), 1172 (2023). https://doi.org/10.3390/brainsci13081172
22. V. Khorev, S. Kurkin, A. Badarin, V. Antipov, E. Pitsik, A. Andreev, V. Grubov, O. Drapkina, A. Kiselev, A. Hramov, Review on the use of brain computer interface rehabilitation methods for treating mental and neurological conditions. J. Integr. Neurosci. **23**(7), 125 (2024). https://doi.org/10.31083/j.jin2307125
23. M. Garcia Pimenta, T. Brown, M. Arns, S. Enriquez-Geppert, Treatment efficacy and clinical effectiveness of EEG neurofeedback as a personalized and multimodal treatment in ADHD: a critical review. Neuropsychiatr. Disease Treat. **17**, 637–648 (2021). https://doi.org/10.2147/NDT.S251547
24. D. Zhao, J. Zhang, The effects of working memory training on attention deficit, adaptive and non-adaptive cognitive emotion regulation of Chinese children with attention deficit/hyperactivity disorder (ADHD). BMC Psychol. **12**(1), 59 (2024). https://doi.org/10.1186/s40359-024-01539-6
25. M.S.H. Al-Saad, B. Al-Jabri, A.F. Almarzouki, A review of working memory training in the management of attention deficit hyperactivity disorder. *Frontiers in Behavioral Neuroscience*, **15**, 686873 (2021). https://doi.org/10.3389/fnbeh.2021.686873
26. M. Guha, Diagnostic and statistical manual of mental disorders: DSM–5. Refer. Rev. **28**(3), 36–37 (2014)
27. H. Musullulu, Evaluating attention deficit and hyperactivity disorder (ADHD): a review of current methods and issues. Front. Psychol. **16**, 1466088 (2025). https://doi.org/10.3389/fpsyg.2025.1466088
28. M. Moses, J. Tiego, D. Demontis, et al., Working memory and reaction time variability mediate the relationship between polygenic risk and ADHD traits in a general population sample. Mol. Psychiatr. **27**, 5028–5037 (2022). https://doi.org/10.1038/s41380-022-01775-5
29. R. Astenvald, M.A. Frick, J. Neufeld, et al., Emotion dysregulation in ADHD and other neurodevelopmental conditions: a co-twin control study. Child Adolescent Psychiatr. Mental Health **16**, 92 (2022). https://doi.org/10.1186/s13034-022-00528-0
30. Y. Zhang-James, E.C. Helminen, J. Liu, et al., Evidence for similar structural brain anomalies in youth and adult attention-deficit/hyperactivity disorder: a machine learning analysis. Trans. Psychiatr. **11**, 82 (2021). https://doi.org/10.1038/s41398-021-01201-4
31. W.S. Choi, S.M. Wang, Y.S. Woo, W.M. Bahk, Therapeutic efficacy and safety of memantine for children and adults with ADHD with a focus on glutamate–dopamine regulation: a systematic review. J. Clin. Psychiatr. **86**(1), 24r15507 (2024). https://doi.org/10.4088/JCP.24r15507

32. Y. Long, N. Pan, S. Ji, et al., Distinct brain structural abnormalities in attention-deficit/hyperactivity disorder and substance use disorders: a comparative meta-analysis. Transl. Psychiatr. **12**, 368 (2022). https://doi.org/10.1038/s41398-022-02130-6
33. C. Ramos-Galarza, D. Brito, B. Rodríguez, B. Guerrero, J. Cruz-Cárdenas, M. Bolaños-Pasquel, Systematic review of executive function stimulation methods in the ADHD population. J. Clin. Med. **13**(14), 4208 (2024). https://doi.org/10.3390/jcm13144208
34. L. Chu, P. Zhu, C. Ma, L. Pan, L. Shen, D. Wu, Y. Wang, G. Yu, Effects of combining group executive functioning and online parent training on school-aged children with ADHD: a randomized controlled trial. Front. Pediatr. **9**, 813305 (2022). https://doi.org/10.3389/fped.2021.813305
35. D. Landinez Martinez, C. Quintero López, V. Gil Vera, Working memory training in children with attention deficit hyperactivity disorder: a systematic review. Rev. Psicol. Clín. Niños Adolescent. **9**, 2022–2023 (2022). https://doi.org/10.21134/rpcna.2022.09.3.7
36. C.G. Lim, C.P. Soh, S.S.Y. Lim, et al., Home-based brain–computer interface attention training program for attention deficit hyperactivity disorder: a feasibility trial. Child Adolescent Psychiatr. Mental Health **17**, 15 (2023). https://doi.org/10.1186/s13034-022-00539-x
37. A. Drigas, A. Sideraki, Brain neuroplasticity leveraging virtual reality and brain–computer interface technologies. Sensors **24**(17), 5725 (2024). https://doi.org/10.3390/s24175725
38. M. Orban, M. Elsamanty, K. Guo, S. Zhang, H. Yang, A review of brain activity and EEG-based brain–computer interfaces for rehabilitation application. Bioengineering **9**(12), 768 (2022). https://doi.org/10.3390/bioengineering9120768
39. B. Abibullaev, A. Keutayeva, A. Zollanvari, Deep learning in EEG-based BCIs: a comprehensive review of transformer models, advantages, challenges, and applications. IEEE Access **11**, 127271–127301 (2023). https://doi.org/10.1109/ACCESS.2023.3329678
40. F. Caiado, A. Ukolov, The history, current state and future possibilities of the non-invasive brain computer interfaces. Med. Novel Technol. Dev. **25**, 100353 (2025). https://doi.org/10.1016/j.medntd.2025.100353
41. B. Abibullaev, J. An, Decision support algorithm for diagnosis of ADHD using electroencephalograms. J. Med. Syst. **36**, 2675–2688 (2012). https://doi.org/10.1007/s10916-011-9742-x
42. A. Gouret, S. Le Bars, T. Porssut, F. Waszak, S. Chokron, Advancements in brain–computer interfaces for the rehabilitation of unilateral spatial neglect: a concise review. Front. Neurosci. **18**, 1373377 (2024). https://doi.org/10.3389/fnins.2024.1373377
43. V. Martínez-Cagigal, J. Thielen, E. Santamaría-Vázquez, S. Pérez-Velasco, P. Desain, R. Hornero, Brain–computer interfaces based on code-modulated visual evoked potentials (c-VEP): a literature review. J. Neural Eng. **18**(6), 061002 (2021). https://doi.org/10.1088/1741-2552/ac38cf
44. A. Maye, D. Zhang, A.K. Engel, Utilizing retinotopic mapping for a multi-target SSVEP BCI with a single flicker frequency. IEEE Trans. Neural Syst. Rehab. Eng. **25**(7), 1026–1036 (2017). https://doi.org/10.1109/TNSRE.2017.2666479
45. M. Abu-Alqumsan, A. Peer, Advancing the detection of steady-state visual evoked potentials in brain–computer interfaces. J. Neural Eng. **13**(3), 036005 (2016). https://doi.org/10.1088/1741-2560/13/3/036005
46. B. Guragain, A. Haider, R. Fazel-Rezai, Hybrid brain–computer interface systems: approaches, features, and trends. InTech (2018). https://doi.org/10.5772/intechopen.75132
47. R. Singla, SSVEP-based BCIs. InTech (2018). https://doi.org/10.5772/intechopen.75693
48. T. Nayak, L.W. Ko, T.P. Jung, Y. Huang, Target classification in a novel SSVEP-RSVP based BCI gaming system. In: *Proceedings of the 2019 IEEE International Conference on Systems, Man and Cybernetics (SMC) Bari, Italy* (2019), pp. 4194–4198. https://doi.org/10.1109/SMC.2019.8914174
49. G.B. Swati, S. Sushmitha, K.S. Srinidhi Bharadwaj, K.B. Niranjana, Steady state visually evoked potential based brain computer interface for game control, in *2018 International Conference on Electrical, Electronics, Communication, Computer, and Optimization Techniques (ICEECCOT), Mysuru, India* (2018), pp. 422–426. https://doi.org/10.1109/ICEECCOT43722.2018.9001590

50. E. Quiles, J. Dadone, N. Chio, E. García, Cross-platform implementation of an SSVEP-based BCI for the control of a 6-DOF robotic arm. Sensors **22**(13), 5000 (2022). https://doi.org/10.3390/s22135000
51. N. Chumerin, N.V. Manyakov, M. van Vliet, A. Robben, A. Combaz, M.M. Van Hulle, Steady-state visual evoked potential-based computer gaming on a consumer-grade EEG device. IEEE Trans. Comput. Intell. AI Games **5**(2), 100–110 (2013). https://doi.org/10.1109/TCIAIG.2012.2225623
52. Y. Liu, X. Li, C. Yang, Design of a control platform for mobile robot with SSVEP–BCI system, in *Proceedings of the 2019 IEEE International Conference on Cybernetics and Intelligent Systems (CIS) and IEEE Conference on Robotics, Automation and Mechatronics (RAM), Bangkok, Thailand* (2019), pp. 198–203. https://doi.org/10.1109/CIS-RAM47153.2019.9095824
53. T. Sakurada, T. Kawase, T. Komatsu, K. Kansaku, Use of high-frequency visual stimuli above the critical flicker frequency in a SSVEP-based BMI. Clin. Neurophysiol. **126**(10), 1972–1978 (2015). https://doi.org/10.1016/j.clinph.2014.12.010
54. U. Chaudhary, N. Birbaumer, A. Ramos-Murguialday, Brain–computer interfaces for communication and rehabilitation. Nat. Rev. Neurol. **12**, 513–525 (2016). https://doi.org/10.1038/nrneurol.2016.113
55. A. Ali, S. Puthusserypady, A 3D learning playground for potential attention training in ADHD: a brain–computer interface approach, in *Proceedings of the Annual International Conference of the IEEE Engineering in Medicine and Biology Society* (2015), pp. 67–70. https://doi.org/10.1109/EMBC.2015.7318302
56. A. Khaleghi, H. Zarafshan, M.R. Mohammadi, Visual and auditory steady-state responses in attention-deficit/hyperactivity disorder. Eur. Arch. Psychiatr. Clin. Neurosci. **269**, 645–655 (2019). https://doi.org/10.1007/s00406-018-0902-6
57. R. Parafita, G. Pires, U. Nunes, M. Castelo-Branco, A spacecraft game controlled with a brain–computer interface using SSVEP with phase tagging, in *Proceedings of the 2013 IEEE 2nd International Conference on Serious Games and Applications for Health (SeGAH), Vilamoura, Portugal* (2013), pp. 1–6. https://doi.org/10.1109/SeGAH.2013.6665309
58. D. Zarka, A. Leroy, A. Cebolla, C. Cevallos, E. Palmero-Soler, G. Cheron, Neural generators involved in visual cue processing in children with attention-deficit/hyperactivity disorder (ADHD). Eur. J. Neurosci. **53**, 1207–1224 (2021). https://doi.org/10.1111/ejn.15040
59. D. Liparas, S.I. Dimitriadis, N.A. Laskaris, A. Tzelepi, K. Charalambous, L. Angelis, Exploiting the temporal patterning of transient VEP signals: a statistical single-trial methodology with implications to brain–computer interfaces (BCIs). J. Neurosci. Methods **232**, 189–198 (2014). https://doi.org/10.1016/j.jneumeth.2014.04.032
60. Y. Chen, Design of a game control system based on brain–computer interface: link to a game, in *Proceedings of the 2021 2nd International Conference on Computing and Data Science (CDS), Stanford* (2021), pp. 255–258. https://doi.org/10.1109/CDS52072.2021.00051
61. C. Pillai, R. Ritch, P. Derr, A. Gonzalez, L.K. Cox, J. Siegfried, J.M. Liebmann, C. Tello, Sensitivity and specificity of short-duration transient visual evoked potentials (SD-tVEP) in discriminating normal from glaucomatous eyes. Invest. Ophthalmol. Visual Sci. **54**(4), 2847–2852 (2013). https://doi.org/10.1167/iovs.12-10097
62. G. Bin, X. Gao, Y. Wang, B. Hong, S. Gao, VEP-based brain–computer interfaces: time, frequency, and code modulations [Research Frontier]. IEEE Comput. Intell. Mag. **4**(4), 22–26 (2009). https://doi.org/10.1109/MCI.2009.934562
63. S. Kim, T. Banaschewski, R. Tannock, Color vision in attention-deficit/hyperactivity disorder: a pilot visual evoked potential study. J. Optometr. **8**(2), 116–130 (2015). https://doi.org/10.1016/j.optom.2014.10.002
64. H. Nezamfar, S.S.M. Salehi, D. Erdogmus, Stimuli with opponent colors and higher bit rate enable higher accuracy for C-VEP BCI, in *Proceedings of the 2015 IEEE Signal Processing in Medicine and Biology Symposium (SPMB), Philadelphia* (2015), pp. 1–6. https://doi.org/10.1109/SPMB.2015.7405476

65. D. Aminaka, S. Makino, T.M. Rutkowski, Chromatic and high-frequency cVEP-based BCI paradigm, in *Proceedings of the 37th Annual International Conference of the IEEE Engineering in Medicine and Biology Society (EMBC), Milan* (2015), pp. 1906–1909. https://doi.org/10.1109/EMBC.2015.7318755
66. M.M. Amini, V. Shalchyan, Designing a motion-onset visual-evoked potential-based brain–computer interface to control a computer game. IEEE Trans. Games **16**(2), 409–418 (2024). https://doi.org/10.1109/TG.2023.3279289
67. A. Vrins, M. Romani, P. Mavromoustakos-Blom, M. Alimardani, NeuroWizard: investigating the effect of gamification on BCI performance among adult users with ADHD, in *Intelligent Human Computer Interaction. IHCI 2024*, ed. by D. Singh, J.W. van 't Klooster, U.S. Tiwary. Lecture Notes in Computer Science, vol. 15558 (Springer, Cham, 2025). https://doi.org/10.1007/978-3-031-88881-6_15
68. E. Perez–Valero, M.A. Lopez–Gordo, M.A. Vaquero–Blasco, An attention-driven videogame based on steady-state motion visual evoked potentials. Exp. Syst. **38**, e12682 (2021). https://doi.org/10.1111/exsy.12682
69. P. Martinez, H. Bakardjian, A. Cichocki, Fully online multicommand brain-computer interface with visual neurofeedback using SSVEP paradigm. Comput. Intell. Neurosci. **2007**, 94561 (2007). https://doi.org/10.1155/2007/94561
70. E.C. Lalor, S.P. Kelly, C. Finucane, et al., Steady-state VEP-based brain-computer interface control in an immersive 3D gaming environment. EURASIP J. Adv. Signal Process. **2005**, 706906 (2005). https://doi.org/10.1155/ASP.2005.3156
71. E. Filiz, R.B. Arslan, Design and implementation of steady state visual evoked potential based brain computer interface video game, in *Proceedings of the 2020 IEEE 20th Mediterranean Electrotechnical Conference (MELECON), Palermo, Italy*, pp. 335–338 (2020). https://doi.org/10.1109/MELECON48756.2020.9140710
72. R. Beveridge, S. Wilson, M. Callaghan, D. Coyle, Neurogaming with motion-onset visual evoked potentials (mVEPs): adults versus teenagers. IEEE Trans. Neural Syst. Rehab. Eng. **27**(4), 572–581 (2019). https://doi.org/10.1109/TNSRE.2019.2904260
73. P. Arpaia, S. Criscuolo, E.D. Benedetto, N. Donato, L. Duraccio, Evaluation of the effectiveness of a wearable, AR-based BCI for robot control in ADHD treatment, in *Proceedings of the 2022 IEEE International Conference on Metrology for Extended Reality, Artificial Intelligence and Neural Engineering (MetroXRAİNE), Rome, Italy* (2022), pp. 630–634. https://doi.org/10.1109/MetroXRAINE54828.2022.9967655
74. P.L. Lee, H.T. Shu, H.C. Chang, Design a brain computer interface gaming system using steady-state visual evoked potential, in *Proceedings of the 2014 IEEE International Conference on Consumer Electronics – Taiwan* (2014), pp. 5–6. https://doi.org/10.1109/ICCE-TW.2014.6904095
75. M. van Vliet, A. Robben, N. Chumerin, N.V. Manyakov, A. Combaz, M.M. Van Hulle, Designing a brain-computer interface controlled video-game using consumer grade EEG hardware, in *Proceedings of the 2012 ISSNIP Biosignals and Biorobotics Conference: Biosignals and Robotics for Better and Safer Living (BRC), Manaus, Brazil* (2012), pp. 1–6. https://doi.org/10.1109/BRC.2012.6222186
76. World Medical Association, World Medical Association Declaration of Helsinki: ethical principles for medical research involving human subjects. J. Am. Med. Assoc. **310**(20), 2191–2194 (2013). https://doi.org/10.1001/jama.2013.281053
77. International Council for Harmonisation, Integrated addendum to ICH E6 (R1): guideline for good clinical practice E6 (R2). Current Step 2, (2015), pp. 1–60
78. Council for International Organizations of Medical Sciences, *International Ethical Guidelines For Health-Related Research Involving Humans* (CIOMS & WHO, Geneva, 2016)
79. V.M. Lema, M. Mbondo, E.N. Kamau, Informed consent for clinical trials: a review. East Afr. Med. J. **86**(3), 133–142 (2009)
80. M. Krzysztofek, *GDPR: General Data Protection Regulation (EU) 2016/679: Post-Reform Personal Data Protection in the European Union* (Kluwer Law International, The Hague, 2018)

81. J.V. Odom, M. Bach, M. Brigell, G.E. Holder, D.L. McCulloch, A. Mizota, International Society for Clinical Electrophysiology of Vision, et al., ISCEV standard for clinical visual evoked potentials: 2016 update. Document. Ophthalmol. **133**, 1–9 (2016). https://doi.org/10.1007/s10633-016-9553-y
82. G.F. Harding, P.M. Jeavons, Photosensitive epilepsy. (Cambridge University Press, Cambridge, 1994)
83. International Organization for Standardization, *ISO/IEC 27001:2013: Information Technology—Security Techniques—Information Security Management Systems—Requirements* (ISO, Geneva, 2013)
84. J. Daemen, V. Rijmen, Specification for the advanced encryption standard (AES). FIPS PUB 197 (2001)
85. L. Sweeney, k-anonymity: a model for protecting privacy. Int. J. Uncertain. Fuzz. Knowl. Based Syst. **10**(5), 557–570 (2002). https://doi.org/10.1142/S0218488502001648
86. S.A. Larsen, L. Klok, W. Lehn-Schiøler, R. Gatej, S. Beniczky, Low-cost portable EEG device for bridging the diagnostic gap in resource-limited areas. Epilept. Disorders **26**(5), 694–700 (2024). https://doi.org/10.1684/epd.2024.1345
87. J.Y. Ng, K.D. Cobey, S. Ahmed, V. Chow, S.G. Maduranayagam, L.J. Santoro, et al., Recommendations and guidelines for creating scholarly biomedical journals: a scoping review. PLOS ONE **18**(3), e0282168 (2023). https://doi.org/10.1371/journal.pone.0282168
88. K.F. Schulz, D.G. Altman, D. Moher, CONSORT 2010 statement: updated guidelines for reporting parallel group randomised trials. J. Pharmacol. Pharmacother. **1**(2), 100–107 (2010). https://doi.org/10.4103/0976-500X.72352
89. T. Ros, S. Enriquez-Geppert, V. Zotev, K.D. Young, G. Wood, S. Whitfield-Gabrieli, et al., Consensus on the reporting and experimental design of clinical and cognitive-behavioural neurofeedback studies (CRED-nf checklist). Brain **143**(6), 1674–1685 (2020). https://doi.org/10.1093/brain/awaa057

Chapter 8
Mind Over Machine: Using Action Observation to Reduce Mental Fatigue in Motor Imagery

Prarthana Saikia, Supriya Bordoloi, and Shyamanta M. Hazarika

Abstract Motor imagery (MI) based mental training has been proven to enhance motor skill learning in both healthy individuals and those with post-stroke hemiparesis. Mentally simulating movements without physically performing them, MI activates the same neural mechanisms that are activated during actual motor performance. The combination of MI and physical training has been shown to improve rehabilitation outcomes, highlighting its potential as a complementary strategy for motor recovery. However, after a prolonged MI, the person feels mentally fatigued affecting cognitive abilities. The efficacy of MI is impaired by mental fatigue (MF), making it challenging to effectively use it for neurorehabilitation. In this study, we investigated the hypothesis that combining action observation (AO) with MI can counteract MF. AO and MI use overlapping brain pathways for motor execution. Combining AO and MI (AO+MI) involves simultaneously observing an action and imagine performing it, enhancing motor representation and engagement. Our research focused on the effects of MF during pure MI and the combination of AO+MI. MF is characterized by an increase in parietal alpha power and a decrease in the N1 component of Event-Related Potential (ERP). Welch's averaged modified

The authors Prarthana Saikia and Supriya Bordoloi contributed equally to this work.

P. Saikia
Biomimetic Robotics and AI Lab, Department of Mechanical Engineering, Indian Institute of Technology Guwahati, Guwahati, India

S. Bordoloi (✉)
Centre for Linguistic Science and Technology, Indian Institute of Technology Guwahati, Guwahati, India
e-mail: bordoloi@iitg.ac.in

S. M. Hazarika
Biomimetic Robotics and AI Lab, Department of Mechanical Engineering, Indian Institute of Technology Guwahati, Guwahati, India

Centre for Linguistic Science and Technology, Indian Institute of Technology Guwahati, Guwahati, India
e-mail: s.m.hazarika@iitg.ac.in

R. Vinjamuri (ed.), *Bridging the Gap between Mind and Machine*,
https://doi.org/10.1007/978-3-032-06713-5_8

periodogram method was used to estimate the power spectral density for all trials. Our results indicated an increase in the average power from the first to the last run of the experiment. Changes in the amplitude of the N1 ERP component and alpha-theta band power were observed in the topographic maps. Low-resolution electromagnetic tomography analysis was used to determine the maximum current density distribution. We found that AO+MI had the highest mean current density for N1 ERP compared to pure MI. In addition to the electrophysiological markers mentioned above, subjective assessment with the visual analogue scale and chalder fatigue scale were also used to assess mental fatigue. Our results showed that AO+ MI can maintain a better level of performance even in the presence of MF.

Keywords Mental fatigue · Action observation · Motor imagery · Attention · Motor rehabilitation

8.1 Introduction

Motor imagery (MI) refers to the mental simulation of movement without any overt physical execution [1]. MI helps people with hemiparesis (after a stroke) improve their motor skills. Such MI training aids in regeneration and allocating a certain brain region to the movement of an arm or leg. When we imagine a movement, our brain generates the same electrical impulses that occur when we actually move a motor limb [2]. Brain-Computer Interface (BCI) uses such electrical measurements of brain activity to translate the user's goals into control signals. Electroencephalography (EEG) combined with such MI-based paradigms provides a simple and non-invasive solution for the BCI system [3]. MI can repeatedly stimulate the injured motor cortex to partially restore patients' motor functions. This is done by reactivating the motor neurons around the injured cells. When motor actions are imagined, the brain motor network gets activated [4]. In addition to cortical regions such as the premotor cortex, supplementary motor area, and primary motor cortex, subcortical regions including cerebellum [5] and the basal ganglia are also engaged [6] during MI. This mental activity in the brain can serve as a control signal for the communication and motor control of prosthetic devices [7]. Thus, MI has attracted considerable interest as a neurorehabilitation technique to improve motor recovery after stroke.

8.1.1 Mental Fatigue and Its Impact on Performance

Prolonged engagement in MI can lead to the onset of mental fatigue (MF), a state characterized by tiredness associated with a decrease in energy, activity levels, and cognitive abilities [8, 9]. The effectiveness of MI-based BCI in neurorehabilitation is hindered by mental states such as attention levels, fatigue, and arousal, which have a substantial impact on the signal quality and reliability of these devices [10]. For MI-based BCI, it is important to investigate the relationships between MF and MI. Self-report ratings are the simplest and most practical method for analyzing and estimating mental exhaustion [11]. Numerous scales and questionnaires have been developed to quantify MF. Studies have shown that increased power in frontal theta (θ) and parietal alpha (α) of EEG rhythms are related to mental exhaustion [12]. Furthermore, increases in spectral power across delta, theta, alpha, and beta frequency bands have been observed with increased levels MF [10]. For instance, EEG spectral parameters to scale MF across various scalp regions were analysed during simulated driving tasks [13]. In the context of MI, the relationship between MI and MI-induced MF using EEG was examined [8]. Five areas of the scalp, viz. frontal, parietal, temporal, central, and occipital, are considered for fatigue analysis. Their findings indicated that prolonged MI can modulate regional brain activity associated with fatigue. Motor commands necessary to carry out actual or imagined movements may be impaired by a prolonged MI. A strong correlation between MF and movement-related cortical potential (MRCP) amplitude was also observed during MI [14]. The effects of prolonged mental fatigue on behavioral and electrophysiological systems were investigated [15]. In addition to EEG rhythms, event-related potential (ERP) is also altered by fatigue. Individuals with MF often report having difficulty maintaining attention to the task at hand. The relationship between MF and attention-related ERP alterations was investigated [16]. A decrease in N1 amplitude over time was reported indicating a decline in attentional processing [17]. In this context, the N1 ERP component serves as a critical neural marker in our study as we examine the effects of MI-induced fatigue on attention during task performance.

8.1.2 Role of Action Observation in Enhancing Motor Imagery

Action observation (AO) is a rehabilitation technique based on the mirror neuron system, which becomes active not only during the execution of movements but also while observing others perform the same action [18]. This neural mirroring mechanism involves key brain regions such as the inferior parietal lobule, premotor cortex, and superior frontal gyrus, collectively known as the mirror neuron system (MNS) [19, 20]. AO engages these regions to internally simulate the observed action, thereby facilitating motor understanding and planning [4]. In clinical contexts, AO enables patients to engage motor-related neural circuits without requiring overt physical movement, making it particularly beneficial during early stages of rehabilitation [21, 22]. Functional neuroimaging studies have demonstrated that AO enhances cortical excitability and induces neuroplastic changes in motor regions [23]. AO-based therapy significantly improves upper limb function in stroke patients, showing its therapeutic relevance [24]. Importantly, AO helps maintain attention and engagement during rehabilitation, which is essential for better outcomes [25]. This suggests that AO can reduce the MF linked to prolonged MI. These cognitive benefits of AO provide a foundation for exploring its combination with MI to further mitigate MF and optimize rehabilitation outcomes.

8.1.3 Combining MI and AO to Combat Mental Fatigue

Building on these cognitive advantages, combining AO with MI presents a promising strategy to address the limitations posed by MF during prolonged MI sessions. The combination of AO and MI, often termed AO+MI, has gained increasing attention in recent years as a cognitively efficient paradigm for motor rehabilitation. AO+MI leverages the overlapping neural circuits engaged by both observation and imagination of movement. Emerging evidence suggests that AO+MI enhances motor performance and recovery, improves attentional engagement and reduces cognitive load compared to MI alone [26]. Studies have shown that AO can preserve motivation and attentional resources throughout extensive mental activity, reducing MF. Visual direction provided by AO serves as a scaffold, facilitating mental simulation and making it easier to perform MI tasks [27–29]. This decreased cognitive load may assist to delay the onset of MF. AO+MI elicited higher and more sustained activation in motor-related cortical areas, suggesting enhanced engagement and potentially lower subjective fatigue during task execution [30]. In the context of attention, the dynamic and externally guided nature of AO is believed to help maintain focus during repetitive imagery tasks. This is especially relevant for BCI applications, where sustained attention is critical to maintaining signal quality and classification accuracy. It was also shown that combining AO with MI led to greater engagement of attentional networks, including the dorsolateral prefrontal cortex, than MI alone, which prevents attentional drift [31]. From a rehabilitation perspective, the AO+MI approach has shown particular promise in stroke recovery and neurorehabilitation. A randomized controlled trial revealed that stroke patients undergoing AO+MI training showed significantly greater improvements in upper limb function and movement kinematics than those performing MI alone [32]. Moreover, AO+MI has been found to increase sensorimotor excitability, even in patients with limited voluntary movement, making it suitable for early-stage rehabilitation [21, 33]. These findings demonstrate the cognitively durable and neurophysiologically robust nature of AO+MI for improving motor learning, MF, and attention, particularly in long-term rehabilitation and BCI training.

Building on the cognitive and neural advantages of AO integrated with MI, the present study aims to examine its potential in counteracting MF induced by prolonged MI. We conceptualize AO as a visual priming mechanism that supports sustained mental engagement by reducing the internal cognitive load associated with MI. Specifically, we hypothesize that AO+MI facilitates a more stable balance of corticomotor excitability compared to MI alone, thereby mitigating fatigue-related neural and cognitive decline. To test this hypothesis, we focused on neurophysiological indicators of MF, particularly the mean spectral power in the theta and alpha EEG bands and the amplitude of the N1 ERP component, which are sensitive to both attentional and fatigue-related changes.

Our study was designed to track the temporal evolution of fatigue by comparing changes in these markers in repeated task runs. We investigated the neural correlates of AO+MI versus pure MI in common motor tasks, aiming to determine whether AO could attenuate the gradual changes typically associated with MF. Topographic EEG maps were also analyzed to visualize any changes in ERP and rhythm activity. Additionally, the maximal current density of N1 ERP responses was explored and investigated using Low-resolution electromagnetic tomography (LORETA) source localization.

8.2 Materials and Methods

8.2.1 Participant Selection

Nine healthy volunteers (aged 25–35 years) with normal or corrected-to-normal vision participated in the study. The experimental objectives were unknown to all participants, and they were not informed that prolonged MI task was designed to induce MF. Prior to participation, written informed consent was obtained from each individual. The study protocol was reviewed and approved by the Institute Human Ethics Committee at the Indian Institute of Technology Guwahati. In addition, participants were required to complete a brief demographic questionnaire asking for their name, gender, age, occupation, and other relevant details.

8.2.2 Experimental Design and Procedure

Each participant had the opportunity to recline comfortably in a chair and place his or her hands on the lap. A short break was taken before the experiment, and then they were prepared for data collection. Participants completed two pre and post-experiment self-report questionnaires: the visual analogue scale for fatigue (VAS-F) [34] and the chalder fatigue scale (CFS) [35]. These questionnaires have been shown to be a viable and reliable tool for quantitative assessment of fatigue in both healthy individuals and patients [36].

The experiment was conducted using E-Prime software [37], with visual stimuli presented on an LCD screen positioned approximately 80 cm from the participant. Each stimulus consisted of a video clip showing an actor performing a specific motor action, i.e. flexion and extension of hand. The experimental session included two task conditions: pure MI and AO+MI.

In the pure MI task, subjects were asked to imagine the action for 6 seconds without requiring motor execution (as shown in the video). A fixation cross and a

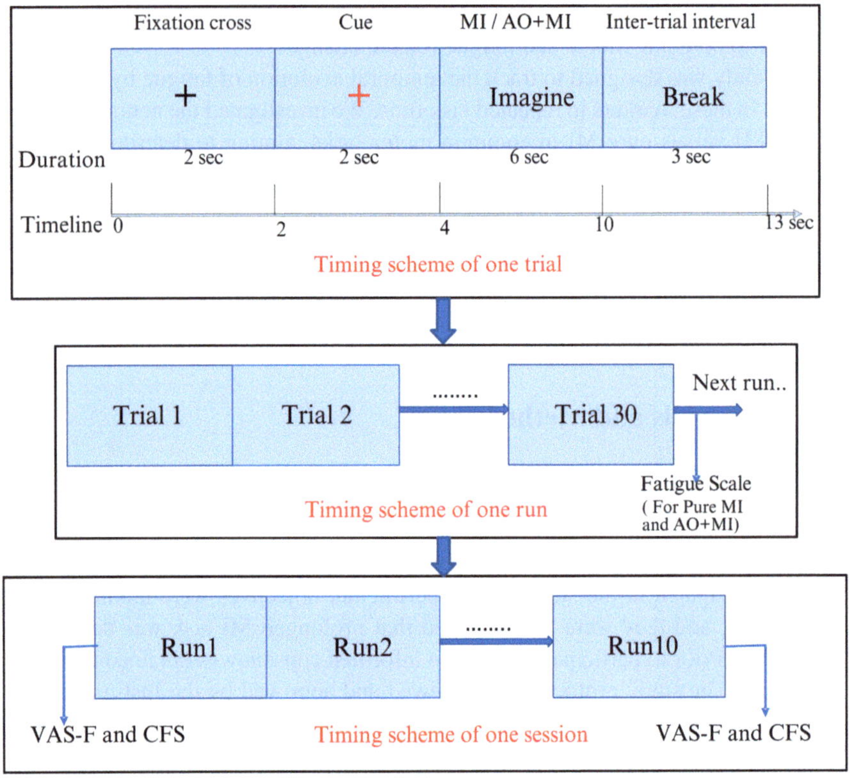

Fig. 8.1 Experimental paradigm and timing details. Each run incorporated 30 trials, each spanning 13 seconds, and a complete session consisted of 10 runs

brief buzzer sound signaled the start of each trial and instructed the participants to be ready for the activity. They mimicked an imaginary version of this event by viewing it at the same speed as the instruction video.

In the combined AO+MI task, a video showing an action was shown for 6 seconds, during which participants were asked to observe and imagine the same action (as shown in the video) at their own preferred pace for 6 seconds. Participants were asked to watch the video and imagine the action from a first-person perspective. The experimental paradigm, illustrated in Fig. 8.1, consists of multiple time blocks. Each block consists of a pure MI task, an AO+MI task, and rest periods.

A 3000 ms pause was given after each trial. A half-minute pause was inserted between the two tasks, MI and AO+MI, in each run, as shown in Fig. 8.2. After each task, subjects were asked a question about the degree of fatigue, motivation, and attention they felt at that time while performing the task. Each condition comprised 10 runs, with each run lasting approximately 13 minutes and containing

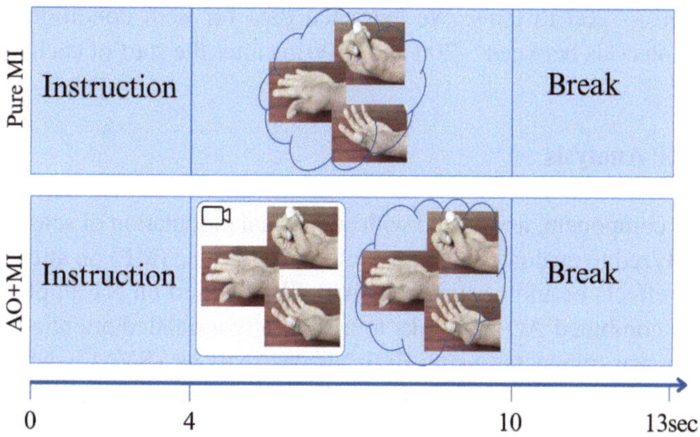

Fig. 8.2 Schematic representation of the experimental conditions. The paradigm includes two task blocks: pure MI and AO+MI

30 randomized trials. A 2-minute rest interval was provided between runs to reduce the potential carry-over effects of fatigue.

8.2.3 EEG and Electrophysiological Measurements

Using the international 10–20 system, the EEG signal was recorded with a V-Amp 16 channel recorder (Brain Products, Germany) [38] from 16 Ag/AgCl surface electrodes (Fp1, Fp2, F3, Fz, F4, C3, Cz, C4, P3, Pz, P4, O1, O2). Throughout the experiment, the electrode impedance was kept below 10 kΩ. The Brain Vision Recorder [38] was used to sample raw signals at a frequency of 500 Hz. The vertical electrooculogram artefact was monitored with an electrode placed under the right eye. The FCz and AFz electrodes were used as reference and ground. Two additional electrodes implanted on TP9 and TP10 were recorded as an offline reference.

To investigate the effects of extended MI on MF, both temporal and frequency domain EEG characteristics were examined in this study. BrainVision Analyzer (BVA) [38] and MATLAB R2023b [39] were used to preprocess the data. Offline data were notch filtered at 50 Hz and bandpass filtered in the passband frequency range from 0.1 to 40 Hz using a Butterworth filter with zero phase shift. The data were then downsampled to 256 Hz and re-referenced to the average of TP9 and TP10 (approximating the mastoid electrodes). Following filtering, raw data were visually inspected, and artefacts were corrected using Independent Component Analysis (ICA) implemented in BVA. Components associated with artefacts were identified and removed based on visual inspection. The continuous data were then segmented into epochs of 6300 ms, starting 200 ms before stimulus onset and extending to 6100 ms post-stimulus. A baseline correction was applied using the pre-stimulus interval from − 200 to 0 ms. We extracted data for both conditions (pure MI, AO+MI) in intervals between −200 and 6100 ms after the start of each trial.

8.2.3.1 ERP Analysis

The N1 ERP component, associated with attentional modulation of sensory processing, was analyzed to evaluate the impact of mental fatigue (MF) on attention and its downstream effects on motor system activity. We focused on N1 amplitude during pure MI and combined AO+MI tasks to assess fatigue-related attentional changes. ERP analysis was conducted using BrainVision Analyzer (BVA), where EEG data were segmented into epochs and averaged for each experimental condition. The N1 component was extracted for all brain regions, and amplitudes were calculated individually for each participant across different time intervals and task conditions.

8.2.3.2 Power Spectral Density

In addition to the ERP analysis, we performed a spectral analysis of the data. Power spectral density for all epochs was estimated using the averaged modified periodogram method of Welch [40] with a segment length of 6.3 seconds and a Hanning window. The power was then calculated for each electrode corresponding to each subject for two different frequency bands, i.e., theta (4–8 Hz) and alpha (8–12 Hz). In conjunction with the extraction of the mean EEG frequency band power, several regions of interest (ROIs) were used to investigate variations of brain activation on the scalp.

8.2.3.3 LORETA Analysis

Source localization techniques were used to estimate the location and dynamics of the underlying neural generators that produce the electrical potentials measured on the scalp [41]. LORETA is one of the source analysis methods used in our study (Pascual-Marqui et al. [42]), performed by the BVA software. LORETA maps the brain activity localized in 3D cortical regions from EEG data recorded on the scalp [43, 44].

8.2.4 Subjective Assessments of Mental Fatigue

To evaluate the subjective perception of MF induced during the MI experiment, two validated fatigue rating scales were employed: VAS-F and CFS. Both scales were administered to the participants before and after the task to assess cognitive fatigue and its associated symptoms, such as drowsiness, reduced alertness, and difficulty in concentration.

The VAS-F questionnaire, comprising 18 items, was used to assess perceived fatigue, concentration difficulty, and alertness levels. In this study, 13 relevant items (1, 2, 3, 4, 5, 11, 12, 13, 14, 15, 16, 17, 18) were selected for our analysis. Participants rated their subjective fatigue symptoms on a continuous scale at two time points: pre-task and post-task. The CFS, consisting of 6 selected items (2, 3, 5, 7, 8, 10), further confirmed the increase in subjective fatigue levels post-task. Each condition in the experiment comprised 10 runs. Following each run, participants were asked to rate their perceived fatigue using a continuous five-point fatigue scale, where 1 represented "no fatigue" and 5 indicated "extreme fatigue". Participants marked the point that best reflected their current level of fatigue, and this score was recorded as the fatigue rating for that run.

8.2.5 Comparative Assessment: Pure MI and AO+MI

To evaluate the differential effects of MI and combined AO+MI, intra-task subjective ratings were collected for four key dimensions: MI quality, fatigue, motivation, and attention. Each participant completed 10 runs under both conditions and mean scores per scale were computed and subjected to paired statistical testing. Prior to analysis, the Shapiro-Wilk test confirmed the normal distribution of the difference scores for all scales ($p > 0.05$) [45]. Therefore, paired t-tests were employed to compare the mean ratings between the two conditions.

8.3 Results

8.3.1 Event Related Potential

We calculated the grand average for each electrode. Then, the peak ERP amplitude of the anterior N1 component at frontal electrodes Fp1, Fp2, F3, Fz, F4 was calculated in the range of 100–180 ms after stimulus onset. The amplitudes of the posterior N1 components were calculated at parietal electrodes P3, P4, and Pz in the time range of 100–200 ms. Since the N1 time course on the scalp varies from anterior to posterior, the posterior electrodes have a longer time interval [46]. It can be seen from Fig.8.3a that the amplitude of N1 ERP component decreases as the time spent on a task increases. From Fig. 8.3b, it is observed that in the last run of the experiment, AO+MI maintains a higher N1 ERP amplitude than MI. This explains that higher levels of the attentional N1 ERP component can be achieved in AO+MI when fatigue is present.

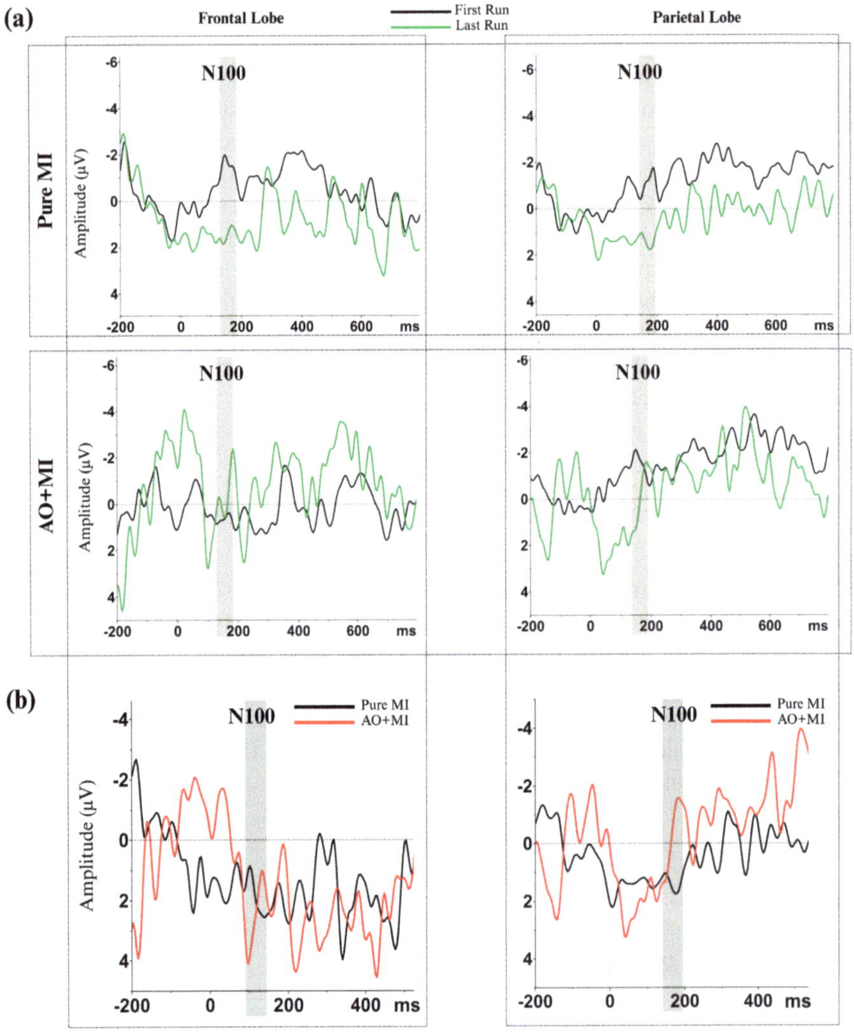

Fig. 8.3 (**a**) ERPs averaged over electrode sites, displaying N1 peak for pure MI and AO+MI; (**b**) Area activity for pure MI and AO+MI during the last experimental run

8.3.2 EEG Spectral Analysis

For spectral analysis, the left frontal lobe, the median frontal lobe, the right frontal lobe, the left posterior lobe, the median posterior lobe, and the right posterior lobe are usually considered. The frontal and parietal regions of the brain were used to calculate alpha band power. Fatigue affects band power values corresponding to both MI and AO+MI as shown by the changes in band power in Fig. 8.4a.

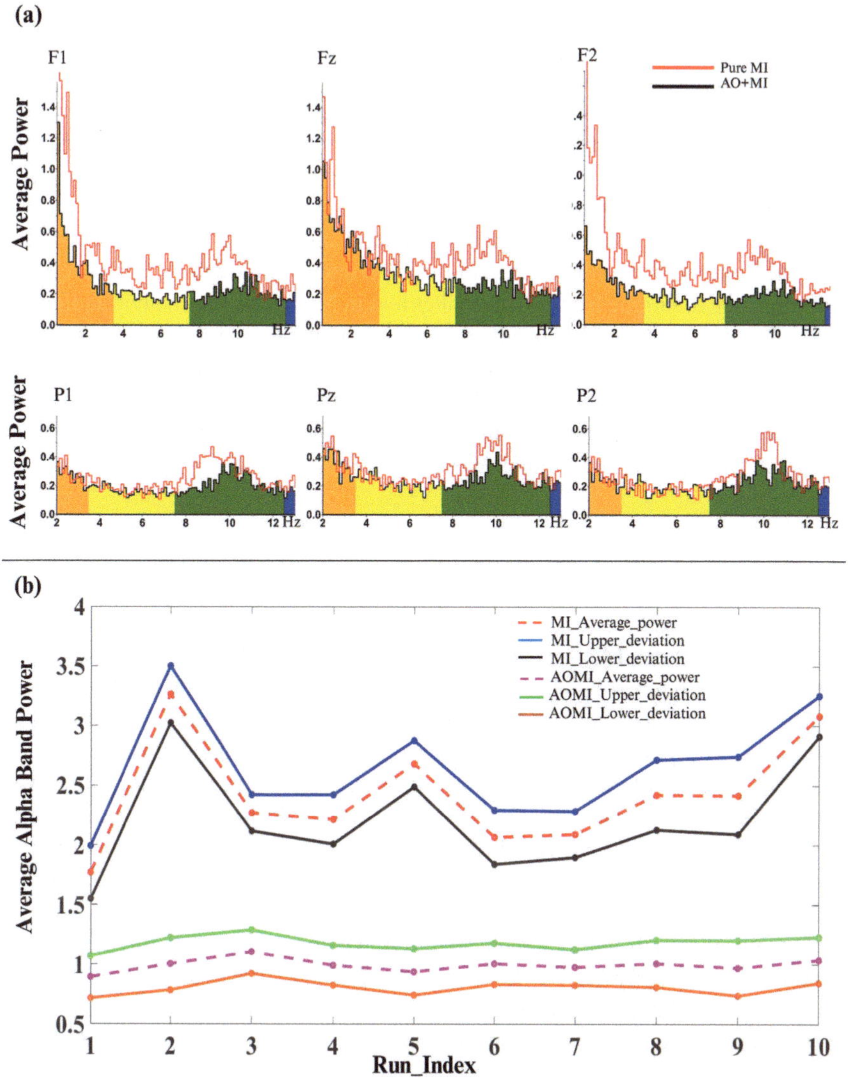

Fig. 8.4 (**a**) Mental fatigue affects average power. (**b**) Change in alpha band power across runs showing an increasing trend in pure MI, indicating rising mental fatigue, while AO+MI remains relatively stable. The solid line on both sides of the average spectrum indicates the standard deviation

It shows the area activity with respect to frequency in the frontal theta band and the parietal alpha band for the two tasks MI and AO+MI corresponding to the last experimental run. The figure shows that MI (plotted with red color) has a higher area activity than AO+MI (plotted with black color).

The average power of alpha bands was calculated and plotted with respect to ten runs. For this purpose, we averaged all power values for the respective frequency band over all trials in one run. The majority of EEG channels showed a shift in band power for both MI and AO+MI activities. Figure 8.4b displays the significant increase in alpha (8–12 Hz) band power from the first to the last run for both MI and AO+MI. Additionally, it is found that the absolute change in band power for MI is greater than that for AO+MI from the first to last run.

Figure 8.5a shows the topographical map of changes in N1 ERP for pure MI and AO+MI from the first to the last run of the experiment. Although N1 ERP decreases with time due to fatigue from MI, a smaller loss in N1 can be obtained by adding AO to MI. The power of the EEG bands varies throughout the session after each run. The incremental variations in theta band power and alpha band power during a prolonged MI condition are shown in Fig. 8.5b. The topographic plots show that AO+MI have a smaller relative incremental change in power in the last run compared to the first run when we compare the MI and AO+MI conditions.

8.3.3 Source Analysis

In our study, LORETA is mainly used to localize the N1 ERP component. It determines the brain areas that were active during MI and AO+MI tasks. All data were analzsed to determine the 3D intensity of participants' brain activity during both tasks. The grand average of the last run of the experiment was used for the LORETA analysis of the N1 component. A time frame between 90 and 150 ms was chosen. The transient view of LORETA in Fig. 8.6 shows the source activation map for the selected time period. The current densities for the ROI from the frontal region corresponding to the highest source activity are 0.00264 $\mu A/mm^2$ for the pure MI and 0.00315 $\mu A/mm^2$ for the AO+MI.

The source signals within the selected ROI, which correlate with the single voxel with the highest activation, are shown in Fig. 8.7. Thus, from both LORETA plots, it can be concluded that the concurrent AO +MI has a greater effect on motor priming in the frontal region of the brain despite the fatigue caused by the sustained MI.

8.3.4 Subjective Assessment of Mental Fatigue

To assess the MF associated with the experimental conditions, we utilized two validated self-report tools: VAS-F and CFS. Participants completed both questionnaires before and after the experimental runs, allowing quantification of subjective fatigue levels induced by the tasks.

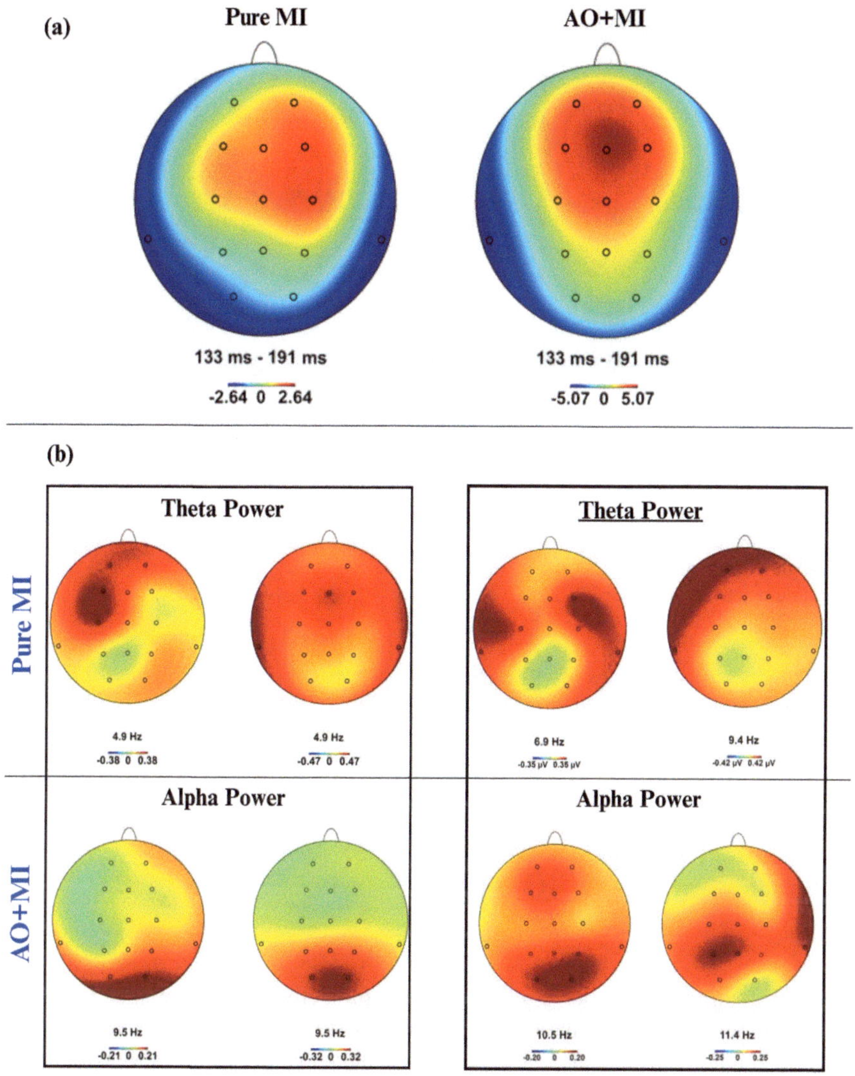

Fig. 8.5 Topographic plots of (**a**) N1 ERP changes for pure MI and AO+MI, (**b**) Changes in frontal theta and the parietal alpha band for MI and AO+MI

8.3.4.1 VAS-F Results

The obtained VAS-F scores reveal that all subjects experienced drowsiness, fatigue, difficulty in concentrating and difficulty in keeping eyes open after the completion of the experiment. Mann-Whitney U-test was conducted on the subjective scores obtained before and after the experiment of each of the 13 items used in the study. The results are shown in Table 8.1. The first column shows the question number,

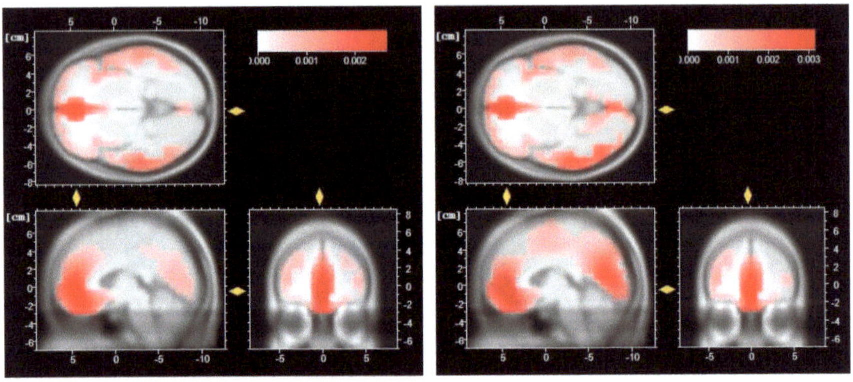

(a)
Current density (magnitude) : 0.00264 μA/mm^2

(b)
Current density (magnitude) : 0.00315 μA/mm^2

Fig. 8.6 LORETA source analysis of N1 component during (**a**) pure MI, (**b**) AO+MI

Fig. 8.7 Source-Time series of single voxel with maximum activation in the frontal lobe

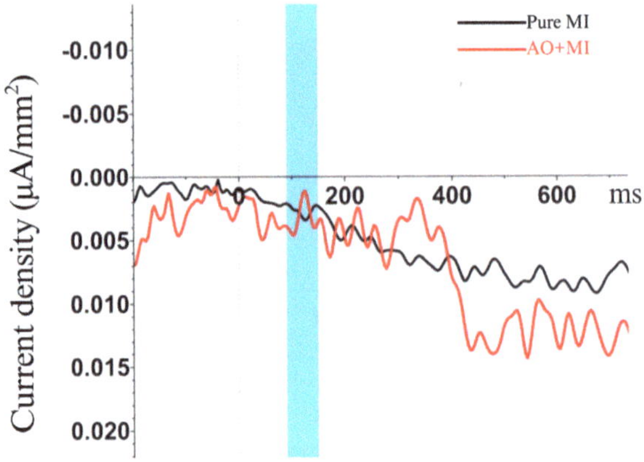

second column shows the p-values on the subjective scores on that particular question before and after the experiment and the third column shows whether it is significant or not. ✓ indicates statistically significant while × indicates insignificant different.

8.3.4.2 CFS Results

The obtained CFS scores reveal that all subjects experienced drowsiness, fatigue, difficulty in concentrating and difficulty in keeping eyes open after the completion of the experiment. Mann-Whitney U-test was conducted on the subjective scores obtained before and after the experiment of each of the 6 items used in the study. The results are shown in Table 8.2. The first column shows the question number, second column shows the p-values on the subjective scores on that particular question

Table 8.1 VAS-F statistical test results on the subjective scores

Item	p-value	Significance
Q1	0.0117	✓
Q2	0.0102	✓
Q3	0.0135	✓
Q4	0.0102	✓
Q5	0.0102	✓
Q11	0.0157	✓
Q12	0.0088	✓
Q13	0.0907	✓
Q14	0.0071	✓
Q15	0.1388	×
Q16	0.0243	✓
Q17	0.1719	×
Q18	0.3976	×

Table 8.2 CFS statistical test results on the subjective scores

Item	p-value	Significance
Q2	0.0930	×
Q3	0.0189	✓
Q5	0.0209	✓
Q7	0.1770	×
Q8	1.0000	×
Q10	0.027	✓

before and after the experiment and the third column shows whether it is significant or not. ✓ indicates statistically significant while × indicates insignificant different.

8.3.5 *Comparative Assessment: Pure MI and AO+MI*

To evaluate the differential effects of MI and combined AO+MI, intra-task subjective ratings were collected for four key dimensions: MI quality, fatigue, motivation, and attention. Each participant completed 10 runs under both conditions and mean scores per scale were computed and subjected to paired statistical testing. Prior to analysis, the Shapiro-Wilk test confirmed the normal distribution of the difference scores for all scales ($p > 0.05$) [45]. Therefore, paired t-tests were employed to compare the mean ratings between the two conditions.

Fatigue ratings showed a significant difference ($p = 0.0402$), indicating a mitigating effect of action observation on mental fatigue as shown in Table 8.3. The motivation scale also showed a significant difference ($p = 0.0063$). No significant differences were observed in attention or imagery quality.

Table 8.3 Statistical test results

Scale	Shapiro-Wilk test (p)	Paired t-test (p)	Significant difference
MI quality	0.3842	0.0657	×
Fatigue	0.4899	0.0402	✓
Motivation	0.3110	0.0063	✓
Attention	0.7399	0.6213	×

8.4 Discussion

8.4.1 AO as a Strategy to Mitigate MF

Our study investigates whether the effects of MF can be mitigated by combining AO with MI. Changes in spectral power are analyzed by examining each subject's EEG data to assess the effects of MF. When a subject was asked to imagine a motor activity while watching a video of an everyday task, it's likely that they were already familiar with that task. And it encourages them to perform the task. Both AO+MI and pure MI are examined for the attention-related N1-ERP component. Figure 8.3 clearly illustrates the variation in N1 amplitude between AO+MI and pure MI. Compared to the initial N1 value, the N1 ERP component for AO+MI shows a small amplitude shift during the last run of the experiment. Thus, it can be concluded that AO+MI increases cortical activity related to motor function more successfully than MI.

8.4.2 Integration of AO with MI

In addition, our results show that the spectral power densities of the theta band and alpha bands vary throughout the session. The power in the alpha bands increases significantly between the beginning and end of the task. The LORETA study shows that the current density is maximal when AO and MI are combined. Thus, these results suggest that the integration of AO and MI helps patients retain higher performance even when they are mentally fatigued.

8.4.3 Subjective Experience of MF During MI and AO+MI

This study examined mental fatigue (MF) associated with pure Motor Imagery (MI) and combined Action Observation with Motor Imagery (AO+MI) using both pre-post subjective scales (VAS-F, CFS) and intra-task comparative ratings. The results showed a consistent increase in fatigue, drowsiness, and attentional decline post-experiment, with several VAS-F and CFS items reaching statistical significance.

These findings are in line with prior research showing that sustained MI practice can induce cognitive fatigue and reduce imagery quality over time [8] [47]. Importantly, intra-task comparisons revealed that AO+MI was associated with significantly lower fatigue and higher motivation ($p = 0.0063$) compared to pure MI. This supports the view that AO provides external visual cues that reduce cognitive demands and enhance affective engagement during mental simulation [27] [26] However, no significant differences were found in attention or imagery quality, suggesting that while AO+MI alleviates mental workload, it may not consistently enhance the subjective vividness of imagery.

8.4.4 Limitations of the Study

While the present study provides promising evidence for the effectiveness of combining AO with MI to counteract mental fatigue, several limitations should be acknowledged. First, the relatively small and homogeneous sample size limits generalization of our results. Future studies with larger and more diverse participant groups are necessary to validate the observed effects across different populations and clinical conditions. Second, the experimental tasks involved only a limited set of simple, common motor actions. While this ensured experimental control, it may not fully represent the complexity of real-world rehabilitation scenarios. Investigating more goal-directed tasks could reveal the practical benefits of AO+MI in applied settings. Third, while LORETA source localization provided insight into the cortical generators of the N1 ERP response, it is limited by its relatively low spatial resolution and reliance on assumptions about source distributions. Interpretation of these results should thus be approached with caution. More precise neuroimaging methods such as fMRI or MEG, or hybrid EEG-fNIRS systems, could enhance spatial localization and cross-validate the cortical regions implicated in AO+MI effects While subjective assessments such as the Visual Analogue Scale (VAS-F) and the Chalder Fatigue Scale (CFS). Reliance on self-reports to capture perceived mental fatigue introduces notable methodological limitations. One notable limitation of self-reports was the difficulty in maintaining sustained engagement throughout pure MI trials, largely due to fatigue and fluctuations in attention. In contrast, while AO+MI was cognitively less demanding, it occasionally resulted in passive observation, where some participants did not actively engage in the imagery. This variation in participant engagement may have influenced the consistency and depth of neural activation, potentially affecting the interpretation of AO+MI's effectiveness in mitigating MF. Furthermore, a gradual increase in sleepiness and MF was consistently reported towards the end of both conditions, indicating possible cognitive overload and lack of task variability. some participants also noted that the repetitive nature of the experiment contributed to monotony, which could affect intrinsic motivation and task adherence. These limitations suggest the need for improved experimental design, such as incorporating more varied and personalized stimuli, optimizing trial duration, and implementing engagement checks to ensure consistent imagery performance throughout the task.

8.5 Conclusion

This study provides evidence that AO+MI enhances neural efficiency and reduces MF more effectively than pure MI. AO+MI led to more stable N1 ERP responses, increased alpha band power, and higher motivation levels, indicating stronger cortical engagement and sustained performance. It's conceivable that visual guidance from AO enhances the accuracy of mental simulations beyond MI alone. Our research findings can be helpful in the development of neurorehabilitation plans for stroke patients.

Acknowledgments The financial support received through the INAE Abdul Kalam Technology Innovation National Fellowship, Govt of India under Grant No.: INAE/121/AKF/40 and the Cognitive Science Research Initiative, Govt. of India under Grant No.: DST/CSRI/2018/301 is gratefully acknowledged. Additionally, all participants' time and willingness to take part in this study are deeply appreciated.

References

1. A. Guillot, F. Di Rienzo, C. Collet, The neurofunctional architecture of motor imagery. *Advanced Brain Neuroimaging Topics in Health and Disease-methods and Applications*, IntechOpen Rijeka, (2014), pp. 433–456
2. M. Jeannerod, Mental imagery in the motor context. Neuropsychologia **33**(11), 1419–1432 (1995)
3. F. Xu, Y. Miao, Y. Sun, D. Guo, J. Xu, Y. Wang, J. Li, H. Li, G. Dong, F. Rong et al., A transfer learning framework based on motor imagery rehabilitation for stroke. Sci. Rep. **11**(1), 1–9 (2021)
4. M. Jeannerod, Neural simulation of action: a unifying mechanism for motor cognition. Neuroimage **14**(1), S103–S109 (2001)
5. M. Lotze, P. Montoya, M. Erb, E. Hülsmann, H. Flor, U. Klose, N. Birbaumer, W. Grodd, Activation of cortical and cerebellar motor areas during executed and imagined hand movements: an fmri study. J. Cogn. Neurosci. **11**(5), 491–501 (1999)
6. E. Bonda, M. Petrides, S. Frey, A. EvANs, Neural correlates of mental transformations of the body-in-space. Proc. Natl. Acad. Sci. **92**(24), 11180–11184 (1995)
7. J. Munzert, B. Lorey, K. Zentgraf, Cognitive motor processes: the role of motor imagery in the study of motor representations. Brain Res. Rev. **60**(2), 306–326 (2009)
8. U. Talukdar, S.M. Hazarika, J.Q. Gan, Motor imagery and mental fatigue: inter-relationship and eeg based estimation. J. Comput. Neurosci. **46**(1), 55–76 (2019)
9. G. Borghini, L. Astolfi, G. Vecchiato, D. Mattia, F. Babiloni, Measuring neurophysiological signals in aircraft pilots and car drivers for the assessment of mental workload, fatigue and drowsiness. Neurosci. Biobehav. Rev. **44**, 58–75 (2014)
10. T. Cao, F. Wan, C.M. Wong, J.N. da Cruz, Y. Hu, Objective evaluation of fatigue by eeg spectral analysis in steady-state visual evoked potential-based brain-computer interfaces. Biomed. Eng. Online **13**(1), 1–13 (2014)

11. A. Pomer-Escher, R. Tello, J. Castillo, T. Bastos-Filho, Analysis of mental fatigue in motor imagery and emotional stimulation based on eeg, in *XXIV Congresso Brasileiro de Engenharia Biomedica-CBEB* (2014)
12. L.J. Trejo, K. Kubitz, R. Rosipal, R.L. Kochavi, L.D. Montgomery et al., Eeg-based estimation and classification of mental fatigue. Psychology **6**(05), 572 (2015)
13. A. Craig, Y. Tran, N. Wijesuriya, H. Nguyen, Regional brain wave activity changes associated with fatigue. Psychophysiology **49**(4), 574–582 (2012)
14. T. Jacquet, R. Lepers, B. Poulin-Charronnat, P. Bard, P. Pfister, B. Pageaux, Mental fatigue induced by prolonged motor imagery increases perception of effort and the activity of motor areas. Neuropsychologia **150**, 107701 (2021)
15. T. Jacquet, B. Poulin-Charronnat, P. Bard, R. Lepers, Persistence of mental fatigue on motor control. Front. Psychol. **11**, 588253 (2021)
16. L.G. Faber, N.M. Maurits, M.M. Lorist, Mental fatigue affects visual selective attention. PloS One **7**(10), e48073 (2012)
17. M.A.S. Boksem, T.F. Meijman, M.M. Lorist, Effects of mental fatigue on attention: an erp study. Cogn. Brain Res. **25**(1), 107–116 (2005)
18. K.A. Garrison, C.J. Winstein, L. Aziz-Zadeh, The mirror neuron system: a neural substrate for methods in stroke rehabilitation. Neurorehabil. Neural Repair **24**(5), 404–412 (2010)
19. G. Rizzolatti, L. Fogassi, V. Gallese, Neurophysiological mechanisms underlying the understanding and imitation of action. Nature Rev. Neurosci. **2**(9), 661–670 (2001)
20. V. Gallese, C. Keysers, G. Rizzolatti, A unifying view of the basis of social cognition. Trends Cogn. Sci. **8**(9), 396–403 (2004)
21. G. Buccino, A. Solodkin, S.L. Small. Functions of the mirror neuron system: implications for neurorehabilitation. Cogn. Behav. Neurol. **19**(1), 55–63 (2006)
22. G. Buccino, A. Molinaro, C. Ambrosi, D. Arisi, L. Mascaro, C. Pinardi, A. Rossi, R. Gasparotti, E. Fazzi, J. Galli, Action observation treatment improves upper limb motor functions in children with cerebral palsy: a combined clinical and brain imaging study. Neural Plast. **2018**(1), 4843985 (2018)
23. L. Fadiga, L. Fogassi, G. Pavesi, G. Rizzolatti, Motor facilitation during action observation: a magnetic stimulation study. J. Neurophysiol. **73**(6), 2608–2611 (1995)
24. D. Ertelt, S. Small, A. Solodkin, C. Dettmers, A. McNamara, F. Binkofski, G. Buccino, Action observation has a positive impact on rehabilitation of motor deficits after stroke. Neuroimage **36**, T164–T173 (2007)
25. E. Pelosin, L. Avanzino, M. Bove, P. Stramesi, A. Nieuwboer, G. Abbruzzese, Action observation improves freezing of gait in patients with parkinson's disease. Neurorehabil. Neural Repair **24**(8), 746–752 (2010)
26. S. Vogt, F. Di Rienzo, C. Collet, A. Collins, A. Guillot, Multiple roles of motor imagery during action observation. Front. Human Neurosci. **7**, 807 (2013)
27. D.L. Eaves, M. Riach, P.S. Holmes, D.J. Wright, Motor imagery during action observation: a brief review of evidence, theory and future research opportunities. Front. Neurosci. **10**, 514 (2016)
28. S. Bordoloi, C.N. Gupta, S.M. Hazarika, Understanding effects of observing affordance-driven action during motor imagery through eeg analysis. Exp. Brain Res. **242**(10), 2473–2485 (2024)
29. S. Bordoloi, P. Saikia, C.N. Gupta, S.M. Hazarika, Neural correlates of motor imagery during action observation in affordance-based actions: preliminary results, in *2022 44th Annual International Conference of the IEEE Engineering in Medicine & Biology Society (EMBC)* (IEEE, 2022), pp. 4088–4092
30. M. Villiger, N. Estévez, M.-C. Hepp-Reymond, D. Kiper, S.S. Kollias, K. Eng, S. Hotz-Boendermaker, Enhanced activation of motor execution networks using action observation combined with imagination of lower limb movements. PloS One **8**(8), e72403 (2013)
31. R. Meers, H.E. Nuttall, S. Vogt, Motor imagery alone drives corticospinal excitability during concurrent action observation and motor imagery. Cortex **126**, 322–333 (2020)
32. U. Marusic, S. Grosprêtre, A. Paravlic, S. Kovač, R. Pišot, W. Taube, Motor imagery during action observation of locomotor tasks improves rehabilitation outcome in older adults after total hip arthroplasty. Neural Plast. **2018**(1), 5651391 (2018)

33. J.R. Emerson, J.A. Binks, M.W. Scott, R.P.W. Kenny, D.L. Eaves, Combined action observation and motor imagery therapy: a novel method for post-stroke motor rehabilitation. AIMS Neurosci. **5**(4), 236 (2018)
34. K.A. Lee, G. Hicks, G. Nino-Murcia, Validity and reliability of a scale to assess fatigue. Psychiatry Res. **36**(3), 291–298 (1991)
35. M. Cella, T. Chalder, Measuring fatigue in clinical and community settings. J. Psychosom. Res. **69**(1), 17–22 (2010)
36. B.Y. Tseng, B.J. Gajewski, P.M. Kluding, Reliability, responsiveness, and validity of the visual analog fatigue scale to measure exertion fatigue in people with chronic stroke: a preliminary study. Stroke Res. Treatment **2010** (2010)
37. Psychology Software Tools, Inc., E-Prime (Version 3.0), Computer Software. Psychology Software Tools, Sharpsburg, PA, 2016
38. Brain Products GmbH, Brain products: Solutions for neurophysiological research, 2025
39. The MathWorks, Inc. MATLAB Version 9.15 (R2023b), Computer software. The MathWorks, Inc., Natick, Massachusetts, 2023
40. P. Welch, The use of fast fourier transform for the estimation of power spectra: a method based on time averaging over short, modified periodograms. IEEE Trans. Audio Electroacoustics **15**(2), 70–73 (2003)
41. S. Dattola, F.C. Morabito, N. Mammone, F. La Foresta, Findings about loreta applied to high-density eeg–a review. Electronics **9**(4), 660 (2020)
42. R.D. Pascual-Marqui, C.M. Michel, D. Lehmann, Low resolution electromagnetic tomography: a new method for localizing electrical activity in the brain. Int. J. Psychophysiol. **18**(1), 49–65 (1994)
43. S. Invitto, C. Faggiano, S. Sammarco, V. De Luca, L.T. De Paolis, Haptic, virtual interaction and motor imagery: entertainment tools and psychophysiological testing. Sensors **16**(3), 394 (2016)
44. M.A. Jatoi, N. Kamel, A.S. Malik, I. Faye, T. Begum, A survey of methods used for source localization using eeg signals. Biomed. Signal Process. Control **11**, 42–52 (2014)
45. S.S. Shapiro, M.B. Wilk, An analysis of variance test for normality (complete samples). Biometrika **52**(3/4), 591–611 (1965)
46. P.J. Rowe, C. Haenschel, N. Khachatoorian, K. Yarrow, Post-stroke object affordances: an eeg investigation. Brain Cogn. **146**, 105639 (2020)
47. A. Guillot, C. Collet, V.A. Nguyen, F. Malouin, C. Richards, J. Doyon, Brain activity during visual versus kinesthetic imagery: an fmri study. Human Brain Mapping **30**(7), 2157–2172 (2009)

Chapter 9
Synergies Between Mind and Machine in Autism Research: An AI-Based Framework for Understanding and Reconstructing Neural Dynamics

Zannate Malik, Mariia Sidulova, and Chung Hyuk Park

Abstract In recent work, artificial intelligence (AI)-driven frameworks in collaboration with neuroscience have proven to overcome challenges in Autism spectrum disorder (ASD) research. Autism presents a unique challenge due to its complexity, variability, and wide range of comorbidities that have room to explore brain-behavior relationships in individuals by utilizing AI. This chapter presents an AI-based framework for individuals with ASD using large-scale, open-access datasets from the Autism Brain Imaging Data Exchange (ABIDE). Leveraging functional MRI and phenotypic data, we apply connectivity and comorbidity analyses to understand better how neural activity patterns relate to behavioral traits and coexisting conditions. The development of computational models that reconstruct high-resolution, heatmap-like plots from fMRI scans and associated metadata enables the visual interpretation of spatial patterns in brain activity by applying machine learning and AI algorithms. Combining brain imaging and behavioral data, the framework supports more interpretable models of ASD heterogeneity. It provides insights into the basic neural mechanisms that may highlight key ASD traits and comorbid descriptions. This work reflects how mind–machine collaboration can enhance personalized understanding of neurodevelopmental conditions, contributing tools that bridge clinical background with computational modeling and digital twin approaches.

Keywords AI-driven neuroimaging frameworks · Autism spectrum disorder (ASD) · Conditional variational autoencoder (CVAEs) · Autism brain imaging data exchange (ABIDE) · Functional connectivity MRI

Z. Malik · M. Sidulova · C. H. Park (✉)
Department of Biomedical Engineering, School of Engineering and Applied Science, The George Washington University, Washington, DC, USA
e-mail: z.malik2@gwu.edu; sidul001@gwu.edu; chpark@gwu.edu

R. Vinjamuri (ed.), *Bridging the Gap between Mind and Machine*,
https://doi.org/10.1007/978-3-032-06713-5_9

9.1 Introduction

Autism spectrum disorder (ASD), or autism, presents significant complexity due to its variability in behavioral traits and symptoms, including “qualitative impairment in social reciprocity, repetitive behaviors, restricted interests, and marked deficits in communication” [13]. The diagnosis of autism was considered rare due to limited knowledge of the disability and is now identified in “more than 1% of children,” highlighting an urgent need to better understand its biological basis [6]. The heterogeneous nature of ASD means there is significant variance in symptoms among individuals, making it especially challenging to pinpoint identifiable neural markers and consistent biological or physiological features for treatment.

Addressing these challenges requires datasets of sufficient size and quality. Single research laboratories typically struggle to gather adequately large or diverse samples to assess and clarify the brain mechanisms underlying ASD. ABIDE consists of two significant collections, ABIDE I and ABIDE II, each providing cross-sectional functional magnetic resonance imaging (fMRI) datasets and detailed phenotypic profiles. This data collection reports consistent neural patterns associated with autism that prove to be reliable and comprehensive for research.

Deep learning (hierarchical neural networks that learn layered features) and generative modeling (techniques that learn to synthesize data resembling their inputs) have opened up new tools and methods for analyzing these complex datasets. One approach involves using conditional variational autoencoders (CVAEs), which have proven effective in interpreting high-dimensional neuroimaging data. The CVAE framework incorporates phenotypic datasets from ABIDE directly into the modeling process, allowing a better representation of individual variability within brain imaging data [15]. This computational method enhances our ability to visualize and interpret nuanced patterns of brain connectivity that relate specifically to behavioral traits and comorbidities seen in individuals with autism.

This chapter applies well-established CVAE and connectivity analysis approaches to the ABIDE dataset collections. Our primary focus is reconstructing clear, interpretable visualizations such as heatmap-like plots depicting brain activity patterns. These plots integrate neuroimaging data and phenotypic metadata to illuminate differences and similarities among ASD subjects. Such visualization tools can significantly enhance clinical and research understanding of ASD heterogeneity by providing intuitive insights into otherwise complex data.

Finally, this chapter demonstrates how integrating advanced computational modeling methods with detailed clinical data bridges the gap between mind and machine with a more refined perception of the brain’s altered connectivity in autism. The combination of clinical resources such as fMRI scans, behavioral assessments, and phenotypic profiles represent the “mind,” and AI tools like autoencoders, graphing neural networks, and reconstructing connectivity maps represent the “machine” which supports more personalized and increases the accuracy of pinpointing neurodevelopmental conditions, demonstrating that cross-disciplinary collaboration leads to impactful advances.

9.2 Background

9.2.1 Functional Connectivity in Autism

Early neuroimaging studies of autism reported widespread differences in how brain regions communicate, fueling debates over whether autistic brains are generally "under-connected" or "over-connected." The under-connectivity theory posits that long-range connections are weaker in subjects with autism, leading to reduced coordination between distant cortical areas ([13, 12]). Indeed, systematic differences in functional connectivity metrics have been consistently observed in autism, with predominantly decreased cortico-cortical connectivity [13]. Many studies have found that ASD is marked by hypoconnectivity, which is reduced synchronized activity, especially between frontal and posterior brain regions [18]. For example, one deep-learning study reported "anterior-posterior underconnectivity in the ASD brain" as a distinguishing pattern [18]. At the same time, other findings complicate this simple model. Some researchers observed regional hyperconnectivity (increased synchronization) in ASD, particularly at short ranges or within local circuits [12]. This led to the notion that autistic brain networks might exhibit local over-connectivity but long-distance disconnection [15, 12]. In other words, frontal cortical regions might be "talking only to themselves" in subjects with autism, with weaker links to other parts of the brain [15].

Today, it is recognized that under and over-connectivity patterns are present in the brains of individuals with ASD [15]. Large-scale analyses have reconciled these seemingly disparate themes. Notably, the first Autism Brain Imaging Data Exchange study (ABIDE I) pooled nearly a thousand resting-state fMRI scans and found both hypo- and hyperconnectivity in ASDs, although with a striking predominance of hypoconnectivity [6]. Di Martino et al., observed that hypoconnectivity dominated their whole-brain analysis, particularly for cortico-cortical and interhemispheric functional connectivity. At the same time, instances of hyperconnectivity were relatively limited (e.g., involving some subcortical circuits) [6]. This observation helps explain why earlier studies sometimes reported contradictory results: depending on the cohort or method, one might detect predominantly reduced long-range connectivity or, in other cases, localized increases in connectivity. Recent evidence suggests that "it is more likely that both over- and under-connectivity patterns are present" across different neural circuits in ASD [15]. For example, one EEG study found long-range under-connectivity alongside short-range over-connectivity in autistic individuals [12], indicating that network alterations in ASD are not monolithic.

A key insight from contemporary research is that findings in autism functional connectivity are highly heterogeneous. There is an ongoing discussion due to different studies reports of "general over-connectivity, under-connectivity, and/or a combination of both" [9]. The heterogeneous nature has made attempts to classify autism using brain connectivity since connectivity patterns can vary widely amongst individuals on the spectrum. Researchers assign differences in study results to factors such as the age of participants, sample composition (sex ratio), analytical methods, and motion artifacts [7, 9]. For instance, childhood studies sometimes report different network abnormalities than adult studies, and many early ASD samples were predominantly male, potentially obscuring sex-specific connectivity profiles [15]. Notably, a persistent under-representation of females with ASD in functional connectivity research is present, which is important given known sex differences in ASD. Because ASD is diverse, there is a broad spectrum of conditions with varying symptom profiles and etiologies, which also contributes to inconsistent connectivity findings [15]. In essence, autism is not a single neural phenotype but a collection of related neurodevelopmental variations, which can manifest as different connectivity "signatures." This variability in brain network organization is a feature of ASD itself. Recent work by Nielsen et al., and others emphasizes that individual autistic participants often have idiosyncratic functional networks that do not conform to a single pattern [9]. One extensive study explicitly tested inter-individual variation and found "robust evidence for increased idiosyncrasy in ASD relative to [typical controls] in [various] networks," meaning that autistic brains show greater inter-individual variability in functional connectivity organization [7]. These authors further showed that such idiosyncratic connectivity might be a principal driver of inconsistency in the autism connectomics literature, as regions with the most variable connectivity in ASD also yielded conflicting results in case-control studies [7]. In summary, the current state of research indicates that while ASD is often associated with overall hypoconnectivity (especially across distant brain regions), there are pockets of hyperconnectivity and a high degree of network variability across individuals, highlighting the importance of large samples and a new analytic framework in capturing the multifaceted neural dynamics of autism.

9.2.2 Overview of ABIDE Datasets

Approaching the heterogeneity of ASD requires datasets that are large and diverse. A landmark effort in this regard has been the Autism Brain Imaging Data Exchange (ABIDE), an open-access data pool that started sharing ASD neuroimaging data. The initial ABIDE project (now referred to as ABIDE I) launched in 2012 as "the first open-data repository for autism neuroimaging," aggregating previously collected resting-state fMRI scans from sites around the world [7]. In total, ABIDE I released 1112 datasets containing 539 individuals with ASD and 573 age-matched typical controls (Zhou et al.) (ages 7–64) along with each subject's structural MRI and a set of phenotypic (clinical and demographic) measures [6]. This unprecedented data-sharing effort provided researchers with a large-scale sample to probe ASD brain connectivity. As Di Martino et al., noted in their description of ABIDE I, "by pooling multiple international data sets, ABIDE [was] expected to accelerate the pace of discovery," enabling both replication and novel discovery in ASD neurobiology [6]. Indeed, analyses of the ABIDE I dataset yielded

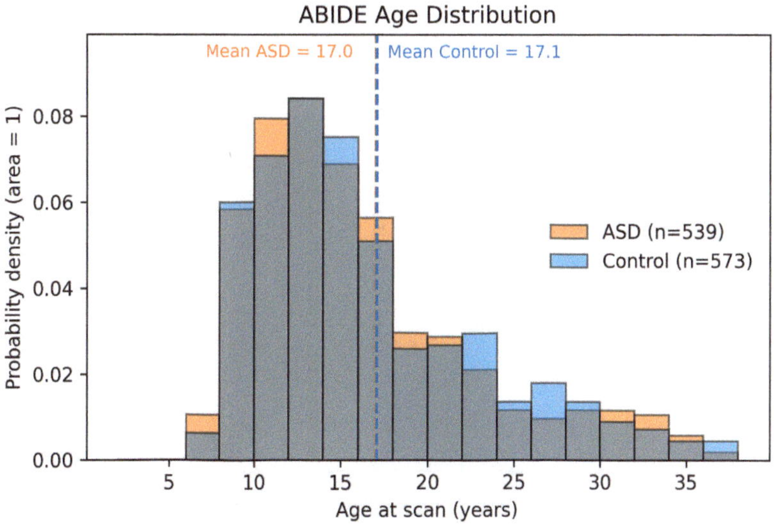

Fig. 9.1 ABIDE Age Distribution. Histogram of age at MRI scan for the ABIDE-I cohort (n = 1112). Orange bars = participants with ASD, blue bars = typically developing controls. Heights are probability densities (area = 1). Vertica dashed lines mark the group means (ASD = 17.0 y, Control = 17.1 y). Bin width = 2 years; 300 dpi image

important insights. For example, using a subset of ABIDE I, Di Martino et al., vreported the concurrent presence of widespread hypoconnectivity and pockets of hyperconnectivity in ASD, reconciling previously inconsistent findings [6]. They found that long-range connections, especially between default-mode network hubs and other regions, tended to be weaker in ASD (supporting the under-connectivity model).

In contrast, fewer connections (often involving subcortical areas) were more substantial in ASD [6]. These results, drawn from hundreds of subjects, helped validate earlier reports and demonstrated the power of multisite data aggregation. Furthermore, ABIDE I showed that pooling data across many research sites is feasible: although scanner models, protocols, and other site effects introduce noise, insightful brain-behavior patterns emerge when datasets are aggregated [13]. "The survey of the ABIDE R-fMRI data sets [provided] unprecedented demonstrations of both replication and novel discovery," Di Martino et al., wrote, emphasizing how combining data can both confirm known findings and reveal new ones [6]. Figure 9.1, 9.2 and 9.3 show the overall age, group, and sex distributions of the ABIDE-I cohort.

Building on this success, a second iteration, ABIDE II, was released in 2017 to expand the resource further. ABIDE II added data from 19 sites, contributing 487 new individuals with ASD and 557 new controls [7]. Combined with ABIDE I, the pooled dataset now comprises over 2100 resting-state fMRI scans (approx. 1026 ASD and 1130 TD individuals), making it one of the largest neuroimaging samples

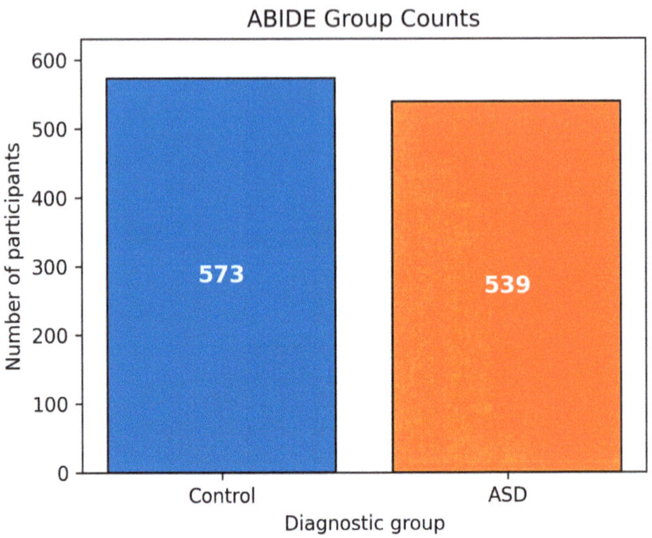

Fig. 9.2 ABIDE Participant Counts by Diagnostic Group Bar chart showing the total number of subjects in the ABIDE dataset, broken down in diagnostic categories (ASD vs Typically Developing Controls), showing the overall sample size and group balance used for our connectivity analyses

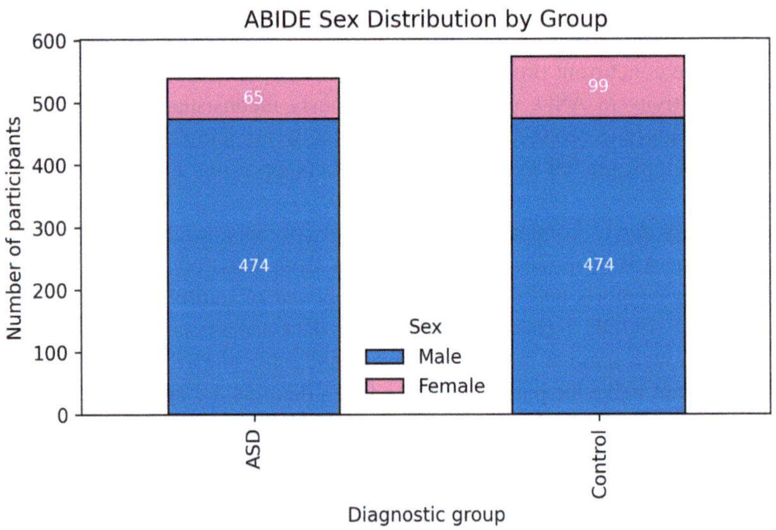

Fig. 9.3 Sex Distribution within ABIDE Diagnostic Groups. Stacked bar plot shows the number of male and female participants in each diagnostic group (ASD vs Typically Developing Controls), highlighting the sex ratio imbalance and shows the importance of integrating phenotypic conditioning in our AI models

in psychiatry. The scale and scope of ABIDE II enhanced research in several ways. First, the enlarged sample offers greater statistical power to detect subtle brain connectivity effects and to examine subgroups within ASD. With ~2156 datasets, researchers can split the data for discovery vs. replication analyses, improving rigor and reproducibility. Second, the multisite, international nature of the combined ABIDE I + II mitigates single-site biases and allows testing the generalizability of models across different scanners and populations. Third, ABIDE II emphasized phenotypic richness and diversity: it contributed a wider age range, more females (addressing the male bias in ABIDE I), and a broader array of behavioral measures and psychiatric assessments for each subject. ABIDE II includes a variety of clinical phenotypes (e.g., measures of autism symptom severity, IQ, and common comorbid symptoms) that researchers can use to link brain connectivity patterns with behavior. It even added diffusion MRI data for 284 subjects, enabling the study of structural and functional connectivity. As the ABIDE II organizers noted, these enhancements intended to "contribute to unraveling key sources of ASD heterogeneity" [7]—for example, by facilitating the search for neuroimaging-based subtypes or by examining how different clinical profiles (such as language delay or ADHD symptoms in ASD) correspond to different connectivity alterations.

The ABIDE I and II repositories have transformed autism neuroimaging by bringing multimodal datasets and resources that allow us to dive deeper into ASD heterogeneity. The datasets show how open science is in neuroimaging, making data available to any researcher and thus modifying access to large samples. Since their release, ABIDE I and II have been used in hundreds of studies, catalyzing discoveries and even challenges to previous findings. Importantly, ABIDE enabled several multisite validation studies. For instance, Nielsen et al., [13] leveraged ABIDE to attempt a machine-learning classification of ASD vs. controls across 16 sites (964 subjects total), which is a follow-up on earlier single-site studies that reported ~80% accuracy [13]. While Nielsen's multisite classifier achieved only modest accuracy (~60% range), it revealed that site-related variability and individual heterogeneity make ASD classification a complex problem, highlighting the need for algorithms to handle such variance [13]. These algorithms exemplify how ABIDE data tests the power of neuroimaging biomarkers.

Additionally, ABIDE's open availability has spurred methodological advances, such as new techniques for data harmonization to reduce side effects [14] and the application of deep learning to brain connectivity. In summary, ABIDE I & II represent a cornerstone of ASD connectomics research, providing the field with large-scale, openly shared datasets that continue to drive multisite analyses, enable comparative and meta-analytic studies, and inspire a new generation of machine learning approaches. By aggregating data across the globe, ABIDE has helped the community move beyond isolated and inconsistent findings toward a more integrated understanding of ASD's neural foundation.

The rise of big data resources like ABIDE concurred with the rise of machine learning and deep learning in neuroimaging. Researchers have increasingly applied advanced computational models to ASD brain data to detect atypical connectivity patterns, classify individuals, and probe the multidimensional nature of the disorder.

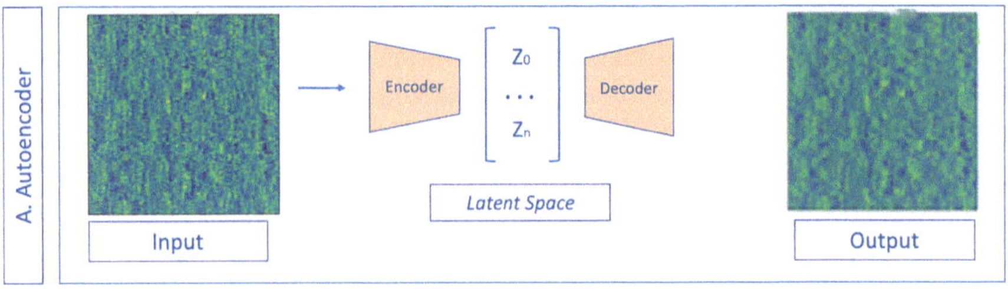

Fig. 9.4 Autoencoder (AE) Architecture. Schematic of a basic autoencoder used for fMRI connectivity data. The 200-ROI x time matrix reconstruction image (left) is passed through an encoder to produce a low dimensional latent representation (latent space), which the decoder then uses to reconstruct the original input to a degraded output (right)

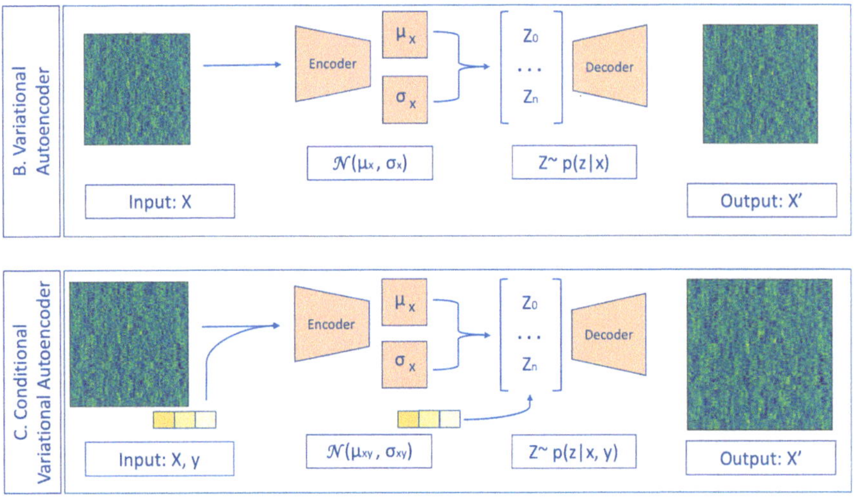

Fig. 9.5 Variational Autoencoder (VAE) (top) and Conditional Variational Autoencoder (CVAE) (bottom) Architectures. Schematic of our VAE model shows the same ROI time matrix is encoded into two vectors (μ and σ) that parameterize a Guassian latent distribution. Samples drawn from this distribution are put into the decoder to generate a fair reconstruction of the input. Whereas a CVAE model at the bottom shows how we concatenate a conditional embedding (phenotypic data y) with the ROI-time input. The encoder produces μ(x,y) and σ(x,y) for joint latent distribution, and the decoder uses both the samples latent z and the conditional y to reconstruct the original fMRI matrix, enabling conditioning through phenotype for connectivity modeling

In Figs. 9.4 and 9.5, we provide an overview of these approaches, focusing on frameworks such as autoencoders (AEs), variational autoencoders (VAEs), conditional VAEs (CVAEs), and connectivity-based analyses that are especially relevant to understanding neural networks in ASD heterogeneity through generative modeling.

Supervised Machine Learning (Classification): Early studies used traditional machine learning algorithms (e.g., support vector machines, random forests) on resting-state connectivity features to discriminate ASD from typical development. Initial results on single-site datasets were promising—for example, some achieved around 80% accuracy in classifying high-functioning autism vs. controls using whole-brain functional connectivity matrices [13]. However, applying these methods to the multisite ABIDE sample led to performance dropping, emphasizing the challenges of generalization. As mentioned, Nielsen et al., attempted to replicate prior results with 16-site data. They found lower accuracy, noting that "large inhomogeneities in acquisition parameters, subject populations, and research protocols limit the sensitivity for detecting abnormalities" in such pooled data [13]. This drawback spurred efforts to improve classification by using more sophisticated models. One landmark study by Heinsfeld et al., employed deep neural networks on ABIDE and reported about 70% classification accuracy—an improvement over conventional approaches and a proof-of-concept that deep learning can "identify autism spectrum disorder using ... fMRI" data to a reasonable extent [9, 18]. Notably, the patterns that emerged from Heinsfeld's trained deep model aligned with neuroscience: the network's features pointed to reduced connectivity between anterior and posterior regions in ASD (consistent with the under-connectivity theory) [18]. Other studies have likewise used deep feed-forward or convolutional neural networks on functional connectivity features to classify ASD, often after applying feature selection or dimensionality reduction (given the high dimensionality of connectomes). For example, a recent multisite study by Kang et al., used multi-view ensemble learning on ABIDE and improved accuracy by combining information across data sites [15]. Despite these advances, classification alone has limitations for understanding ASD: an accuracy of ~70% indicates substantial overlap between ASD and typical connectivity patterns, and optimized models for prediction do not always yield easily interpretable insights about the brain. Researchers now explore more unsupervised and generative modeling techniques that can reveal how ASD brains differ without necessarily reducing the problem to binary classification.

Deep Generative Models (VAE/CVAE): In recent years, variational autoencoders have gained traction as a powerful tool to study neuroimaging data, including in ASD research [15]. VAEs are probabilistic generative models that learn a low-dimensional latent representation of the data and can generate (or reconstruct) data samples from this latent space. Their appeal lies in capturing the underlying distribution of complex data (like brain connectivity patterns) in a structured way. "Generative models, such as Variational Autoencoders (VAEs), are increasingly employed for atypical pattern detection in brain imaging," note Sidulova and Park, "during training, these models learn to capture the underlying patterns within 'normal' brain images..." In practice, one can train a VAE on neurotypical connectivity data (learning a normative latent space) and then examine how well the model can reconstruct ASD data; deviations in reconstruction may highlight unusual connectivity patterns in ASD. Sidulova and Park [15] leveraged VAEs in precisely this way to investigate ASD: they report that their model "aimed to uncover atypical interconnectivity between brain regions" by measuring differences between reconstructed and input connectivity matrices for ASD individuals. An advantage of VAEs is that they can model both "low- and high-level features of fMRI data" [15]. Unlike techniques such as ICA or seed-based correlation focusing on pairwise connections or components, a VAE can encode holistic connectivity patterns (potentially capturing distributed network anomalies characteristic of ASD). Several studies have utilized VAEs or related autoencoder models to extract features for ASD. For example, Choi et al., employed a deep autoencoder to learn representations of functional connectivity in ASD. They found that the model's latent features reflected known aberrant patterns—their VAE identified co-occurring under-connectivity and over-connectivity in ASD, and these network disruptions correlated with participants' IQ scores [15]. Generative models can not only detect ASD vs. control differences but also link them to behavioral metrics, shedding light on ASD's cognitive heterogeneity.

Researchers have experimented with different VAE architectures in the fMRI domain, including convolutional neural network (CNN) VAEs, recurrent neural network (RNN) VAEs, and hybrids [15]. Each architecture has strengths: CNN-based VAEs excel at capturing spatial structure in connectivity maps, while RNN or LSTM-based VAEs can model temporal dynamics of fMRI time series. In their comparative study, Sidulova and Park evaluated a CNN-VAE, an RNN-VAE, and a parallel CNN + RNN VAE on ABIDE data. They found that the CNN-based VAE was most effective for functional connectivity analysis in ASD, outperforming the recurrent variants in reconstruction accuracy and identifying group differences, suggesting that preserving topographical information (e.g., network layout) was crucial for their task. Notably, Sidulova and Park did not stop at unsupervised learning: they also introduced Conditional VAEs (CVAEs), which incorporate additional information (covariates) into the generative process. They added phenotypic data (such as sex) as a conditional input to the VAE, motivated by whether including such attributes could improve the model's ability to detect meaningful patterns. The rationale is that ASD exhibits sex-related prevalence differences and possibly sex-specific neural features, so a CVAE could learn separate latent distributions for males and females and thus reduce bias or reveal sex-by-diagnosis interactions. Indeed, their results indicated that adding conditions (like group labels or demographic factors) helped the VAE "reduce sex-related bias" and slightly enhanced performance in capturing ASD-specific connectivity anomalies [15]. More broadly, CVAEs and related approaches allow for more targeted generative modeling—for instance, one can condition on diagnosis to generate synthetic ASD vs. control connectivity maps for comparison or condition on age to examine developmental effects in ASD. Recent studies in other domains have similarly used CVAEs for connectivity network detection and to factor out confounds: e.g., [15] employed an adversarial CVAE to identify brain network differences in Alzheimer's disease while controlling for age and sex, demonstrating that conditioning can help "reduce the effect of age and sex bias in the latent vectors." In the ASD context, [8] used a CVAE with an attention mechanism to integrate age/sex information when classifying brain connectivity of children with neurodevelopmental disorders. They found that incorporating such phenotypic attributes improved the model's ability to disentangle disease-specific patterns from other variations [15]. These examples underscore a trend in ASD neuroimaging toward multimodal and semi-supervised models that blend imaging data with auxiliary information (diagnosis, demographics, cognitive scores) to characterize the spectrum's complexity better.

Other Computational Approaches: Besides deep neural networks, researchers have applied various computational models to ASD data. Graph theory-based analyses remain popular for examining network properties (such as nodal centrality or network efficiency) in ASD [15]. Such studies often report altered network topology in autism (e.g., decreased network integration or abnormal hub connectivity), complementing the pairwise connectivity findings. Clustering and subtype discovery methods have also been explored. For instance, diagnosis-informed connectivity subtyping has been used to group ASD individuals into subgroups based on their functional connectivity patterns, revealing that specific subsets of ASD show more pronounced connectivity disruptions than others Choi et al., [3]. Another cutting-edge development uses computational neuroscience models (beyond pure machine learning) to bridge neural activity and behavior in ASD. For example, recent work by Takahashi et al. introduced a computational psychiatry framework combining neural network simulations with empirical fMRI data [16]. Their 2023 study hypothesized that functional connectivity (FC) and neural excitability interact to affect ASD symptoms. They first trained a hierarchical recurrent neural network (inspired by predictive coding theory) to perform a cognitive task. They manipulated its connectivity and excitability parameters to produce "ASD-like" behavior in the simulation. They found that certain conditions (e.g., very homogeneous neuronal excitability combined with high functional connectivity in the model) led to performance deficits analogous to those seen in ASD (such as reduced generalization in emotion recognition). Intriguingly, when they mapped these simulated parameters onto real ABIDE subjects, the model predicted which subgroup of participants would have an ASD diagnosis. Their results suggest a novel explanatory model: hyper-connectivity and atypical neural excitability interaction could underlie specific ASD cognitive profiles [16]. This work exemplifies how AI-based frameworks (here, recurrent neural nets with biologically inspired design) can simulate neural dynamics and test those simulations against empirical data. Our CVAE model pulls out hidden connectivity features that simple correlation maps often miss, they graph theoretic metrics then reveal how key hubs in the brain reconfigure in autism; and our hybrid CNN-RNN-VAE model simulates the brain's evolving activity over time. Together, these methods handle the brain's high dimensional, non linear complexity in ways that basic linear analyses simply cannot. For example, VAEs can untangle non-linear interactions amongst hundreds of ROIs like in the prefrontal, parietal, and subcortical hubs. In practice one latent dimension might capture variance and another latent dimension might align with the clinical assessment. Both of these latent axes can be used as biomarkers to be used as neurofeedback or predicting clinical and behavioral responses which is an ongoing challenge.

9.3 Methods and Computational Approach

9.3.1 *Dataset*

To effectively analyze functional connectivity and comorbidities in ASD, we utilized the Autism Brain Imaging Data Exchange (ABIDE) initiative's datasets. ABIDE provides comprehensive open-access datasets collected from multiple laboratories across the US and parts of Europe, aiming to advance our understanding of autism's neural foundations by facilitating large-scale analysis [6]. Specifically, ABIDE I includes data from 1112 individuals (539 ASD, 573 controls) across 17 international sites.. In contrast, ABIDE II expands on this with an additional 1114 individuals (521 ASD, 593 controls) collected from independent laboratories ([6]; see Fig. 9.1 for age distribution and Fig. 9.2 for diagnostic group counts). The datasets include resting-state functional MRI (rs-fMRI) scans alongside detailed phenotypic profiles such as clinical diagnoses, cognitive test scores, demographics, and comorbid conditions.

9.3.2 *Preprocessing*

Before applying any computational methods, fMRI data requires thorough preprocessing such as correcting slice-timing offsets, realigning frames to account for head motion, and regressing out physiological artifacts. This step helps to reduce noise, motion artifacts, and other external interferences so that our CVAE models can see true brain signals. Using Python-based neuroimaging tools, particularly Nilearn and Nipype libraries, we carried out standard preprocessing routines such as slice-time correction, realignment for head motion correction, normalization to standard anatomical space, spatial smoothing, and temporal filtering [1]. This comprehensive preprocessing ensures that subsequent connectivity analyses yield reliable and reproducible results.

Functional connectivity is measuring correlations between brain regions based on resting-state fMRI scans. In order to quantify this connectivity, we utilized region-of-interest (ROI)-based analysis using widely recognized brain atlases, such as Schaefer's atlas with 200 ROIs and the Harvard-Oxford atlas with 96 ROIs. Based on functional and anatomical characteristics, these atlases segment the brain into distinct regions (parcellations) (see Fig. 9.6 as an example). For example, we might correlate the prefrontal cortex parcel's signal with the parietal lobe parcel, filling out the 200x200 matrix using Schaefer's atlas that will lay out the brain's resting state networks. The time-series data extracted from these ROIs allowed us to compute pairwise correlation matrices representing functional connectivity patterns across brain regions ([4]: see Figs. 9.6 and 9.7).

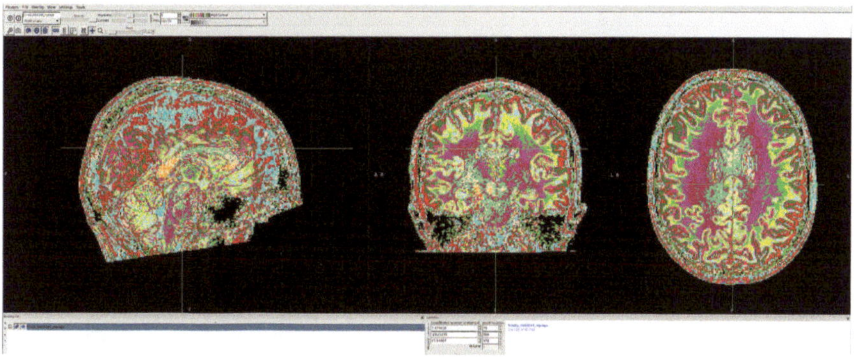

Fig. 9.6 Massachusetts General Hospital (MGH) Cortical Parcellation Overlay (Harvard-Oxford Atlas). Sagittal, coronal, and axial views of an individual ABIDE subject's resting state fMRI in FSLeyes, overlaid with the MGH cortical atlas labels. The color parcellations show morphometric routes in the cortex into functional regions for downstream connectivity reconstruction and heatmap visualization

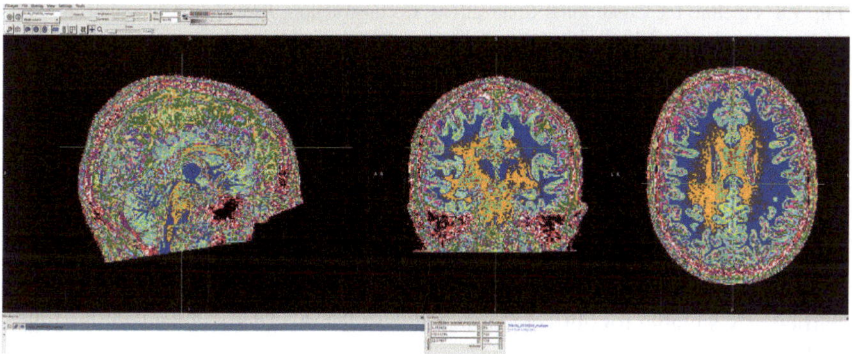

Fig. 9.7 Massachusetts General Hospital (MGH) Subcortical Parcellation Overlay (Harvard-Oxford Atlas). The same subject as Fig. 9.6, now overlaid with the subcortical atlas. By combining cortical and subcortical segmentations, our CVAE model can capture whole-brain connectivity patterns which can be interpreted to ASD-related connectivity in large-scale networks and deep nuclei

9.3.3 Computational Modeling: Conditional Variational Autoencoders (CVAEs)

To handle the complex and high-dimensional nature of neuroimaging data, we utilized conditional variational autoencoders (CVAEs). CVAEs represent an advanced class of deep generative models that encode complex data variability patterns into interpretable latent representations [10]. Specifically, CVAEs incorporate conditional information like phenotypic metadata (age, sex, social communication score, etc.) alongside the brain scans during the learning process, enabling better modeling

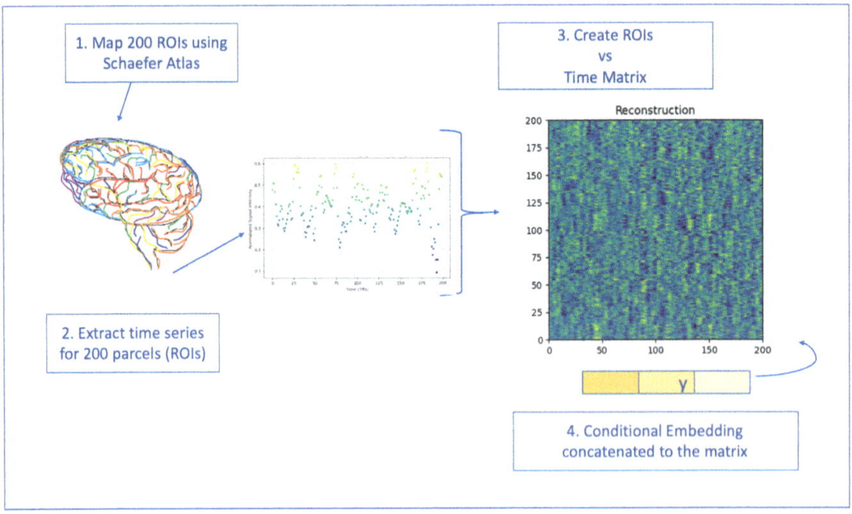

Fig. 9.8 Pipeline for Conditional Connectivity Reconstruction. Parcellation mapping in Step 1 where we apply the Schaefer's atlas to delineate cortical parcels on the subject's brain. Step 2, show the time series extraction where we pull the fMRI signal from each of the 200 regions, producing a region by time matrix. Step 3 shows matrix reconstruction where the autoencoders (AE, VAE, CVAE) learn to reconstruct this ROIxTime matrix, recovering the original connectivity patterns. Step 4, finally concatenates a learned conditional embedding (age, sex, diagnosis, etc.) to the matrix, enabling the CVAE to model individual variability in neural dynamics

of individual variability. These conditions allow the network to take apart age-related maturation or diagnosis that impact connectivity patterns. This method is particularly suitable for autism research due to the disorder's considerable clinical and biological backgrounds [15].

In practical terms, our Python-based pipeline (see Fig. 9.8) involved training CVAE models on preprocessed ABIDE datasets. We input each participant's functional connectivity matrix (derived from their fMRI scans), and phenotypic data (e.g., age, gender, clinical scores, comorbidity status) were input to the CVAE models. The output is a reconstructed, denoised connectivity matrix that captures the fundamental neural dynamics specific to each individual. This reconstruction capability makes CVAEs particularly effective for visualizing and interpreting subtle connectivity differences among ASD subgroups.

We specifically utilized convolutional neural networks (CNN) and recurrent neural networks (RNN) individually and in hybrid configurations (CNN-RNN). CNNs effectively capture spatial connectivity patterns, while RNNs are suitable for capturing temporal or sequential data dependencies [2]. Our hybrid CNN-RNN approach thus leverages strengths from both architectures, enhancing reconstruction accuracy and interpretability of neural dynamics.

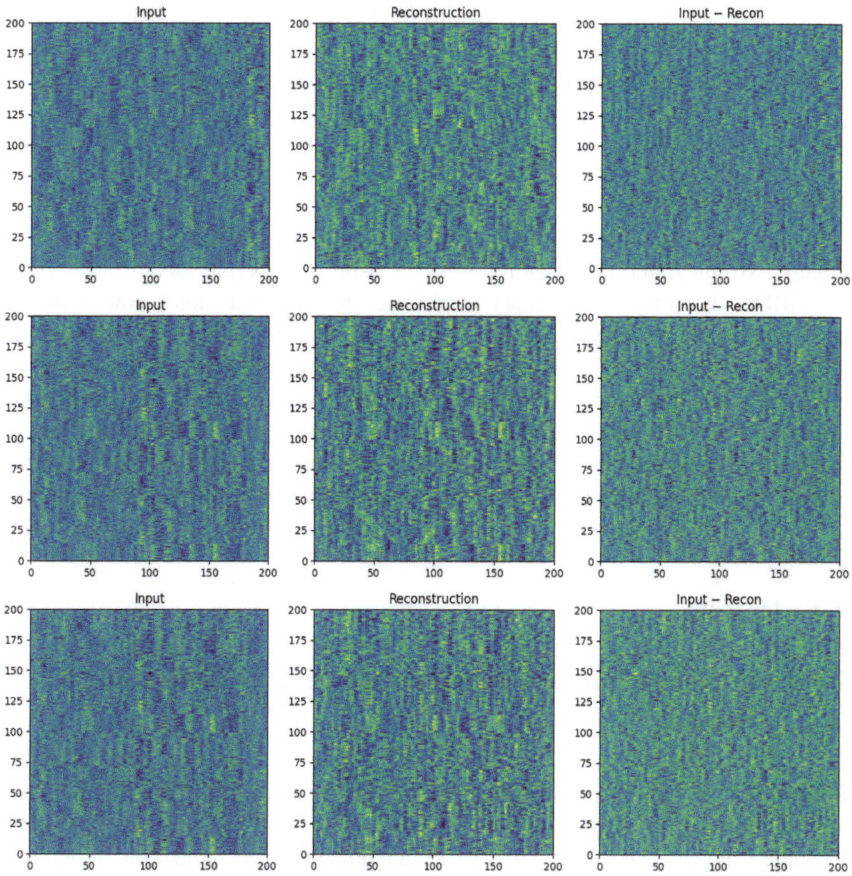

Fig. 9.9 Three Individual Subjects Reconstructed Functional Connectivity Heatmaps. Using the original 200x200 Roi-time matrix we show a comparison of the training input, reconstructed output, and validation output by our CNN-RNN-VAE

9.3.4 Generation of Connectivity Heatmap Visualizations

The final component of our computational pipeline involves translating the reconstructed connectivity matrices into intuitive visualizations (see Fig. 9.9) with high-resolution heatmaps sized 200 $\times$ 200 pixels. These visualizations facilitate clear and direct interpretation of spatial patterns within the brain. Using custom Python scripts (developed with matplotlib and seaborn libraries), we transformed the reconstructed matrices into colored heatmap plots. In these plots, color intensity indicates the strength of functional connectivity between pairs of brain regions, providing an immediate visual representation of neural dynamics across individuals and groups.

Comparing visualizations generated from different atlases (e.g., Schaefer versus Harvard-Oxford) further enriched our interpretations, highlighting how the choice of brain parcellation might influence observed connectivity patterns.

9.3.5 *Comorbidity Analysis Integration*

Finally, we integrated comorbidity analyses using phenotypic metadata provided in ABIDE datasets. The primary goal here was to understand how different comorbidities (e.g., ADHD, anxiety disorders, epilepsy) correlate with specific connectivity patterns in ASD brains. To achieve this, participants were grouped based on the reported comorbid conditions in addition to diagnoses they already had like ASD with ADHD. Visual maps were made of each group's brain-connectivity patterns and then compared with other subgroups and mapping, we could see how each comorbid condition was linked to its own unique pattern of brain connections.

9.4 Results

9.4.1 *Functional Connectivity Patterns in ASD Using ABIDE Datasets*

Before looking into group level connectivity patterns, we first display how our hybrid CNN-RNN-VAE reconstructs individual connectivity matrices for three example subjects (see Fig. 9.9). To systematically understand how functional connectivity differs among individuals with Autism Spectrum Disorder (ASD), we generated connectivity matrices from the ABIDE datasets using conditional variational autoencoders (CVAEs). These matrices capture intricate neural interactions, representing how different brain regions are functionally related.

Our computational pipeline (Fig. 9.8) generated reconstructed connectivity matrices for each individual. We produced high-resolution heatmap-like visualizations from these matrices to intuitively highlight differences in brain connectivity patterns between ASD and control groups. Figures 9.9 and 9.10 are visualizations showing clear comparative examples between ASD and typically developing individuals.

From visual inspection of these heatmaps, distinct connectivity differences were readily apparent, particularly in the Default Mode Network (DMN) and Salience Network. Previous studies have highlighted these networks as consistently implicated in ASD-related alterations [9]. Our visualizations consistently supported these findings, with visible differences in connectivity strengths among frontal, parietal, and temporal regions.

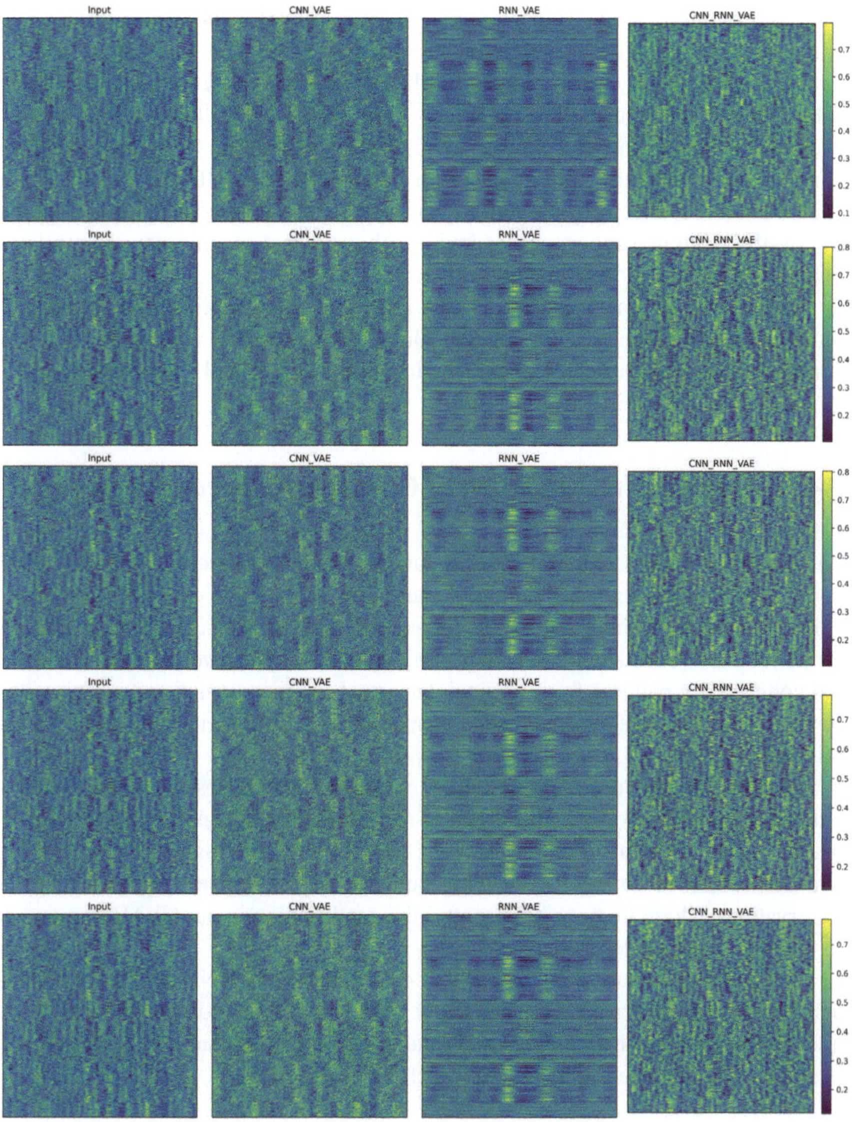

Fig. 9.10 Comparative Reconstructions Across Autoencoder Architectures (Training Input vs CNN-VAE vs RNN-VAE vs CNN-RNN-VAE). Each of the five rows shows one ABIDE validation subject's normalized fMRI snippet by 200 ROIs x 200 TRs

9.4.2 Comorbidity-Specific Connectivity Patterns

An essential part of our analysis involved examining how different comorbidities in ASD populations correlated with distinct connectivity patterns. Comorbidities such as Attention Deficit Hyperactivity Disorder (ADHD), anxiety disorders, and epilepsy are frequent in ASD and may influence observed neural dynamics [11].

We grouped ABIDE participants by these comorbid conditions and generated subgroup-specific connectivity heatmaps (Fig. 9.10). Visually comparing these heatmaps revealed unique connectivity signatures associated with each comorbid condition. For instance, individuals with co-occurring ADHD exhibited distinct frontal-striatal connectivity alterations compared to those without this comorbidity. These observations support earlier findings that suggest specific neural correlates underlying ASD comorbid conditions [9].

These subgroup analyses demonstrate the value of integrating phenotypic data into functional connectivity modeling. Our CVAE approach effectively captured complex relationships between neural connectivity and clinical characteristics.

9.4.3 Quantitative Validation of Connectivity Reconstruction

To objectively assess reconstruction accuracy, we quantitatively evaluated CVAE performance using standard statistical measures, specifically mean squared error (MSE) between original and reconstructed connectivity matrices. Reconstruction errors remained consistently low, indicating that CVAE effectively preserved essential connectivity patterns from original datasets. Low reconstruction error provides further confidence in our generated visualizations' interpretability and clinical relevance.

Additionally, we used unsupervised clustering techniques, such as k-means clustering, to further validate subgroup distinctions derived from reconstructed connectivity patterns. Apparent clustering of connectivity patterns by ASD diagnosis and comorbidities emerged, providing additional quantitative support for visual observations. These findings help control the conditioning in our CVAE framework to take apart ASD heterogeneity.

Quantitative evaluation confirmed that our hybrid CNN-RNN-VAE consistently achieves low reconstruction error (MSE < 0.05 across all validation subjects) with a level of reliability that ensures the visualized heatmaps meaningfully reflect true neural dynamics rather than modeling artifacts.

9.5 Discussion

Our analysis of functional connectivity in autism spectrum disorder (ASD) using conditional variational autoencoders (CVAEs) provided clear, visual, and interpretable insights into neural connectivity patterns. Leveraging the robust datasets provided by the Autism Brain Imaging Data Exchange (ABIDE), our computational approach confirmed previously established neural differences and identified connectivity signatures linked specifically to common comorbidities.

9.5.1 Contextualizing Connectivity Patterns in ASD

Functional connectivity alterations in networks such as the Default Mode Network (DMN) and the Salience Network are reported frequently in autism literature. The DMN, crucial for social cognition and self-referential processing, showed clear connectivity reductions among individuals with ASD compared to typically developing controls [9]. Our visualizations strongly supported these previous findings. Moreover, differences in the Salience Network, known for its role in processing salient environmental stimuli, were also apparent and consistent with prior studies [17]. These visual results validate existing literature and emphasize the reliability of CVAE-based reconstruction methods for interpreting nuanced connectivity patterns in neurodevelopmental conditions.

One important factor from our results was the significant impact of brain atlas on observed connectivity patterns (see Figs. 9.6 and 9.7). When using Schaefer's atlas, we saw a tiny patch of over-connectivity between the right dorsal anterior insula and nearby prefrontal parcels which simply vanished when switching to the Harvard Oxford atlas. These findings indicate that future ASD studies might benefit from simultaneously working with multiple brain atlases to help capture hidden, and not-so-hidden signals from larger to small scale parcellations.

Integrating phenotypic data, specifically comorbid conditions, into our connectivity analyses was an important detail to our approach. ASD commonly co-occurs with conditions such as ADHD, anxiety disorders, and epilepsy, each potentially influencing neural connectivity [11]. Our comorbidity-specific connectivity heatmaps revealed clear and distinct signatures corresponding to these co-occurring conditions. For instance, the frontal-striatal connectivity alterations we identified among individuals with ADHD align with previously documented functional anomalies related to attention and executive functioning deficits [5].

These findings solidify the clinical relevance of integrating detailed phenotypic data and behavioral assessments with neuroimaging analyses. Furthermore, this supports our approach where the potential for these computational methods with AI algorithms to inform personalized interventions tailored to individual neurodevelopmental profiles.

A significant strength of our computational approach was the high interpretability and reproducibility of generated visualizations, evidenced by low CNN-RNN VAE reconstruction errors. The application of CVAEs provided usable results that clarified complex neural patterns otherwise difficult to perceive through traditional analysis methods. More importantly, the visual clarity of these reconstructions could aid clinicians and researchers by providing built-in representations of neural dynamics associated with ASD and its comorbidities, which could potentially improve diagnosis and facilitation of targeted intervention strategies.

Unsupervised clustering analyses like K-means clustering further supported the distinct profiles of ASD only, ASD + ADHD, and ASD + anxiety of our subgroup characterizations, clearly separating the connectivity patterns among ASD diagnostics and comorbidity groups. These machine learning algorithms provide reliable tools for targeted, reproducible clinical research in quantifiable subgrouping which were proven to be not random.

Although ABIDE's 2100 clinical multisite scans give us unique statistical potential, residual site effects like differences in TR/TE, field strength, and head-coil may still create bias in connectivity measures. While extensive preprocessing mitigated some variability, residual inter-site differences may still affect the generalizability of findings.

Second, our cross-sectional analysis with the 7–64 age in years at a single time point cannot reveal how these networks evolve like in longitudinal cohorts where we can map the progression in developmental trajectories in ASD. Future research incorporating longitudinal datasets could reveal dynamic connectivity patterns and help navigate key markers that help us understand neurodevelopmental trajectories and influence reinforcement learning models for personalized interventions.

Third, although CVAE reconstructions provide strong interpretability, the method is data-driven, dependent on underrepresented subgroups for example, females with rare comorbidities, in data quality and size. In future studies, enhancing dataset diversity and quantity would further refine these models, improving accuracy and clinical relevance.

Future research directions could focus on combining multimodal imaging data (such as EEG or diffusion MRI) to strengthen interpretations of neural dynamics along with fMRI scans with phenotypic data. Furthermore, by applying these computational approaches with AI tools to clinical trials and treatment outcomes, the research could further validate their clinical utility and potentially lead to precision medicine approaches tailored to individual neural profiles and more accurate readings of subnetwork connectivity in relation to comorbidities.

Overall, our results highlight how computational modeling can clarify complex neural connectivity in ASD, by bridging clinical phenotypic data and neuroimaging analyses through CVAE and VAE-based reconstruction. It also demonstrates intricate connectivity in the brain through networking and how they interact or shift with the presence of comorbidities. This AI-driven research give us an idea of the practicality of mind–machine collaborations with advanced computational tools can deepen our understanding of ASD heterogeneity and ultimately guide clinical interventions toward personalized, precision-driven care in autism.

9.6 Conclusion

In understanding brain functional connectivity in autism and exploring through data-driven perspectives, we wanted to design a transparent and interpretable method to bridge computational insights with clinical contexts. By using the ABIDE datasets and integrating subject-specific factors like age, sex, and diagnostic scores into a CVAE model, we could reconstruct and visually interpret functional connectivity patterns that reflect neurodevelopmental traits and ASD heterogeneity. Rather than treating machine learning as a black-box solution, this approach treats AI as a tool for uncovering latent structures in neural data where structures that may help researchers, clinicians, and engineers better understand the diversity within the autism spectrum. This chapter emphasizes the need for mind-machine synergy: models that not only compute but also communicate, learn from, and adapt to the complexity of human neurobiology.

References

1. A. Abraham, M.P. Milham, A. Di Martino, R.C. Craddock, D. Samaras, B. Thirion, G. Varoquaux, Deriving reproducible biomarkers from multi-site resting-state data: An autism-based example. NeuroImage **147**, 736–745 (2017). https://doi.org/10.1016/j.neuroimage.2016.10.045
2. H. Choi et al., Deep learning in autism spectrum disorder: Classification and implications for functional connectivity. Front. Neuroinform. **14** (2020). https://doi.org/10.3389/fninf.2020.609927
3. H. Choi et al., Diagnosis-informed connectivity subtyping discovers subgroups of autism with reproducible symptom profiles. NeuroImage **256**(119212), 119212 (2022). https://doi.org/10.1016/j.neuroimage.2022.119212. pubmed.ncbi.nlm.nih.gov/35430361/
4. C. Craddock et al., The Neuro Bureau preprocessing initiative: Open sharing of preprocessed neuroimaging data and derivatives. Front. Neuroinform. (2013). https://doi.org/10.3389/fninf.2013.00041
5. A. Di Martino, X.N. Zuo, C. Kelly, R. Grzadzinski, M. Mennes, A. Schvarcz, J. Rodman, C. Lord, F.X. Castellanos, M.P. Milham, Shared and distinct intrinsic functional network centrality in autism and attention-deficit/hyperactivity disorder. Biol Psychiatry **74**(8), 623–632 (2013). https://doi.org/10.1016/j.biopsych.2013.02.011
6. A. Di Martino et al., The autism brain imaging data exchange: Towards a large-scale evaluation of the intrinsic brain architecture in autism. Mol. Psychiatry **19**, 659–667 (2014). https://doi.org/10.1038/mp.2013.78
7. A. Di Martino et al., Enhancing studies of the connectome in autism using the autism brain imaging data exchange II. Sci. Data **4**(170010), 1–15 (2017)
8. M.S. Gao et al., Learning a phenotypic-attribute attentional brain connectivity embedding for ADHD classification using rs-fMRI, in *Proceeding of the IEEE EMBC*, (2020), pp. 5472–5475
9. J.V. Hull et al., Resting-state functional connectivity in autism spectrum disorders: A review. Front. Psych. **7**, 205 (2017). https://doi.org/10.3389/fpsyt.2016.00205
10. D.P. Kingma, M. Welling, Auto-encoding variational Bayes, in *Proceedings of the International Conference on Learning Representations (ICLR)*, (2014)
11. M.-C. Lai, M.V. Lombardo, S. Baron-Cohen, Autism. Lancet **383**(9920), 896–910 (2014). https://doi.org/10.1016/S0140-6736(13)61539-1
12. J.M. Lee, S. Kyeong, E. Kim, K.-A. Cheon, Abnormalities of inter- and intra-hemispheric functional connectivity in autism spectrum disorders: A study using the autism brain imaging data exchange database. Front. Neurosci. **10**, 191 (2016). https://doi.org/10.3389/fnins.2016.00191
13. J.A. Nielsen et al., Multisite functional connectivity MRI classification of autism: ABIDE results. Front. Hum. Neurosci. **7**, 599 (2013). https://doi.org/10.3389/fnhum.2013.00599
14. S. Saponaro et al., Multi-Site Harmonization of MRI Data Uncovers Machine-Learning Discrimination Capability in Barely Separable Populations: An Example from the ABIDE Dataset. NeuroImage: Clin. **35**, 103082–103082 (2022). https://doi.org/10.1016/j.nicl.2022.103082. Accessed 23 Sept 2024

15. M. Sidulova, C.H. Park, Conditional variational autoencoder for functional connectivity analysis of autism spectrum disorder functional magnetic resonance imaging data: A comparative study. Bioengineering **10**(10), 1209 (2023). https://doi.org/10.3390/bioengineering10101209
16. Y. Takahashi et al., Interaction between functional connectivity and neural excitability in autism: A novel framework for computational modeling and application to biological data. Comput. Psychiatry **7**(1), 14–29 (2023)
17. L.Q. Uddin, K. Supekar, C.J. Lynch, A. Khouzam, J. Phillips, C. Feinstein, S. Ryali, V. Menon, Salience network-based classification and prediction of symptom severity in children with autism. JAMA Psychiatry **70**(8), 869–879 (2013). https://doi.org/10.1001/jamapsychiatry.2013.104
18. L. Xu, Z. Sun, J. Xie, J. Yu, J. Li, J. Wang, Identification of autism spectrum disorder based on short-term spontaneous hemodynamic fluctuations using deep learning in a multi-layer neural network. Science Direct Retrieved June 18, 2025, from https://www.sciencedirect.com/science/article/pii/S1388245720306088#bib113 (2021)

Chapter 10
Human Intent into Motion: Neuro-Motor Controlled Prosthetic Hand

Maibam Pooya Chanu, Ramana Vinjamuri, and Nayan M. Kakoty

Abstract In the rapidly evolving world of human-machine interaction, the ability to translate human thoughts into physical movements is gaining popularity in the area of rehabilitation robotics. This breakthrough marks a significant step forward in assistive technology. At the core of this progress, neuro-motor control is a promising approach that enables individuals with limb impairments to regain lost hand movements and experience a renewed sense of functionality through advanced prosthetic hands.

This chapter explores how human intention can be decoded and transformed into physical action for control of a prosthetic hand through advances in neuro-motor control. A deeper understanding of the fundamental principles of neural pathways involved in motor control including the role of motor planning, action initiation and execution will be discussed. Further, this chapter will delve into the recent advances in neuro-motor control using brain-machine interface with the integration of sensory feedback. By synthesizing the current trends in research, this chapter presents the state-of-the-art in neuro-motor control for translating human intentions to prosthetic hand movements for improving the quality of life of individuals with upper limb amputations.

Keywords Neuro-motor control · Prosthetic hand · Human intention decoding · Sensory feedback

M. P. Chanu · N. M. Kakoty (✉)
Embedded Systems and Robotics Laboratory, School of Engineering, Tezpur University, Assam, India
e-mail: nkakoty@tezu.ernet.in

R. Vinjamuri
Vinjamuri Lab, Computer Science and Electrical Engineering, University of Maryland, Baltimore, MD, USA
e-mail: rvinjam1@umbc.edu

R. Vinjamuri (ed.), *Bridging the Gap between Mind and Machine*,
https://doi.org/10.1007/978-3-032-06713-5_10

10.1 Introduction

Prosthetic hand technology represents an important field at the intersection of biomechanics, neuroscience and engineering to restore the functionalities of individuals who have loss upper limb. Furthermore, it significantly affect daily living activities (dla) [1, 2]. The human hand is an exceptional example of evolutionary process, with unmatched dexterity and adaptability in the act of grasping and handling objects. For example, the hand exhibits diverse grasping patterns in activities like precision grasp involving the index finger and thumb, power grasp involving all fingers and the palm. These skilled behavior involve in hand movements is term as dexterity. Hand dexterity plays a crucial role during dla. In the history of science fiction, one of the most iconic portrayals is that of a bionic prosthetic hand that closely mimic the natural hand in both appearance and function. However, despite these futuristic depictions, real-world prosthetic technology still lags behind in achieving such natural integration and performance. Additionally, the response of the most advanced bionic limbs are slow and has limited of number of degrees of freedom (DoF) without sensory feedback.

Globally, it is estimated that over 57.7 million people are living with limb amputations due to traumatic causes as of 2017 [3]. According to the National Sample Survey 76th Round (2018) report, 1.4% of India's population has locomotor disabilities [4]. Among these individuals, 3.9% have acquired artificial hands i.e. approximately 0.737 million hand amputees in India are using prosthetic hands. However, these prosthetic hands are mostly for cosmetic appeal and lack of human hand motor functionalities. Further, the growing demand for the replacement of upper limb motor function is raising a significant concern in the healthcare community. In addition, the projected global market growth shows a compound annual growth rate (CAGR) of 9.15% reaching USD $\approx$753.37 million during 2024–2032. Furthermore, the global prosthetics market is experiencing rapid growth driven by increasing demand for better prosthetic solutions.

Nevertheless, there is a steady progress in the design of prosthetic hand to mimic the human hand as presented in Fig. 10.1. This evolution began with the non-functional iron-clad hand in the sixteenth century, continued in the twentieth century with body-powered hook devices that were capable of simple grasping. Further, development has advanced with sophisticated prosthetic hand driven by synergies with embodied intelligence. Throughout this progress, biomimetic design has played an important role in development of prosthetic hand with the application of human machine interface (HMI) and brain computer interface (BCI) with artificial sensory feedback system.

Understanding the mechanisms underlying neuro-motor control is crucial for developing interventions to rehabilitate movement disorders, enhance movement performance and design advanced prosthetic hand. BCI encompass a system design to interpret the recorded activity of the brain and convert them into control

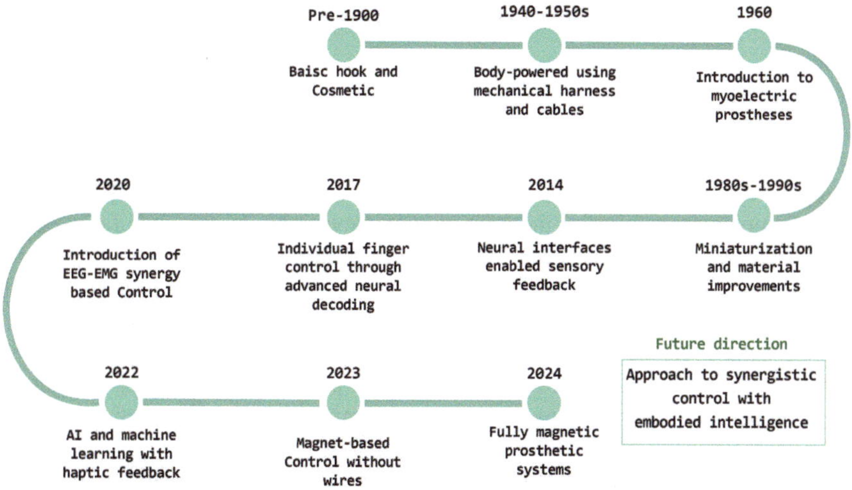

Fig. 10.1 Timeline flow chart illustrating the evolution of prosthetic technology. Key milestones include: Pre-1900, basic hook and cosmetic; 1940-1950s, body-powered using mechanical harness and cables; 1960, introduction to myoelectric prostheses; 1980s- 1990s, miniaturization and material improvements; 2014, neural interfaces enabled sensory feedback; 2017, individual finger control through advanced neural decoding; 2020, introduction of EEG-EMG synergy-based control; 2022, AI and machine learning with haptic feedback; 2023, magnet-based control without wires; 2024, fully magnetic prosthetic systems. Future direction: approach to synergistic control with embodied intelligence

commands for prosthetic hand. These system enable interfaces between neural activity and prosthetic hand. Within this domain, neuro-motor hand prostheses is a subset that focuses on capturing signals from brain regions responsible for voluntary movement. By studying these brain activities, the control commands are generated for control of actuators embedded in the prosthetic hand.

At the intersection of thought and action, there lies one of the most profound question in human experience: *"how does intention become movement?"* This chapter begins by exploring the complex neural mechanisms that allow the human brain to coordinate precise finger movements. The chapter focus on investigating how the coordinated brain signals are being applied to develop advanced prosthetic hands that respond according to human intent. Through detailed examination of the interaction in the major regions of the brain, and how the alteration in pathways affect the finger movement will be elucidated. This is followed by the discussion on fundamental neuroscience underlying neural pathways, motor planning, action execution and sensorimotor integration. This chapter ends with the discussion on the technological advancements that enable real-time prosthetic hand control using brain signals and open research issues.

10.2 The Neural Foundations of Movement

Imagine reaching for a cup of coffee, feeling its warmth as the fingers curl around the ceramic cup. This simple act often taken for granted depends on a highly coordinated flow of neural signals from the brain to the muscles. However, for individuals with limb amputation, this type of dla are an unattainable goal–until now. Neuro-motor control lies at the core of such movement, the invisible language of movement that our brains speak. Neuro-motor control refers to the complex neural mechanisms by which the brain plans, initiates and executes the movement with the integration of sensory feedback. In the context of assistive technologies, neuro-motor control serves as the foundational framework for decoding human intent and translating it into functional movement in prosthetic hands. For individuals with upper limb amputations, neuro-motor controlled hand prosthetics offers a revolutionary means for restoring lost functionality.

The human hand is a multifaceted organ of human beings capable to different types of movements and manipulations with precision and dexterity. This dexterity is governed by a complex neural control system involving the hierarchical motor planning, precise sensorimotor integration and adaptive learning mechanisms. Therefore, understanding these neural processes is foundational for developing neuroprosthetics that restore hand functionalities for individuals with limb amputation.

10.2.1 The Role of Brain in Human Hand Control

The motor planing and execution in human is a hierarchically organized process involving various regions of the brain coordinating to convert intention to physical action. This motor coordination involves the central nervous system of the brain including the motor cortex, somatosensory cortex, basal ganglia (BG), thalamus, cerebellum, and sensory feedback systems as shown in Fig. 10.2. The motor cortex regions are responsible for the intent generation, execution and regulation of hand movements [5]. The BG and cerebellum regions are responsible for modulating and refining hand movements, learning and habit formation [6]. The BG is a canonical structure that integrates striatum (STR), globous pallidus externus (GPe), globous pallidus internus (GPi), subthalamic nucleus (STN) and substantia nigra (SN). The SN is divided into substantia nigra pars compacta (SNc) and substantia nigra pars reticulara (SNr). These neurons in the cortical basal ganglia cerbello thalamus (CBCT) network are interconnected with the synaptic inputs in direct and indirect pathways namely, glutamate, gamma-aminobutyric acid (GABA) and α-amino-3-hydroxy-5-methyl-4-isoxazolepropionic acid (AMPA). The STR receives glutamatergic projections from both the motor cortex and the thalamus and contains D1 and D2 dopaminergic receptors. In the direct pathway, the motor cortex sends the glutamatergic input to STR. The dopaminergic input from the SNc regulates the activity of STR. The GABAergic neuron from STR inhibit the GPi and GPe in which

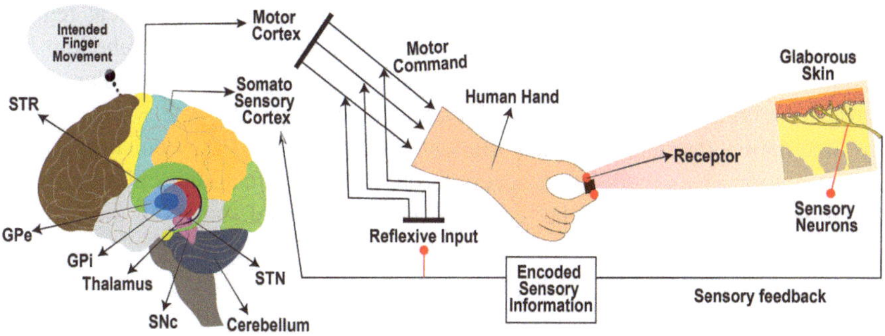

Fig. 10.2 Diagram illustrating the neural pathway of intended finger movement. The process begins in the brain with the intended movement signal in the striatum (STR), passing through the motor cortex and somatosensory cortex. Motor commands are sent to the human hand, where a receptor on the glabrous skin detects sensory input. Sensory neurons transmit feedback, completing the loop with encoded sensory information and reflexive input. Key brain regions labeled include the thalamus, cerebellum, and others

the direct pathway is excited as the D2 receptors are inhibited by the dopamine from SNc. Additionally, to directly inhibit the output regions of BG, the GPe also inhibits the STN. When the STN is suppressed, the GPi and SNr are inhibited. This means the GPe reduces activity in the output nuclei in two ways. As a result, the GPi and SNr excites the thalamus. This enable the thalamic neurons to become excited and send excitatory signals to the cortex. This excitation of the thalamus enhance the initiation of voluntary movements [7].

In the indirect pathway, the GPe receives inhibitory input from STR. As a result, the GPe becomes less active and can no longer inhibit the STN, GPi and SNr. With reduced inhibition, the STN is excited and receives additional excitatory input from the cortex via hyperdirect pathway [8]. Once activated, the STN excites GABAergic neurons in the GPi and SNr which in turn suppress the thalamus and thereby inhibiting motor cortex [9]. This helps to suppress unwanted movements. Besides from the direct and indirect pathways, the hyperdirect pathway consist of a single, direct connection from the frontal cortex to the STN and employs glutamate as its neurotransmitter. It is believed that the hyperdirect pathway enable quick suppression of motor activity, helping the brain to rapidly inhibit movements when needed [10, 11].

The regulation of motor activity within the direct and indirect pathways is influenced by the dopaminergic system. Dopamine released from the SNc modulates the STR neurons by binding to D1 and D2 receptors. Activation of D1 receptors initiate movements in the direct pathway. Activation of D2 receptors inhibits the movement in the indirect pathway. This dual modulatory effect of dopamine enable to achieve the precision and efficacy of voluntary motor control. Beyond the execution of motor output, the brain actively processes and integrates sensory inputs from the hand with additional modalities including visual, auditory and vestibular. This help to maintain spatial and temporal accuracy in hand movements. This multisensory integration is mediated by interactions among different cortical regions such as the parietal cortex involved in spatial representation. The cerebellum and thalamus relays information to STR. Further, STN projects to cerebellar cortex through pontine nuclei. Together, BG and cerebellum influence motor control by modulating the activity of key cortical areas involved in movement planning and execution. This modulation occurs through their shared connections with the thalamus, which acts as a relay center that passes refined signals to the cerebral cortex for coordinated movement output [13].

10.2.2 Mechanism for Sensory Feedback

The human brain synthesizes sensory inputs for unified and accurate representations of both the external environment and the body itself [14]. The brain exhibits adaptive modulation, adjusting the weight and relevance of multiple sensory cues based on the incoming sensory information [15]. Sensory integration within the brain primarily occurs in the parietal cortex for processing tactile and visual inputs. The primary somatosensory cortex is responsible for encoding tactile input from the hand and the secondary somatosensory cortex in merging the tactile information with signals from other sensory modalities. The posterior parietal cortex is involved in combining sensory feedback with motor planning and spatial awareness. The intraparietal sulcus helps in representing object characteristics and hand configurations. These areas work in coordination with key motor-related cortical regions to facilitate precise hand function and control.

The thalamus transmit the sensory information from peripheral receptors to corresponding cortical areas. The cerebellum fine-tune and coordinate hand movements through sensory feedback processing. The human brain orchestrates hand control through a distributed and hierarchical architecture that processes sensory, cognitive and emotional cues in parallel. For individuals with limb amputation, these natural control systems are disrupted resulting in reduced functional capabilities. Although current prosthetic hands replicate the physical form and movement of a natural hand; limitations persist in sensory feedback, intuitive control and user adaptability. To overcome these challenges, advancing prosthetic design by emulating the coordination of brain regions and implementing layered control strategies offers a promising approach to restore more naturalistic hand function.

10.3 Hierarchical Control Strategy in Hand Prosthetics

The control of prosthetic hand follows the hierarchical control from high level movement planning and intentions in human brain to low level motor command

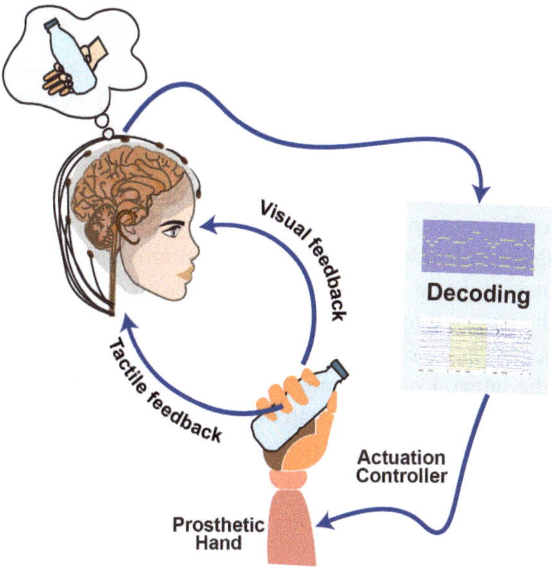

Fig. 10.3 Diagram illustrating a feedback loop for a prosthetic hand. The process begins with a brain signal, depicted as a thought bubble with a hand holding a bottle. Arrows indicate the flow of information to a decoding device, labeled "Decoding," which processes the signal. The decoded information is sent to an actuation controller, which directs the prosthetic hand to grasp a bottle. Feedback loops labeled "Visual feedback" and "Tactile feedback" connect the prosthetic hand back to the brain, completing the cycle. The diagram emphasizes the integration of neural signals and feedback in controlling the prosthetic hand

and execution in muscular skeletal system. However, the control of prosthetic hand remains a challenging problem as it involves the synergistic integration of signal processing, actuations and sensory feedback mechanism. Neuro-motor control of prosthetic hand has three main components as shown in Fig. 10.3.

Neural activity associated with intended movements were recorded, preprocessed and decode to extract movement intention. This signals were then translated into control commands. The resulting commands are used to actuate the neuro-motor prosthetic hand. Based on the shape, size and weight of the object being manipulated, the prosthetic hand can select the type of grasp to be performed. Further, the motor command to control the prosthetic hand can be adjusted by using sensor or feedback from the users residual limb to prevent slipping and misalignment while grasping objects.

However, this hierarchical control strategy of prosthetic hand has limitations due to complex and high dimensional nature of human hand movement. This type of control requires numerous input and output to control each DoF. Additionally, the variability of the sensory feedback system can impair the reliability in the control of prosthetic hand [16].

10.4 BCI in Prosthetic Hand Control

In order to entirely emulate human hand motor functionality, advancements in signal processing, micro-electronics and machine learning have been in core for BCI based prosthetic hand control. However, the complete understanding of how the human brain coordinate the complex movements and grasping activities of the human hand remains elusive. This presents a numerous challenges in the realms of neuroscience and engineering for realization of prosthetic hand mimicking human hand. From the perspective of neuroscience, elucidating the mechanism of the brain governing hand control holds the potential to shed light in neural foundation of motor skills, action perception, integration of sensorimotor and object manipulation [18, 19]. These processes are the integral part to human evolution for advancement and recovery of motor functions through rehabilitation. From the perspective of engineering, understanding the complex mechanism of hand control in human brain provides a valuable insights in design and control aspects of prosthetic hands [20, 21].

10.4.1 Brain-inspired Prosthetic Hand Control Strategies

BCI systems based on brain signals using advancements in neuroscience and engineering have considered as effective methodology for control of prosthetic hand. The initiation and fine tuning of grasping-related actions are represented in specific parts of the cerebral cortex. Furthermore, the electroencephalogram (EEG) is a good method to capture this activity, as it has high temporal resolution. Studies have shown that EEG patterns can distinguish between hand movements [22]. Non-invasive BCIs are preferred due to their efficacy, safety and affordability. Typically, EEG measures originating from the sensorimotor cortex [23] are a central factor use by BCI-controlled prosthetics to decipher intention for hand movements.

EEG-based decoding of hand movement has been extensively studied to reconstruct hand synergies and control prosthetic devices. The primary motor cortex particularly channels C3, CZ and C4 has neural signals associated with the voluntary hand movement [24, 25]. Beta rebound and alpha-range task-related coherence have further highlighted the significance in channels CZ, CP1, FC1, FC2, CP2, P3 and frontal regions such as FZ, F3, and F7 during movement execution [26, 27]. Additionally, increased activation in contralateral regions P4, C4, and F4 occurs just before movement onset. The inclusion of low-frequency time-domain EEG signals have been found with enhance decoding performance [28]. Over the past three decades, numerous studies have explored the use of EEG for hand movement recognition. The classification accuracy achieved in these studies varies depending on the extracted features and the classification algorithms employed [29, 30].

A study utilizing a linear classifier based on discriminant analysis combined with spectral domain weighted common spatial pattern features reported a classification accuracy of 73.2% for distinguishing hand grasp and release actions [31]. Several time-domain features including slope sign changes, Willison amplitude, variance, and skewness have been effectively applied in EEG-based hand movement classification tasks [32–34]. Further, work in [35] employed non-invasive EEG combined with optimized machine learning models for control of prosthetic hands. 11 time domain features including minimum value, maximum value, median amplitude, mean amplitude, standard deviation, variance, waveform length, mean absolute value, root mean square, skewness and kurtosis were employed for classification using optimizable support vector machine (SVM). The classifier distinguished between hand open and close intentions with an average testing accuracy of 93.4% demonstrating reliability. The optimized model used a Gaussian kernel with tuned parameters for improved classification accuracy. The classification output was interfaced with a prosthetic hand. On detection of intended movements, the system trigger corresponding motor actuation to execute hand open or close operations.

Additionally, the movement-related neural synergy was examined by analyzing the coherence in spatial power distribution patterns and the power spectral density of independent EEG components [36]. Using independent component analysis, 15 EEG channels from the frontal, central, and parietal regions were identified as most relevant for decoding motor-related brain activity. Both time-domain and synergy-based features were extracted from these selected channels and used to train a SVM classifier. A Bayesian optimizer was employed to fine-tune the SVM hyperparameters. The resulting optimized classifier achieved an average testing accuracy of $94.39 \pm 0.84\%$ using the extracted synergistic features. The output of the classifier was employed to control a prosthetic hand demonstrating the feasibility of synergy-based decoding for intuitive neuroprosthetic control. Further, several studies also explored how the brain orchestrates complex hand movements by coordinating and activating groups of muscles in terms of synergy for efficient and coordinated control of various hand movement [37–40].

10.4.2 Sensory Feedback in Prosthetic Hand Control

The information that human body receives through senses like touch, vision, and hearing during physical movement are referred to as sensory feedback in the context of prosthetic hand control. This sensory input plays an important role in enabling the brain to regulate and execute movements during complex motor tasks. In BCI systems, the absence of such feedback results in deviation of human hand like natural movement by prosthetic hand. To address this, recent research has focused on integrating multi-modal sensory feedback including tactile, force and position into BCI systems to enhance the control of the prosthetic hand.

Emerging studies have increasingly emphasized improving movement control in neuro-motor prosthesis through enhanced feedback mechanisms. Visual feedback is an essential element in neuro-motor control loop for prosthetic hand. This will provide the users with real time information about the activity of the brain and the interpretation of the prosthetic hand. By observing visual feedback such as changes in the interface or object position, the users can adjust their mental strategies to improve the control. Studies have shown that visual feedback increases the accuracy and learning speed in BCI based control [41]. Additionally, person with hand amputation typically retain intact auditory systems. This leads to auditory feedback as a promising method for sensory feedback. However, it remains uncertain whether auditory feedback effectively supports adaptation to dynamic and kinematic changes, or how it compares to visual feedback in terms of control performance. Non-invasive sensory feedback using vibrotactile, mechanotactile, and electrotactile stimulation techniques were also employed in prosthetic hands [42].

Currently, researchers have focus on employing feedback to one type of sensation at a time. There is a challenge in communicating more than one type of sensory feedback leading to high computational load to the controller of the prosthetic hand. This makes the prosthetic hand control systems away from that of the human hand in terms of cognitive abilities. To address these challenges, combined multiple feedback methods wherein each can delivers different type of sensation in parallel is under research.

10.5 Discussions and Open Issues

This chapter discusses the underlying brain mechanisms involved in natural hand control and their relevance to the development of prosthetic hands that interface with the human brain. It emphasizes the need for prosthetic systems to incorporate sensory feedback and develop control interfaces capable of translating external stimuli into neural-like signals. Additionally, adaptive algorithms that can integrate information from multiple sensory sources is essential for multiplexing sensory feedback.

However, current prosthetic hand control strategies continue to face notable challenges when compared to the capabilities of the human hand. Unlike natural hand, which process a rich variety of somatosensory feedback such as texture, weight, shape, pain and proprioception; most prosthetic hands rely on a limited set of sensors. These typically measure contact force, position or temperature and fall short in reproducing the depth and realism of natural touch. Although this chapter outlines several solutions, important challenges related to neuro-motor control of prosthetic hand remain unresolved and require further investigations. Some of the particular directions are:

1. To design a prosthetic hand that replicate the neuro-motor functions of the natural hand with the integration of advanced materials and sensors capable of detecting and responding to motor commands and sensory feedback.
2. To implement intelligent control algorithms that fuse information from different neural and sensory channels leveraging machine learning and artificial intelligence techniques to decode motor intent and adjust movements dynamically based on user interaction.

3. To incorporate adaptive learning mechanisms that mirror the plasticity of brain using models from computational neuroscience allowing the prosthetic hand to refine control through experience, feedback and reinforcement learning.

In conclusion, human brain inspired control strategy of prosthetic hand is promising with multiple research challenges. A biomimetic and adaptive prosthetic hand that meet the needs of the persons with hand amputation can be developed by replicating the synergistic hierarchical control strategies of the human hand. Although progress has been made, there is still the need of better prosthetic hand that could seamlessly interact with the human nervous system. More complex and sophisticated algorithm to control the prosthetic hand are required for multiple sensory feedback realization. Utilizing this strategy, a neuro-motor controlled prosthetic hands can be developed to accurately translate the intention of a user into movement improving the quality of life.

Acknowledgments The support received under project number BBF/BITS/FY/2023-24/ART/103, BITS BioCyTiH Foundation and project number GP/2025/AR/048, I-Hub Foundation for COBOTICS, Technology Innovation Hub of IIT Delhi are gratefully acknowledged.

References

1. D.M. Desmond, Coping, affective distress, and psychosocial adjustment among people with traumatic upper limb amputations. J. Psychosomatic Res. **62**(1), 15–21 (2007)
2. S. Aas, Prosthetic embodiment. Synthese **198**(7), 6509–6532 (2021)
3. C.L. McDonald, S. Westcott-McCoy, M.R. Weaver, J. Haagsma, D. Kartin, Global prevalence of traumatic non-fatal limb amputation. Prosthetics and Orthotics Int. **45**(2), 105–114 (2021)
4. The Ministry of Statistics and Programme Implementation (GOI), India - Survey of Persons with Disabilities NSS 76th Round: July 2018–December 2018—Overview, 5 2022.
5. M. Husain, Neural control of hand movement. Brain **145**(4), 1191–1192 (2022)
6. L.H. Kim, D.H. Heck, R.V. Sillitoe, Cerebellar functions beyond movement and learning. Ann. Rev. Neurosci. **47**(1), 145–166 (2024)
7. A. Ramot, F.H. Taschbach, Y.C. Yang, Y. Hu, Q. Chen, B.C. Morales, X.C. Wang, A. Wu, K.M. Tye, M.K. Benna et al., Motor learning refines thalamic influence on motor cortex. Nature 1–10 (2025)
8. Z. Polyakova, S. Chiken, N. Hatanaka, A. Nambu, Cortical control of subthalamic neuronal activity through the hyperdirect and indirect pathways in monkeys. J. Neurosci. **40**(39), 7451–7463 (2020)
9. A. Nambu, S. Chiken, H. Sano, N. Hatanaka, J. A. Obeso, Dynamic activity model of movement disorders: a unified view to understand their pathophysiology. Clini. Neurol. **64**(6), 390–397 (2024)
10. A. Nambu, S. Chiken, H. Sano, N. Hatanaka, J.A. Obeso, Dynamic activity model of movement disorders: the fundamental role of the hyperdirect pathway. Movement Disorders **38**(12), 2145–2150 (2023)
11. W. Chen, C. de Hemptinne, A.M. Miller, M. Leibbrand, S.J. Little, D.A. Lim, P.S. Larson, P.A. Starr, Prefrontal-subthalamic hyperdirect pathway modulates movement inhibition in humans. Neuron **106**(4), 579–588 (2020)

12. A.A. Moustafa, R.D. McMullan, B. Rostron, D.H. Hewedi, H.H. Haladjian, The thalamus as a relay station and gatekeeper: relevance to brain disorders. Rev. Neurosci. **28**(2), 203–218 (2017)
13. T.M. Florio, Emergent aspects of the integration of sensory and motor functions. Brain Sci. **15**(2), 162 (2025)
14. H.C. Dijkerman, E.H. De Haan, Somatosensory processing subserving perception and action: Dissociations, interactions, and integration. Behav. Brain Sci. **30**(2), 224–230 (2007)
15. E. Nemati, D.B. Grayden, A.N. Burkitt, P.Z. Eskikand, Balancing prior knowledge and sensory data in a predictive coding model of coherent motion detection. PLOS Comput. Biol. **21**(5), e1013116 (2025)
16. B. Rostamian, M. Koolani, P. Abdollahzade, M. Lankarany, E. Falotico, M. Amiri, N.V. Thakor, Texture recognition based on multi-sensory integration of proprioceptive and tactile signals. Sci. Rep. **12**(1), 21690 (2022)
17. J. Kühn, T. Hu, A. Tödtheide, E. Pozo Fortunić, E. Jensen, S. Haddadin, The synergy complement control approach for seamless limb-driven prostheses. Nature Mach. Intell. **6**(4), 481–492 (2024)
18. R. De-la Torre, E. D. Oña, C. Balaguer, A. Jardón, Robot-aided systems for improving the assessment of upper limb spasticity: a systematic review. Sensors **20**(18), 5251 (2020)
19. S. Bhattacharjee, R. Kashyap, T. Abualait, S.-H. Annabel Chen, W.-K. Yoo, S. Bashir, The role of primary motor cortex: more than movement execution. J. Motor Behav. **53**(2), 258–274 (2021)
20. L. Cao, G. Li, Y. Xu, H. Zhang, X. Shu, D. Zhang, A brain-actuated robotic arm system using non-invasive hybrid brain–computer interface and shared control strategy. J. Neural Eng. **18**(4), 046045 (2021)
21. S. Khalid, F. Alnajjar, M. Gochoo, A. Renawi, S. Shimoda, Robotic assistive and rehabilitation devices leading to motor recovery in upper limb: a systematic review. Disabil. Rehabil. Assist. Technol. **18**(5), 658–672 (2023)
22. J. Shin, A.S.M. Miah, M.H. Kabir, M.A. Rahim, A. Al Shiam, A methodological and structural review of hand gesture recognition across diverse data modalities. IEEE Access **12**, 142606–142639 (2024)
23. M. Stępień, J. Conradi, G. Waterstraat, F.U. Hohlefeld, G. Curio, V.V. Nikulin, Event-related desynchronization of sensorimotor EEG rhythms in hemiparetic patients with acute stroke. Neurosci. Lett. **488**(1), 17–21 (2011)
24. N. Yoshimura, H. Tsuda, T. Kawase, H. Kambara, Y. Koike, Decoding finger movement in humans using synergy of EEG cortical current signals. Sci. Rep. **7**(1), 1–11 (2017)
25. M. Teplan et al., Fundamentals of EEG measurement. Meas. Sci. Rev. **2**(2), 1–11 (2002)
26. E. Formaggio, S.F. Storti, I. Boscolo Galazzo, M. Gandolfi, C. Geroin, N. Smania, A. Fiaschi, P. Manganotti, Time-frequency modulation of ERD and EEG coherence in robot-assisted hand performance. Brain Topography **28**(2), 352–363 (2015)
27. J. Wang, Y.-H. Chen, J. Yang, M. Sawan, Intelligent classification technique of hand motor imagery using EEG beta rebound follow-up pattern. Biosensors **12**(6), 384 (2022)
28. P. Ofner, A. Schwarz, J. Pereira, D. Wyss, R. Wildburger, G.R. Müller-Putz, Attempted arm and hand movements can be decoded from low-frequency EEG from persons with spinal cord injury. Sci. Rep. **9**(1), 1–15 (2019)
29. K. Wang, M. Xu, Y. Wang, S. Zhang, L. Chen, D. Ming, Enhance decoding of pre-movement EEG patterns for brain–computer interfaces. J. Neural Eng. **17**(1), 016033 (2020)
30. P. Gaur, H. Gupta, A. Chowdhury, K. McCreadie, R.B. Pachori, H. Wang, A sliding window common spatial pattern for enhancing motor imagery classification in EEG-BCI. IEEE Trans. Instrum. Meas. **70**, 1–9 (2021)
31. G. Lange, C.Y. Low, K. Johar, F.A. Hanapiah, F. Kamaruzaman, Classification of electroencephalogram data from hand grasp and release movements for BCI controlled prosthesis. Proc. Technol. **26**, 374–381 (2016)
32. A. Khorshidtalab, M.-J.E. Salami, M. Hamedi, Robust classification of motor imagery EEG signals using statistical time–domain features. Physiol. Meas. **34**(11), 1563 (2013)

33. S.C. Machavarapu, M.K. Mukul, D. Kumar, EEG classification based on variance, in *2014 International Conference on Green Computing Communication and Electrical Engineering*, pp. 1–4
34. C. Mühl, C. Jeunet, F. Lotte, EEG-based workload estimation across affective contexts. Front. Neurosci. **8**, 114 (2014)
35. M. Pooya Chanu, D. Pei, P. Olikkal, R.K. Vinjamuri, N.M. Kakoty, Electroencephalogram based control of prosthetic hand using optimizable support vector machine, in *Proceedings of the 2023 6th International Conference on Advances in Robotics* (2023), pp. 1–6
36. P.C. Maibam, D. Pei, P. Olikkal, R.K. Vinjamuri, N.M. Kakoty, Enhancing prosthetic hand control: a synergistic multi-channel electroencephalogram. Wearable Technol. **5**, e18 (2024)
37. A. Bicchi, M. Gabiccini, M. Santello, Modelling natural and artificial hands with synergies. Philos. Trans. R. Soc. B Biol. Sci. **366**(1581), 3153–3161 (2011)
38. C. Alessandro, I. Delis, F. Nori, S. Panzeri, B. Berret, Muscle synergies in neuroscience and robotics: from input-space to task-space perspectives. Front. Comput. Neurosci. **7**, 43 (2013)
39. M. Santello, G. Baud-Bovy, H. J¨orntell, Neural bases of hand synergies. Front. Comput. Neurosci. **7**, 23 (2013)
40. V. Patel, M. Burns, R. Chandramouli, R. Vinjamuri, Biometrics based on hand synergies and their neural representations. IEEE Access **5**, 13422–13429 (2017)
41. Y. Zhang, C. Zhang, R. Jiang, S. Qiu, H. He, A distribution adaptive feedback training method to improve human motor imagery ability. IEEE Trans. Neural Syst. Rehabil. Eng. **33**, 380–390 (2025)
42. J. Zhang, C.-H. Chou, M. Hao, W. Liang, Z. Zhang, A. Xie, J.L. Patton, W. Pei, N. Lan, Somatotopically evoked tactile sensation via transcutaneous electrical nerve stimulation improves prosthetic sensorimotor performance. IEEE Trans. Neural Syst. Rehabil. Eng. **32**, 2815–2825 (2024)

Chapter 11
Design of a Human Machine Interface for Assisted Speech on an EOG-based Voice Speller

Diego Caro-López, Claudia R. De Stefano-Ramos, Aranza Carrillo-Márquez, Pablo E. Herrera-Zamora, Jorge de J. Lozoya-Santos, José L. Contreras-Vidal, and Mauricio A. Ramírez-Moreno

Abstract Individuals with severe motor and speech disabilities often face significant barriers to engage in communication. The purpose of this study is to propose a non-invasive Human–Machine Interface that enables such individuals to convey messages using electrooculography signals. The system is composed of three main components: an eye gesture detection method, a customizable virtual keyboard, and a voice assistance module. The eye gesture module recognizes five distinct actions: four directional eye movements (right, left, up, and down) and voluntary blinks. These gestures are captured through surface electrodes and processed using signal filtering and thresholding techniques. A calibration test was designed to record individual eye dynamics and generate personalized control parameters for accurate gesture detection and interaction. The virtual keyboard includes adaptive features such as text prediction and completion, implemented through Natural Language Processing to optimize communication speed. Keyboard navigation is controlled with eye movements and letter selection is confirmed via double blinking, and finalized messages are converted into speech through an integrated text-to-speech engine within the voice assistance module. This project contributes to assistive

D. Caro-López
School of Engineering and Sciences, Tecnologico de Monterrey, Monterrey, Mexico

Noninvasive Brain–Machine Interface System Laboratory, Department of Electrical and Computer Engineering, NSF IUCRC BRAIN, University of Houston, Houston, TX, USA
e-mail: A00833057@tec.mx

C. R. De Stefano-Ramos · A. Carrillo-Márquez · P. E. Herrera-Zamora · J. de J. Lozoya-Santos · M. A. Ramírez-Moreno (✉)
School of Engineering and Sciences, Tecnologico de Monterrey, Monterrey, Mexico
e-mail: A01382749@tec.mx; A00832698@tec.mx; A01198045@tec.mx; jorge.lozoya@tec.mx; mauricio.ramirezm@tec.mx

J. L. Contreras-Vidal
Noninvasive Brain–Machine Interface System Laboratory, Department of Electrical and Computer Engineering, NSF IUCRC BRAIN, University of Houston, Houston, TX, USA
e-mail: jlcontr2@central.uh.edu

R. Vinjamuri (ed.), *Bridging the Gap between Mind and Machine*,
https://doi.org/10.1007/978-3-032-06713-5_11

neurotechnology by presenting a scalable, user-centered electrooculography-based communication solution for individuals with severe motor and speech impairments.

Keywords Human-machine interface · Electrooculography · Augmentative and alternative communication

11.1 Introduction

Motor disabilities include a wide range of conditions involving impairments in gross and fine motor function, affecting individuals of all ages. The World Health Organization estimates that approximately 1.3 billion people—nearly 16% of the global population—live with significant disabilities [1]. Motor impairments often hinder communication, as deficits in movement can delay speech production and reduce communicative effectiveness. In children with cerebral palsy, gross motor severity explains over 38% of the variability in communication outcomes [2]. Additionally, systematic reviews in neurodevelopmental disorders such as autism spectrum disorder report moderate correlations between motor deficits and social-communication impairments [3]. These findings highlight the impact of motor disabilities on social and communicative development.

Human-Computer interaction is the field of study that focuses on optimizing how users interact with computer systems called Human-Computer Interfaces (HCIs), through intuitive and responsive interfaces that satisfy users' needs. Human-Machine interaction focuses on the design, development, and evaluation of interfaces between humans and machines. Closely related is the field of Human-Machine Interfaces (HMIs), which extends the principles of HCIs beyond traditional computers to include a wide variety of machines such as autonomous vehicles, robots, and smart devices [4].

At the intersection of HMIs and the medical field lies the development of Assistive Technology (AT), especially for individuals with a wide variety of motor disabilities such as cerebral palsy, amyotrophic lateral sclerosis (ALS) or muscular dystrophy. These technologies aim to enhance or restore autonomy by enabling individuals to control elements of their environment, including communication. One significant category within AT is Augmentative and Alternative Communication (AAC), which includes tools and systems that replace or supplement spoken language [5].

AAC systems often rely on various types of input signals, including imaging, mechanical, electromechanical, biosignals, touch, breath, and Brain-Computer Interfaces (BCIs) [5–8]. Among biosignal-based systems, Electrooculography (EOG) has emerged as a promising control modality for AAC applications due to several advantages. EOG measures the corneo-retinal standing potential that occurs with eye movement, and it features a linear relationship with gaze direction, high signal amplitude, and waveform characteristics that are relatively easy to detect and process [5, 9–11]. These features make EOG especially suitable for applications such as mouse emulation and virtual keyboard control in assistive communication systems.

Common approaches to implementing EOG-based AAC systems include virtual keyboards combined with gaze-based navigation and a selection method. For instance, Dhillon et al. [9] proposed a QWERTY keyboard controlled by horizontal and vertical eye movements, with key selection triggered by Electromyography (EMG) signals from eyebrow movements. Similarly, Nathaniel et al. [11] introduced a QWERTY keyboard that employed a regression model to estimate gaze displacement and used a dwell-based validation technique, wherein a detected gaze fixation selected the desired key. Their system achieved an average character selection accuracy of 99.23% and a typing speed of 11.89 characters per minute, demonstrating the viability of EOG-based systems for communication.

The present project aims to contribute to this field through the proposal of an HMI that enables users to communicate using a virtual keyboard integrated with a text-to-speech module. The system operates as a closed-loop interface controlled entirely through EOG signals. Specifically, it is capable of detecting and classifying five distinct ocular gestures: four directional eye movements (right, left, up, and down), and eye blinks. The selection mechanism is based on a double-blink pattern, which provides a reliable and intuitive method for key confirmation.

The virtual keyboard used in this system introduces a novel circular layout designed to optimize selection time and user accessibility. Unlike traditional linear QWERTY keyboards, the circular configuration reduces gaze travel distance and aligns more naturally with horizontal eye movement patterns. Furthermore, the keyboard incorporates a Natural Language Processing (NLP) component based on a GPT-2 model trained in Spanish, which enables predictive text input and word autocompletion. These features enhance both the speed and fluidity of user communication, helping to mitigate the inherently slower input rates of gaze-controlled systems. Additional functionalities are included to improve user experience and adaptability. A customization button allows users or caregivers to modify system settings and extend the keyboard's capabilities to suit individual preferences.

Additionally, electroencephalography (EEG) signals are intended to be recorded simultaneously during the use of the proposed system. These data will be preprocessed and analyzed to locate brain activity sources through Independent Component Analysis and inverse source localization.

11.2 Methods

11.2.1 Calibration Test

A calibration test (Fig. 11.1) was developed in Python (v3.12.3, Python Software Foundation) to estimate EOG thresholds for specific eye gestures. The test begins

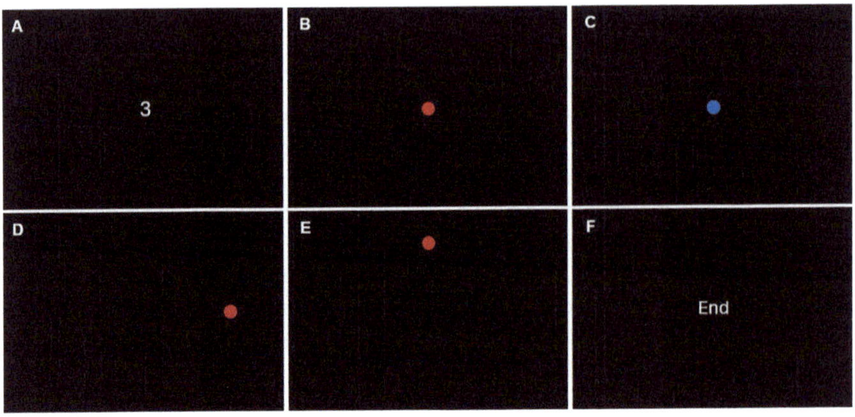

Fig. 11.1 Key frames of the calibration test animation. **(a)** Start of the initial countdown. **(b)** Rest phase at the center. **(c)** Blink cue. **(d)** Cardinal right target. **(e)** Cardinal top target. **(f)** End of the animation

with a countdown from 3 to 1 on a black background. After the countdown, a red fixation point appears at the center of the screen for a random duration between 2 and 4 seconds, followed by a red dot presented for 1 second in one of four cardinal directions (right, left, up, or down) in random order .

The central fixation point turns blue at two random times during calibration. On each occasion, participants will perform a double blink while maintaining gaze on the blue dot. A final red dot appears at the center for 2 seconds, followed by the word "End" to indicate calibration was complete.

11.2.2 Virtual Keyboard Design

The virtual keyboard was developed in Python (v3.12.3, Python Software Foundation) and implemented for real-time use with commands activated through eye gestures. A circular interface was chosen because it offers some advantages over traditional linear layouts. The keys are arranged in concentric rings around a central fixation point at the center of the screen (Fig. 11.2a, c). A traditional QWERTY keyboard was also developed (Fig. 11.2b, d) and can be accessed via a customization button included in the keyboard's graphical user interface (GUI). This customization button also allows the user to switch between light and dark modes by changing the key and background colors.

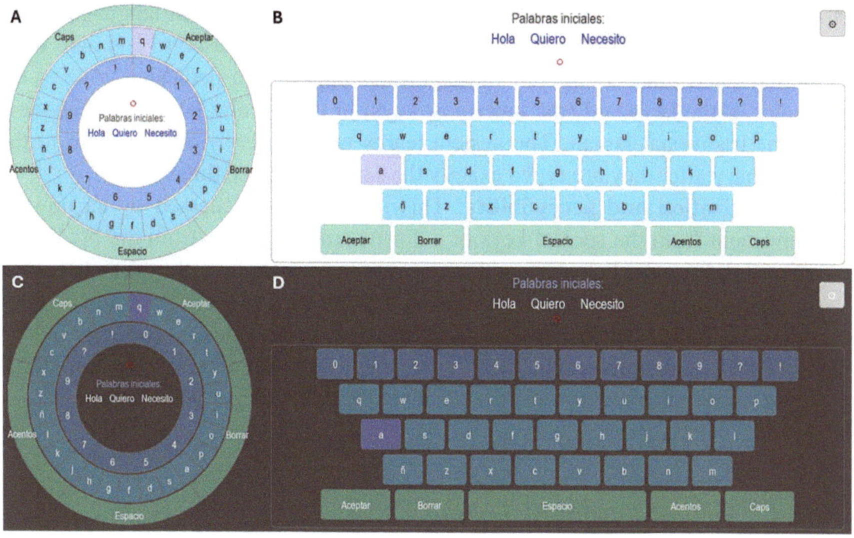

Fig. 11.2 (**a**) Circular virtual keyboard developed with its respective traditional QWERTY layout in light (**b**) and dark mode (**c**), (**d**)

11.2.3 Virtual Keyboard Algorithm

The virtual keyboard provides a fully navigable GUI using the Pygame library (https://www.pygame.org/), designed for flexible inputs and assistive features, including predictive text and voice feedback. It consists of four main blocks listed below:

1. **Navigation and Selection:** Users navigate between zones (e.g., letters, numbers, special keys) using eye gestures. In the circular layout, horizontal eye movements switch between keys within a ring, while vertical movements change the depth of the selected key (i.e., from the outer to the inner rings). In the QWERTY layout, eye movement direction determines the navigation direction across the grid of keys. The currently selected zone is highlighted. Alternatively, users can navigate using keyboard arrows or the mouse. Keys can be selected via double blink, the Enter key, or a mouse click. A confirmation sound is played upon each key press.
2. **Special Keys and Actions:** These are the outermost keys, located on the external ring of the circular keyboard interface, and at the bottom of the QWERTY keyboard. *Aceptar* (Accept) uses text-to-speech to read aloud the typed text, then clears the input field. *Borrar* (Delete) removes the last character from the input. *Espacio* (Space) inserts a space to finish a word. If a word was completed exactly with the same characters three times, it is added to a custom dictionary for word autocompletion. *Acentos* (Accents) activates accent mode, allowing vowels to be typed with accents. *Caps* toggles between uppercase and lowercase letters.

3. **Predictive text and autocompletion:** A GPT-2-based Spanish language model suggests the next word to be typed. A word dictionary is used to compare word prefixes and offer completions. Suggestions appear in a dedicated zone at the center of the keyboard (the innermost keys) in the circular layout, and above it in the traditional layout. The system learns new words by adding them to the dictionary after they are typed exactly the same way three times.
4. **Customization and accessibility:** This menu can be accessed by an upward movement from the special keys or a downward movement from the prediction zone in the circular layout, or with an upward movement from the prediction zone in the traditional keyboard. Users can increase or decrease the keyboard size for better visibility, switch between light and dark color palettes, choose between QWERTY and alphabetical letter order, toggle between circular and traditional layouts, and enable or disable the predictive text feature.

11.2.4 Virtual Keyboard Testing

To evaluate the performance of the proposed virtual keyboard designs, a simulation-based analysis was conducted using five representative communication phrases, shown in Table 11.1. The phrases were typed using both the circular layout and the QWERTY layout, with simulated eye movements represented by directional key presses. Each horizontal or vertical movement was counted as a navigation step, and double blinks were simulated with a selection key. The total number of required movements and selections was recorded for each phrase on both layouts, allowing a comparative assessment of navigation efficiency.

11.3 Results

Table 11.2 summarizes the number of directional movements required to type each phrase, providing a comparative view of the navigation demands imposed by each interface. The results show that the total number of keystrokes required to type each phrase was comparable between the circular and traditional QWERTY layouts. However, a notable difference was observed in the distribution of horizontal

Table 11.1 Phrases used for the simulated typing analysis, along with their English translations

Phrase	Spanish	English
1	Me siento bien	I feel good
2	Quiero agua	I want water
3	Sí	Yes
4	No	No
5	Necesito ir al baño	I need to go to the bathroom

Table 11.2 Comparison of simulated directional commands required to compose and output each phrase using each virtual keyboard layout

	Circular			QWERTY		
Phrase	Horizontal	Vertical	Total	Horizontal	Vertical	Total
1	37	12	**49**	24	25	**49**
2	16	8	**24**	10	13	**23**
3	15	1	**16**	11	5	**16**
4	12	1	**13**	12	7	**19**
5	45	14	**59**	36	21	**57**

Total refers to the sum of horizontal and vertical commands.

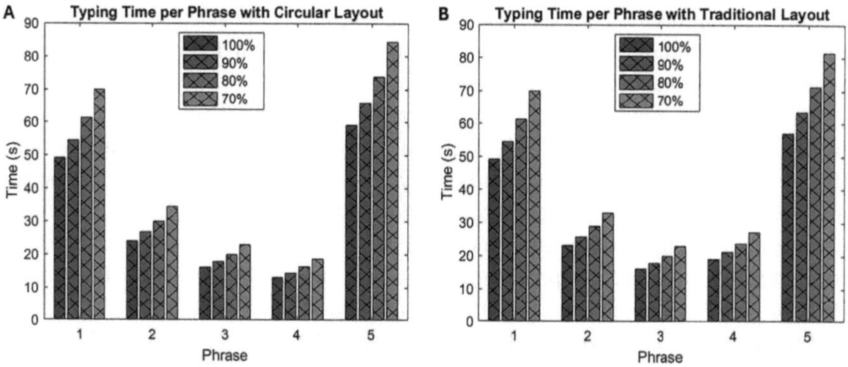

Fig. 11.3 Typing time for each phrase at different gesture detection accuracies for the circular **(a)** and traditional **(b)** virtual keyboard layouts

and vertical movements. The circular layout consistently required fewer vertical keystrokes, which were offset by an increase in horizontal navigation. This suggests that the circular interface may reduce the need for complex or effortful vertical eye gestures, potentially offering improved comfort or usability for users with limited vertical gaze control.

Assuming a one-second delay per command detection, simulated typing times were estimated using the data from Table 11.2. Additionally, different gesture detection accuracy levels were considered. Figure 11.3 presents a comparison of typing times for each phrase across varying accuracies using both virtual keyboard layouts. This one-second delay limits the maximum interaction speed to 60 commands per minute at 100% gesture detection accuracy. At 70% accuracy, the effective rate decreases to 42 commands per minute due to repeated commands needed for recognition errors.

11.4 Discussion

This work presents the development of an HMI utilizing EOG signals to enable communication for individuals with severe motor and speech disabilities. The primary contributions include the proposal of a modular HMI system integrating a responsive eye gesture detection module, a customizable virtual keyboard, and a text-to-speech voice assistance module. By focusing on user adaptability, interface flexibility, and predictive text features, the system aims to bridge a gap for a user group facing significant communication barriers.

A key finding, illustrated in Table 11.2 and Fig. 11.3, is the difference in gaze navigation demands between the circular and QWERTY keyboard layouts. Simulated tests showed that while both layouts required a similar total number of commands to type the phrases, the circular layout consistently reduced the number of vertical eye movements, replacing them with horizontal commands. As shown in Fig. 11.3, overall typing times remained comparable across different gesture recognition accuracies. This reduced reliance on vertical saccades may provide ergonomic benefits, especially for users with limited vertical gaze control or increased fatigue during vertical eye movements.

Favoring horizontal saccades is supported not only by these findings but also by oculomotor research. A cross-sectional study across a wide age range found that vertical saccades exhibit significantly longer latencies, greater hypometria, and lower peak velocities compared to horizontal ones [12]. These limitations can negatively impact selection speed and accuracy, particularly during large eye movements. Therefore, emphasizing horizontal navigation—such as that in the circular layout—can enhance performance and reduce user fatigue.

Further support comes from findings showing that both eye and hand tracking of horizontal motion are more accurate and less delayed than their vertical counterparts, even with unpredictable trajectories [13]. This directional asymmetry reflects more efficient neural and cognitive mechanisms for horizontal movement prediction and control. Leveraging these physiological advantages can improve the reliability of gesture detection and overall system responsiveness.

Additional features such as Natural Language Processing for word prediction and autocompletion, as well as customization options like layout switching, color palettes, and predictive text toggling, contribute to a user-centered design. These elements support personalization and may enhance long-term usability and user satisfaction. The integration of a voice output engine enables real-time communication in AAC scenarios.

Nonetheless, the study has limitations. The system was validated through simulation, and further testing is needed with clinical populations. Gesture detection accuracy, may benefit from improved signal processing or machine learning techniques. Practical deployment may also present challenges related to electrode stability, environmental noise, and interface robustness.

Overall, the results align with current trends in AT, which emphasize adaptability, multimodal control, and biosignal integration. This project provides a flexible, modular foundation for EOG-based communication systems and paves the way for future development, including hybrid EEG–EOG platforms and broader clinical applicability.

11.5 Conclusion

The developed EOG-based HMI system demonstrates the feasibility of enabling assisted speech and communication for individuals with severe motor and speech impairments through non-invasive eye gesture detection and a virtual keyboard interface. Features such as adaptive layout customization, predictive text, and integrated voice output contribute to an enhanced user experience and improved communication efficiency.

A comparative simulation of the circular and QWERTY keyboard layouts showed a similar total number of required gestures, but the circular layout consistently reduced vertical eye movements—an advantage for users with limited vertical gaze control. Future work should focus on evaluating the system with clinical populations, and refining the interface for real-world applications.

Acknowledgments The authors would like to acknowledge the NSF IUCRC BRAIN Center at the University of Houston and its affiliate site at Tecnológico de Monterrey and for their support in the development of this work.

Author Contributions Conceptualization, D.C.-L., C.R.D.S.-R., and M.A.R.-M.; methodology, D.C.-L., C.R.D.S.-R., A.C.-M., and P.E.H.-Z.; software, D.C.-L., A.C.-M., and P.E.H.-Z.; validation, D.C.-L., C.R.D.S.-R., M.A.R.-M., and J.L.C.-V.; formal analysis, D.C.-L., M.A.R.-M., J.L.C.-V., and C.R.D.S.-R., investigation, D.C.-L., and C.R.D.S.-R.; data curation, D.C.-L., and C.R.D.S.-R.; writing—original draft preparation, D.C.-L.; writing—review and editing, D.C.-L., and M.A.R.-M.; visualization, D.C.-L., and M.A.R.-M.; supervision, M.A.R-M.; ; funding acquisition, J.J.L.-S.; resources, J.J.L.-S.; project administration, M.A.R.-M. All authors have read and agreed to the published version of the manuscript.

Competing Interests The authors declare no competing interests.

References

1. World Health Organization, Disability and health (2023). Accessed Jul 12, 2025
2. A. Coleman, K.A. Weir, R.S. Ware, R.N. Boyd, Relationship between communication skills and gross motor function in preschool-aged children with cerebral palsy. Arch. Phys. Med. Rehabil. **94**(11), 2210–2217 (2013)
3. L.A. Wang, V. Petrulla, C.J. Zampella, R. Waller, R.T. Schultz, Gross motor impairment and its relation to social skills in autism spectrum disorder: a systematic review and two meta-analyses. Psychol. Bull. **148**(3–4), 273 (2022)
4. D. Mourtzis, J. Angelopoulos, N. Panopoulos, The future of the human–machine interface (hmi) in society 5.0. Future Internet **15**(5), 162 (2023)
5. C.G. Pinheiro, E.L. Naves, P. Pino, E. Losson, A.O. Andrade, G. Bourhis, Alternative communication systems for people with severe motor disabilities: a survey. Biomed. Eng. Online **10**, 1–28 (2011)
6. Y. Elsahar, S. Hu, K. Bouazza-Marouf, D. Kerr, A. Mansor, Augmentative and alternative communication (aac) advances: a review of configurations for individuals with a speech disability. Sensors **19**(8), 1911 (2019)

7. B. Peters, B. Eddy, D. Galvin-McLaughlin, G. Betz, B. Oken, M. Fried-Oken, A systematic review of research on augmentative and alternative communication brain-computer interface systems for individuals with disabilities. Front. Human Neurosci. **16**, 952380 (2022)
8. M.A. Hernández-Mustieles, E. Magaña-Cruz, J. Machado-Trejo, C.E. Corona-González, J.D.J. Lozoya-Santos, M.A. Ramírez-Moreno, Design of a virtual chatbot platform for basic needs communication through imagined speech bci, in *Discovering the Frontiers of Human-Robot Interaction: Insights and Innovations in Collaboration, Communication, and Control* (Springer, Berlin, 2024), pp. 381–399
9. H.S. Dhillon, R. Singla, N.S. Rekhi, R. Jha, Eog and emg based virtual keyboard: a brain-computer interface, in *2009 2nd IEEE International Conference on Computer Science and Information Technology* (IEEE, 2009), pp. 259–262
10. D. Borghetti, A. Bruni, M. Fabbrini, L. Murri, F. Sartucci, A low-cost interface for control of computer functions by means of eye movements. Comput. Biol. Med. **37**(12), 1765–1770 (2007)
11. N. Barbara, T.A. Camilleri, K.P. Camilleri, Eog-based eye movement detection and gaze estimation for an asynchronous virtual keyboard. Biomed. Signal Process. Control **47**, 159–167 (2019)
12. E.L. Irving, L. Lillakas, Difference between vertical and horizontal saccades across the human lifespan. Exp. Eye Res. **183**, 38–45 (2019)
13. F.R. Danion, J. Mathew, N. Gouirand, E. Brenner, More precise tracking of horizontal than vertical target motion with both the eyes and hand. Cortex **134**, 30–42 (2021)

Chapter 12
Enhancing Hand Exoskeleton Grasping Performance with Robust Force Controller Development

Kaushik Halder and M. Felix Orlando

Abstract Object grasping is a crucial aspect of performing activities of daily living, as it facilitates individuals to interact effectively with their environment. In elderly human subjects, this capability often becomes challenging due to factors such as reduced muscular strength, diminished motor coordination, and neurodegenerative disorders that impair hand functionality. Additionally, age-related declines in grasping strength and joint flexibility further complicate the ability to precisely and accurately grasp objects. Therefore, this study demonstrates the design and development of an assistive device aimed at improving object grasping performance and overall quality of life in the elderly population, which contributes to bridging the gap between mind and machine. This work focuses on improvement in grasping assistance tasks while developing a two-fingered hand exoskeleton along with a robust grasping force control strategy. Real-time force feedback, obtained through embedded FSR-402 sensors on fingertips, the proposed Sliding Mode Control (SMC)-based approach enables continuous regulation of actuator output, thereby maintaining precise and stable grasping force across different-sized objects. The effectiveness of the proposed system was evaluated based on three distinct groups of object grasping experiments (i. constant grasping force, ii. multi-step grasping force tracking, and iii. disturbance rejection) for three different-sized objects while involving two elderly human subjects. The promising findings indicate that the combination of the proposed robust SMC- based grasping force control with a developed two-fingered hand exoskeleton constitutes a significant advancement toward functional human-machine integration, particularly in the domain of rehabilitation, assistive technology, and activities of daily living.

Keywords Sliding mode control · Hand exoskeleton · Grasping force control · Human-machine integration · Assistive technology

---

K. Halder · M. Felix Orlando (✉)
Department of Electrical Engineering, Indian Institute of Technology Roorkee, Roorkee, Uttarakhand, India
e-mail: k_halder@ee.iitr.ac.in; m.orlando@ee.iitr.ac.in

R. Vinjamuri (ed.), *Bridging the Gap between Mind and Machine*,
https://doi.org/10.1007/978-3-032-06713-5_12

12.1 Introduction

Assistive robotic technologies are gaining increasing global attention for the rehabilitation and support of individuals with orthopedic and neuromuscular impairments. Elderly individuals, in particular, often experience diminished motor function, which severely affects their ability to perform daily activities, especially those requiring precise hand movements. Impaired hand function compromises the ability to grasp and manipulate objects effectively [1–3]. Consequently, the development of hand exoskeletons equipped with advanced force control mechanisms is essential in assistive technology to improve grasping performance in Activities of Daily Living (ADL). Numerous studies have explored the design and implementation of such devices for grasping assistance [4–26]. Within this context, achieving robust and reliable force control for object manipulation remains a critical and emerging area of research.

In recent years, considerable research efforts have been directed toward the development of various hand exoskeleton systems aimed at delivering appropriate grasping forces to support individuals with weakened or impaired hand function in performing ADL. The design of these devices varies depending on the specific application and user requirements. As outlined in prior studies [4, 5], several key design considerations are critical for the effective implementation of hand assistive exoskeletons. These include: (i) ensuring the compactness of the mechanism to avoid interference with environmental interactions during ADL tasks; (ii) achieving functional grasping that closely replicates the natural motion and safety characteristics of a healthy human hand; (iii) maintaining a lightweight and portable structure to enhance user comfort and usability; and (iv) incorporating size adjustability to accommodate varying finger dimensions, which is essential for broader user compatibility.

Hand assistive exoskeletons can be broadly classified into three categories based on their force transmission mechanisms: tendon-driven systems, pneumatic actuator-based designs, and linkage-based mechanisms. Tendon-driven exoskeletons offer several advantages, such as compact form factors and reduced weight, making them suitable for wearable applications [6–8]. However, these systems often suffer from friction-induced nonlinearities between the cable and sheath, which can result in considerable tension loss during force transmission, thereby affecting grasping efficiency. For instance, a single-actuator cable-driven exoskeleton for hand assistance was proposed in [7]. Nonetheless, for effective support during ADL, it is essential that each digit of the user's impaired hand be equipped with an independent force transmission unit to ensure precise and reliable assistance. Butzer et al. [8] developed a spring-assisted, cable-driven hand exoskeleton designed to facilitate four different types of grasping tasks commonly encountered in ADL. The system incorporated a remotely located actuation unit to minimize the weight borne by the user. However, the maximum grasping force generated by the device was limited to 5.2 N, which falls short of the force typically required for effective ADL performance. Another class of assistive exoskeletons utilizes pneumatic actuators, offering advantages such as low weight, enhanced user comfort, and increased compliance due to the low elastic modulus of soft actuators [9–13]. Despite these benefits, the inherent elasticity and pronounced hysteresis effects of pneumatic actuators pose significant challenges for precise force control. In contrast, rigid linkage-based exoskeletons generally offer improved reliability, more predictable mechanical behavior, and simpler maintenance when compared to tendon-driven or pneumatic systems. An additional notable advantage of rigid linkage-based designs is their ability to transmit high and stable forces to the fingertips, which is essential for effective grasping assistance. These mechanisms are also capable of reliably replicating the flexion and extension movements of human fingers. Owing to these benefits, numerous researchers have developed hand assistive systems utilizing direct-driven rigid linkage architectures. However, integrating actuators directly onto the linkage structure introduces several challenges, including increased control complexity and reduced compactness, often resulting in a heavier and bulkier overall design [14–19]. For instance, M. Felix Orlando et al. [19] proposed a novel two-fingered exoskeleton based on a direct-driven four-bar linkage mechanism, optimized for performing object translation tasks. In an effort to reduce overall weight and improve the portability of hand exoskeletons, several researchers have explored the use of under-actuated mechanisms in assistive device design [20–23]. Recently, Li et al. [20] introduced an under-actuated linkage-based exoskeleton specifically designed to assist the index finger, aiming to improve grasping performance in elderly individuals with diminished muscle strength. The proposed system is characterized by its lightweight structure, mechanical compatibility, portability, and inherent back-drivability, making it well-suited for daily assistive applications. Hong et al. [21] proposed a novel under-actuated exoskeleton for the index finger, specifically designed to facilitate power grasping, in combination with a passive linkage structure for the thumb. The design incorporates a spherical four-bar mechanism, with actuators positioned laterally along the hand to enhance wearability. Additionally, a three-fingered under-actuated exoskeleton was introduced in [22], aimed at assisting elderly individuals in grasping tasks. This system was developed with a strong emphasis on ensuring kinematic compatibility between the exoskeleton and the natural motion of the user's hand. Furthermore, a 3 degree of freedom (3 DOF) under-actuated linkage-based hand exoskeleton has been presented in [23], targeting rehabilitation applications by supporting functional hand movements during therapeutic exercises.

12.1.1 Related Works

The majority of existing research in the field has primarily concentrated on the design and mechanical development of hand assistive devices for individuals with impaired or weakened hand function. In contrast, relatively few studies [18–20, 24, 25] have addressed the integration of force control strategies alongside exoskeleton development for object grasping applications. Within the current state-of-the-art in hand assistive technology, most proposed control approaches have focused on position control, joint angle tracking, torque trajectory regulation, and admittance control, with limited emphasis on robust and precise force control for functional grasping. Despite advancements in hand assistive technologies, there remains a critical need for robust force control strategies to enhance grasping performance and operational efficiency. Within the domain of hand exoskeleton force control, only a limited number of studies have proposed effective control schemes. These include Proportional-Derivative (PD)-based control [17, 23], intention-based control strategies [18], Proportional-Integral-Derivative (PID) admittance control [20], and simple proportional control methods [19]. Notably, Jo and Bae [17] introduced a compact, force-controllable hand assistive device incorporating a series elastic actuator. Their approach employed a PD controller, optimized via Linear Quadratic Regulator (LQR) tuning, in conjunction with a disturbance observer to improve control robustness and accuracy. In recent years, H. Li et al. [20] incorporated a robustness-oriented analysis within their control framework by implementing a PID-based admittance control strategy for grasping assistance. However, their study lacks a comprehensive evaluation of disturbance rejection capabilities and does not include controller validation under varying force trajectory tracking scenarios. Guotao et al. [23] implemented a hybrid control strategy combining position control and admittance control in their developed hand exoskeleton system. Specifically, a Proportional-Derivative (PD) controller was utilized for regulating finger position, while the interaction between the exoskeleton and the external environment was managed through an admittance control scheme. However, their approach did not include a robustness analysis with respect to external disturbances, leaving the system's performance under unpredictable conditions unaddressed. In the context of hand exoskeleton control, only a limited number of studies [25–27] have adopted Sliding Mode Control (SMC) to enhance robustness. For instance, Wege et al. [25] employed SMC for position control in a hand exoskeleton system, while Babaiasl et al. [26] applied SMC for joint angle trajectory tracking in an upper-limb rehabilitation robot. Notably, to improve grasping performance using hand exoskeleton, the implementation of a robust control scheme is essential. In this context, this study proposes a force feedback-based SMC strategy for direct force control in hand assistive devices. The primary objective is to design and implement a robust sliding mode force control approach, with optimally tuned parameters, for an own developed underactuated two-fingered hand exoskeleton, aimed at improving grasping performance.

12.2 Kinematic Model and Force Analysis

12.2.1 Forward Kinematics

In this study, a two-fingered hand exoskeleton was developed to support grasping tasks in elderly individuals. The device features a lightweight design (200 g), adjustable sizing to accommodate various hand dimensions, force control capability, and an underactuated linkage mechanism. The kinematic model along with CAD model of the hand exoskeleton is depicted in Fig. 12.1. Based on the Fig. 12.1b, the

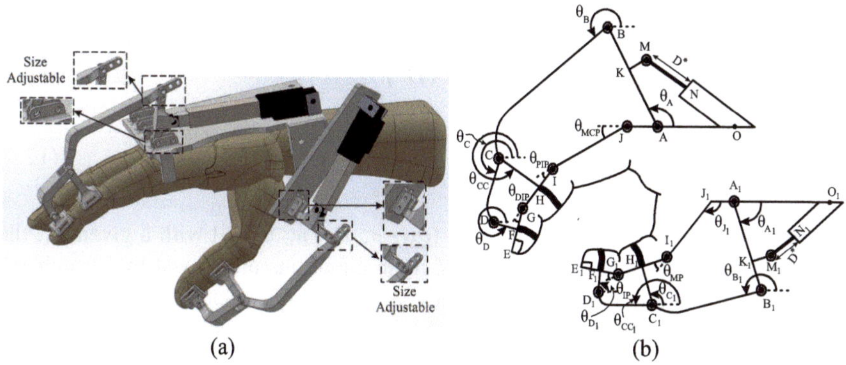

Fig. 12.1 Designed hand exoskeleton, (**a**) CAD model of the prototype and (**b**) Kinematic model of the hand exoskeleton [28]

base frame for index finger exoskeleton is placed at O. Notably, the linear actuator is also attached at O point to actuate the index finger. This arrangement provides the revolute joint at point A as active joint with corresponding rotation angle θ_A. The index finger exoskeleton system has eight revolute joints at points A, B, C, D, and M, including three biological joints at points G, I, and J. As per kinematic model of the index finger exoskeleton, G, I, and J points are indicated as distal interphalangeal (DIP), proximal interphalangeal (PIP) and metacarpophalangeal (MCP) joints, respectively, and corresponding joint angles are θ_{DIP}, θ_{PIP}, and θ_{MCP}. The constraints equations for MCP, PIP, and DIP joints of index finger exoskeleton are considered as, $\theta_{PIP} = \frac{3}{4}\theta_{MCP}$ and $\theta_{DIP} = \frac{2}{3}\theta_{PIP}$ [20]. The flexion and extension of fingers can be achieved by the linear motion of actuator, that insists the fingers to have the desired postures for grasping different sized objects. The index finger exoskeleton comprises two closed loop chains and associated equations are as follows [28],

$$L_{AB}e^{j\theta_A} + L_{BC}e^{j\theta_B} + L_{CG}e^{j\theta_C} = L_{AJ} + L_{JI}e^{j(\pi+\theta_{MCP})} + L_{IH}e^{j(\pi+\theta_{MCP}+\theta_{PIP})} \tag{12.1}$$

$$L_{AB}e^{j\theta_A} + L_{BC}e^{j\theta_B} + L_{CG}e^{j\theta_C} = L_{AJ} + L_{JI}e^{j(\pi+\theta_{MCP})} + L_{IH}e^{j(\pi+\theta_{MCP}+\theta_{PIP})} \tag{12.2}$$

$$L_{CD}e^{j\theta_{CC}} + L_{DE}e^{j\theta_D} = L_{HG}e^{j(\pi+\theta_{MCP}+\theta_{PIP})} + L_{GF}e^{j(\pi+\theta_{MCP}+\theta_{PIP}+\theta_{DIP})} \tag{12.3}$$

where, L_{ik} is considered as length between two points i and k. L_{AJ}, L_{JI}, L_{IG}, and L_{GE} represent the metacarpal, proximal, middle, and distal phalanges' length, respectively. Six passive joints at B, C, D, J, I, and G points are having dependent joint angles, which are depending on active joint at point A (joint angle θ_A). Here, θ_A is considered as an active joint angle which has a direct relation (given in Eq. (12.4)) with the displacement of linear actuator (D^*).

$$\theta_A = \arccos(\frac{L_{OA}^2 + L_{OK}^2 - (D^* + L_{ON})^2}{2L_{OA}L_{OK}}) + \arctan(\frac{L_{KM}}{L_{AK}}) \tag{12.4}$$

To obtain a numerical solution of $\theta = [\theta_{MCP}, \theta_{PIP}, \theta_{DIP}]$ with a given θ_A, the aforementioned closed-loop kinematics chain equations are solved by Levenberg-Marquardt algorithm. Forward kinematics equations of the index finger exoskeleton is expressed in Eq. (12.5), where, x_{in} and y_{in} are index fingertip coordinates.

$$\begin{aligned} x_{in} &= L_{OJ} + L_{JI}\cos(\pi + \theta_{MCP}) + L_{IG}\cos(\pi + \theta_{MCP} + \theta_{PIP}) \\ &\quad + L_{GE}\cos(\pi + \theta_{MCP} + \theta_{PIP} + \theta_{DIP}) \\ y_{in} &= L_{JI}\sin(\pi + \theta_{MCP}) + L_{IG}\sin(\pi + \theta_{MCP} + \theta_{PIP}) \\ &\quad + L_{GE}\sin(\pi + \theta_{MCP} + \theta_{PIP} + \theta_{DIP}) \end{aligned} \tag{12.5}$$

Similarly, Eqs. (12.6)–(12.7) illustrate the closed-loop kinematics chain equations for thumb exoskeleton. A linear actuator is also mounted at point O_1 to facilitate the motion of thumb digit.

$$\begin{aligned} L_{A_1B_1}e^{-j\theta_{A_1}} + L_{B_1C_1}e^{j\theta_{B_1}} + L_{C_1G_1}e^{j\theta_{C_1}} &= L_{A_1J_1} + L_{J_1I_1}e^{j\theta_{J_1}} \\ &\quad + L_{I_1H_1}e^{j(\pi - \theta_{MP} + \theta_{J_1})} \end{aligned} \tag{12.6}$$

$$L_{C_1D_1}e^{j\theta_{CC_1}} + L_{D_1E_1}e^{j\theta_{D_1}} = L_{H_1G_1}e^{j(\pi - \theta_{MP} + \theta_{J_1})} + L_{G_1F_1}e^{j(\pi - \theta_{IP} - \theta_{J_1})} \tag{12.7}$$

The kinematic model of thumb exoskeleton highlights five passive joints at B_1, C_1, D_1, I_1, and the joint angle at G_1 is dependent and governed by the motion of active joint at A_1 (expression for joint angle θ_{A_1} is similar to the expression of θ_A, given in Eq. (12.4)). Two biological joints metacarpophalangeal (MP) and interphalangeal (IP) are represented at points I_1 and G_1, respectively while considering the biological constraint as $\theta_{IP} = \frac{1}{2}\theta_{MP}$ [20]. Given the values of θ_{A_1} and θ_{J_1} Eqs. (12.6)–(12.7) can be numerically solved by Levenberg-Marquardt algorithm, to obtain θ_{MP} and θ_{IP}. The forward kinematic equations for thumb digit are presented below, where, x_{th} and y_{th} represent the coordinates of thumb digit tip [28].

$$\begin{aligned} x_{th} &= L_{O_1J_1} + L_{J_1I_1}\cos(2\pi - \theta_{J_1}) + L_{I_1G_1}\cos(\pi + \theta_{MP} - \theta_{J_1}) \\ &\quad + L_{G_1E_1}\cos(\pi - \theta_{IP} - \theta_{MP} + \theta_{J_1}) \\ y_{th} &= L_{J_1I_1}\sin(2\pi - \theta_{J_1}) + L_{I_1G_1}\sin(\pi + \theta_{MP} - \theta_{J_1}) \\ &\quad + L_{G_1E_1}\sin(\pi - \theta_{IP} - \theta_{MP} + \theta_{J_1}) \end{aligned} \tag{12.8}$$

12.2.2 Actual Grasping Force Analysis

Cartesian position of the fingertip can be computed using the forward kinematics. If we consider the fingertip position of index finger or thumb digit at a particular instant as X_{tip} and Y_{tip} for a specific active angle θ_A, which is given in the Eqs. (12.5) and (12.8) with respect to the base frame, differential kinematics allows to develop a direct relationship between the Cartesian velocity and $\dot{\theta}_A$ through Jacobian $\boldsymbol{J}$ as,

$$\dot{\boldsymbol{X}} = \boldsymbol{J}_{2\times 1}\dot{\theta}_A, \tag{12.9}$$

where, $\boldsymbol{X} = [X_{tip} \quad Y_{tip}]^T$. Now, applying the concept of calculating actual force, the relationship between joint torque and tip force is given by,

$$\boldsymbol{\tau} = \boldsymbol{J}^T \boldsymbol{F} \tag{12.10}$$

where, $\boldsymbol{F}$ is actual force at the fingertip and $\boldsymbol{\tau}$ is torque at the hinge (active angle). Before going for the force calculation using Jacobian transpose procedure, it is considered that the proposed linkage based hand exoskeleton digits are having ideal dynamics (zero mass and unity viscous friction) [19, 29]. So, the torque can be written by, $\boldsymbol{\tau} = \dot{\theta}_A$. The above relation enables to find the fingertip force as,

$$\boldsymbol{F} = (\boldsymbol{J}^T)^{-1}\boldsymbol{\tau} = (\boldsymbol{J}^T)^{-1}\dot{\theta}_A \tag{12.11}$$

12.3 Robust Force Controller Design

SMC represents a robust state-feedback control strategy particularly effective for nonlinear systems subjected to model uncertainties and external disturbances. Its principal attributes include: (i) the ability to reduce the effective system dynamics by transforming higher-order system behavior into lower-order dynamics via sliding motion on a predefined manifold; and (ii) inherent robustness, as the control law maintains system performance in the presence of bounded disturbances and parameter variations. Force control in hand exoskeleton systems presents significant challenges, primarily due to the mechanical complexity and high degrees of freedom inherent in their structure. Conventional control approaches, which typically rely on parameter tuning within limited operating ranges, often exhibit degraded performance when faced with variations in system parameters or external loading conditions. To address these limitations, this study employs a grasping force control strategy based on SMC, which is robust to uncertainties and unknown external disturbances, enabling precise force control. The schematic diagram of the proposed SMC-based grasping force control strategy is depicted in Fig. 12.2. To implement the force control strategy, a relationship between the actuator's input control signal

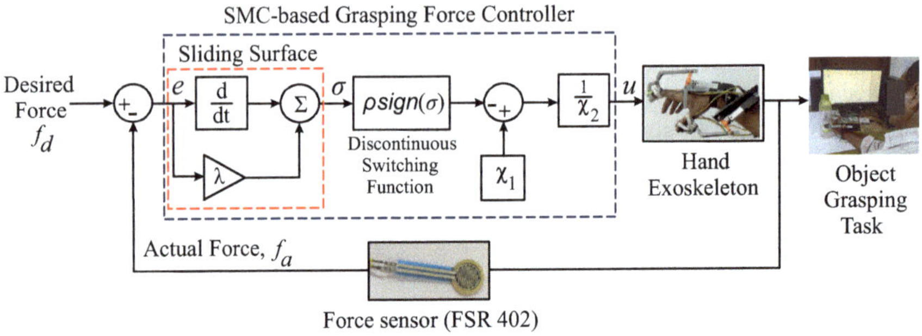

Fig. 12.2 The concept of the control scheme involving SMC

and its corresponding velocity is established, as presented in Eq. (12.12). This formulation is derived based on specifications provided in the linear actuator datasheet.

$$\dot{D}^* = a_1 - a_2 u \tag{12.12}$$

where, a_1 and a_2 can be found from the data sheet of the linear actuator. Equation (12.4) can be rewritten as,

$$\theta_A = \Xi(D^*) \tag{12.13}$$

where, $\Xi(D^*) = \arccos(\frac{L_{OA}^2 + L_{OK}^2 - (D^* + L_{ON})^2}{2L_{OA}L_{OK}}) + \arctan(\frac{L_{KM}}{L_{AK}})$. The time derivative of the previously defined expression in Eq. (12.13) is obtained as follows:

$$\dot{\theta}_A = \Xi^{'}(D^*)\dot{D}^* \tag{12.14}$$

As established in the subsection concerning the analysis of actual grasping force, it is known that:

$$\boldsymbol{F}_{act} = (J^T)^{-1}\dot{\theta}_A \tag{12.15}$$

Therefore, force at the fingertip is as follow,

$$f_a = \|\boldsymbol{F_{act}}\| = \Phi\dot{\theta}_A = \Phi\,\Xi^{'}(D^*)\dot{D}^* \tag{12.16}$$

where, $\Phi = \|(J^T)^{-1}\|$. The sliding surface can be expressed as a function of the system error as,

$$\sigma = \dot{e} + \lambda e \tag{12.17}$$

where, $\lambda > 0$ and the control errors e and $\dot{e}$ are defined as,

$$e = f_d - f_a, \quad \dot{e} = \dot{f}_d - \dot{f}_a \tag{12.18}$$

The variables f_a and f_d represent the measured force from the sensor and the object grasping force, respectively, which the exoskeleton system aims to achieve. Equation (12.17) can be interpreted as a first-order filter, where e serves as the input and σ as the output. Within this framework, the influence of uncertainties in σ can be attenuated based on the selection of the filter bandwidth λ. Notably, when $\sigma = 0$,

$$e(t) = e(0)\exp(-\lambda t) \tag{12.19}$$

$$\dot{e}(t) = (-\lambda)e(0)\exp(-\lambda t) \tag{12.20}$$

By combining Eqs. (12.13) and (12.14), the time derivative of sliding variable can be obtained as:

$$\begin{aligned} \dot{\sigma} &= \frac{\partial \sigma}{\partial D^*}\dot{D}^* = \lambda\frac{\partial e}{\partial D^*} + \frac{\partial \dot{e}}{\partial D^*} \\ &= (-\lambda\Phi\Xi''(D^*)\dot{D}^* - (2\dot{\Phi}\Xi'(D^*)\Xi''(D^*) + \Phi\Xi''')\dot{D}^{*2})\dot{D}^* \end{aligned} \tag{12.21}$$

The dynamics of the sliding variable can now be expressed as follows:

$$\dot{\sigma} = -(\chi_1 - \chi_2 u)(a_1 - a_2 u)^2 \tag{12.22}$$

where, $\chi_1 = \lambda\Phi\Xi''(D^*) + 2\dot{\Phi}\Xi'(D^*)\Xi''(D^*)a_1 + \Phi\Xi'''(D^*)a_1$ and $\chi_2 = 2\dot{\Phi}\Xi'(D^*)\Xi''(D^*)a_2 + \Phi\Xi'''(D^*)a_2$. Control law can be assumed as [28],

$$u = \frac{1}{\chi_2}(\chi_1 - \rho sign(\sigma)) \tag{12.23}$$

To analyze the stability on the sliding surface, a Lyapunov function is proposed as follows:

$$V = \frac{1}{2}\sigma^2 \tag{12.24}$$

(Asymptotically stable $\dot{V} \leq 0$ for $\sigma \neq 0$). It is fundamentally required that the system states converge toward and remain on the sliding manifold governed by the sliding mode dynamics [30, 31].

$$\dot{V} = \sigma\dot{\sigma} = \sigma(\chi_1 - \chi_2 u)(a_1 - a_2 u)^2 \tag{12.25}$$

$$\begin{aligned} \dot{V} &= -\sigma\ \rho\ sign(\sigma)(a_1 - \frac{a_2}{\chi_2}(\chi_1 - \rho sign(\sigma)))^2 \\ &= -\ \rho|\sigma|\gamma \end{aligned} \tag{12.26}$$

where, $\gamma = (a_1 - \frac{a_2}{\chi_2}(\chi_1 - \rho sign(\sigma)))^2$. Thus, ρ should be defined such that $\dot{V} \leq -\rho|\sigma|\gamma$. However, issue of chattering on the sliding surface can be mitigated by replacing the discontinuous switching function $sign(\sigma)$ with a continuous time-dependent function. The continuous function is selected as follows:

$$
\begin{aligned}
sat(\frac{\sigma}{\phi}) &= \frac{\sigma}{\phi} \qquad if\ |\frac{\sigma}{\phi}| < 1 \\
&= sign(\frac{\sigma}{\phi})\ otherwise
\end{aligned}
\tag{12.27}
$$

where, ϕ = boundary layer thickness and it is time varying.

Furthermore, an optimization strategy based on Particle Swarm Optimization (PSO) is implemented to identify the optimal control parameters. The algorithm minimizes the Root Mean Square Error (RMSE) between the desired and experimentally measured fingertip force, which serves as the objective function. In this implementation, a swarm of 60 particles updates positions based on individual and global bests to minimize the objective function. The velocities and positions of the particles are updated iteratively in accordance with the formulations presented in Eqs. (12.28) and (12.29).

$$
\begin{aligned}
Vel_j(k+1) = {} & wVel_j(k) + c_1r_1(Pos_{Best\,j}(k) - Pos_j(k)) \\
& + c_2r_2(G_{Best}(k) - Pos_j(k))
\end{aligned}
\tag{12.28}
$$

$$
Pos_j(k+1) = Pos_j(k) + V_j(k+1) \tag{12.29}
$$

Here, $Pos_j(k+1)$ and $Vel_j(k+1)$ are the position and velocity of the jth particle at $(k+1)$th iteration, respectively. $Pos_{Best\,j}(k)$ indicates personal best position of jth particle at kth iteration. The global best of the swarm at kth iteration, is expressed as $G_{Best}(k)$. r_1 and r_2 are the random numbers in the range [0,1] whereas, c_1 and c_2 are the coefficient of acceleration and w is the inertia weight. Position and velocity of each particle have been randomly initialized. Figures 12.3a, b illustrate positions of the particle of swarm at the last iteration and from the figure it is clear that all the particles of the swarm converges to a particular value ($\lambda = 0.273$ and $\rho = 0.189$). Figure 12.3c illustrates the reduction of the cost function throughout the simulation process.

12.4 Results and Discussions

12.4.1 Experimental Setup

In this study, three sets of experiments were performed: (i) maintaining a constant grasping force, (ii) disturbance rejection, and (iii) force trajectory tracking. The schematic representation of the hardware setup used for the validation of grasping

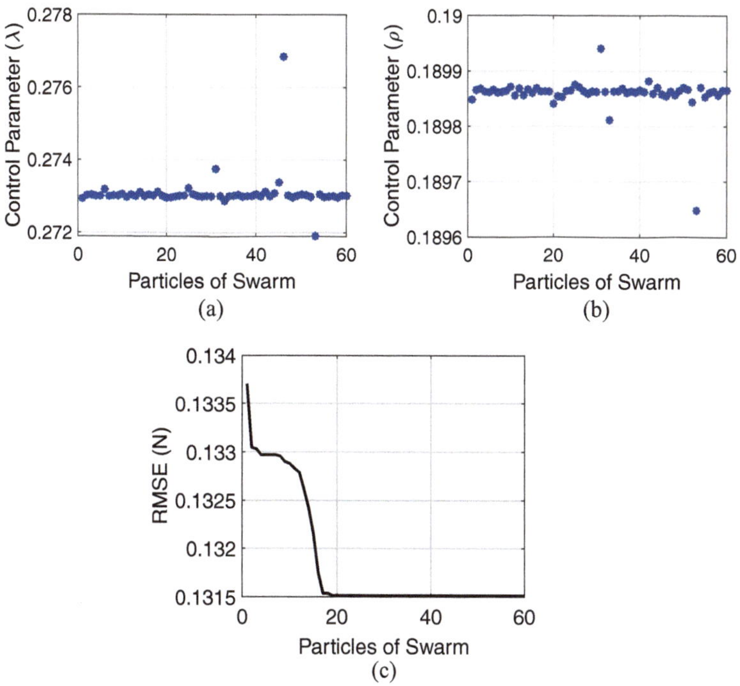

Fig. 12.3 Results of PSO, (**a**) Optimal value of λ, (**b**) Optimal value of ρ, and (**c**) Error convergence curve

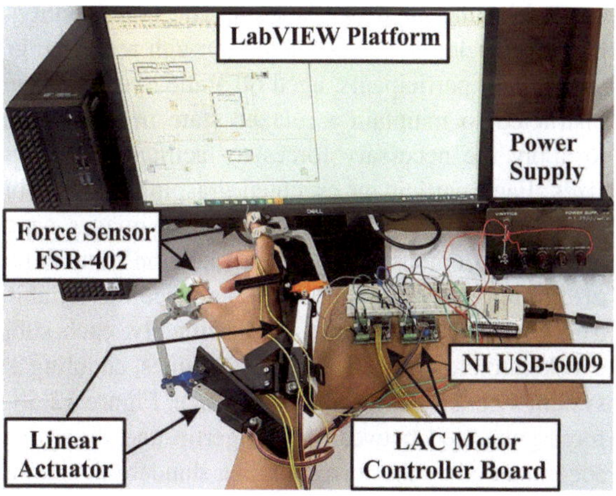

Fig. 12.4 Experimental setup

force control strategy is depicted in Fig. 12.4. The experimental setup comprises two linear actuators, force sensors, a PC-based hardware interface, a 0–24 V DC power supply, and a workstation computer, as shown in Fig. 12.4. The hand assistive device

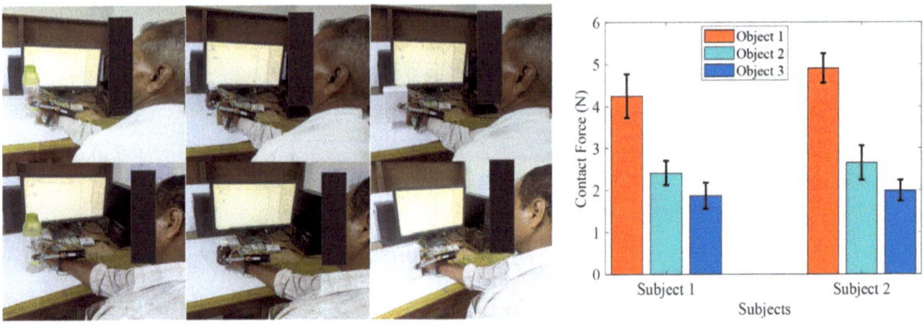

(a) (b)

Fig. 12.5 Real-time experimental results for constant grasping force experiment of index finger, (**a**) Experimentation with two elderly human subjects for the grasping task and (**b**) Mean contact force with standard deviation for each object grasping

incorporates Actuonix Linear Actuators (model L12-50-210-12P & L12-30-210-12P) paired with a dedicated motor controller board (LAC board) [32]. The actuators are integrated with the custom LAC microcontroller for precise control. Force measurement is achieved using a force sensing resistor (FSR 402). Data acquisition is facilitated by the NI USB-6009 DAQ card, which serves as the communication interface between the hardware components and a Windows 10-based desktop computer. The system is controlled via LabVIEW real-time platform featuring a graphical user interface, which serves as the assistive pilot setup for the present experimental investigation. Figure 12.5 illustrates the experimental snapshots while grasping three different sized objects along with the bar chart of the contact forces exerted by the hand exoskeleton. The functionality of the proposed two-fingered hand exoskeleton was evaluated through real-time grasping experiments involving two elderly participants, aged 60 and 62 years. During the trials, the subjects were instructed to maintain a relaxed state in their fingers, allowing the exoskeleton to apply the necessary forces to facilitate object grasping across various object sizes. Each participant executed grasping tasks involving three objects of varying sizes. Figure 12.5b presents sequential images captured during the experimental procedure, demonstrating the exoskeleton's capability to assist in grasping objects of different dimensions [33]. To enhance the reliability of the findings, a repeated trial analysis was performed. Specifically, each subject was instructed to use the exoskeleton to grasp each object ten times, enabling a more robust assessment of the system's consistency and performance. Figure 12.5b illustrates the average contact force generated between the fingertip and three distinct objects during grasping, accompanied by the corresponding standard deviations for each case. The bar graph analysis indicates that the developed hand exoskeleton is capable of effectively transmitting force to the fingertip across varying object sizes.

12.4.2 *Experimental Results*

12.4.2.1 Real-Time Constant Grasping Force Experiments

The most common task in daily living is Reach-Grasp, which requires coordination between the wearer's hand and the assistive device. With this viewpoint, a robust force control strategy is designed for elderly human subjects to enhance grasping performance. Therefore, we have performed a real-time constant grasping force experiment on two elderly human subjects with the objective of minimizing force error between actual and desired grasping force. Figures 12.6 and 12.7 depict the experimental results for one human subject while grasping an object. In this experiment, 5 N force is considered the desired force for grasping an object. Figure. 12.6 illustrates the force response for constant grasping force experiments for index finger exoskeleton with control signal and force error trajectory. Figure 12.6a shows the step response for the index finger exoskeleton, where the exoskeleton takes significantly less time (1.7 sec) to reach the object, and then the grasping force reaches the desired value within 4 to 5 sec. Furthermore, Fig. 12.7 presents the constant grasping force experiment outcomes for thumb digit exoskeleton. Figures 12.6b and 12.7b depict the control signals of the constant grasping force experiment for the index and thumb exoskeleton, respectively. In Fig. 12.6b, the maximum control

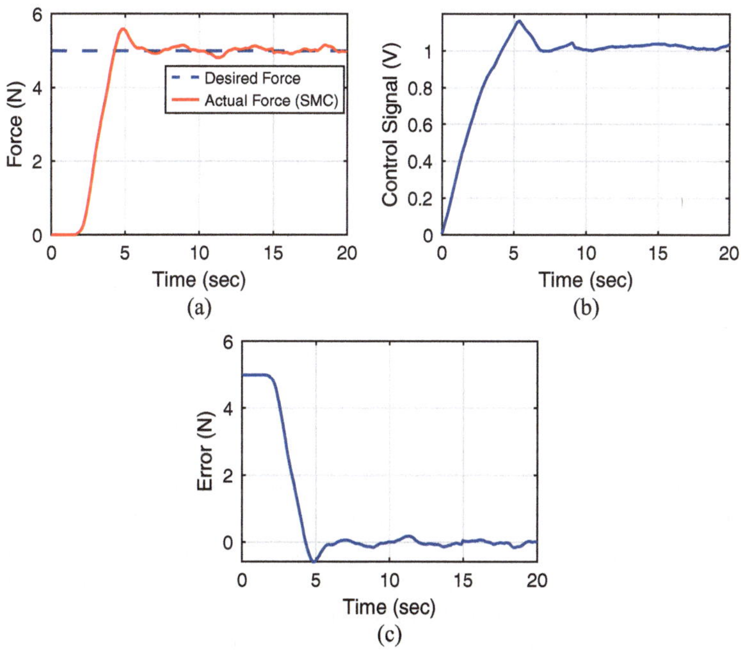

Fig. 12.6 Real-time experimental results for constant grasping force experiment of index finger, (**a**) Force response, (**b**) Control signal, and (**c**) Error profile

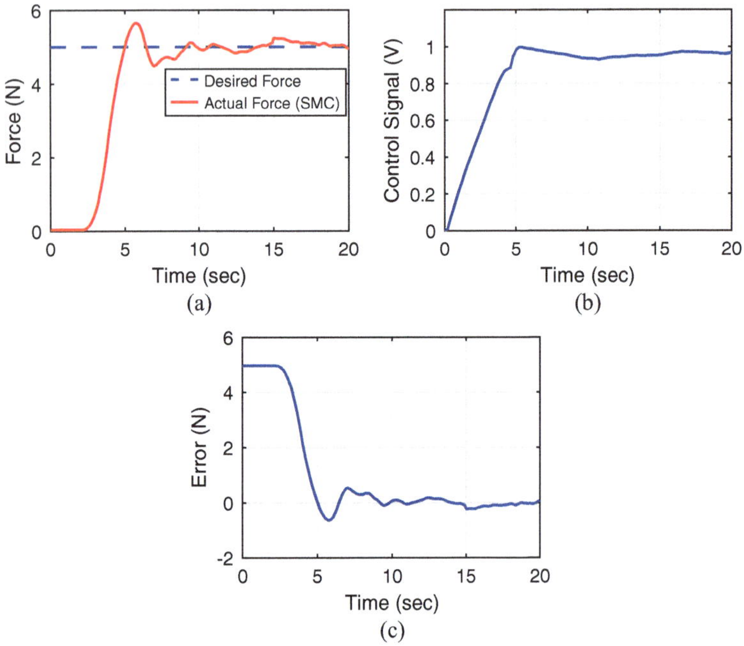

Fig. 12.7 Real-time experimental results for constant grasping force experiment of thumb digit, (**a**) Force response, (**b**) Control signal, and (**c**) Error profile

signal is 1.16 V and after 10 sec, the control signal converges towards a value ranging from 1 to 1.03 V (with order of difference O(10^{-2})). Similarly, Fig. 12.7b shows the maximum control signal for the thumb exoskeleton is 0.99 V and after a significant time (10–12 sec) it converges within a range of 0.94 to 0.96 V. From the viewpoint of error trajectory (shown in Fig. 12.6c), the performance parameters are observed as the final absolute error: 0.0158 N, Mean Absolute Error (MAE) in the last 10 sec: 0.0598 N, and Root Mean Square Error (RMSE) in last 10 sec: 0.07664 N. Notably, Fig. 12.7a depicts the constant force experiment results for thumb exoskeleton, where, it is observed that the exoskeleton takes 2.3 sec to reach the grasped object and then the force control scheme helps to achieve the desired grasping force in less time. Figure 12.7c shows the error trajectory for the thumb exoskeleton. The performance parameters for thumb exoskeleton constant force experiments are final absolute force error: 0.0615 N, MAE, and RMSE in the last 10 sec: 0.0889 N and 0.1077 N, respectively. To evaluate the effectiveness of the proposed grasping force control strategy, experiments were conducted involving two elderly participants and three different sized objects (experimental snapshots are presented in Fig. 12.5a. Table 12.1 presents the performance metrics of constant grasping force experiment, including the mean RMSE values and corresponding standard deviations of ten trails for each object and participant. The results indicate that the RMSE values range between 0.5344 N and 0.6437 N.

Table 12.1 Grasping performance analysis involving two subjects

Subject	Finger	Performance metric (RMSE)		
		Object-1	Object-2	Object-3
Subject-1	Index	0.5640 (0.1830)	0.6347 (0.1986)	0.5649 (0.1610)
	Thumb	0.5344 (0.1969)	0.6419 (0.1203)	0.5755 (0.1322)
Subject-2	Index	0.6437 (0.1542)	0.6012 (0.1945)	0.5944 (0.2019)
	Thumb	0.6273 (0.2016)	0.6125 (0.1912)	0.6084 (0.1251)

12.4.2.2 Disturbance Rejection Experiments

To evaluate the robustness of the proposed force control strategy, a real-time disturbance rejection experiment was performed involving an elderly participant. The experiment aimed to maintain a target grasping force of 5 N for both the index finger and thumb through the corresponding segments of the exoskeleton. During the initial phase, the control system successfully achieved the desired force output within 5 seconds. Subsequently, at the 7 sec mark, an abrupt external disturbance was introduced by applying a pulling force to the exoskeleton while the subject maintained a firm grasp on an object, thereby testing the system's capacity for disturbance compensation. Figures 12.8a and 12.9a illustrate the force response profiles of the index and thumb digit of the exoskeleton during the disturbance rejection experiment. Analysis of the force trajectories indicates that the robust SMC-based force controller successfully reaches the target grasping force within approximately 4 to 5 seconds. Following the application of the external disturbance, transient undershoot responses were observed, with reductions of approximately 37% for the index finger and 21% for the thumb exoskeleton, respectively. The corresponding control signals recorded during the experiment are depicted in Figs. 12.8b, c. Examination of these plots reveals that the external disturbance has a noticeable impact on the control behavior. Nonetheless, the proposed robust force control scheme effectively compensates for the disturbance by adjusting the control inputs to ensure convergence to the desired grasping force. In this experimental trial, the control signals peaked at 1.21 V for the index finger and 1.17 V for the thumb segment of the exoskeleton. Figures 12.8c and 12.9c illustrate the error trajectories corresponding to the index and thumb exoskeletons, respectively. The grasping force trajectories indicate RMSE of 1.5054 N for the index finger and 1.2113 N for the thumb. Additionally, the MAE observed after the application of the external disturbance were 0.3169 N for the index and 0.2247 N for the thumb. These results demonstrate that the proposed robust force control algorithm is capable of accurately tracking a constant target grasping force, even in the presence of sudden external disturbances.

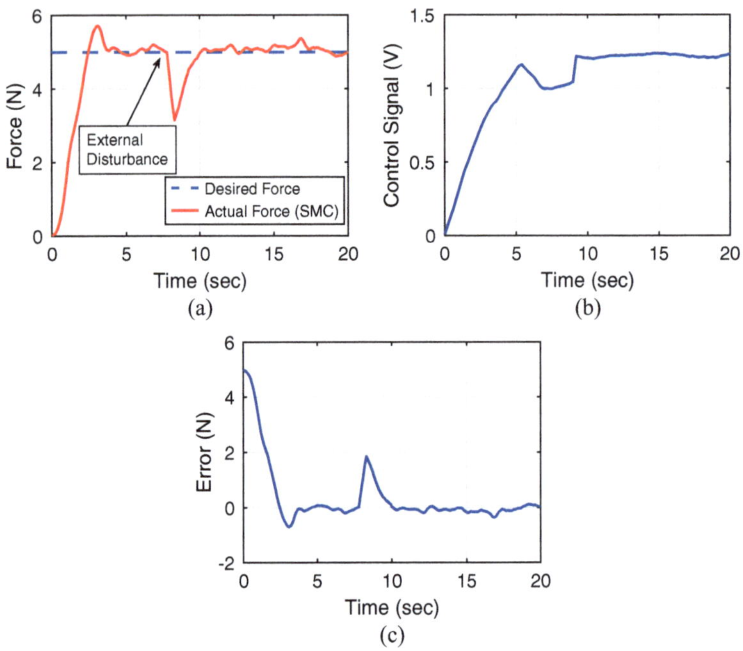

Fig. 12.8 Real-time disturbance rejection experimental results of index finger, (**a**) Force response, (**b**) Control signal, and (**c**) Error profile

12.4.2.3 Force Control Strategy Validation

In practical grasping scenarios, variations in the object's weight, such as the addition or removal of external loads, are common. When an individual maintains a specific grasping force and the load changes, the required grasping force must be adjusted accordingly to ensure a stable grip. To address this condition, the present study evaluates the performance of the proposed controller using a multi-step force trajectory tracking task, thereby validating its ability to adapt to dynamic force requirements during object manipulation. Figure 12.10a presents the force response of the index finger exoskeleton during a multi-step input tracking task implemented using the SMC strategy. The force output exhibits peak values ranging from 8.5 N to 9 N across the steps. Correspondingly, the error trajectory for this experiment is shown in Fig. 12.10b and the performance metric (RMSE) was evaluated as 0.8026 N, indicating a high level of tracking accuracy under varying force demands. During transitions between different force step levels, noticeable transient errors occur around the 20th and 30th seconds, as indicated by the error trajectory in Fig. 12.10b. However, the controller rapidly compensates for these deviations, and the force output converges to the desired values with minimal steady-state error shortly after each transition. Figure 12.11 illustrates the grasping performance of the thumb exoskeleton in response to multi-step force inputs regulated by the SMC

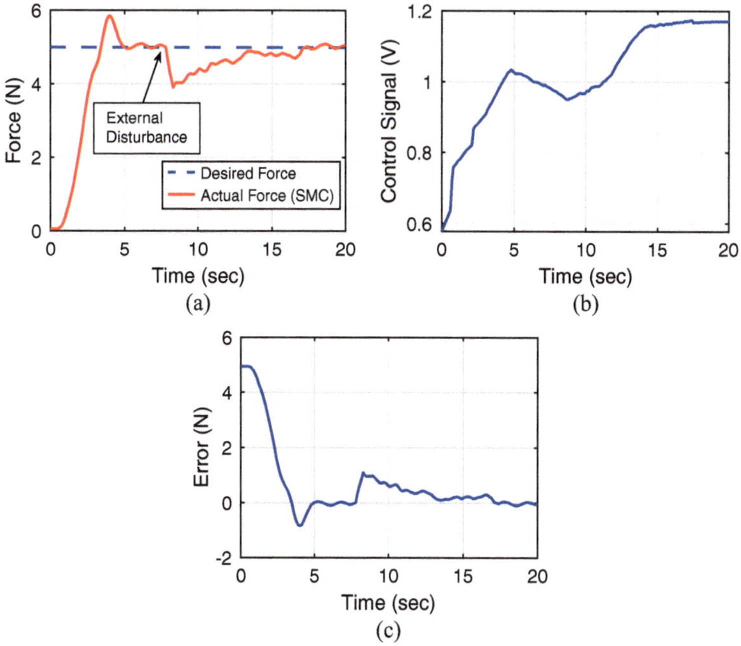

Fig. 12.9 Real-time disturbance rejection experimental results of thumb digit, (**a**) Force response, (**b**) Control signal, and (**c**) Error profile

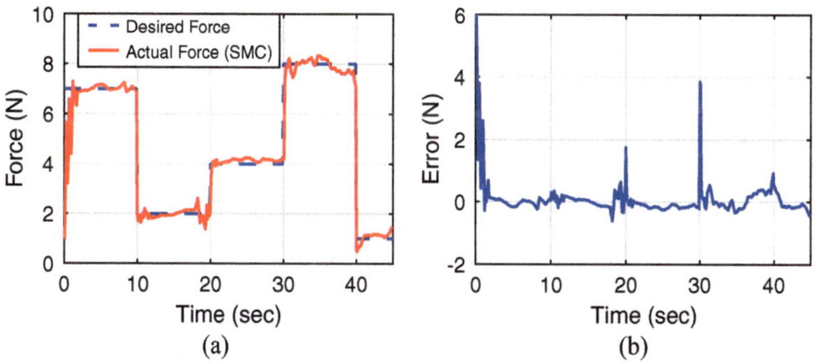

Fig. 12.10 Multi-step response for the index finger during an object grasping with SMC strategy, (**a**) Force response and (**b**) Error profile

strategy. The force response demonstrates peak values ranging between 7.8 N and 8.5 N. The associated error trajectory is presented in Fig. 12.11b, the RMSE was determined as 0.6482 N, indicating effective tracking performance under varying force demands.

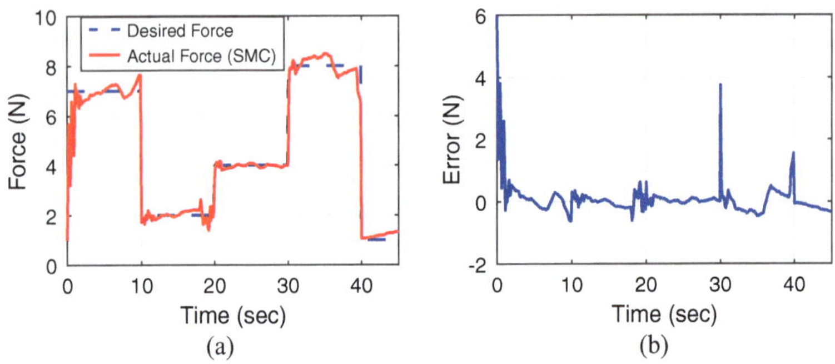

Fig. 12.11 Multi-step response for the index finger during an object grasping with SMC strategy, (**a**) Force response and (**b**) Error profile

12.4.3 Discussions

In this study, a robust force control approach utilizing SMC with optimally tuned parameters is proposed for a two-fingered assistive robotic system, aimed at improving the grasping capabilities of elderly individuals. In comparison with previously developed hand assistive devices designed for object grasping and utilizing linkage-based mechanisms, such as those with a wearable mass of 127 g per finger (maximum grasping force: 12.3 N) [20], 172 g per finger (maximum grasping force: 10 N) [21], 300 g for three fingers (maximum grasping force: 14.9 N) [34], and 185 g per finger (maximum grasping force not reported) [35], the hand exoskeleton presented in this study offers significant improvements. It features a reduced overall weight (200 g), employs an underactuated design, and is capable of generating sufficient normal force (maximum fingertip force of 14 N) required for effective grasping tasks associated with activities of daily living (ADLs). Although numerous hand exoskeletons have been developed within the domain of assistive technologies, comprehensive analyses of force control strategies remain limited. Therefore, the primary objective of this work is to implement a robust force control framework to support grasping tasks in elderly individuals. To evaluate the effectiveness of the proposed control system, three distinct real-time grasping experiments were conducted with two elderly participants: (i) constant force maintenance, (ii) disturbance rejection, and (iii) tracking of dynamic force trajectories.

In the first set of experiments, constant-force grasping tasks were performed using objects of three different sizes. For stable grasping, it is essential that the exoskeleton consistently applies a uniform contact force. In a comparable study, Li et al. [20] evaluated their single-finger exoskeleton under similar conditions and reported a steady-state force error of approximately 0.1 N. In contrast, our proposed dual-fingered system demonstrated significantly improved performance, with final absolute errors of 0.0158 N for the index finger and 0.0615 N for the thumb, corresponding to a 61.35% reduction in steady-state error. Additionally, effective disturbance rejection remains a critical factor in ensuring reliable grasping assistance during real-world interactions. In existing literature, Li et al. [20] have also conducted disturbance rejection experiments; however, their study lacks a comprehensive control performance analysis. To evaluate the robustness of the proposed force control strategy, a real-time disturbance rejection experiment was carried out. The results demonstrated an overall RMSE of 1.3583 N (mean of index finger and thumb digit). Moreover, following the introduction of an abrupt external disturbance, MAE was observed to be 0.2708 N (mean of index finger and thumb digit), indicating the controller's effectiveness in restoring force stability under perturbations. To further validate the performance of the proposed force control strategy, force trajectory tracking experiments were conducted using different reference profiles. A similar study by M. Felix Orlando et al. [19] involved tracking an irregular sinusoidal force trajectory using a two-fingered exoskeleton, reporting RMSE values of 0.35 N for the index finger and 0.65 N for the thumb. In comparison, the results from our system demonstrate improved tracking accuracy. Specifically, for a multi-step force trajectory, the system attained RMSEs of 0.8026 N for the index finger and 0.6482 N for the thumb. These results highlight the effectiveness of the proposed control approach in tracking dynamic force profiles.

Table 12.2 presents a comparative analysis highlighting the advantages of the proposed exoskeleton system relative to the design introduced by Felix Orlando et al. [19]. Overall, the performance evaluation of the implemented force control strategy demonstrates that the developed exoskeleton is a promising candidate for assistive applications, particularly in scenarios requiring consistent grasping force regulation. Overall, the promising results obtained in this study suggest that the developed two-fingered exoskeleton, integrated with the proposed control scheme, holds significant potential as an assistive device to improve grasping performance in users requiring hand support.

12.5 Conclusions

This book chapter presents the comprehensive design and development of a hand assistive device with grasping force control strategy tailored to enhance object grasping capabilities and support daily living activities in the elderly population. By integrating advanced control strategies with user-centric design, this work contributes meaningfully to narrowing the interface gap between human intent and robotic assistance, thereby promoting improved functional independence and quality of life. Grasping experiments were conducted with two elderly participants using three distinct objects to evaluate the efficacy of the proposed system, including real-time disturbance rejection tests to assess control robustness. The satisfactory experimental outcomes demonstrate that the developed hand exoskeleton, integrated with a SMC-based grasping force control strategy, effectively enhances grasping performance during ADL. These findings underscore the potential of the proposed

Table 12.2 Comparison table between our designed hand exoskeleton with proposed SMC-control scheme with existing hand exoskeleton [19]

Proposed Exoskeleton	**M. Felix Orlando et al. [19]**
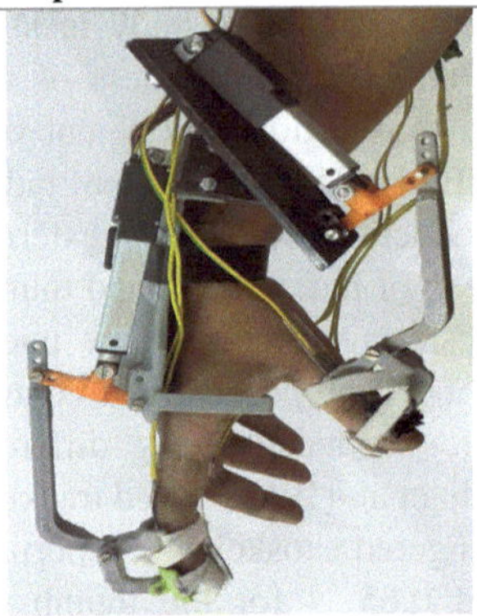	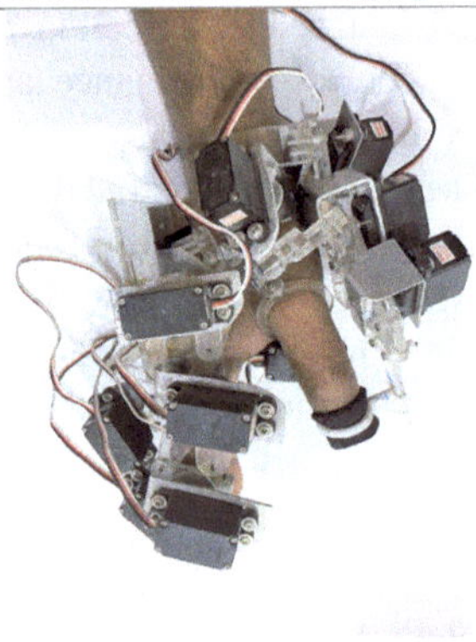
Underactuation mechanism based and lightweight (200 g)	Direct driven system bulky and heavy in weight (700 g)
All the five digits can be individually associated by the proposed compact device	Meant only for index finger and thumb: Not suitable for middle and pinky fingers as they interfere between digits
Appropriate for grasping-task by exerting normal force on fingertips	Force applicability and control towards object- grasping suffer
Wearing time 7 sec and removing time 5 sec	Wearing time 40 sec and removing time 30 sec
SMC-based grasping force controller with closed-loop stability analysis	Proportional controller without closed-loop control stability

device to provide reliable and adaptive assistance for users with diminished hand function. To comprehensively assess the robustness and real-world applicability of the proposed system, future research will involve a broader group of elderly participants. This extended study will be conducted under the ethical committee and approval of the Medical Society of India, ensuring adherence to the highest standards of research integrity and participant safety.

References

1. C. Fallaha, M. Saad, J. Ghommam, Y. Kali, Sliding mode control with model-based switching functions applied on a 7-DOF exoskeleton arm. IEEE/ASME Trans. Mechatron. **26**(1), 539–550 (2021)
2. M.R. Haghjoo, H. Lee, M.R. Afzal, A. Eizad, J. Yoon, Mech-walker: a novel single-DOF linkage device with movable frame for gait rehabilitation. IEEE/ASME Trans. Mechatron. **26**(1), 13–23 (2021)
3. D. Esposito, et al., Design of a 3D-printed hand exoskeleton based on force-myography control for assistance and rehabilitation. Machines **10**(1), 1–16 (2022)

4. D.J. Hyun, H. Lim, S. Park, S. Nam, Singular wire-driven series elastic actuation with force control for a waist assistive exoskeleton, H-WEXv2. IEEE/ASME Trans. Mechatron. **25**(2), 1026–1035 (2020)
5. D. Huamanchahua, P. Toledo-Garcia, J. Aguirre, S. Huacre, Hand exoskeletons for rehabilitation: a systematic review, in *2022 IEEE International IOT, Electronics and Mechatronics Conference (IEMTRONICS)* (2022), pp. 1–6
6. B.B. Kang, H. Choi, H. Lee, K.J. Cho, Exo-glove poly II: a polymer based soft wearable robot for the hand with a tendon-driven actuation system. Soft Robot. **6**(2), 214–227 (2019)
7. S. Park, L. Weber, L. Bishop, J. Stein, M. Ciocarlie, Design and development of effective transmission mechanisms on a tendon driven hand orthosis for stroke patients, in *IEEE International Conference on Robotics and Automation (ICRA)* (2018), pp. 2281–2287
8. T. Butzer, O. Lambercy, J. Arata, R. Gassert, Fully wearable actuated soft exoskeleton for grasping assistance in everyday activities. Soft Robot. **8**(2), 128–143 (2020)
9. J. Wang, Y. Fei, W. Pang, Design, modeling, and testing of a soft pneumatic glove with segmented PneuNets bending actuators. IEEE/ASME Trans. Mechatron. **24**(3), 990–1001 (2019)
10. H. Zhang, A.S. Kumar, F. Chen, J.Y.H. Fuh, M.Y. Wang, Topology optimized multimaterial soft fingers for applications on grippers, rehabilitation, and artificial hands. IEEE/ASME Trans. Mechatron. **24**(1), 120–131 (2019)
11. K.H.L. Heung, R.K.Y. Tong, A.T.H. Lau, Z. Li, Robotic glove with soft-elastic composite actuators for assisting activities of daily living. Soft Robot. **6**(2), 289–304 (2019)
12. H. Li, L. Cheng, Z. Li, W. Xue, Active disturbance rejection control for a fluid-driven hand rehabilitation device. IEEE/ASME Trans. Mechatron. **26**(2), 841–853 (2021)
13. L. Ge, et al., Design, modeling, and evaluation of fabric-based pneumatic actuators for soft wearable assistive gloves. Soft Robot. **7**(5), 583–596 (2020)
14. T. Chen, P.S. Lum, Hand rehabilitation after stroke using a wearable, high DOF, spring powered exoskeleton, in *Proceedings of 38th Annual International Conference of the IEEE Engineering in Medicine and Biology Society* (2016), pp. 578–581
15. J. Arata, et al., A new hand exoskeleton device for rehabilitation using a three-layered sliding spring mechanism. *IEEE International Conference on Robotics and Automation (ICRA)* (2013), pp. 3902–3907
16. C.J. Nycz, T. Butzer, O. Lambercy, J. Arata, G.S. Fischer, R. Gassert, Design and characterization of a lightweight and fully portable remote actuation system for use with a hand exoskeleton. IEEE Robot. Autom. Lett. **1**(2), 976–983 (2016)
17. I. Jo, J. Bae, Design and control of a wearable hand exoskeleton with force-controllable and compact actuator modules, in *Proceedings of IEEE International Conference on Robotics and Automation* (2015), pp. 5596–5601
18. C. Xie, Q. Yang, Y. Huang, S. Su, T. Xu, R. Song, A hybrid arm-hand rehabilitation robot with EMG-based admittance controller. IEEE Trans. Biomed. Circuits Syst. **15**(6), 1332–1342 (2021)
19. M. Felix Orlando, L. Behera, A. Dutta, A. Saxena, Optimal design and redundancy resolution of a novel robotic two-fingered exoskeleton. IEEE Trans. Med. Robot. Bionics **2**(1), 59–75 (2020)
20. H. Li, L. Cheng, N. Sun, R. Cao, Design and control of an underactuated finger exoskeleton for assisting activities of daily living. IEEE/ASME Trans. Mechatron. **27**, 2699–2709 (2022)
21. M.B. Hong, S.J. Kim, Y.S. Ihn, G.C. Jeong, K. Kim, KULEX-hand: an underactuated wearable hand for grasping power assistance. IEEE Trans. Robot. **35**(2), 420–432 (2019)
22. I. Sarakoglou, A. Brygo, D. Mazzanti, N.G. Hernandez, D.G. Caldwell, N.G. Tsagarakis, HEXOTRAC: a highly under-actuated hand exoskeleton for finger tracking and force feedback, in *IEEE/RSJ International Conference on Intelligent Robots and Systems (IROS)* (2016), pp. 1033–1040
23. G. Li, L. Cheng, Z. Gao, X. Xia, J. Jiang, Development of an untethered adaptive thumb exoskeleton for delicate rehabilitation assistance. IEEE Trans. Robot. **38**(6), 3514–3529 (2022)

24. P. Mondal, M. Mondal, S. Ghosh, M. Felix Orlando, Force control of an index finger exoskeleton for rehabilitation purpose, in *IEEE Applied Signal Processing Conference (ASPCON)* (2018), pp. 183–187
25. A. Wege, G. Hommel, Development and control of a hand exoskeleton for rehabilitation of hand injuries, in *IEEE International Conference on Intelligent Robots and Systems* (2005), pp. 3046–3051
26. M. Babaiasl, S.N. Goldar, M.H. Barhaghtalab, V. Meigoli, Sliding mode control of an exoskeleton robot for use in upper-limb rehabilitation, in *RSI International Conference on Robotics and Mechatronics* (2015), pp. 694–701
27. A.B.W. Miranda, A. Forner-Cordero, Upper limb exoskeleton control based on sliding mode control and feedback linearization, in *ISSNIP Biosignals and Biorobotics Conference: Biosignals and Robotics for Better and Safer Living (BRC)* (2013), pp. 1–6
28. K. Halder, M. Felix Orlando, Design and development of robust force control strategy of a hand exoskeleton for grasping assistance, in *28th International Conference on System Theory, Control and Computing (ICSTCC)* (2024), pp. 539–544
29. L. Sciavicco, B. Siciliano, *Modelling and control of robot manipulators* (Springer, London, 2000)
30. V. Utkin, J. Guldner, J. Shi, *Sliding Mode Control in Electromechanical Systems* (Taylor and Francis, Milton Park, 1999)
31. J.J.E. Slotine, *Applied Nonlinear Systems* (Prentice-Hall, Hoboken, 1991)
32. ACTUONIX: https://www.actuonix.com/l12-50-210-6-p
33. K. Halder, A. Roy, M. Felix Orlando, Neural network-based indirect adaptive force control of a two-fingered hand exoskeleton toward robust grasping assistance. IEEE Trans. Syst. Man Cybern. Syst. **55**(5), 3614–3629 (2025)
34. M. Sarac, M. Solazzi, M.A. Otaduy, A. Frisoli, Rendering strategies for underactuated hand exoskeletons. IEEE Robot. Autom. Lett. **3**(3), 2087–2092 (2018)
35. I.H. Ertas, E. Hocaoglu, D.E. Barkana, V. Patoglu, Finger exoskeleton for treatment of tendon injuries, in *IEEE International Conference on Rehabilitation Robotics* (2009), pp. 194–201

Chapter 13
A Pairing-Free Approach for End-to-End Mapping from Human to Kinematically-Dissimilar Robotic Hands

Matthew Stanley and Xiaoli Zhang

Abstract End-to-end mapping from human to robot hand poses offers a compelling, non-invasive approach for translating visual hand data into robot joint commands. However, existing methods rely heavily on labor-intensive, hand-structure-specific paired datasets, hindering scalability and cross-device adaptability. To address these limitations, we propose End-to End Partial Supervision Mapping (EPSM), a novel framework that decouples pose information from structure-specific information through a shared latent grasp feature space. EPSM employs a unified encoder to extract low-dimensional grasp representations from depth images, while separate decoders reconstruct robot joint angles tailored to specific hand structures. By sharing encoder weights across models and isolating decoder training to individual robotic hands, EPSM enables cross-structure mapping via decoder swapping–without requiring any human-robot paired training data across different configurations. We demonstrate the effectiveness of EPSM by mapping human hand poses to three robotic hands with varying finger counts (three, four, and five), showcasing its potential as a generalizable and scalable solution for human-robot hand interaction.

Keywords Machine learning · Computer vision · Human-robot interface · Teleoperations · Control · Mapping · Control architecture

13.1 Introduction

Natural, effective, and intuitive human-to-robot hand pose mapping is essential for both (1) teleoperation, in which a human operator executes a remote task through teleoperating a robot hand [1–3], and (2) autonomous grasp/manipulation tasks, in which a robot learns from human demonstrations [4, 5]. Both scenarios

M. Stanley · X. Zhang (✉)
Intelligent Robotic And Systems Lab, Colorado School of Mines, Golden, CO, USA
e-mail: mdstanley@mines.edu; xlzhang@mines.edu

R. Vinjamuri (ed.), *Bridging the Gap between Mind and Machine*,
https://doi.org/10.1007/978-3-032-06713-5_13

are common in applications such as manufacturing, space exploration, underwater operations, and assistive robotics. Recently, end-to-end human-to-robot hand pose mapping has attracted researchers' attention for being intuitive and non-invasive during operation. End-to-end mapping [6–8] uses a convolutional neural network (CNN) to extract the comprehensive human hand pose information in a non-linear form, from the rich information in images and map them to the robot hand joints.

The difficulty in obtaining end-to-end mapping models lies within the data-hungry training process. Training requires large quantities of human-robot pairwise hand pose data that are generated manually. Various sources of uncertainties, including human pose errors, robot hand control tolerances, and pose-matching errors, can compound during the manual pose-pairing process. Thus, considerable manual effort is required during the pose-pairing process to keep data uncertainty low to train a more accurate model.

The most common method for generating the paired data for structurally similar hands is kinematic mapping [6, 9, 10]; however, this becomes difficult for cross-structure mapping as ambiguities exist in how to accommodate the differences. Additionally, ensuring robust mapping for cross-structure applications among robotic hands is especially challenging. [11, 12] Although various mapping methods like finger-tip mapping [13, 14], joint mapping [11, 14–16], and virtual-object-based mapping [17–19] have been developed that can ignore specific discrepancies between the human and robot structure, they require great attention to generate pairs based on virtual-objects.

With a heavy dependency on meticulously collected paired data, the limitations of the current end-to-end approaches are:

1. Dependency on the large quantity of manually paired data. While most regression models require labeled data to learn the desired output for a set of inputs, teleoperation models also require data to be paired, meaning that multiple datasets for each structure in the model must be created consisting of both structures in the same pose. The training process requires many meticulously paired human-robot hand poses that are challenging and time-consuming to produce. For example, Teachnet requires 400k pairs of manually paired depth images between humans and robots for training. The paring process is so labor-intensive that just generating the paired dataset for end-to-end mapping can be a significant contribution to this field.
2. Compounding uncertainty. The manual pairing process compounds many sources of uncertainties that ultimately affect the model's performance. Variations in input image resolutions, measurement uncertainty in joint angles, and errors during the pairing process can affect the model's training performance concurrently due to the supervised end-to-end mapping scheme.
3. Structure specific dataset. When generated, a dataset is labeled and paired for only a specific hand structure and cannot be generalized to others. For example, a dataset created to map human pose to a 4-fingered robot cannot be adapted or modified for mapping to a 3-fingered robot. A new dataset needs to be collected from scratch to achieve the mapping to 3-fingered robots.

To address these challenges, our vision is that:

1. The model needs to operate on non-paired data to reduce the data collection effort and the uncertainty compounding.
2. The model needs to be structure-independent to be generalizable to various hand structures without additional training effort

Towards this vision, the goal of this research is to develop a training scheme that only depends on joint-labeled, non-pairwise data. We term this approach End-to-End Partial Supervision Mapping (EPSM) as it requires only joint angle data to train the encoder-decoder network, eliminating the need for pairwise labels during the mapping process. Because joint angle labeling can be automated with existing computer version packages, such as Google MediaPipe, or directly reading joint encoders from the robot structures, the labeling process only requires minimal effort compared with the original pairing process. By implementing the encoder-decoder structure with weight-sharing encoders, the model is capable of learning synergistic latent features inspired by synergy extraction methods [11, 12, 20–26] to provide a common framework for the model to learn end-to-end pose mapping without paired data.

13.2 Related Works

13.2.1 End-to-End Data-Driven Mapping

Data-driven mapping uses machine learning techniques to perform cross-domain conversion. One approach to creating the mapping between domains is through autoencoders. These neural networks reduce the dimensionality of the features in the first domain and then reconstruct them in the second domain. There are several types of autoencoders, but they all follow the same principles: human hand input is mapped to a set of latent features, which are then mapped to robot joint angles [4, 27, 28]. The most common type is joint-to-joint, where human inputs are joint angles; however, this requires the human to wear a motion capturing device, such as marker tracking or inertial sensors, to obtain human joint positions [13, 14, 18, 21, 24, 27, 29]. Having wearable sensors is not ideal, as they are obtrusive and can reduce the natural workspace that each finger can reach.

An approach that has become increasingly studied is end-to-end mapping [6, 8]. Rather than mapping from human joints to robot joints, end-to-end mapping uses images as input and outputs joint information based on the images. The benefit of this approach is that it does not require additional equipment, such as a tracking suit, to collect human joint information to use as input. Since images are used as input, only a camera is required to operate an end-to-end model. As powerful as end-to-end mapping can be in converting images into robot joint angles, a major drawback to these models is the reliance on large quantities of training data. Due to the high dimensionality that comes from using images, end-to-end mapping has the extra condition of discerning which features in the images are important to determining each joint angle.

13.2.2 *Importance of Paired Data*

The most common concern with data-driven pose mapping, both for direct joint mapping and end-to-end mapping, is the collection of pairwise data, as the quality of the mapping depends on the quality of the pairs. This, however, requires a pre-existing method of pair generation. Therefore, various methods have been generated to create pairs to train teleoperation models.

Methods like [6] provide methods for creating human-to-robot pairs by finding similarities between the two hand structures. While these approaches work great for anthropomorphic robots, they also create ambiguity for non-anthropomorphic robot hands.

Another approach for determining pairs is to project the grasping surface across various objects to determine similar grasping poses between various hand structures [12, 19]. These approaches ensure that each hand structure can grasp multiple objects in a similar way. While this ensures successful grasping on equally shaped objects, methods trained using these models cannot guarantee paired grasping when grabbing objects not in the dataset or attempting to make non-grasping poses.

Developing approaches to generate these paired datasets is imperative for data-driven mapping, as the performance of data-driven models are directly influenced by the quality of the pairs in the dataset. This means that, for the best mapping quality, a custom data between the two hand structures being mapped must be generated. In short, the majority of one's time spent on creating a data-driven mapping model is spent on generating high-quality paired datasets.

13.3 Methods

Going from the image space (I) to the joint space (θ) produces complications since an image is a high-dimensional feature set with significantly more information than necessary to describe the joint space. It is therefore necessary to extract a set of latent features from the image that can easily be collected from the images and constructed back to the joint space. For this reason, the EPSM model adopts the autoencoder structure to focus on two tasks: the encoder performs structure-independent feature representation, while the decoder performs feature reconstruction to structure-specific joint angles.

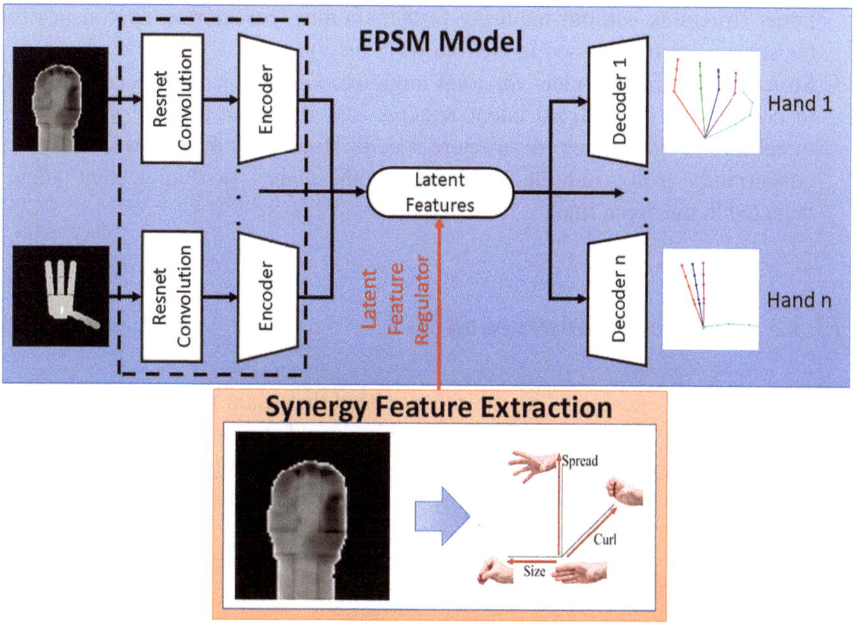

Fig. 13.1 An overview of the architecture of the ESPM model that demonstrates the capability for mapping to multiple hand structures. The model (*blue region*) consists of a weight-shared encoder (left box) and multiple decoders (right) for each hand structure. The weight-sharing encoder is capable of achieving cross-structure mapping by regulating the latent feature with a known synergy domain space (*orange region*). This domain is shown to exist for hand structures, meaning this approach is capable of mapping various hand models

13.3.1 EPSM Architecture

The EPSM model is designed to circumvent the paired data constraints of traditional supervised learning and is visualized in the blue box from Fig. 13.1. This is achieved through the following:

1. A shared weight encoder to construct a hand-structure-independent subspace. By implementing shared weights encoders to process images from all hand structures, the generated subspace contains generalizable abstract, low-dimensional feature descriptors independent of input hand structures. The encoder uses ResNet architecture for primary feature extraction, identifying key physical features from hand images. A custom fully connected layer is then used as a secondary to encode the physical features into the generalizable, structure-independent grasping subspace.
2. A structure-independent grasping subspace. This subspace contains common abstract features of various grasping poses regardless of the hand structure. This common subspace ensures proper decoder mapping for joint angles in

target structures without requiring further training. Structuring approaches for the subspace are discussed in the following section.

3. Structure-specific decoders for joint angle reconstruction. The decoders learn to convert the generalized latent features into the robot-specific joint angles to achieve desired poses. A structure-specific decoder is trained with encoders concurrently using joint-angle-labeled depth images to output joint angles depicted in the depth image.

13.3.2 Subspace Construction

The latent features generated by the encoder need to contain structure-independent pose information. One method for describing hand poses in low dimensional features is synergy [11, 17, 18, 21, 26, 29–31]. In physiology, synergy describes complex tasks that the body performs in a low-dimensional state. The current standard for generating synergy is through PCA [17, 18, 21, 26, 29, 30]; however, this information has led to the development of more advanced empirical methods for generating synergy [11, 18, 29].

The traditional PCA approach is difficult to leverage, as it is highly dependent on the dataset used to extract synergy, making it difficult to reproduce the exact same values for another set of hand poses. Empirical methods used for mapping purposes are also limited to being structure-dependent, requiring mapping in the synergy domain to convert it into the form of another hand structure. To make the EPSM model work without needing paired datasets between each hand structure, a new method of extracting synergy is required. The method needs to have the following constraints:

1. The subspace should be independent of the structure of the hand.
2. The synergy values should be normalized within a reasonable range. For training purposes, the units of all the losses should be the same. The most common approach to training a model is to normalize the variable between a range of [0, 1].
3. The synergy values should describe the operational needs in the hand pose.

This approach can leverage any set of synergy features necessary for operation. For this paper, the synergy domain is designed based on the grasping synergy domain from [26] and is a modification to the approach from [11]; where most grasping poses could be presented by describing the average spread, bend, and curl of the fingers, as demonstrated in Fig. 13.2.

The method begins with a projection matrix $P_{3\times N}$, where each column represents each finger and each row represents a synergy value. Each value in the matrix is either 0 or 1 to represent whether the joint angle influences movement along the corresponding synergy axis.

A scaling factor $K_{3\times 1}$ is developed to preserve the units from the input joints angle to the synergy values. Each element in this vector is defined as:

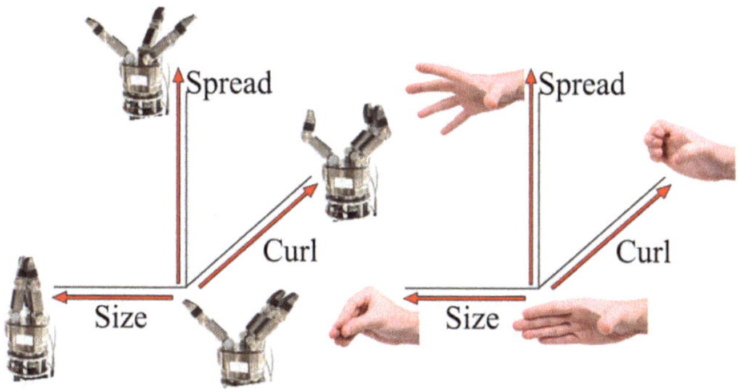

Fig. 13.2 A Figure from [11] illustrating the synergy subdomain. This domain simplifies the hand grasping poses into three vectors: sizing, spreading, and curling. This synergy domain works for many grasping tasks and can be used for various hand structures

$$k_i = \sum_{n=0}^{N} P_{i,n} \tag{13.1}$$

With this, the formula for extracting the synergy subdomain is defined as:

$$S = (P \cdot \Theta) \odot K \tag{13.2}$$

Since this formulation preserves the input units and scale, synergy values are normalized by normalizing the joint angles before plugging them into the equation. This is ideal, as in most instances the latent feature outputs should be normalized in the same manner as the decoder outputs.

13.3.3 EPSM Training Routine

The training can be split into two parts. The first is the latent feature regulation. In this step, the goal is to ensure that the latent features resemble the synergy subdomain generated in Sect. 13.3.2. This is important, as the goal of the latent features is to learn the structure-independent features that best describe the desired output poses. Denoting the latent feature prediction as Λ', the desired joint angle as Θ, and the predicted joint angle as Θ', the regulation loss can be defined as:

$$L_{reg} = ||\Lambda' - (P \cdot \Theta) \odot K||_2 \tag{13.3}$$

The second step is to ensure the reconstruction of the desired joint angles. This is required to ensure that each decoder learns the desired joint angles. This is easily achieved by the following:

$$L_{rec} = ||\Theta' - \Theta||_2 \quad (13.4)$$

The final loss becomes:

$$L = L_{rec} + \alpha L_{reg} \quad (13.5)$$

Where the weight α is used to balance learning the final synergy subdomain compared to the final joint space. This value can be adjusted to balance between self-mapping and cross-mapping. Values between 1.0 and 0.8 were tested with slight trade-off between joint angle reconstruction and latent features generation; ultimately, the α is set to 0.9 as it provided the good trade-off for this report. For each epoch, the model is trained on all hand structures simultaneously. This ensures that during each training loop, the EPSM model does not learn to prefer one hand structure over another. The total losses for each structure are applied to the shared-weight encoder and the corresponding decoder.

13.4 Experimental Design

13.4.1 Validation Method

All of these evaluation metrics are compared across three parts of the EPSM model. The first part evaluates the representation quality of latent features, which are outputs from weight-sharing encoders, in terms of capturing pose synergies. The representation quality is assessed by comparing latent feature outputs from encoders to the synergy values calculated from ground truth joint angles with Eq. (13.2). A good-performing model is expected to output the latent features that closely resemble the ones calculated from ground truth joint angles. The second part evaluates the accuracy of joint angle mapping. Because the pairwise poses for other hand structures are unknown and thus no ground truth joint angles are known, the mapping accuracy can only be calculated for the self-mapping case. The model predicts joint angles from images with labeled human hand poses, then the errors between the estimates and the labels are evaluated. Third, the cross-structure accuracy is evaluated. While no ground truth is known for the other hand structures, the mapping can be evaluated in the synergy subdomain, since the joint angles for the known ground truth and predicted joint angle from the EPSM model can be converted into synergy values using Eq. (13.2).

To properly validate the processes, the EPSM model will be evaluated by calculate the mean error and standard deviation of the error. The errors are calculated between the output of the model and the expected values, being joint angles in radians for the hand reconstruction mapping evaluations and the normalized synergy vector for the latent feature and cross-structure mapping evaluations. These methods are selected for the following reasons:

1. The mean error shows how accurately the model can predict the testing set. This value is expected to be close to 0, indicating that on average the model predicts the correct value.
2. The standard deviation shows the spread of the error. This value is expected to be slightly above the total uncertainty in the dataset, as it will represent the uncertainty the EPSM model has in its predictions.
3. The mean error and standard deviation together can be used to calculate confidence in the model's performance. Using the Student's t-test, the model will validated by comparing the results to a 90% confidence interval. 90% is chosen as it demonstrates reliability in the model with needing significantly precise equipment.

13.4.2 Data Collection

EPSM requires the dataset to consist of labeled image and joint angle data, though does not require the data to be paired between each hand structure's datasets. Each hand pose image is labeled with a vector of true joint angles for every finger of the pose depicted. Images of the hand poses are depth images captured with an Xbox Kinect camera. The joint angle labeling for robot hands is acquired through the built-in encoders. The human joint angles are acquired by collecting the joint poses from Google MediaPipe [32] and implementing inverse kinematics to calculate the joint angles.

To create the dataset, a list of desired poses is presented to the human subjects and robots. Both are instructed to select a random pose to start, then slowly move between poses at random while the Kinect collected samples of the movement. Instructing the robot to random poses ensures that the order of samples in each hand's dataset is different, making the datasets unpaired. The robot hands used in this paper are the 3 Shadow Dexterous Hand models: the 5-fingered model, the 4-fingered model, and the 3-fingered model. This process is illustrated in Fig. 13.3.

The tolerance for Shadow hand control is 1° (0.017 radians), as stated in the control manual. The human data tolerance is not stated, as it is affected by the resolution of the camera and the distance the hand is from the camera. To approximate the uncertainty, the human hand is set at a constant open-hand pose with multiple joint angles calculated. On average, the uncertainty in calculating the hand pose is about 0.09 radians, however, the uncertainty is likely higher since the open-hand pose produced the poses with the least uncertainty. Additional uncertainties are introduced due to the resolution of the camera, but is difficult to calculate the effects on the mapping. This means that, at best, the total uncertainty in the mapping training data is about 0.107 radians.

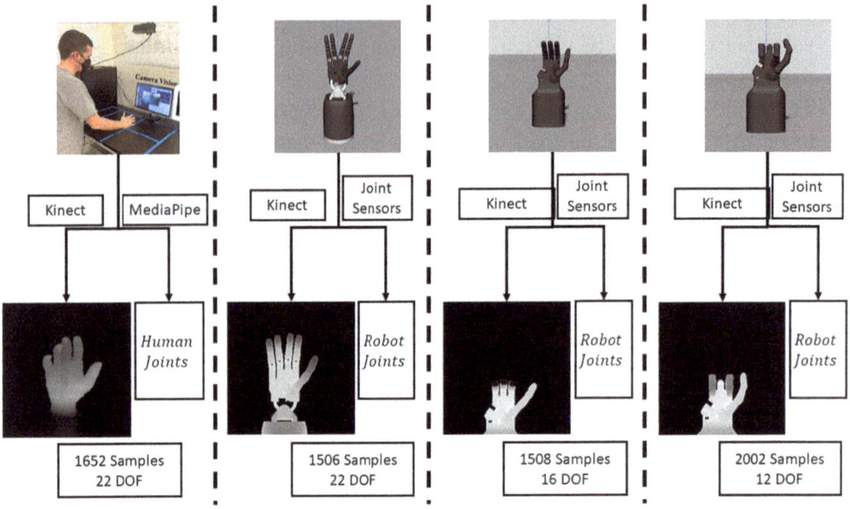

Fig. 13.3 The data collection process for each hand structure. The dashed lines symbolize the fact that the dataset for each hand structure is collected in complete isolation from the other hand structures. The human hand structure is collected using the method on the left, with a Kinect camera to capture depth image and the Google MediaPipe [32] to extract the joint angles. The other robot hands (the remaining right columns) also use the Kinect camera to capture depth images, but the joint angles are measured from the robot's encoders

13.5 Result

13.5.1 Synergy Representation in Latent Features

The first thing to analyze is the latent feature generated by the encoder. The latent features should exist in the subdomain created in Sect. 13.3.2. Representing this subdomain is critical, as it concisely summarizes the hand poses in the grasping dataset and thus can be reconstructed into any pose in the testing set.

The encoder accurately generates similar vectors to the subdomain, with an average difference compared to the expected value of −0.0028 with a standard deviation of 0.147. The expectation is that the encoder's predictions have a large uncertainty since it learns to represent the subdomain from multiple structures and inherits the tolerances from each dataset. Despite this, it is also expected to converge on an accurate subdomain generation since one must exist to accurate mapping across multiple hand structures. The results from the latent feature mapping are visualized in Fig. 13.4, where the patterns generated by the latent features directly align with the synergy subdomain features.

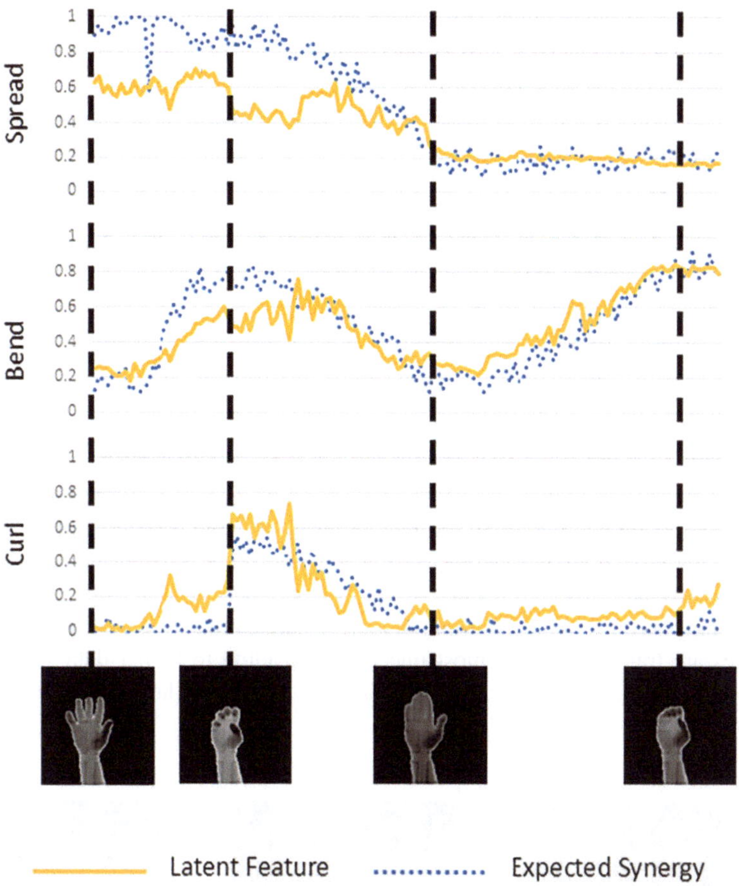

Fig. 13.4 Demonstration of latent feature generated across various human hand poses. The expected synergy values are shown with the dashed lines while the latent features are shown with the solid lines. The input images begin with an open-hand pose, then move towards grabbing a bottle cap, back to open-hand, then finally to a cylinder grab. The latent features progress with the same variance as the expected synergy values and have very similar values across each frame. The greatest difference in value is seen when the finger spread is maximized, resulting from the robot hands being capable of more spreading than the human hand

13.5.2 Same Structure Joint Mapping

The second round of tests involved using human hand images and reconstructing the human joint angles. By remapping to the same structure, these tests provide an estimate of the amount of information lost during the mapping process, as the exact joint angle is known for the image and joint pair. In the joint domain, the average error is about −0.0431 radians with a standard deviation of 0.199 radians. Using the Student t-test, the claim that the EPSM model accurately predicts the joint angle is validated. All of these values are in the joint-space domain, though for evaluating the cross-structure mapping, it is important to also perform the tests in the synergy domain to produce a valid comparison.

As expected, the error decreases in the synergy domain, having a mean error of −0.0137 and a standard deviation of 0.140. This error reduction is expected, since (1) the synergy domain is constrained between [0,1] while most joints are restrained between [0, 1.5707], giving an expected 36.3% error reduction, and (2) the mapping relies on the low dimensional restraints of the same synergy domain, resulting in higher accuracy in this domain.

13.5.3 Cross-Structure Mapping

First, the human hand poses are mapped to the Shadow Dexterous hand, a five-fingered robot hand with very similar anatomy to the human hand. Mapping to the robot hand results in a synergy error of about 0.0102 and a standard deviation of 0.160. This is a significant decrease in average error for a slight increase in data spread. These are expected results, as (1) the robot decoders are trained with data containing less uncertainty, resulting in more accurate predictions, and (2) the latent features are trained with the most uncertainty, leading to higher data spread. As demonstrated in Fig. 13.5, the generated hand poses are similar to the various hand poses in the testing images.

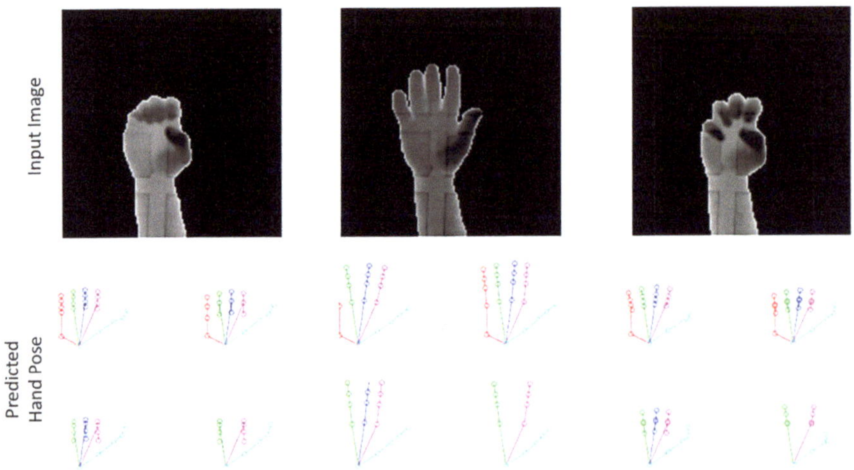

Fig. 13.5 Image demonstrating four unique poses in the testing dataset along with the predicted joint angles for each hand structure. The first row contains the input images to the encoder, the second row contains the plots of each hand structure. Each figure consists of the human hand (top-left), the 5-finger robot hand (top-right), the 4-finger robot hand (bottom-left), and the 3-finger robot hand (bottom-right). The final pose for each hand matches the hand pose in the image

For non-anthropomorphic hands, a similar pattern occurs despite the physical differences. For the 4-fingered model of the Shadow Dexterous hand, the mapping results in an average synergistic error of −0.0091 with a standard deviation of 0.179. For the 3-fingered model, the mapping results in an average synergistic error of −0.0049 with a standard deviation of 0.159. These results align with the assumptions made for the 5-finger Shadow Dexterous hand. While the 3-fingered hand's mean error is significantly smaller than the value for the 5-fingered hand (being less than 50% of the original value), this is due to the 3-fingered robot having significantly fewer joints impacting the synergy value. As stated in Fig. 13.3, the 3-fingered hand has 12 DOF compared to the 22 DOF seen in the 5-fingered robot hand. Synergy values have approximately 50% fewer joints to align with for the 3-fingered hand, meaning that during training these joint angles contribute twice as much to the corresponding synergies, thus decreasing the average error while the uncertainty remains fairly similar.

13.6 Conclusion

This work presents the EPSM architecture as a data-driven, end-to-end approach for partial supervised human-to-robot hand pose mapping across multiple hand structures. By introducing a synergy domain that enhances the latent feature space within the autoencoder design, EPSM effectively balances input feature reconstruction with accurate cross-structure mapping. Empirical evaluation using mean error and standard deviation metrics demonstrates that EPSM achieves reliable grasp pose prediction while accounting for variability inherent in human hand datasets.

The EPSM model produces intuitive mapping, though it does have notable trade-offs. The main concern with the EPSM model is the limited range of motion it can map due to the latent features and the cross-structure mapping. The synergy features implemented in this paper are grasping synergies, which means that the model in this paper can only produce grasping poses; however, the model can be trained with another set of synergy features to produce different poses. Additionally, the EPSM model is limited by the dexterity of all hand structures to which it can map.

In exchange, the EPSM model shines in its ability to map to multiple hand structures simultaneously with easier training data. Since it is trained with all hand structures at once, the model can convert any hand pose into the poses of all hand structures, meaning it can produce instruct multiple hand types in performing the same task with needing to substitute mapping models. Additionally, the partial supervised learning method deployed by EPSM means that it requires more available training data to fully train it. Since it does not require pairwise data, the training data are easier to collect, as it does need to be collected at the same time as other datasets.

Future work will focus on enhancing the latent feature space to further mitigate these limitations and to generalize the method for a broader range of manipulation tasks. Overall, EPSM offers a scalable and efficient framework for unsupervised human-to-robot pose mapping, contributing a valuable tool for applications in robotic manipulation and human–robot interaction.

References

1. A. Torabi, M. Khadem, K. Zareinia, G.R. Sutherland, M. Tavakoli, Manipulability of teleoperated surgical robots with application in design of master/slave manipulators, in *2018 International Symposium on Medical Robotics (ISMR)* (IEEE, Atlanta, 2018), pp. 1–6. https://doi.org/10.1109/ISMR.2018.8333307
2. A. Khasawneh, H. Rogers, J. Bertrand, K.C. Madathil, A. Gramopadhye, Human adaptation to latency in teleoperated multi-robot human-agent search and rescue teams. Automation Constr. **99**, 265–277 (2019). https://doi.org/10.1016/j.autcon.2018.12.012
3. G.G. Muscolo, S. Marcheschi, M. Fontana, M. Bergamasco, Dynamics modeling of human-machine control interface for underwater teleoperation. Robotica **39**, 618–632 (2020). https://doi.org/10.1017/S0263574720000624
4. G. Du, P. Zhang, J. Mai, Z. Li, Markerless kinect-based hand tracking for robot teleoperation. Int. J. Adv. Robot. Syst. **9** (2012). https://doi.org/10.5772/50093
5. A. Handa, K.V. Wyk, W. Yang, J. Liang, Y.W. Chao, Q. Wan, S. Birchfield, N. Ratliff, D. Fox, Dexpilot: vision-based teleoperation of dexterous robotic hand-arm system. Proc. IEEE Int. Conf. Robot. Automation, 9164–9170 (2020). https://doi.org/10.1109/ICRA40945.2020.9197124
6. S. Li, X. Ma, H. Liang, M. Görner, P. Ruppel, B. Fang, F. Sun, J. Zhang, Vision-based teleoperation of shadow dexterous hand using end-to-end deep neural network, in *2019 International Conference on Robotics and Automation (ICRA)* (2019), pp. 416–422
7. V. Patraucean, A. Handa, R. Cipolla, Spatio-temporal video autoencoder with differentiable memory. *ICLR 2016 - Workshop track* (2015). abs/1511.0
8. A. Singh, L. Yang, K. Hartikainen, C. Finn, S. Levine, End-to-end robotic reinforcement learning without reward engineering. Robot. Sci. Syst. **73** (2019). https://www.roboticsproceedings.org/rss15/p73.html
9. D. DeMers, K. Kreutz-Delgado, Inverse kinematics of dextrous manipulators. Neural Syst. Robot. 75–116 (1997). https://doi.org/10.1016/B978-0-08-092509-7.50008-7
10. S. Bhattacharya, M. Pivtoraiko, A classification of configuration spaces of planar robot arms for a continuous inverse kinematics problem. Acta App. Math. Acta Applicandae Mathematicae **139**(1), 133–166 (2015). https://doi.org/10.1007/s10440-014-9973-1
11. C. Meeker, M. Haas-Heger, M. Ciocarlie, A continuous teleoperation subspace with empirical and algorithmic mapping algorithms for nonanthropomorphic hands. IEEE Trans. Automation Sci. Eng. 1–14 (2020). https://doi.org/10.1109/TASE.2020.3035156
12. G. Gioioso, G. Salvietti, M. Malvezzi, D. Prattichizzo, Mapping synergies from human to robotic hands with dissimilar kinematics: an approach in the object domain. IEEE Trans. Robot. **29**, 825–837 (2013). https://doi.org/10.1109/TRO.2013.2252251
13. J. Liu, Y. Zhang, Mapping human hand motion to dexterous robotic hand, in *2007 IEEE International Conference on Robotics and Biomimetics (ROBIO)* (2008), pp. 829–834. https://doi.org/10.1109/ROBIO.2007.4522270
14. C. Stanton, A. Bogdanovych, E. Ratanasena, Teleoperation of a humanoid robot using full-body motion capture, example movements, and machine learning, in *Australasian Conference on Robotics and Automation, ACRA* (2012)
15. R. O'Flaherty, P. Vieira, M.X. Grey, P. Oh, A. Bobick, M. Egerstedt, M. Stilman, Humanoid robot teleoperation for tasks with power tools, in *IEEE Conference on Technologies for Practical Robot Applications, TePRA* (2013). https://doi.org/10.1109/TePRA.2013.6556362

16. R. Meattini, D. Chiaravalli, L. Biagiotti, G. Palli, C. Melchiorri, Combined joint-cartesian mapping for simultaneous shape and precision teleoperation of anthropomorphic robotic hands. IFAC-PapersOnLine **53**, 10052–10057 (2020). https://doi.org/10.1016/J.IFACOL.2020.12.2726
17. G. Gioioso, G. Salvietti, M. Malvezzi, D. Prattichizzo, An object-based approach to map human hand synergies onto robotic hands with dissimilar kinematics. Robot. Sci. Syst. **8**, 97–104 (2013). https://doi.org/10.15607/rss.2012.viii.013
18. J. Romero, T. Feix, C.H. Ek, H. Kjellström, D. Kragic, Extracting postural synergies for robotic grasping. IEEE Trans. Robot. **29**, 1342–1352 (2013). https://doi.org/10.1109/TRO.2013.2272249
19. G. Salvietti, L. Meli, G. Gioioso, M. Malvezzi, D. Prattichizzo, Object-based bilateral telemanipulation between dissimilar kinematic structures, in *IEEE International Conference on Intelligent Robots and Systems* (2013), pp. 5451–5456. https://doi.org/10.1109/IROS.2013.6697145
20. C.I. Mavrogiannis, C.P. Bechlioulis, M.V. Liarokapis, K.J. Kyriakopoulos, Task-specific grasp selection for underactuated hands, in *Proceedings - IEEE International Conference on Robotics and Automation* (2014), pp. 3676–3681. https://doi.org/10.1109/ICRA.2014.6907391
21. J. Starke, C. Eichmann, S. Ottenhaus, T. Asfour, Synergy-based, data-driven generation of object-specific grasps for anthropomorphic hands, in *IEEE-RAS International Conference on Humanoid Robots* (2019), pp. 327–333. https://doi.org/10.1109/HUMANOIDS.2018.8624990
22. M. Bianchi, P. Salaris, A. Bicchi, Synergy-based hand pose sensing: reconstruction enhancement. Int. J. Robot. Res. **32**, 396–406 (2013). https://doi.org/10.1177/0278364912474078
23. D. Pei, V. Patel, M. Burns, R. Chandramouli, R. Vinjamuri, Neural decoding of synergy-based hand movements using electroencephalography. IEEE Access **7**, 18155–18163 (2019). https://doi.org/10.1109/ACCESS.2019.2895566
24. W. Chen, C. Xiong, S. Yue, Mechanical implementation of kinematic synergy for continual grasping generation of anthropomorphic hand. IEEE/ASME Trans. Mechatron. **20**, 1249–1263 (2015). https://doi.org/10.1109/TMECH.2014.2329006
25. M. Santello, M. Bianchi, M. Gabiccini, E. Ricciardi, G. Salvietti, D. Prattichizzo, M. Ernst, A. Moscatelli, H. Jörntell, A.M.L. Kappers, K. Kyriakopoulos, A. Albu-Schäffer, C. Castellini, A. Bicchi, Hand synergies: integration of robotics and neuroscience for understanding the control of biological and artificial hands. Phys. Life Rev. **17**, 1–23 (2016). https://doi.org/10.1016/j.plrev.2016.02.001
26. G. Salvietti, Replicating human hand synergies onto robotic hands: a review on software and hardware strategies. Front. Neurorobot. **12**, 1–6 (2018). https://doi.org/10.3389/fnbot.2018.00027
27. T. Tang, J. Jia, H. Mao, Dance with melody: an lstm-autoencoder approach to music-oriented dance synthesis, in *MM 2018 - Proceedings of the 2018 ACM Multimedia Conference* (2018), pp. 1598–1606. https://doi.org/10.1145/3240508.3240526
28. M. Spüler, N. Irastorza-Landa, A. Sarasola-Sanz, A. Ramos-Murguialday, in *Artificial Neural Networks and Machine Learning – ICANN 2016*, ed. by A. E.P. Villa, M. Paolo, P. Rivero, A. Javier. Extracting Muscle Synergy Patterns from EMG Data Using Autoencoders (Springer International Publishing, Cham, 2016), pp. 47–54
29. S. Katyara, F. Ficuciello, D.G. Caldwell, B. Siciliano, F. Chen, Leveraging Kernelized synergies on shared subspace for precision grasping and dexterous manipulation. IEEE Trans. Cogn. Develop. Syst. **15**(4), 2064–2076 (2023). https://doi.org/10.1109/TCDS.2021.3110406
30. E. Battaglia, M. Kasman, A.M. Fey, Moving past principal component analysis: Nonlinear dimensionality reduction towards better hand pose synthesis, in *2022 International Symposium on Medical Robotics, ISMR 2022* (2022) https://doi.org/10.1109/ISMR48347.2022.9807580
31. M.L. Lafash, *Synergy* (Oxford University Press, Oxford, 2008)
32. C. Lugaresi, J. Tang, H. Nash, C. McClanahan, E. Uboweja, M. Hays, F. Zhang, C.-L. Chang, M. Yong, J. Lee, W.-T. Chang, W. Hua, M. Georg, M. Grundmann, in *Third Workshop on Computer Vision for AR/VR at IEEE Computer Vision and Pattern Recognition (CVPR) 2019*. MediaPipe: A Framework for Perceiving and Processing Reality (2019). https://mixedreality.cs.cornell.edu/s/NewTitle_May1_MediaPipe_CVPR_CV4ARVR_Workshop_2019.pdf

Chapter 14
Human-Centered Shared Autonomy for Motor Planning, Learning, and Control Applications

M. H. Farhadi, Ali Rabiee, Sima Ghafoori, Anna Cetera, Wei Xu, and Reza Abiri

Abstract With recent advancements in AI and computation tools, intelligent paradigms have emerged to empower different fields such as shared autonomy and human-machine teaming in healthcare with new capabilities. Advanced AI algorithms (e.g., reinforcement learning) can be trained and developed to autonomously make individual decisions to achieve desired plan and motion goals. However, such independent decisions and goal achievements might not be ideal within healthcare, where human intent plays a pivotal and crucial role in guiding human-machine paradigms. This chapter presents a comprehensive review of human-centered shared autonomy AI frameworks, particularly by focusing on upper limb biosignal-based machine interfaces and their associated end-effector based motor control systems, including computer cursors, robotic arms, and planar robotic platforms. Our primary focus is delving into motor planning, learning (rehabilitation), and controls, specifically covering conceptual foundations of human-machine teaming in reach-and-grasp tasks, analyzing both the theoretical principles and practical implementations from both human and machine perspectives. We extended the discussion in each section to elaborate on how human and machine inputs can be blended as shared autonomy paradigms, with the healthcare applications. The chapter examines topics on human factors, biosignal processing for intent detection, shared autonomy approaches in brain-computer interfaces (BCI), rehabilitation, robots, assistive robotics, and finally, Large Language Models (LLM) as the next frontier. With a foundation based on human-centered factors, we also proposed adaptive shared autonomy AI as a potential high-performance paradigm for the two interactive human and AI agents. We identified current challenges in human-centered shared autonomy implementation and proposed future directions, particularly examining

M. H. Farhadi · A. Rabiee · S. Ghafoori · A. Cetera · R. Abiri (✉)
Translational Neurorobotics Laboratory, Department of Electrical, Computer, and Biomedical Engineering, University of Rhode Island, Kingston, RI, USA
e-mail: mh.farhadi@uri.edu; ali.rabiee@uri.edu; sima.ghafoori@uri.edu; annacetera@uri.edu; reza_abiri@uri.edu

W. Xu
Human Centered Ai labs at Los Angles, CA, USA

R. Vinjamuri (ed.), *Bridging the Gap between Mind and Machine*,
https://doi.org/10.1007/978-3-032-06713-5_14

the roles of AI reasoning agents in advancing these systems. Through this analysis, we bridge neuroscientific understanding with robotics approaches to develop more intuitive, effective, and ethical human-machine teaming frameworks.

Keywords Shared autonomy · Human-centered AI · Biosignal processing · Brain-computer interfaces · Rehabilitation robotics · Assistive technology · Reach-and-grasp · Human-machine teaming · Large language models

14.1 Introduction

The symbiotic integration of human capabilities with intelligent machines is rapidly transforming our approach to complex motor tasks, particularly in domains requiring nuanced interaction and assistance, such as healthcare and robotics [1, 2]. Within this evolving landscape, human-centered shared autonomy has emerged as a pivotal paradigm. This chapter embarks on a comprehensive exploration of shared autonomy (traditionally known as shared control) frameworks. Our focus is primarily on end-effector based systems, encompassing a range of applications from intuitive computer cursor control and sophisticated robotic arm manipulation to advanced rehabilitation and assistive robots within the healthcare sector. The core tenet of these systems lies in creating a collaborative partnership where the human user and the semi-autonomous system work together by leveraging their respective strengths to achieve a common goal [3].

The imperative for human-centered approaches stems from the need to ensure that autonomous systems remain aligned with user intentions, adapt to individual user needs and capabilities, and maximize tasks performance all while maintaining a feeling of agency and trust in users [4]. This is particularly critical when dealing with biosignals (such as electromyography (EMG) or electroencephalography (EEG)) as the primary communication channel between the human and the machine [5]. These signals, while rich in information about user intent, are often complex, noisy, and highly variable, posing significant challenges for robust interpretation and seamless translation into machine actions [6]. Shared autonomy addresses these challenges by not seeking to fully automate tasks, nor to place the entire burden of control on the human, but rather to intelligently arbitrate and blend control inputs from both ends [1].

This chapter will delve into the conceptual foundations of human-machine teaming, examining how principles from cognitive science, neuroscience, and robotics converge to inform the design of effective shared control systems. We will analyze both the theoretical underpinnings and the practical implementations of these frameworks, tracing their evolution and impact. A significant portion of our discussion will be dedicated to the critical role of biosignal processing in accurately detecting human intentions [7]. This involves exploring advanced algorithms for signal acquisition, feature extraction, and pattern recognition that enable the machine to infer the user's desired actions and goals.

We will survey the diverse applications of shared autonomy across various domains. In Brain-Computer Interfaces (BCIs), shared control offers a pathway to more intuitive and reliable operation of assistive devices for individuals with severe motor impairments [8]. Shared autonomy facilitates personalized therapy by dynamically adjusting the level of assistance based on the patient's engagement and performance, promoting motor learning and recovery [9], ultimately empowering users by enhancing their ability to perform activities of daily living with greater independence and precision.

The burgeoning field of Large Language Models (LLMs) is also beginning to intersect with shared autonomy, offering new avenues for more naturalistic interaction, high-level task understanding, and even AI-driven reasoning to support this collaborative process [10].

Despite significant advancements, the implementation of truly human-centered shared autonomy faces ongoing challenges. These include ensuring seamless and intuitive interaction, developing robust co-adaptation mechanisms, addressing ethical considerations related to user privacy, autonomy, and potential for algorithmic bias [11], and validating system efficacy in real-world scenarios [12]. This chapter will identify these critical challenges and propose directions for development.

Our goal is to provide researchers, practitioners, and students with a thorough understanding of the principles, technologies, applications, and future horizons of human-centered shared autonomy. This chapter provides a unified analysis of shared autonomy for motor control by connecting three distinct applications: Brain-Computer Interfaces (BCIs), rehabilitation, and assistive robotics [2]. Our contribution is to establish a common technical framework for these historically separate fields, one grounded in the mapping of user biosignals to control policies via adaptive arbitration [7, 13]. By examining methods–from Bayesian inference to modern deep learning architectures–through this unified lens, we identify shared challenges and opportunities for creating models that generalize across these human-in-the-loop systems [14, 15].

14.2 Human-Centered Shared Autonomy Principles

The effective integration of intelligent systems into human endeavors necessitates a paradigm shift from automation-centric designs towards those that prioritize a robust and adaptive synergy with the human user. Human-centered shared autonomy (HCSA) is at the forefront of this change, representing a field dedicated to the development of systems wherein human and artificial intelligence establish a cohesive and symbiotic partnership. The principles underpinning HCSA are not merely design heuristics; they constitute a foundational philosophy critical for realizing the transformative potential of autonomous technologies, particularly in domains with significant human consequence, such as healthcare, assistive robotics, and complex operational environments. Adherence to these principles is paramount for the creation of systems that are not only technically proficient but also intuitive, trustworthy, responsive, and ultimately, empowering for the individuals they are intended to assist [1].

This section distills the core design principles of human-centered shared autonomy (HCSA), detailing (i) why the human remains team-leader, (ii) how assistance arbitration paradigms govern control blending, and (iii) how intent inference closes the loop.

14.2.1 Human as the Leader in Human-AI Teams

A central principle within numerous shared-autonomy paradigms, particularly those involving critical decision-making or operation in unpredictable environments, is placing the human as the leader within this human-AI teamwork [3, 16]. This principle acknowledges that while artificial intelligence can display remarkable proficiency in perception and execution, autonomous systems often lack the generalized intelligence, common-sense reasoning, and nuanced comprehension of human values needed to navigate real-world complexities. Accordingly, the human's leadership typically involves defining high-level objectives, making strategic decisions at critical junctures, and actively monitoring the system's performance and behavioral patterns–rather than issuing continuous low-level control commands [1, 17, 18]. Positioning the human in this capacity ensures the autonomous system functions as a capable collaborator whose actions remain congruent with the human's objectives and ethical precepts [19]; it also preserves accountability and the user's sense of agency –the experience of volition and control over one's actions and outcomes [4]–while enabling timely intervention to rectify or override system behavior when it deviates from expectations or when unforeseen circumstances arise [20].

To facilitate this, systems frequently incorporate adjustable autonomy (also termed adaptive automation [21]). This functionality allows the level of AI assistance to be dynamically modulated, either via explicit user directives, context-aware algorithms that adapt to shifting task demands, or implicitly through the inference of the user's cognitive state (e.g., fatigue, engagement levels). The user must be able to formulate an accurate mental model of the system's capabilities, its current operational status, its decision-making rationale (to an appropriate degree of abstraction), and its probable future actions. Simultaneously, the system adapts to the user's learning curve and evolving understanding. This understanding empowers the user to supervise effectively and anticipate the necessity for intervention.

14.2.2 Interaction Paradigms and Levels of Automation

The level of automation in shared autonomy emerges from the combination of interaction paradigm and arbitration scheme. An interaction paradigm describes the fundamental roles and collaborative structure within the human-robot team, as outlined in several works [3, 22]. To implement these different paradigms, we need different arbitration schemes for blending the input of each party. Table 14.1 provides a comprehensive overview of arbitration policies across various shared autonomy implementations, highlighting the diversity of approaches in current research. In co-activity, the human and robot work within a shared environment, often on independent tasks that contribute to a broad objective, requiring spatial and temporal coordination. The master-slave paradigm places one agent, typically the human, in a position to dictate actions or trajectories, which the other agent (the robot) executes or supports–a common mode in teleoperation applications. In the teacher-student paradigm, one agent guides or adjusts the other to promote learning or rehabilitation; for example, a robot therapist might adaptively guide a patient's movements. Finally, collaboration in a partner-partner sense implies that human and robot act as peers, dynamically sharing responsibilities and control to achieve tasks that neither could perform as effectively in isolation, often involving mutual adaptation.

To implement these different paradigms, we need different arbitration schemes for blending the input of each party. Arbitration schemes specify the mechanisms for allocating or sharing control during task execution, determining "who controls what and when". These schemes range from fixed autonomy, where control authority is assigned a priori and remains static within task segments, to mode-switching, where control authority is explicitly transferred between agents at discrete junctions such as clear phase boundaries or specific triggers. Parallel schemes allow the human and robot to act simultaneously but on different, separable aspects of the task, often corresponding to disjoint Degrees of Freedom–for instance, the human might manage the positioning of an end-effector while the robot handles its orientation or grasp force. At the most fluid end of the spectrum, continuous policy-blending combines inputs or control policies from both the human (u_{human}) and the robot (u_{AI}) continuously to generate a single command u_{final} for the system. A common mathematical formulation for this is a weighted sum [13, 30]:

$$u_{final} = (1 - \alpha)u_{human} + \alpha u_{AI} \tag{14.1}$$

The blending coefficient, $\alpha \in [0, 1]$, dictates the relative contribution of each agent, with $\alpha = 0$ signifying full human control and $\alpha = 1$ full robot control. Intermediate values represent varying degrees of shared control. The specific terms being blended (u_{human}, u_{AI}) can represent various control aspects such as forces, velocities, or position commands, depending on the application (e.g., as in [25] where control velocities from EMG and robot assistance are blended).

In continuous policy-blending, the coefficient α can be static or, more dynamically in HCSA, adaptive. An adaptive α allows the system to modulate the level of robot autonomy based on real-time factors such as confidence in intent inference [13] or safety metrics like proximity to obstacles [23, 27]. Game-theoretic arbitration [21, 23, 27] models the human and robot as players with distinct cost functions, determining dynamic arbitration strategies through evaluation of safety indices and task progress metrics.

Table 14.1 Overview of arbitration policies, evaluation environments, and functionalities

Reference	Arbitration policy	Evaluation environment & functionality tested
Losey et al. [3]	Review: Co-activity, master-slave, teacher-student, collaboration; Arbitration definition.	Conceptual review; Case studies: Upper-limb robotic rehabilitation, haptic feedback for prosthetics.
Franceschi et al. [23]	Role Arbitration (RA) via Game Theory (Coop. & Non-Coop.); Dynamic α (blending) via Fuzzy Logic or force.	pHRI with UR5 robot; Co-transportation/manipulation: obstacle avoidance, via-point reaching.
Madduri et al. [18]	Review: Co-adaptation in User-Machine Interfaces (UMI); Leader-follower; Game theory for user-decoder.	Conceptual review: Co-adaptive motor control interfaces (BCI, myoelectric, kinematic).
McFarland et al. [24]	BCI control mapping (EEG to cursor). (Level of automation in BCI input translation).	Noninvasive BCI; 2D cursor movement & target selection on screen; Emulates mouse operation.
Sarasola-Sanz et al. [25]	Hybrid BMMI; Blended control (EEG-triggered EMG + robot assist).	7-DoF Upper limb exoskeleton; Stroke rehabilitation: functional tasks (grasping, pointing).
Jain and Argall [26]	Probabilistic intent recognition; Blended shared control based on intent confidence.	6-DoF MICO robotic arm; Assistive robotics: object retrieval/manipulation with joystick & Sip-N-Puff.
Franceschi et al. [27]	RA (Cooperative Game Theory); Dynamic α via interaction force.	UR5 robot for pHRI; Planar circular trajectory following with human intervention.
Nikolaidis and Shah [16]	Shared Mental Models for H-R Teaming; Robot adapts plan based on human's state/actions. (Arbitration via plan adaptation).	Simulated and Baxter robot; Physical assembly tasks (IKEA furniture).
Ajoudani et al. [28]	Shared control in HRI; Control sharing levels (human/robot lead, cooperation).	Broad review: pHRI, industrial co-manipulation, assistive robotics.
Li et al. [21]	Continuous Role Adaptation via Game Theory and Policy Iteration; Variable impedance based on Nash equilibrium.	pHRI experimental setup; Cooperative manipulation tasks (trajectory following).
Mörtl et al. [29]	Leader-Follower role allocation analysis in pHRI; Energy exchange based role identification.	1-DoF haptic knob, virtual crank, Dyadic co-manipulation (human-human and human-robot).

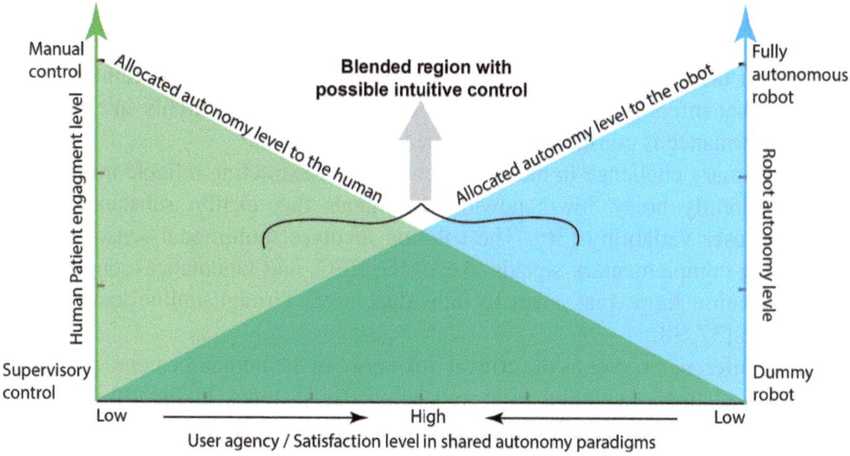

Fig. 14.1 Conceptual autonomy–agency continuum for shared autonomy paradigms

The choice of interaction paradigm, arbitration scheme, and resultant level of automation profoundly influences the human user's experience. Systems that are opaque in their decision-making, unpredictable in their actions, or those that frequently override user input without clear justification can escalate frustration and erode user trust [12]. This directly impacts user agency and cognitive workload [31, 32], a trade-off conceptually illustrated in Fig. 14.1. Optimal shared autonomy systems balance performance gains with preservation of user empowerment, often requiring adaptive arbitration that responds to individual user capabilities and preferences [17].

14.2.3 Decoding Human Intent for Adaptive Assistance

Intent inference, where an autonomous system determines a human user's objectives or actions from limited observations, is fundamental to shared autonomy as it enables systems to anticipate user needs and provide timely assistance without explicit commands [2].

Three primary methodologies address intent inference in shared autonomy systems. Probabilistic methods explicitly model uncertainty in human behavior and noisy sensory data, maintaining probability distributions over possible user goals that update with new observations, as exemplified by recursive Bayesian filtering and POMDP frameworks [33]. These approaches excel in interpretability but depend heavily on the accuracy of their models and priors. Data-driven methods leverage deep learning architectures like CNNs and RNNs to learn complex mappings from behavioral cues (EMG, gaze, kinematics) to user intentions, achieving high accuracy when sufficient training data is available [34]. However, they often lack transparency and struggle with novel intents outside their training distribution. Model-based strategies employ explicit representations of human cognition, biomechanics, or task structures to interpret user behavior, such as inverse reinforcement learning that infers goals from observed actions [35]. While highly interpretable, their performance is constrained by model fidelity.

The primary challenge in intent inference lies in extracting reliable information from inherently noisy, low-bandwidth biosignals that exhibit substantial inter- and intra-user variability [36]. The solution involves multimodal sensor fusion–combining complementary signals like EMG, EEG, and kinematics–coupled with personalization loops that adapt to individual users through online learning and calibration [37, 38].

Intent inference serves as the critical link between the human-centered principles and practical implementation of shared autonomy systems. It directly influences how arbitration schemes adjust the blending coefficient α in real-time, with higher confidence in intent inference typically leading to increased autonomous assistance [13]. The quality of intent inference fundamentally determines whether a shared autonomy system enhances or hinders the user's sense of agency. Accurate inference enables seamless assistance that feels like an extension of the user's will, while poor inference leads to frustrating conflicts between human and machine intentions [39]. This challenge becomes particularly acute in biosignal-based systems, and robotic applications where neural and behavioral signals provide the primary communication channel, as we will explore in Sect. 14.3.

14.2.4 Towards Human-AI Joint Cognitive Systems

The trajectory of shared autonomy is pointing towards a paradigm that is more deeply integrated with human-AI teaming. The ultimate goal is the creation of a human-AI joint cognitive system, in which the human and machine function as a single, cohesive cognitive unit to achieve goals that are intractable for either partner alone [40]. This represents a shift in focus from purely optimizing physical task execution to improving the system's cognitive ergonomics, where the interactions are explicitly designed to minimize the user's cognitive load and mental fatigue [41]. The aim is to create a fluid, low-effort partnership. Rather than the user constantly directing the AI, the system begins to proactively assist based on task context and a predictive understanding of the user's high level goals [42, 43].

Achieving this requires moving beyond reactive intent inference. For instance, instead of only decoding a user's immediate motor command, the system can learn to anticipate the next likely sub-task in a sequence (e.g., after picking up a cup, the next goal is likely bringing it to the mouth). This is a core challenge being explored in human-in-the-loop reinforcement learning, where the AI can be rewarded for actions that not only advance the task but also minimize the need for user corrections, thereby reducing cognitive burden [44]. Future biosignal-based interfaces will play a critical role, not by attempting to "read the user's mind," but by providing a real-time estimate of cognitive states like workload or surprise, which can serve as a feedback signal to modulate the AI's level of autonomy [45].

The practical path towards this tighter integration also depends on improving the AI's transparency. The development of more explainable AI for robotics is crucial, as it allows the system to signal its intentions to the user, making its behavior more predictable and trustworthy [46]. This creates a positive feedback loop: a more predictable system is easier to work with, which lowers cognitive load and builds the trust necessary for truly symbiotic human-AI partnership.

Having established the foundational principles of human-centered shared autonomy, we now examine their application across three critical domains: brain-computer interfaces, rehabilitation, and assistive technologies.

14.3 Shared Autonomy in Brain-Computer Interface Domain

14.3.1 Overview and Gaps

Brain-Computer Interfaces (BCIs) have evolved from direct, low-level control systems toward more autonomous architectures that aim to reduce cognitive load, improve reliability, and enhance user experience [47]. Shared autonomy in BCIs involves dynamically blending user intention with system intelligence–commonly using biosignal-driven cues (e.g., EEG, EMG) and decision-making algorithms to assist users in completing motor or communication tasks. Current approaches fall into three categories: Direct control involves users providing continuous or discrete commands (e.g., left/right movement) through modulation of specific brain rhythms or event-related potentials. While this offers high user agency, it is mentally demanding and slow, especially for multidimensional tasks [5, 48]. Goal selection systems infer the user's intended goal from high-level signals and autonomously execute the corresponding action (e.g., selecting a target object). This reduces cognitive burden but requires accurate intent recognition and is often limited by low information transfer rates [8, 49]. Predictive shared control systems combine user input with predictive models (e.g., probabilistic inference, deep learning, or reinforcement learning) to collaboratively plan and execute actions. These systems can resolve ambiguity and correct errors, improving safety and efficiency [50]. However, these systems face key limitations: Signal variability presents challenges as EEG signals are highly non-stationary and subject to intra- and inter-subject variability, leading to unreliable decoding [51]. Calibration and adaptation issues arise because many systems require long calibration periods and fail to adapt dynamically to changes in neural patterns, fatigue, or user intent over time [52]. Trust and transparency become problematic as autonomy increases, since users may struggle to understand or trust the system's decisions, especially if they lack feedback or explanation mechanisms [32]. Cognitive load remains a concern because direct control BCIs impose significant mental effort, while poorly calibrated shared control models can increase user frustration due to erratic or incorrect assistance [53]. Lack of standardization in evaluation and benchmarking of shared

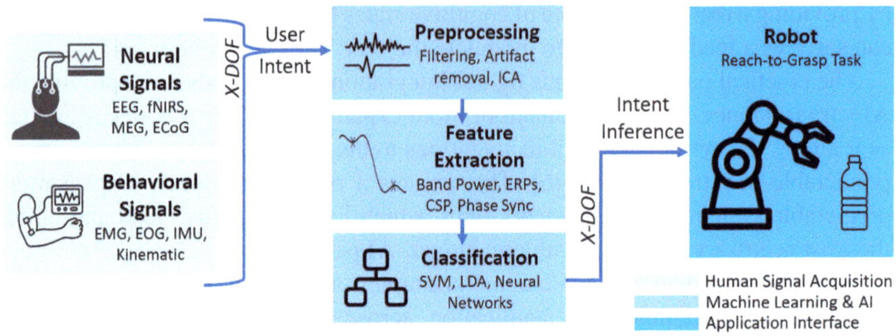

Fig. 14.2 Traditional BCI pipeline for a reach-to-grasp task, illustrating intent decoding from neural and behavioral signals through preprocessing, feature extraction, and classification. Traditional BCIs operate under a sequential paradigm, preserving user input (X degrees of freedom, DOF)

autonomy approaches creates inconsistency, making it difficult to compare methods across studies. Additionally, most systems rely on static control policies rather than adaptive models that learn and evolve with the user (Fig. 14.2). These constraints highlight the need for co-adaptive, user-in-the-loop models that continuously learn from user behavior and feedback in BCI applications [54, 55].

14.3.2 Neural and Behavioral Signals

Biosignal processing in shared autonomy systems combines neural and behavioral signals to infer user intention, monitor cognitive state, and support flexible human-robot interaction. These signals provide direct access to brain activity, muscle dynamics, and movement behavior, which together improve the robustness and reliability of human-in-the-loop systems.

Neural signals commonly used in non-invasive BCIs include electroencephalography (EEG), functional near-infrared spectroscopy (fNIRS), magnetoencephalography (MEG), and electrocorticography (ECoG). EEG is most commonly used because of its affordability, high temporal resolution, and portability [51, 52]. EEG measures brain activity by recording electrical signals from neurons in the cortex and researchers analyze EEG signals across different frequency ranges (delta, theta, alpha, beta, gamma) associated with different cognitive or motor states. It can also capture event-related potentials (ERPs), such as the P300 or N200, which indicate how the brain processes information and makes decisions [56]. MEG, while offering better spatial resolution than EEG, is costly and non-portable. ECoG, recorded directly from the cortical surface, provides high-fidelity signals but is invasive and thus limited to clinical contexts.

Behavioral signals provide physical expressions of brain activity and help decode user intention in shared autonomy. Electromyography (EMG) records electrical activity from skeletal muscles and is used to detect residual motor function or anticipate motion intention in prosthetics and exoskeletons [57]. Electrooculography (EOG) captures eye movement potentials and is commonly applied to gaze-based interfaces, providing directional cues or confirmation signals. Inertial measurement units (IMUs), motion capture systems, and pressure sensors record kinematic data–such as joint angles, limb acceleration, or grip force–that reflect motor intent and biomechanics. These modalities can be used alone or combined with neural signals to build hybrid control systems [58, 59].

Multimodal integration of neural and behavioral signals strengthens shared autonomy by improving signal redundancy and reducing ambiguity in intention decoding. For instance, combining EEG with EMG can distinguish between imagined and attempted movements [60, 61] , while combining EOG with EEG allows gaze-contingent BCI control [62, 63]. However, multimodal systems also create problems with synchronization, computational complexity, and user burden. Creating interfaces that combine these signals effectively while maintaining speed and comfort remains a challenge for developing better BCI systems. Key areas for future work include adaptive feature fusion methods, wearable multi-sensor integration, and intelligent signal prioritization to unlock the capabilities of hybrid biosignal frameworks in autonomous BCI systems.

14.3.3 Signal Processing and Feature Extraction

Transforming raw biosignals into actionable information requires a series of preprocessing, feature extraction, and classification steps. Preprocessing techniques such as bandpass filtering, artifact rejection (e.g., ICA for ocular and muscle artifacts), and baseline normalization are fundamental to improving signal quality and minimizing noise [64, 65]. High-quality preprocessing is especially important in shared autonomy systems, where unreliable input can lead to incorrect system responses, ultimately reducing user trust and shared task success. Feature extraction techniques identify relevant information from the preprocessed signals for downstream classification or regression tasks. In EEG, common features include band power (e.g., delta, theta, alpha, beta), event-related potentials (ERPs), phase synchrony [66], and spatial filters such as Common Spatial Patterns (CSP) or Filter Bank CSP (FBCSP) [67, 68]. For behavioral signals such as EMG or kinematic data, time-domain features (e.g., root mean square, variance) [69] and frequency-domain features (e.g., FFT, wavelet transforms) [70] are frequently used [71].

Recent developments in deep learning allows for end-to-end feature learning from raw biosignals using architectures such as convolutional neural networks (CNNs), recurrent neural networks (RNNs), and attention-based models [72, 73]. These approaches can automatically capture spatiotemporal dependencies across channels and time [74], eliminating the need for manually computed features. However, deep learning models present challenges, particularly in shared autonomy scenarios where transparency, user interpretability, and real-time responsiveness are crucial. Black-box models may yield high accuracy but can be difficult to trust or troubleshoot in human-in-the-loop systems.

Additionally, the accuracy and performance of feature extraction pipelines is often user-dependent. Variability in neural and muscular signals between users–and even within the same user over time [75, 76] –can reduce generalizability, leading to performance degradation without adaptive learning mechanisms. This is especially problematic in shared autonomy systems, where continuous co-adaptation between the user and machine is necessary to maintain harmonious user-system integration [77]. Systems that rely on static models or poorly calibrated features may either fail to assist the user effectively or override user input inappropriately, disrupting the collaborative balance. To address these challenges, future feature extraction approaches in shared autonomy should aim for reliability to inter- and intra-user variability, interpretability and feedback integration for user understanding, low-latency and computational efficiency for real-time use, and adaptive and co-learning frameworks that evolve with the user.

14.3.4 Classification and Real-Time Intent Inference

Real-time intent decoding forms the backbone of shared autonomy brain-computer interface (BCI) systems, where human neural signals must be continuously translated into actionable commands while maintaining seamless human-machine collaboration. In these systems, the BCI component captures and decodes user intent from neural signals, while the autonomous component provides intelligent assistance to compensate for decoding uncertainties and reduce cognitive load. This collaborative framework requires robust real-time classification that can operate under the demanding constraints of continuous data streaming, minimal latency, and adaptive performance.

The process of real-time intent inference begins with continuous EEG signal acquisition, followed by rapid preprocessing, feature extraction, and classification within tight temporal windows–typically requiring complete processing cycles of 100–200 milliseconds to maintain natural interaction flow [78, 79]. However, this pipeline faces significant technical challenges that are particularly pronounced in shared autonomy applications. EEG signals suffer from inherently low signal-to-noise ratios [80, 81], with neural activity of interest often buried beneath artifacts from muscle movement, eye blinks, and environmental interference [51, 82]. Motion artifacts become especially problematic in assistive applications where users may have involuntary movements or tremors. Additionally, maintaining sub-200ms latency is essential for preserving user trust and task performance, as delays in decoding disrupt the natural flow of shared control and can lead to user frustration or system abandonment [53]. These systems must also contend with high variability both within individual users across sessions and between different users, requiring robust classification approaches that can adapt to changing neural patterns while operating within computational constraints of real-time embedded systems [31].

Modern classification approaches for real-time intent decoding have evolved from traditional machine learning methods to sophisticated deep learning architectures. While support vector machines (SVM) and linear discriminant analysis (LDA) remain valuable for their computational efficiency and interpretability in resource-constrained applications [83, 84], deep learning models have demonstrated superior performance in capturing the complex, nonlinear relationships inherent in neural signals. Convolutional neural networks (CNNs) excel at extracting spatial patterns from multi-channel EEG data [85, 86], while recurrent neural networks (RNNs) and their variants capture temporal dependencies essential for decoding dynamic motor intent. Graph neural networks (GNNs) have emerged as particularly promising for modeling the complex connectivity patterns between brain regions during motor planning and execution [87, 88]. These advanced architectures enable more nuanced decoding of user intent, which is vital for shared autonomy systems that must distinguish between different levels of user engagement and confidence to appropriately modulate autonomous assistance.

The integration of these technical advances allows for shared autonomy capabilities that define modern assistive BCI systems. Real-time classifiers combined with transfer learning and adaptive filtering, allow systems to not only decode explicit user commands but also infer user confidence, attention levels, and intended goals [89, 90]. This multi-layered understanding enables proactive assistance where the autonomous component can anticipate user needs, provide corrective guidance, and smoothly transition between different levels of autonomy based on decoded neural states. Such capabilities are particularly valuable in assistive and rehabilitative applications, where users may have varying levels of motor control and cognitive load, requiring systems that can dynamically adapt their assistance strategies. Future developments should focus on co-adaptive algorithms that simultaneously optimize both human neural patterns and machine learning models [91, 92], hybrid sensor integration to complement EEG with other modalities, and energy-efficient edge computing architectures to support deployment in portable, long-term use scenarios [14].

14.3.5 Human-Centered Performance Metrics in BCI Shared Autonomy

Evaluating autonomy in BCIs requires a multidimensional performance assessment framework that captures both system-level efficiency and user-centered outcomes. Traditional performance metrics–such as task success rate, classification accuracy, and information transfer rate (ITR)–remain widely used in BCI literature [5, 51]. However, these metrics alone are insufficient for evaluating the quality of shared autonomy, which involves both machine performance and human experience. Recent studies emphasize the importance of human-centered metrics such as user workload, typically assessed through instruments like NASA Task Load Index (NASA-TLX) [93], perceived control and trust, which influence user engagement and acceptance [32, 53], system adaptability and responsiveness, which reflect how well the system co-adapts to user behavior [14], and user satisfaction or usability, often quantified using tools like the System Usability Scale (SUS) [94]. A major challenge lies in developing standardized benchmarking protocols that account for the dynamic interaction between user and system. For example, systems that assist more aggressively may improve task success but reduce user agency–a trade-off that must be captured through joint autonomy-performance metrics [53]. To holistically evaluate shared autonomy, future work should consider hybrid frameworks that combine quantitative performance (e.g., accuracy, speed), behavioral indicators (e.g., error correction patterns), and qualitative feedback (e.g., user-reported metrics).

14.3.6 *Case Study: Graph-Based AI Translation for Shared Control in BCI Grasping Systems*

The role of AI as a translator of neural intent is demonstrated in our bio-inspired, multimodal BCI system for assistive grasping (Fig. 14.3). The pipeline begins with Stage 1 (S1), a vision-based grasping platform that collects synchronized RGB-D and EEG (tEEG) data as users observe and plan reach-to-grasp movements. This multimodal approach combines neural signals from users wearing a 16 channel EEG recording system (g.tec) with RGB-D visual information for reconstructing arm and hand kinematics through camera-based pose estimation. Advanced signal processing modules extract relevant spatial, spectral, and temporal features from the EEG data while simultaneously processing the behavioral signals captured through computer vision. These preprocessed biosignals then feed into Stage 2 (S2), where computational neuroscience approaches utilize graph-based neural decoding models that capture dynamic functional connectivity patterns across EEG channels–

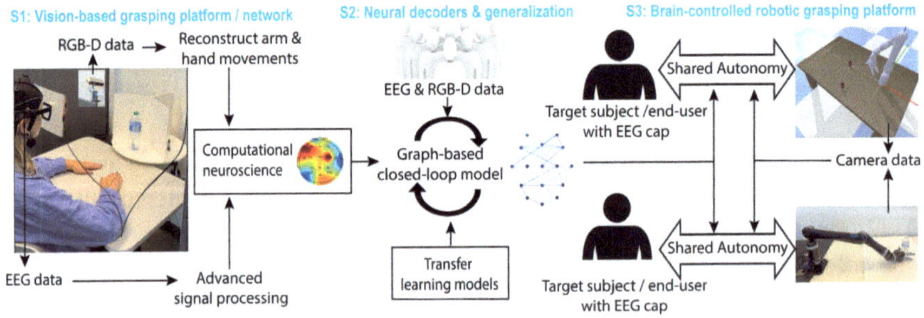

Fig. 14.3 Graph-based AI translation pipeline for BCI grasping system with three components: S1 collects RGB-D and EEG data during reach-to-grasp movements, S2 uses graph-based neural decoders with transfer learning for motor intent classification, and S3 implements shared control between decoded neural signals and robotic autonomy via Kinova JACO arm

modeling the brain as a network to decode motor intent more accurately than traditional sequential processing methods.

In S2, we utilize transfer learning to generalize across subjects, where pretrained models learn shared neural representations related to specific grasp types and are fine-tuned for individual users with minimal data to improve usability and reduce calibration time. The decoded intent is then integrated in Stage 3 (S3), the brain-controlled robotic grasping platform, where AI continuously blends user-driven neural input with autonomous robotic feedback to control a Kinova JACO robotic arm under a shared autonomy framework. This final stage enables both target subjects and end-users to participate in smooth, adaptive interaction through the bidirectional shared autonomy paradigms shown in the figure. The co-adaptive loop–driven by real-time inference from the graph-based model and camera data integration–exemplifies the vision of bio-inspired AI, where human and machine collaborate to complete complex tasks like object grasping in a way that mimics natural sensorimotor coordination while reducing the cognitive burden inherent in traditional BCI approaches. This shared autonomy BCI framework represents a paradigm shift from direct control BCIs toward user-in-the-loop systems collaborative systems that intelligently balance user agency with autonomous assistance.

14.4 Shared Autonomy in Rehabilitation Domain

14.4.1 Overview and Gaps

Stroke is a leading cause of death and often results in long-lasting upper limb motor disabilities [95]. Physical therapy is crucial for helping survivors regain motor control, but traditional methods and clinical systems alone can't meet the high demand due to the resource-intensive nature of therapy [96]. As rehabilitation shifts towards enabling patients to practice and regain strength independently, upper limb robotic rehabilitation has gained prominence [96]. These platforms are increasingly becoming more portable and cost-efficient, catering to various upper limb activities ranging from finger manipulation to shoulder movements [97–101]. Studies comparing robotic rehabilitation with traditional methods have shown positive impacts [96, 102, 103]. However, the widespread adoption of these devices is hindered by several factors. Complex designs often lead to higher costs, making them less affordable. Additionally, usability is a concern, especially for individuals with severe motor impairments who need simpler tasks and user-friendly interfaces. Beyond hardware considerations like cost and portability, a more significant challenge lies in refining the control strategies and adaptive algorithms that mediate the robot's real-time interaction with the patient.

14.4.2 Applications and Mechanisms

Upper limb rehabilitation robots are primarily designed as either end-effector-based or exoskeleton-based systems [97–100]. Exoskeleton-based robots, with multiple degrees of freedom, allow for a wider range of movements due to more connection points. In contrast, end-effector-based robots are predominantly adopted due to their simplicity, ease of production and control, and adaptability to different arm sizes of subjects. Additionally, they have the lowest cost and the greatest potential for commercialization [104]. The study [105] has proven the effectiveness of end-effector-based devices over exoskeleton-based robots. These end-effector robots are further categorized into three types: robots for manipulation [106–109], for reaching [110–133], and for both manipulating and reaching [134, 135]. While robots designed for reaching activities are often prioritized in rehabilitation due to their focus on training upper body movements prior to hand manipulation, they also cater to a wide range of patient needs [136]. Starting with a simple design of just one degree of freedom (DOF) for those with severe conditions, these robots can be adapted to include up to three DOFs, facilitating movement training across three axes. Additionally, by primarily operating within a two-dimensional plane, these robots maintain ease of operation, implementation, and user-friendliness, further enhancing their utility across different phases of rehabilitation therapy.

There are various platforms for upper limb extremity rehabilitation. they differ from their hardware designs to their control strategies and treatment modes. These platforms could generally be categorized into two major groups; bilateral, where both the affected and unaffected arms are involved, and unilateral, where only the affected arm participates in training [104, 137]. Additionally, the physical design of these robots can be categorized into various forms, such as H-shaped, triangular, linkage-based, and mobile, among others. Equally crucial is how these robots are programmed and controlled to adapt to the diverse requirements of different therapeutic processes.

14.4.3 Control Strategies

Developing control strategies plays a crucial role in designing rehabilitation robots, particularly for assisting stroke survivors in restoring limb functions [98, 100, 136, 138, 139]. The choice of control strategy is tailored to the individual needs of the patient and the severity of their motor control impairment, forming the foundation for shared autonomy frameworks where human intent and robotic assistance must be seamlessly integrated. These strategies fall into two main categories: high-level control and low-level control (Fig. 14.4). High-level control concentrates on triggering motor plasticity, whereas low-level control focuses on overseeing and adjusting a range of factors such as position, velocity, force or torque, impedance, admittance, and processing the adaptive feedback from high-level control. This

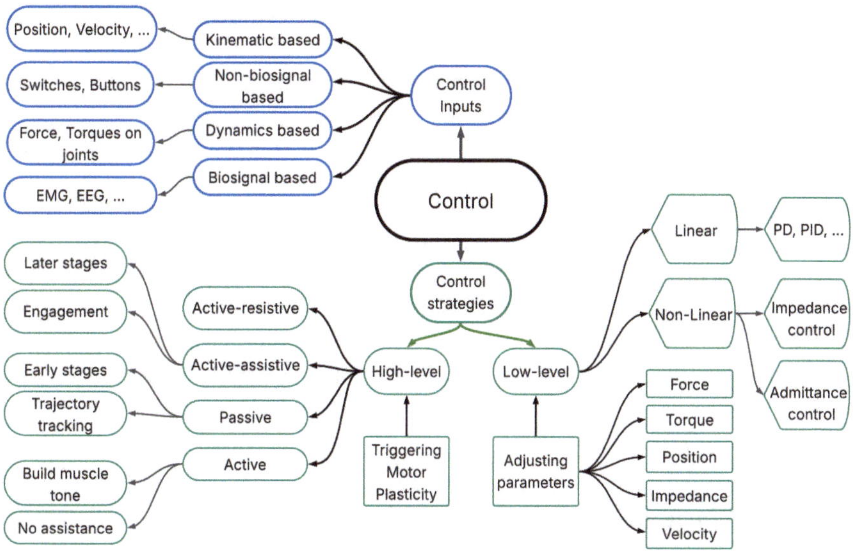

Fig. 14.4 Hierarchical taxonomy of control inputs and strategies for upper-limb rehabilitation robots

hierarchical structure enables dynamic allocation of control authority between the patient and robot. High-level control strategies can be further classified into four main types: passive, active-assistive, active, and active-resistive, each offering different levels of patient autonomy and robot intervention.

In the early stages of rehabilitation, the passive mode is predominantly used, focusing on enhancing the range of motion and reducing spasticity in the affected limb [139]. In this phase, the rehabilitation robot is programmed to assist patients in smoothly following a predetermined path. The initial control strategy often implemented is motion control or trajectory tracking, typically in its passive form. This method ensures that the limb follows a set trajectory within the robot's capabilities, without necessarily requiring active effort from the patient.

Active-assistive control strategies become prominent as therapy advances and patients start regaining control of their limbs [138]. This approach, crucial in later stages of rehabilitation, provides just enough assistance for patients to complete tasks, thus encouraging them to actively participate and exert effort in exercises. This method is tailored to promote better engagement and more effective rehabilitation outcomes, especially as patients' engagement in the exercise becomes increasingly important.

Active, or "transparent," modalities in rehabilitation robots allow patients to move without assistance or resistance from the robot, turning it into a measurement tool during assessments [98]. It's crucial to maintain this transparency to avoid influencing the patient's performance. In later recovery stages, when patients need to build muscle tone, resistive modalities are used. These don't follow preset paths but instead challenge the patient with resistance proportional to their movement speed, enhancing engagement and muscle strengthening in robot-assisted exercises. Generally, these platforms primarily utilize passive and semi-active control strategies [136], often avoiding fully active or resistive modes that are intended for patients who have largely regained their abilities.

Low-level control groups in rehabilitation merge to support high-level objectives [98, 100, 138]. These groups are mainly divided into linear and nonlinear control approaches. Linear control encompasses basic methods such as Proportional-Derivative (PD), Proportional-Integral-Derivative (PID), and their variants. However, due to the nonlinearity in the dynamics of upper extremity exoskeletons, arising from flexible links, joints, and actuators, nonlinear control methods are preferred to enhance system robustness and compatibility with external factors. Nonlinear control includes techniques like computed torque control, impedance control, sliding mode control, adaptive control, and admittance control.

14.4.4 Biosignal Integration for Shared Control

Control inputs are a vital element in defining the control architecture of rehabilitation devices, and they are divided into four main categories [100, 140] (Fig. 14.4). Kinematic-based inputs record orientations, positions, velocities, and accelerations of specific parts of the device or limb. Dynamic-based inputs focus on the force or torque applied by the user on the joints or ends of rehabilitation devices. Biological-based inputs use EMG (and sometimes EEG) signals in an on/off or proportional fashion to align device movements with the user's motor intent. Finally, non-biological-based inputs involve switches or buttons to detect human intentions and initiate limb movements. While kinematic and dynamic sensors remain the most common in clinical platforms, integrating biosignals enables a richer, more intuitive human–robot interaction.

14.4.5 Adaptive Algorithms

Building on the control strategies and input modalities described above, adaptive algorithms serve as the critical mechanism that customizes robotic assistance to each patient's real-time needs. By integrating kinematic data (e.g., joint angles and velocities), dynamic feedback (interaction forces or torques), and biosignals (such as EMG or EEG indicators of muscle engagement), these algorithms continuously recalibrate low-level control loops–whether impedance, admittance, or direct torque control–to deliver the appropriate high-level mode of assistance (passive, active-assistive, active, or active-resistive). In model-reference adaptive schemes, the controller continuously compares the actual limb trajectory to a predefined reference model, adjusting control gains in real time to accommodate uncertainties in patient dynamics and human–robot interaction. For repetitive tasks, an iterative learning component retains tracking errors from previous trials and refines feedforward commands to promote faster convergence toward smooth, accurate movements. Adaptive impedance and admittance controllers modulate virtual stiffness and damping based on real-time assessments of muscle effort–providing minimal resistance during passive or assistive phases and gradually increasing challenge as voluntary engagement intensifies. More advanced neuro-adaptive and fuzzy-logic approaches utilize neural networks or fuzzy estimators to model complex, nonlinear patient–robot interactions without relying on explicit system representations. These techniques enable progressive refinement of control strategies across therapy sessions, facilitating personalized interventions that adapt to each patient's evolving motor capabilities and promote more effective rehabilitation [100, 139].

14.4.6 *Case Study: Stroke Rehabilitation Framework*

The work from [133] presents a novel robotic rehabilitation framework designed specifically for passive stroke rehabilitation using a one-dimensional magnetic-based actuation mechanism (Fig. 14.5). The system targets transitional shoulder movements (flexion-extension) and employs a non-contact magnetic interface to

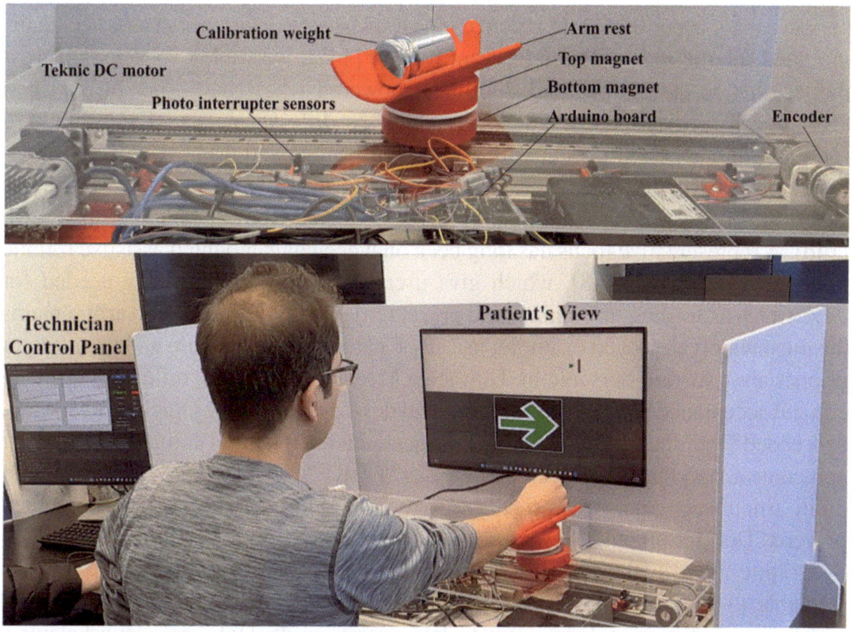

Fig. 14.5 Experimental setup demonstrating user interaction with the magnetic-based robotic rehabilitation platform. The designed planar robotic platform (top), a preview of running a trial with a participant (bottom)

ensure smooth, safe, and user-friendly interaction. Replacing conventional mechanical linkages, the platform minimizes physical strain and risk while enhancing motion quality and weight compensation. The setup includes a magnetically actuated end-effector and an ergonomic armrest, with motion tracked and stabilized via an Extended Kalman Filter (EKF) for robust control, even under sensor failure. In trials with 12 participants, users reported high satisfaction in terms of safety, comfort, and system responsiveness. The platform demonstrated stable, synchronized motion and low tracking error, validating its potential for home-based early rehabilitation. While this initial design supports passive therapy along a single linear axis, future development aims to expand to two-dimensional movements and incorporate assistive control strategies, enabling more active participation and broader functional training for stroke survivors.

14.5 Shared Autonomy in Assistive Technologies Domain

14.5.1 Overview and Gaps

Assistive technologies encompass a broad range of domains aimed at supporting individuals with disabilities in daily activities. These systems can be categorized into mobile servant robots, physical assistant robots, wearable suits/sensors, and person carrier robots [141]. In practice, this translates to devices such as wheelchair-mounted robotic manipulators (servant robots) [142], wearable exoskeletons/prosthetics/sensors [143, 144], and intelligent powered wheelchairs (person carriers) [145]. Each of these assistive devices addresses different needs—for example, a robotic arm can help a person with quadriplegia grasp and drink from a cup independently [15], whereas a smart wheelchair with obstacle avoidance can enhance mobility for those who cannot safely drive a standard power chair on their own [146]. A representative focus has long been on wheelchair-mounted assistive robotic manipulators [2, 147, 148], which give users the reach and dexterity needed for ADLs such as feeding, grooming, and object retrieval. The Kinova Jaco arm remains the most widely deployed example; its 6-DoF carbon-fiber structure and three-finger gripper are now reimbursable by U.S. state Medicaid programs, reflecting growing clinical acceptance [149]. Other commercial devices include the Dutch iARM [150], descended from the original MANUS project and designed for intuitive joystick or app control, and the U.S.-developed DEKA "LUKE" prosthetic arm [151], the first multi-articulated upper-limb prosthesis to receive FDA clearance, offering up to ten powered DoF for amputees.

Despite the availability of such assistive systems, because the control channels available to people with motor impairments are inherently low-bandwidth, noisy, and fatigue-inducing, asking users to tele-operate multi-DoF arms, exoskeletons, or power chairs in a purely manual fashion is often impractical and slow [13]. Conversely, turning the devices into fully autonomous robots can leave users feeling disempowered and vulnerable when the autonomy misinterprets goals or environmental nuances [30]. Shared-autonomy paradigms integrate intelligent assistance directly into these devices to bridge this gap: the human provides strategic intent and retains final authority, while the onboard autonomy supplies continuous low-level stabilization, obstacle avoidance, and motion refinement. Empirical studies consistently show that this collaborative division of labor lowers task time and error rates, preserves the user's sense of agency, and increases overall satisfaction compared with either manual or fully automated control schemes [13, 30]. For these reasons, shared control has become a foundational design principle across assistive domains, ensuring that advanced robotics enhances, rather than supplants, human capability.

14.5.2 Shared Autonomy Architectures

In the context of assistive technologies, shared autonomy (also called shared control) refers to control approaches that combine the user's input with the robot's autonomous actions to achieve the user's intended goals [38]. In these systems, the user typically provides high-level commands or coarse guidance, and the autonomy module refines and executes low-level actions, effectively amplifying the user's capabilities [15] (Fig. 14.6). Early traditional shared control systems often relied on manual mode switching [152, 153], requiring the user to sequentially control different degrees of freedom (e.g., selecting "move arm base" vs "rotate wrist" modes) to cope with limited input dimensionality. However, frequent mode switches drastically slow down task execution and impose a high cognitive load on the user [154].

To relieve that burden, researchers introduced task-aware architectures that embed contextual knowledge of the environment and of likely user goals [2, 155]. These architectures use contextual and goal information to assist the user accordingly, rather than blindly executing pre-programmed motions [156]. For example, if a user is teleoperating a robot arm to pick up a cup, a task-aware system will incorporate knowledge of the cup's location and the goal of grasping when deciding how to assist. A central element of these architectures is the arbitration mechanism, the policy that decides how to blend or switch between user control and autonomous control at each moment [157]. Designing this arbitration is non-trivial: too little autonomy yields high user workload, while too much can erode the user's sense of control [158]. Modern approaches, therefore, strive for user-centric, adaptive arbitration, adjusting the level of assistance based on the context and the individual user's needs [44]. Many frameworks formalize this as a partially observable decision problem [159, 160]: the system maintains a belief over possible user goals and selects autonomous actions that best advance the likely goal. When the user's goal is known or explicitly specified, assistance can be planned via standard Markov Decision Processes (MDPs) for the task. If the goal is uncertain, algorithms based on Partially Observable MDPs (POMDPs) or Bayesian inference are employed to

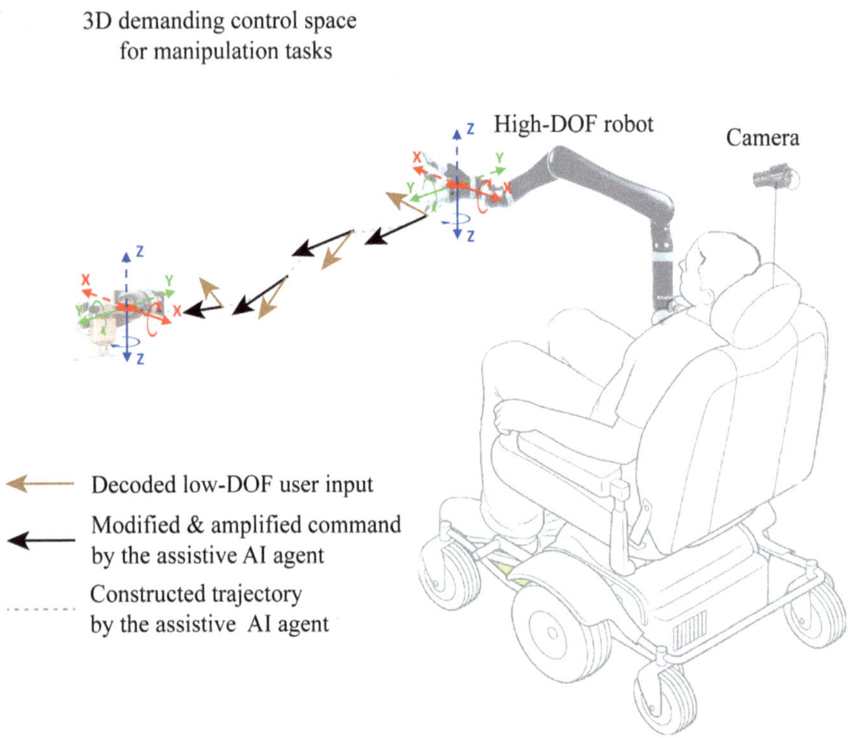

Fig. 14.6 A conceptual design of an assistive AI-powered robotic manipulation system amplified low-dimensional user inputs for high-DoF tasks

infer intent and plan accordingly [30]. This approach was demonstrated in prior works [161] that predict the user's intended target from their initial joystick motions or gaze, then autonomously guide the robot toward that target. The result is an intelligent assistive policy that, for instance, can start moving the arm toward a likely object of interest even before the user provides a perfectly precise command [162].

Building on these task-aware arbitration frameworks, recent research has begun to replace hand-engineered heuristics with end-to-end learning components, often deep reinforcement learning (RL) agents [15, 44, 147], trained in high-fidelity simulation, that can discover when to yield, when to assist, and how to interpolate smoothly between those extremes. By optimizing over millions of synthetic episodes, such agents learn to embed intent inference, motion refinement, and obstacle avoidance inside a single policy, eliminating the need for explicit mode switches or manually tuned sub-modules.

14.5.3 User-Intent Inference and Personalization Pipeline

At the heart of shared autonomy lies the challenge of inferring the user's intent from limited and noisy inputs, and then personalizing the assistance to the user's abilities and preferences [33]. Users with severe motor impairments may control assistive robots through a variety of interfaces: examples include low-dimensional inputs like a two-axis joystick, sip-and-puff switches, head tilts, eye-gaze trackers, or brain-computer interface signals (EEG/EMG) [163–168]. These input channels typically provide far fewer degrees of control than the robot has degrees of freedom—for instance, controlling a 7-DoF arm with a 2-DoF joystick—and the signals can be noisy or intermittent. Therefore, the system must decode the user's intended action or goal from whatever cues it can observe. In non-invasive setups, this inference is complicated by artifacts and user fatigue: surface EEG/EMG signals require extensive user training to be reliable, and even then, they carry limited information bandwidth [5, 169]. Similarly, a user who can only issue simple binary commands (e.g. a head switch for left/right) cannot directly teleoperate a robotic arm to reach and grasp an object without intelligent mediation [170].

To tackle this, assistive shared autonomy systems employ a pipeline of intent inference algorithms. One common approach is goal prediction: as the user begins an action, the system compares the motion or commands to a set of possible targets or tasks [161, 171]. For example, if a robotic wheelchair sees several doorways and the user's initial joystick push is roughly toward one of them, a Bayesian estimator or classifier can infer a high probability that the user intends to go through that doorway. In robotic manipulators, techniques like inverse reinforcement learning [172] and Bayesian intent prediction [173] have been used to guess which object a user is reaching for, after observing a portion of the motion trajectory. Notably, when no explicit goal selection is provided, the system can maintain a probability distribution over possible goals and continuously update it as more user input comes in. Javdani et al. [160] implemented this via a POMDP-based shared autonomy framework where the human's control input is treated as observations about the goal, and the robot's autonomous policy aims to maximize expected progress towards the inferred goal. This allows the robot to assist proactively, for example, moving toward the most likely target while still being ready to switch if the user's input indicates a different intention.

Beyond goal inference, the personalization aspect of the pipeline adapts the system's behavior to individual users. Users have varying reaction times, precision, strength, and preferences for how much help they want. An adaptive shared-control system can learn these characteristics over time. One strategy is to adjust the assistance level (autonomy gain) based on performance [157]: if the user is struggling (e.g., taking too long or issuing erratic commands), the system can increase autonomy; if the user seems adept, the system can relinquish more control. Atan et al. [38] introduced an arbitration method that monitors the user's task performance and expertise level, and dynamically tunes the autonomy assistance at runtime. This user-centric approach led to higher efficiency in a reaching and grasping task by giving novices more assistance and experts more manual control, aligning with each user's abilities. Another facet of personalization is intent interpretation tuning: the system may learn to interpret a particular user's input idiosyncrasies. For example, one person's neck movements might consistently overshoot targets, so the system's mapping of head tilt to cursor movement can be customized (through calibration or learning) to be less sensitive. Similarly, a shared autonomy framework can incorporate a model of the user's reaction or comfort. Some recent work proposes using reinforcement learning with human feedback to adjust how aggressively or conservatively the robot assists, based on the user's satisfaction signals [44].

14.5.4 Evaluation, Validation and Regulatory Path

Evaluating a shared-autonomy device starts with technical metrics such as success rate, completion time, motion precision, and the reduction in user effort. Studies of wheelchair-mounted arms, for example, routinely compare how quickly and reliably users can grasp objects under direct control versus shared control [15]. Equally important are subjective workload and agency measures: tools such as the NASA-TLX questionnaire quantify perceived cognitive load in human-robot collaboration [174], while recent sense of agency experiments show a predictable trade off, greater robot autonomy boosts task performance but can diminish the operator's feeling of control [4]. Hence, optimal systems aim for adaptive, mid-spectrum autonomy that lifts efficiency yet preserves user empowerment.

Validation then progresses to end-user trials, safety certification, and long-term reliability checks. Because recruiting clinical populations can be difficult, developers often prototype in simulation frameworks such as Assistive Gym, which model ADL scenarios and human biomechanics before moving to real users [175]. When devices reach the clinic or home, they must satisfy standards like ISO 13482 for personal-care robots, demonstrating compliant forces, failsafe behaviors, and predictable fault handling [141]. Cybersecurity, data privacy, and wear-and-tear robustness add further regulatory hurdles, which partly explains why many laboratory prototypes never reach the market. Nevertheless, evolving testbeds and clearer standards are lowering these barriers, fostering a new generation of assistive robots that are not only capable but also trusted for everyday use.

14.5.5 Case Study: Adaptive Reinforcement Learning for Amplification of Limited Inputs in Shared Autonomy (ARAS)

As a concrete illustration, Rabiee et al. [15] introduced the Adaptive Reinforcement learning for Amplification of limited inputs in Shared autonomy (ARAS) frame-

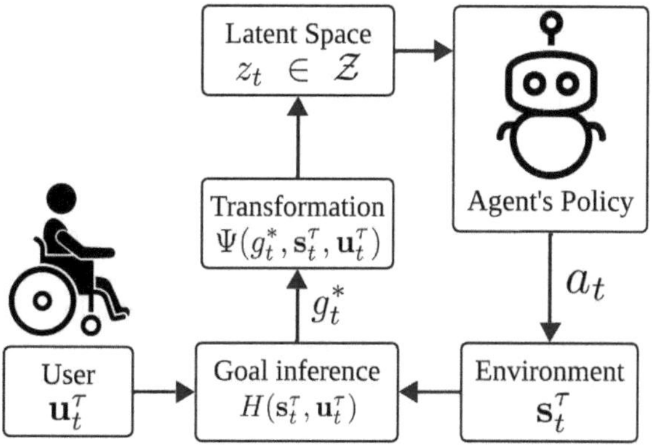

Fig. 14.7 ARAS formulation diagram

work, whose high-level dataflow is sketched in Fig. 14.7. At each control cycle, the system receives a short window of low-bandwidth user commands u_t^τ together with scene information s_t^τ from an overhead RGB camera. A goal inference module $H(s_t^\tau, u_t^\tau)$ maintains a Bayesian belief over the user's intention and returns the most likely goal g_t^*. These signals are fused by a transformation $\Psi(g_t^*, s_t^\tau, u_t^\tau)$ that filters irrelevant pixels and embeddings, producing a compact latent state z_t that captures only task-relevant features. A deep Q-network then maps z_t to a high-level robot action a_t, which the physical arm executes; the resulting state feeds back into the loop, allowing the policy to amplify limited inputs into smooth 7-DoF manipulation without mode switching or explicit mid task queries. In essence, ARAS lets the human supply strategic intent while the autonomy layer handles continuous motion planning and goal refinement in real time.

Extensive evaluation confirms the benefits of this formulation. After 50,000 simulated pick-and-place episodes, the learned policy transferred zero-shot to hardware, where 23 participants completed 90 trials apiece. Compared with manual teleoperation, ARAS cut average completion time, reduced required input keystrokes, and lifted overall success from 86% to 92.8 while preserving user agency. In simulation, ARAS maintained near-perfect performance (99.8% success, 31 s completion) under fixed goals and still exceeded 96% success when users changed intentions midtask, outperforming both a vanilla DQN and a hindsight-optimization baseline across all scenarios. Subjective NASA-TLX scores and Likert surveys showed significant drops in perceived workload and the highest ratings for satisfaction, flexibility, and future use likelihood, indicating that users valued the system's blend of help and control. Together, these results demonstrate that a single latent-space RL policy can infer, amplify, and adapt low-DoF inputs at interactive rates, delivering high-precision shared autonomy that meets both technical and experiential benchmarks.

14.6 Emerging Trends

14.6.1 Large Language Models in Human-Centered Shared Autonomy

Large-language-model (LLM) techniques are rapidly reshaping shared autonomy by turning natural language into a control modality and by supplying robots with onboard reasoning. This development allows LLMs to function as adaptive, dialogue-native co-pilots rather than post-hoc interfaces, due to advancements in their reasoning capabilities and multimodal interfaces.

14.6.1.1 Reasoning Capabilities

Large Language Models (LLMs) are evolving from simple command-followers into reasoning engines for shared autonomy, enabling them to handle ambiguous goals, plan complex tasks, and apply commonsense knowledge. Early demonstrations such as SayCan [176] grounded a frozen LLM's high-level plans in low-level affordance scores, allowing a mobile manipulator to execute multistep kitchen tasks from free-form instructions with a 74% real-world success rate (double that of a language-only baseline). More recent models push this idea further; PaLM-E [177] interleaves text tokens with raw images and proprioceptive states inside a single transformer, allowing one network to generalize across navigation, grasping, and visual question-answering. This reasoning process can be made more robust through closed-loop deliberation. For instance, the Inner Monologue framework [178] keeps the LLM "thinking out loud," injecting real-time success detection and scene descriptions back into its context so the robot can re-plan or query the user if execution goes off course.

Beyond direct execution, LLMs are also used as synthesis engines to make robust policies. Code-generation approaches such as RobotGPT [179] treat multiple drafts from ChatGPT as demonstrations, then distill them into a reinforcement-learning policy that boosts manipulation success from 38 to 91% on pick-and-place tasks. A key aspect of advanced reasoning is also understanding the model's own limitations. To address the failure modes and uncertainty of LLM planners, the KNOWNO framework [180] uses conformal prediction to calibrate an LLM's confidence. When multiple high-probability next steps exist, the robot "knows it doesn't know" and explicitly asks the human for guidance, cutting unnecessary interventions by up to 24% while preserving task-completion guarantees.

This capacity for nuanced reasoning is being applied to decompose complex task sequences. Some frameworks combine LLMs with classical planners to guarantee logical soundness [181], while others leverage external knowledge. The AdaptBot framework, for instance, integrates an LLM with a Knowledge Graph to break down generic commands into specific actions, using human-in-the-loop dialogue to resolve ambiguities [182]. More recent work uses LLMs for object-level planning, where the model reasons about object state transitions to generate sequences of subgoals for traditional task and motion planners [183]. This approach is enhanced by the LLM's embedded commonsense knowledge, which proves critical for inferring unstated user needs. The Guide-LLM framework [184] demonstrated this by planning safe indoor navigation paths for patients with visual impairments, inferring potential hazards using text-based maps. Similarly, the SAVOR framework [185] uses a VLM's commonsense priors about food properties (e.g., softness, shape) to select appropriate utensils and actions for robot-assisted feeding while refining its plan through visual and haptic feedback.

14.6.1.2 Multimodal Interfaces

Advanced reasoning must be paired with interfaces that support effective communication. Therefore, parallel progress in multimodal interfaces is critical, focusing on deeply integrating vision and language, generating expressive and explainable feedback, and exploring novel fusion with human biosignals to create a more seamless human-robot partnership.

Recent Vision-Language-Action (VLA) models are moving beyond single-step commands to enable long-horizon, visually-grounded tasks. Foundational models like RT-2 [186] have demonstrated that VLM-based models can be finetuned with robotic data to acquire secondary robotic capabilities, directly translating visual and language commands into actions. The Vision-Language Model Predictive Control framework [187] exemplifies the integration of these models into the robot's control loop, using a VLM's understanding to continuously sample and predict the outcome of potential action sequences. To scale the training of such systems, the field is also developing large-scale, primitive-level datasets that break down complex tasks into a vocabulary of fundamental skills to create more robust agents [188].

To foster user trust and facilitate effective collaboration, the system must also be able to explain its reasoning. Recent LLMs are capable of generating context-aware, natural language explanations [189–191]. Some systems can now generate contrastive explanations, clarifying not only why a certain action was taken, but why an alternative was not, which is crucial for debugging and building accurate end-user mental models [192]. Beyond text, by combining a VLM with Grad-CAM heatmaps, a robot can also visually highlight the part of the scene that influenced its navigation decision while verbally explaining its reasoning [193].

14.6.2 Ethical Considerations and Human Factors

The practical success of shared autonomy hinges on more than just technical execution and an integration of ethical and human factors is required to mitigate the risk of severe, real-world consequences. These risks extend beyond system abandonment [12] to include physical harm, psychological damage, and the reinforcement of societal inequities. The black box nature of many AI models creates challenges for transparency [19], but it also may foster misaligned trust. Users can develop automation complacency, where they over-trust a system that is mostly reliable and become blind to the edge cases where it can critically fail, leading to physical harm [194]. The risk is not simply that the AI fails, but that the system's design failed to foster calibrated trust, leaving the user unaware of the system's operational boundaries.

Furthermore, a poorly designed system can even cause psychological harm. Systems that are overly assertive or that frequently override user commands can erode the user's sense of agency [4], leading to a state of learned helplessness, which is against the goals of rehabilitation systems [195, 196]. This can cause a degradation of the user's own skills(also known as skill atrophy), which is a crucial concern in therapeutic applications [197]. The user's experience can also be degraded by the "frustration of the excluded middle", where the system handles all simple tasks but disengages during complex scenarios, leaving the user with high cognitive burden precisely when they need the most support [198].

Finally, unexamined systems risk amplifying societal biases. Algorithmic bias [11] can lead to discriminatory outcomes when intent recognition or computer vision models, trained on data from dominant demographic, economic, or cultural groups, fail to perform reliably for minority users [199]. This can render a device ineffective or unsafe, creating scenarios where the benefits of the system are only available to a selective group, thereby limiting the systems accessibility. Addressing these factors is a pragmatic necessity for creating safe, impartial, and effective shared autonomy systems.

14.7 Summary

This chapter has provided a comprehensive analysis of human-centered shared autonomy, arguing for a common technical framework to connect the distinct applications of Brain-Computer Interfaces (BCIs), rehabilitation, and assistive robotics. The proposed framework is grounded in a core challenge: mapping noisy user biosignals to effective control policies through adaptive arbitration. Our review detailed the foundational principles for such systems, including maintaining the human as the leader in the human-AI team, implementing varied interaction paradigms, and the critical role of robust intent inference to translate user goals into action.

Across the three domains, we examined how these principles are applied to address specific needs, such as mitigating the cognitive load of direct BCI control , enabling personalized, adaptive therapy in rehabilitation , and amplifying low-bandwidth commands for high-dimensional assistive tasks. The case studies presented serve as examples of these concepts in practice. Throughout this analysis, we identified shared challenges, including the high variability of biosignals , the difficulty of robustly decoding intent , and the fundamental trade-off between task performance and the user's sense of agency.

Finally, we explored emerging trends that will shape the future of this field. While Large Language Models show promise for enabling more complex reasoning and multimodal interaction, their integration, like all advancements in this area, must be navigated with careful attention to ethical and human factors. As the paper highlights, issues of automation complacency, skill atrophy, and algorithmic bias are critical safety and usability concerns that must be proactively addressed. The trajectory of this research is toward more deeply integrated human-AI joint cognitive systems, but achieving this goal requires a steadfast focus on developing technologies that are not merely capable, but also transparent, trustworthy, and empowering for the end-user.

Acknowledgments The materials reported in this chapter were supported by the NSF CAREER under award ID 2441496 and the NSF grant under award ID 2245558. Additionally, the projects were supported by URI Foundation Grant on Medical Research and the Rhode Island INBRE program from the National Institute of General Medical Sciences of the NIH under grant number P20GM103430.

References

1. A. Ajoudani, A.M. Zanchettin, S. Ivaldi, A. Albu-Schäffer, K. Kosuge, O. Khatib, Progress and prospects of shared control in human-robot interaction. IEEE Trans. Robot. **34**(6), 1430–1447 (2018)
2. R. Abiri, A. Rabiee, S. Ghafoori, A. Cetera, Toward human-centered shared autonomy ai paradigms for human-robot teaming in healthcare (2024). arXiv:2407.17464
3. D.P. Losey, C.G. McDonald, E. Battaglia, M.K. O'Malley, A review of intent detection, arbitration, and communication aspects of shared control for physical human-robot interaction. Appl. Mech. Rev. **70**(1), AMR-17-1021 (2018)
4. M.A. Collier, R. Narayan, H. Admoni, The sense of agency in assistive robotics using shared autonomy (2025). arXiv:2501.07462
5. J.R. Wolpaw, N. Birbaumer, D.J. McFarland, G. Pfurtscheller, T.M. Vaughan, Brain–computer interfaces for communication and control. Clin. Neurophysiol. **113**(6), 767–791 (2002)
6. D. Farina, N. Jiang, H. Rehbaum, A. Holobar, B. Graimann, H. Dietl, O.C. Aszmann, The extraction of neural information from the surface EMG for the control of upper-limb prostheses: emerging avenues and challenges. Physiol. Measur. **31**(11), R25–R53 (2010)
7. J. He, C.G. Siong, A. Nagarajan, L.E. Miller, H.S. Liew, C. Gao, Decoding human motor intent from electromyography for shared human-robot control. Annu. Rev. Control Robot. Auton. Syst. **3**, 349–373 (2020)
8. J.D.R. Millán, R. Rupp, G. Mueller-Putz, R. Murray-Smith, C. Giugliemma, M. Tangermann, C. Vidaurre, F. Cincotti, R. Leeb, J. del Millán, Combining brain–computer interfaces and assistive technologies: state-of-the-art and challenges. Front. Neurosci. **4**, 161 (2010)
9. L. Marchal-Crespo, D.J. Reinkensmeyer, Review of control strategies for robotic movement training after neurologic injury. J. NeuroEng. Rehab. **6**(1), 20 (2009)
10. A. Brohan, N. Brown, J. Carbajal, Y.C. Chebotar, J. Dabis, C. Finn, K. Gopalakrishnan, K. Hausman, A. Herzog, D. Ho, J. Hsu, J. Hughes, B. Ichter, A. Irpan, E. Jang, R. Kevorkyants, S. Lee, S. Levine, Y. Lin, B. Liu, H. Liu, M. Laskin, I. Lu, Q. Luo, L. Metz, K. Oslund, J. Paina, C. Parada, P. Peng, S. Peri, K. Pichai, I. Popov, J. Puckering,

C. Quiambao, K.R. Raju, K. Rao, C. Rihn, T. Romaszko, S.M. Jeong, N. Schrader, E. Sieb, A. Skillman, S. Sniecinski, J. Sorokin, R. Stone, D. Su, Y. Suri, N. Tirumala, F. Tung, Q. Van Der Marel, V. Vanhoucke, A. Vezhnevets, N. Watters, T. Xiao, P. Xu, S. Xu, M. Yan, A. Yao, J. Yee, J. Zitkovich, K. Zolna, RT-2: vision-language-action models transfer visual knowledge to robotic control (2023). arXiv:2307.15818. https://arxiv.org/abs/2307.15818
11. E. Fosch-Villaronga, H. Drukarch, On the EU artificial intelligence act and the need to protect people with disabilities from algorithmic discrimination: mind the gap. Comput. Law Secur. Rev. **43**, 105624 (2021)
12. B. Hayes, B. Scassellati, Challenges in shared autonomy. AI Mag. **34**(4), 63–70 (2013)
13. S. Jain, A. Farshchiansadegh, A. Broad, F. Abdollahi, F. Mussa-Ivaldi, B. Argall, Assistive robotic manipulation through shared autonomy and a body-machine interface, in *2015 IEEE International Conference on Rehabilitation Robotics (ICORR)*. (IEEE, Piscataway, 2015), pp. 526–531
14. F. Lotte, W. Samek, C. Vidaurre, Co-adaptive brain–computer interfaces: a review of algorithms and methods. IEEE Trans. Neural Syst. Rehab. Eng. **30**, 134–145 (2022)
15. A. Rabiee, S. Ghafoori, M. Farhadi, R. Beyer, X. Bai, D.J. Lin, S. Ostadabbas, R. Abiri, Learning multimodal ai algorithms for amplifying limited user input into high-dimensional control space (2025). arXiv:2505.11366
16. S. Nikolaidis, J.A. Shah, Human-robot cross-training: a human factors approach to human-robot teaming, in *Proceedings of the 8th ACM/IEEE International Conference on Human-Robot Interaction (HRI)* (IEEE, Tokyo, 2013), pp. 355–362
17. J.M. Beer, A.D. Fisk, W.A. Rogers, Toward a framework for levels of robot autonomy in human-robot interaction. J. Hum. Robot Interact. **3**(2), 74–99 (2014)
18. M.M. Madduri, S.A. Burden, A.L. Orsborn, Biosignal-based co-adaptive user-machine interfaces for motor control. Curr. Opin. Biomed. Eng. **27**, 100462 (2023)
19. M.L. Cummings, Automation and accountability in complex systems, in *Proceedings of the Human Factors and Ergonomics Society Annual Meeting*, vol. 48, no. 3 (Sage, Los Angeles, 2004), pp. 432–436
20. T.B. Sheridan, R. Parasuraman, Human-automation interaction. Rev. Hum. Factors Ergon. **1**(1), 89–129 (2005)
21. Y. Li, K.P. Tee, W.L. Chan, R. Yan, Y.S. Chua, D.K. Limbu, Continuous role adaptation for human-robot shared control. IEEE Trans. Robot. **31**(3), 672–681 (2015)
22. N. Jarrassé, T. Charalambous, E. Burdet, A framework to describe, analyze and generate interactive motor behaviors. PLoS ONE **7**(11), e49945 (2012)
23. P. Franceschi, N. Pedrocchi, M. Beschi, Human-robot role arbitration via differential game theory. IEEE Trans. Autom. Sci. Eng. **21**(4), 5953–5968 (2024)
24. D.J. McFarland, D.J. Krusienski, W.A. Sarnacki, J.R. Wolpaw, Emulation of computer mouse control with a noninvasive brain-computer interface. J. Neural Eng. **5**(2), 101–110 (2008)
25. A. Sarasola-Sanz, A.M. Ray, A. Insausti-Delgado, N. Irastorza-Landa, W.J. Mahmoud, D. Brötz, C. Bibián-Nogueras, F. Helmhòld, C. Zrenner, U. Ziemann, E. López-Larraz, A. Ramos-Murguialday, A hybrid brain-muscle-machine interface for stroke rehabilitation: usability and functionality validation in a 2-week intensive intervention. Front. Bioeng. Biotechnol. **12**, 1330330 (2024)
26. S. Jain, B. Argall, Probabilistic human intent recognition for shared autonomy in assistive robotics. ACM Trans. Hum. Robot Interact. **9**(1), 1–23 (2019)
27. P. Franceschi, N. Pedrocchi, M. Beschi, Adaptive impedance controller for human-robot arbitration based on cooperative differential game theory, in *2022 IEEE International Conference on Robotics and Automation (ICRA), Philadelphia* (2022), pp. 7888–7894
28. A. Ajoudani, A.M. Zanchettin, S. Ivaldi, A. Albu-Schäffer, K. Kosuge, O. Khatib, Progress and prospects of shared control in human-robot interaction. IEEE Trans. Robot. **34**(5), 1378–1399 (2018)
29. A. Mörtl, M. Lawitzky, A. Kucukyilmaz, T.M. Sezgin, C. Basdogan, S. Hirche, The role of roles: physical cooperation between humans and robots. Int. J. Robot. Res. **31**(13), 1656–1674 (2012)

30. A.D. Dragan, S.S. Srinivasa, A policy-blending formalism for shared control. Int. J. Robot. Res. **32**(7), 790–805 (2013)
31. X. Zhang, G. Xu, L. Xie, M. Xu, Y. Yang, A survey on real-time biosignal processing systems for BCI applications. IEEE Access **8**, 123678–123692 (2020)
32. J.Y. Chen, M.J. Barnes, M. Harper-Sciarini, Supervisory control of multiple robots: human-performance issues and user-interface design. IEEE Trans. Syst. Man Cybern. Syst. **44**(2), 181–195 (2014)
33. A. Jain, B.D. Argall, Probabilistic human intent recognition for shared autonomy in assistive robotics. J. Hum. Robot Interact. **8**(2), 1–29 (2019). Also appears as: Probabilistic Human Intent Recognition for Shared Autonomy in Assistive Robotics. Proc. ACM Hum.-Robot Interact. 8, 2, Article 7 (May 2019), 29 pages. https://doi.org/10.1145/3322279. And related to work presented at RSS 2018, and PMC7233691 from IEEE Trans Neural Syst Rehabil Eng. 2020 Apr;28(4):897–906
34. M.M. Islam, M.I. Hossain, S. Roy, M.B. Hossain, J. Shin, Machine learning for human activity recognition: a comprehensive review. Algorithms **11**(3), 91 (2024). Article states Vol 11, Issue 3, but published date Feb 2024. Using the provided info. If actual is Vol 17, 2024, that would be more typical
35. B.D. Argall, A.D. Dragan, Inferring human intent and predicting human action in human–robot collaboration. Annu. Rev. Control Robot. Auton. Syst. **7**, 35–57 (2024)
36. Z.M. Wang, T.X. Zhao, H.Y. Wang, Y.L. Li, L. Zhu, W. Wu, L.W. Zhang, A survey on eeg signal processing: preprocessing, feature extraction, and classification. Front. Neurosci. **17**, 1203059 (2023)
37. H. Wang, T. Yuan, Y. Zhao, D. Bai, Multimodal fusion of emg and vision for human grasp intent inference in prosthetic hand control. Biomed. Signal Process. Control **95**, 106437 (2025). Pre-proof available on ResearchGate March 2024. Publication date listed as April 2025
38. U. Atan, V.R. Bharadwaj, C. Jiang, Assistive control of robot arms via adaptive shared autonomy, in *2024 IEEE International Conference on Advanced Intelligent Mechatronics (AIM)* (IEEE, Piscataway, 2024), pp. 1096–1102
39. D.E. Gopinath, B.D. Argall, Active intent disambiguation for shared control robots. IEEE Trans. Neural Syst. Rehab. Eng. **28**(4), 907–916, (2020). Corresponds to arXiv:2005.03652
40. E. Hollnagel, D.D. Woods, *Joint Cognitive Systems: The Foundations of Cognitive Systems Engineering* (CRC Press, Boca Raton, 2005)
41. P.A. Hancock, I. Nourbakhsh, B. Scassellati, The future of work with robots and AI. Hum. Factors **63**(6), 921–928 (2021)
42. G. Hoffman, O. Zuckerman, The proactive agent: a new metaphor for human–AI interaction. ACM Trans. Interact. Intell. Syst. **9**(2–3), 1–25 (2019)
43. C.M. Huang, B. Mutlu, Anticipatory robot control for efficient human-robot collaboration, in *2016 11th ACM/IEEE International Conference on Human-Robot Interaction (HRI)* (2016), pp. 83–90
44. S. Reddy, A.D. Dragan, S. Levine, Shared autonomy via deep reinforcement learning (2018). arXiv:1802.01744
45. C. Mühl, B. Allison, P. Gerjets, H. Hinrichs, A survey of affective brain-computer interfaces: principles, state-of-the-art, and challenges. Brain-Comput. Interfaces **1**(2), 66–84 (2014)
46. S. Anjomshoae, A. Najjar, D. Calvaresi, K. Främling, Explainable agents and robots: a survey (2019). arXiv:1909.07243
47. R. Abiri, S. Borhani, E.W. Sellers, Y. Jiang, X. Zhao, A comprehensive review of EEG-based brain–computer interface paradigms. J. Neural Eng. **16**(1), 011001 (2019)
48. R. Abiri, S. Borhani, J. Kilmarx, C. Esterwood, Y. Jiang, X. Zhao, A usability study of low-cost wireless brain-computer interface for cursor control using online linear model. IEEE Trans. Hum. Mach. Syst. **50**(4), 287–297 (2020)
49. J. Kilmarx, R. Abiri, S. Borhani, Y. Jiang, X. Zhao, Sequence-based manipulation of robotic arm control in brain machine interface. Int. J. Intell. Robot. Appl. **2**(4), 149–160 (2018)

50. A. Ghasemi, M.J. Johnson, S. Karnam, D. Gopinath, H. Zhang, N. Sarkar, Shared autonomy for assistive robotic manipulation with a brain-computer interface: theory and experiments. Sci. Robot. **7**(66), eabf8026 (2022)
51. F. Lotte, M. Congedo, A. Lécuyer, F. Lamarche, B. Arnaldi, A review of classification algorithms for eeg-based brain–computer interfaces: a 10 year update. J. Neural Eng. **15**(3), 031005 (2018)
52. D.J. McFarland, J.R. Wolpaw, EEG-based brain–computer interfaces. Curr. Opin. Biomed. Eng. **4**, 194–200 (2017)
53. E.A. Felton, J.C. Williams, G.C. Vanderheiden, R.G. Radwin, Mental workload during brain–computer interface training. Ergonomics **55**(5), 526–537 (2012). https://doi.org/10.1080/00140139.2012.662526
54. D.B. Silversmith, R. Abiri, N.F. Hardy, N. Natraj, A. Tu-Chan, E.F. Chang, K. Ganguly, Plug-and-play control of a brain–computer interface through neural map stabilization. Nat. Biotechnol. **39**(3), 326–335 (2021)
55. N. Natraj, S. Seko, R. Abiri, R. Miao, H. Yan, Y. Graham, A. Tu-Chan, E.F. Chang, K. Ganguly, Sampling representational plasticity of simple imagined movements across days enables long-term neuroprosthetic control. Cell **188**(6), 1208–1225 (2025). https://doi.org/10.1016/j.cell.2025.02.001
56. G. Pfurtscheller, F.H. Lopes da Silva, Event-related EEG/MEG synchronization and desynchronization: basic principles. Clin. Neurophysiol. **110**(11), 1842–1857 (1999)
57. E.A. Clancy, E.L. Morin, R. Merletti, Processing of myoelectric signals for controlling powered limb prostheses. IEEE Trans. Biomed. Eng. **50**(6), 743–754 (2002)
58. G. Schalk, D.J. McFarland, T. Hinterberger, N. Birbaumer, J.R. Wolpaw, BCI2000: a general-purpose brain-computer interface (BCI) system. IEEE Trans. Biomed. Eng. **51**(6), 1034–1043 (2004)
59. G.R. Müller-Putz, R. Scherer, C. Brunner, R. Leeb, G. Pfurtscheller, Combining EEG and EMG for improved intention recognition in upper-limb prosthesis control. IEEE Trans. Biomed. Eng. **62**(9), 2173–2184 (2015)
60. A.H. Do, P.T. Wang, C.E. King, S.N. Chun, Z. Nenadic, Brain-computer interface controlled robotic gait orthosis. J. Neuroeng. Rehab. **10**(1), 1–9 (2013)
61. S. Tortora, S. Ghidoni, C. Chisari, F. Artoni, S. Micera, Hybrid human-machine interface for gait decoding through bayesian fusion of EEG and EMG classifiers. Front. Neurorobot. **14**, 89 (2020)
62. T.O. Zander, C. Kothe, Towards passive brain–computer interfaces: applying brain–computer interface technology to human–machine systems in general. J. Neural Eng. **8**(2), 025005 (2011)
63. B.Z. Allison, E.W. Wolpaw, J.R. Wolpaw, BCI demographics: how many (and what kinds of) people can use an SSVEP BCI? IEEE Transa. Neural Syst. Rehab. Eng. **18**(2), 107–116 (2010)
64. A. Delorme, S. Makeig, Eeglab: an open source toolbox for analysis of single-trial EEG dynamics including independent component analysis. J. Neurosci. Methods **134**(1), 9–21 (2004)
65. M.K. Islam, A. Rastegarnia, Z. Yang, Methods for artifact detection and removal from scalp EEG: a review. Neurophysiol. Clin. **46**(4–5), 287–305 (2016)
66. J.P. Lachaux, E. Rodriguez, J. Martinerie, F.J. Varela, Measuring phase synchrony in brain signals. Hum. Brain Mapp. **8**(4), 194–208 (1999)
67. B. Blankertz, R. Tomioka, S. Lemm, M. Kawanabe, K.R. Muller, Optimizing spatial filters for robust EEG single-trial analysis. IEEE Signal Process. Mag. **25**(1), 41–56 (2008)
68. K.K. Ang, Z.H. Chin, H. Zhang, C. Guan, Filter bank common spatial pattern (FBCSP) in brain–computer interface, in *IEEE International Joint Conference on Neural Networks (IJCNN)* (IEEE, Piscataway, 2012), pp. 2390–2397
69. B. Hudgins, P. Parker, R.N. Scott, A new strategy for multifunction myoelectric control. IEEE Trans. Biomed. Eng. **40**(1), 82–94 (1993)

70. K. Englehart, B. Hudgins, P.A. Parker, Classification of the myoelectric signal using time-frequency based representations. Med. Eng. Phys. **21**(6–7), 431–438 (1999)
71. A. Phinyomark, P. Phukpattaranont, C. Limsakul, Feature extraction and selection for myoelectric control based on wearable emg sensors. Sensors **12**(9), 12402–12440 (2012)
72. V.J. Lawhern, A.J. Solon, N.R. Waytowich, S.M. Gordon, C.P. Hung, B.J. Lance, Eegnet: a compact convolutional neural network for EEG-based brain–computer interfaces. J. Neural Eng. **15**(5), 056013 (2018)
73. S. Roy, I. Kiral-Kornek, S. Harrer, Chrononet: a deep recurrent neural network for abnormal EEG identification. Comput. Biol. Med. **113**, 103395 (2019)
74. R.T. Schirrmeister, J.T. Springenberg, L.D.J. Fiederer, M. Glasstetter, K. Eggensperger, M. Tangermann, F. Hutter, W. Burgard, T. Ball, Deep learning with convolutional neural networks for EEG decoding and visualization. Hum. Brain Mapp. **38**(11), 5391–5420 (2017)
75. P. Shenoy, M. Krauledat, B. Blankertz, R.P. Rao, K.R. Müller, Towards adaptive classification for BCI. J. Neural Eng. **3**(1), R13 (2006)
76. C. Vidaurre, B. Blankertz, Towards a cure for BCI illiteracy. Brain Topograph. **23**(2), 194–198 (2010)
77. S. Perdikis, J.D.R. Millán, Brain-machine interfaces: a tale of two learners. IEEE Syst. Man Cybern. Mag. **6**(3), 12–19 (2020)
78. S.G. Mason, A. Bashashati, M. Fatourechi, K.F. Navarro, G.E. Birch, A real-time interface based on the p300 speller. IEEE Trans. Neural Syst. Rehab. Eng. **15**(4), 540–548 (2007)
79. R. Scherer, G.R. Müller, C. Neuper, B. Graimann, G. Pfurtscheller, Toward self-paced brain–computer communication: navigation through virtual worlds. IEEE Trans. Biomed. Eng. **55**(2), 675–682 (2008)
80. P.L. Nunez, R. Srinivasan, *Electric Fields of the Brain: The Neurophysics of EEG* (Oxford University Press, Oxford, 2006)
81. M.X. Cohen, *Analyzing Neural Time Series Data: Theory and Practice* (MIT Press, Cambridge, 2014)
82. A. Rezeika, M. Benda, P. Stawicki, F. Gembler, A. Saboor, I. Volosyak, Brain–computer interface spellers: a review. Brain Sci. **8**(4), 57 (2018)
83. D.J. Krusienski, E.W. Sellers, F. Cabestaing, S. Bayoudh, D.J. McFarland, T.M. Vaughan, J.R. Wolpaw, Toward enhanced p300 speller performance. J. Neurosci. Methods **167**(1), 15–21 (2008)
84. B. Blankertz, L. Acqualagna, S. Dähne, S. Haufe, M. Schultze-Kraft, I. Sturm, M. Uşümüş, M.A. Wenzel, G. Curio, K.R. Müller, The berlin brain–computer interface: non-medical uses of BCI technology. Front. Neurosci. **5**, 9 (2011)
85. P. Bashivan, I. Rish, M. Yeasin, N. Codella, Learning representations from EEG with deep recurrent-convolutional neural networks (2015). arXiv:1511.06448
86. R. Manor, A.B. Geva, Convolutional neural network for multi-category rapid serial visual presentation BCI. Front. Comput. Neurosci. **9**, 146 (2015)
87. T. Song, W. Zheng, P. Song, Z. Cui, EEG emotion recognition using dynamical graph convolutional neural networks. IEEE Trans. Affect. Comput. **11**(3), 532–541 (2020)
88. W.L. Zheng, B.L. Lu, Investigating critical frequency bands and channels for EEG-based emotion recognition with deep neural networks. IEEE Trans. Auton. Mental Develop. **7**(3), 162–175 (2015)
89. V. Jayaram, M. Alamgir, Y. Altun, B. Scholkopf, M. Grosse-Wentrup, Transfer learning in brain-computer interfaces. IEEE Comput. Intell. Mag. **11**(1), 20–31 (2016)
90. H. Kang, Y. Nam, S. Choi, Composite common spatial pattern for subject-to-subject transfer. IEEE Signal Process. Lett. **16**(8), 683–686 (2009)
91. A.L. Orsborn, S. Dangi, H.G. Moorman, J.M. Carmena, Closed-loop decoder adaptation shapes neural plasticity for skillful neuroprosthetic control. Neuron **82**(6), 1380–1393 (2014)
92. Z.C. Danziger, E. Raffin, A. Tabbal, J. Signoret, M.A. Maier, A. Schmied, Learning algorithms for human–machine interfaces. IEEE Trans. Biomed. Eng. **56**(5), 1502–1511 (2009)

93. S.G. Hart, L.E. Staveland, Development of NASA-TLX (task load index): results of empirical and theoretical research, in *Advances in Psychology*, vol. 52 (Elsevier, Amsterdam, 1988), pp. 139–183
94. J. Brooke, Sus-a quick and dirty usability scale, in *Usability Evaluation in Industry* (Taylor & Francis, Milton Park, 1996), pp. 189–194
95. S. Anwer, A. Waris, S.O. Gilani, J. Iqbal, N. Shaikh, A.N. Pujari, I.K. Niazi, Rehabilitation of upper limb motor impairment in stroke: a narrative review on the prevalence, risk factors, and economic statistics of stroke and state of the art therapies, in *Healthcare*, vol. 10, no. 2 (MDPI, Basel, 2022), p. 190
96. S. Khalid, F. Alnajjar, M. Gochoo, A. Renawi, S. Shimoda, Robotic assistive and rehabilitation devices leading to motor recovery in upper limb: a systematic review. Disabil. Rehab. Assist. Technol. **18**(5), 658–672 (2023)
97. H.M. Qassim, W. Wan Hasan, A review on upper limb rehabilitation robots. Appl. Sci. **10**(19), 6976 (2020)
98. P. Maciejasz, J. Eschweiler, K. Gerlach-Hahn, A. Jansen-Troy, S. Leonhardt, A survey on robotic devices for upper limb rehabilitation. J. Neuroeng. Rehab. **11**(1), 1–29 (2014)
99. A. Demofonti, G. Carpino, L. Zollo, M.J. Johnson, Affordable robotics for upper limb stroke rehabilitation in developing countries: a systematic review. IEEE Trans. Med. Robot. Bion. **3**(1), 11–20 (2021)
100. J. Narayan, B. Kalita, S.K. Dwivedy, Development of robot-based upper limb devices for rehabilitation purposes: a systematic review. Augment. Hum. Res. **6**, 1–33 (2021)
101. M.J. Johnson, R. Rai, S. Barathi, R. Mendonca, K. Bustamante-Valles, Affordable stroke therapy in high-, low-and middle-income countries: from theradrive to rehab cares, a compact robot Gym. J. Rehabil. Assist. Technol. Eng. **4**, 2055668317708732 (2017)
102. P. Lum, D. Reinkensmeyer, R. Mahoney, W.Z. Rymer, C. Burgar, Robotic devices for movement therapy after stroke: current status and challenges to clinical acceptance. Top. Stroke Rehab. **8**(4), 40–53 (2002)
103. S. Mansour, K.K. Ang, K.P. Nair, K.S. Phua, M. Arvaneh, Efficacy of brain–computer interface and the impact of its design characteristics on poststroke upper-limb rehabilitation: a systematic review and meta-analysis of randomized controlled trials. Clin. EEG Neurosci. **53**(1), 79–90 (2022)
104. B. Sheng, Y. Zhang, W. Meng, C. Deng, S. Xie, Bilateral robots for upper-limb stroke rehabilitation: state of the art and future prospects. Med. Eng. Phys. **38**(7), 587–606 (2016)
105. S.H. Lee, G. Park, D.Y. Cho, H.Y. Kim, J.Y. Lee, S. Kim, S.B. Park, J.H. Shin, Comparisons between end-effector and exoskeleton rehabilitation robots regarding upper extremity function among chronic stroke patients with moderate-to-severe upper limb impairment. Sci. Rep. **10**(1), 1806 (2020)
106. J.C. Metzger, O. Lambercy, D. Chapuis, R. Gassert, Design and characterization of the rehapticknob, a robot for assessment and therapy of hand function, in *2011 IEEE/RSJ International Conference on Intelligent Robots and Systems* (IEEE, Piscataway, 2011), pp. 3074–3080
107. G. Devittori, R. Ranzani, D. Dinacci, D. Romiti, A. Califfi, C. Petrillo, P. Rossi, R. Gassert, O. Lambercy, Automatic and personalized adaptation of therapy parameters for unsupervised robot-assisted rehabilitation: a pilot evaluation, in *2022 International Conference on Rehabilitation Robotics (ICORR)* (IEEE, Piscataway, 2022), pp. 1–6
108. M.A. Baniasad, F. Farahmand, N.N. Ansari, Wrist-robohab: a robot for treatment and evaluation of brain injury patients, in *2011 IEEE International Conference on Rehabilitation Robotics* (IEEE, Piscataway , 2011), pp. 1–5
109. R. Buschfort, J. Brocke, A. Hess, C. Werner, A. Waldner, S. Hesse, Arm studio to intensify the upper limb rehabilitation after stroke: concept, acceptance, utilization and preliminary clinical results. J. Rehab. Med. **42**(4), 310–314 (2010)
110. S. Micera, M.C. Carrozza, E. Guglielmelli, G. Cappiello, F. Zaccone, C. Freschi, R. Colombo, A. Mazzone, C. Delconte, F. Pisano, et al., A simple robotic system for neurorehabilitation. Auton. Robot. **19**, 271–284 (2005)

111. H.I. Krebs, M. Ferraro, S.P. Buerger, M.J. Newbery, A. Makiyama, M. Sandmann, D. Lynch, B.T. Volpe, N. Hogan, Rehabilitation robotics: pilot trial of a spatial extension for mit-manus. J. Neuroeng. Rehab. **1**, 1–15 (2004)
112. E.C. Lu, R. Wang, R. Huq, D. Gardner, P. Karam, K. Zabjek, D. Hébert, J. Boger, A. Mihailidis, Development of a robotic device for upper limb stroke rehabilitation: a user-centered design approach. Paladyn J. Behav. Robot. **2**(4), 176–184 (2011)
113. M. Sivan, J. Gallagher, S. Makower, D. Keeling, B. Bhakta, R.J. O'Connor, M. Levesley, Home-based computer assisted arm rehabilitation (HCAAR) robotic device for upper limb exercise after stroke: results of a feasibility study in home setting. J. Neuroeng. Rehab. **11**, 1–17 (2014)
114. S. Mazzoleni, P. Sale, M. Tiboni, M. Franceschini, M.C. Carrozza, F. Posteraro, Upper limb robot-assisted therapy in chronic and subacute stroke patients: a kinematic analysis. Am. J. Phys. Med. Rehab. **92**(10), e26–e37 (2013)
115. J.V. Kopke, M.D. Ellis, L.J. Hargrove, Feasibility of two different EMG-based pattern recognition control paradigms to control a robot after stroke–case study, in *2020 8th IEEE RAS/EMBS International Conference for Biomedical Robotics and Biomechatronics (BioRob)* (IEEE, Piscataway, 2020), pp. 833–838
116. J.S. Sulzer, M.A. Peshkin, J.L. Patton, Design of a mobile, inexpensive device for upper extremity rehabilitation at home, in *2007 IEEE 10th International Conference on Rehabilitation Robotics* (IEEE, Piscataway, 2007), pp. 933–937
117. C.K. Chang, E.P. Washabaugh, A. Gwozdziowski, C.D. Remy, C. Krishnan, A semi-passive planar manipulandum for upper-extremity rehabilitation. Ann. Biomed. Eng. **46**, 1047–1065 (2018)
118. R. Colombo, F. Pisano, A. Mazzone, C. Delconte, S. Micera, M.C. Carrozza, P. Dario, G. Minuco, Design strategies to improve patient motivation during robot-aided rehabilitation. J. Neuroeng. Rehab. **4**, 1–12 (2007)
119. M. Ceccarelli, M. Russo, D. Cafolla, B.D. Chaparro-Rico, Operation safety of a 2-DoF planar mechanism for arm rehabilitation. Inventions **6**(4), 85 (2021)
120. Q. Wu, B. Chen, H. Wu, Adaptive admittance control of an upper extremity rehabilitation robot with neural-network-based disturbance observer. IEEE Access **7**, 123807–123819 (2019)
121. D. Campolo, P. Tommasino, K. Gamage, J. Klein, C.M. Hughes, L. Masia, H-man: a planar, h-shape cabled differential robotic manipulandum for experiments on human motor control. J. Neurosci. Methods **235**, 285–297 (2014)
122. C. Freeman, A.M. Hughes, J. Burridge, P. Chappell, P. Lewin, E. Rogers, A robotic workstation for stroke rehabilitation of the upper extremity using FES. Med. Eng. Phys. **31**(3), 364–373 (2009)
123. B. Ghannadi, R. Sharif Razavian, J. McPhee, Configuration-dependent optimal impedance control of an upper extremity stroke rehabilitation manipulandum. Front. Robot. AI **5**, 124 (2018)
124. S. Guillén-Climent, A. Garzo, M.N. Muñoz-Alcaraz, P. Casado-Adam, J. Arcas-Ruiz-Ruano, M. Mejías-Ruiz, F.J. Mayordomo-Riera, A usability study in patients with stroke using merlin, a robotic system based on serious games for upper limb rehabilitation in the home setting. J. Neuroeng. Rehab. **18**(1), 1–16 (2021)
125. J. Whitall, S.M. Waller, K.H. Silver, R.F. Macko, Repetitive bilateral arm training with rhythmic auditory cueing improves motor function in chronic hemiparetic stroke. Stroke **31**(10), 2390–2395 (2000)
126. E. Vergaro, M. Casadio, V. Squeri, P. Giannoni, P. Morasso, V. Sanguineti, Self-adaptive robot training of stroke survivors for continuous tracking movements. J. Neuroeng. Rehab. **7**(1), 1–12 (2010)
127. S. Hesse, H. Schmidt, C. Werner, C. Rybski, U. Puzich, A. Bardeleben, A new mechanical arm trainer to intensify the upper limb rehabilitation of severely affected patients after stroke: design, concept and first case series. Eur. Medicophys. **43**(4), 463–468 (2007)

128. R.M. Mahoney, H.M. Van der Loos, P.S. Lum, C. Burgar, Robotic stroke therapy assistant. Robotica **21**(1), 33–44 (2003)
129. E. Rashedi, A. Mirbagheri, B. Taheri, F. Farahmand, G. Vossoughi, M. Parnianpour, Design and development of a hand robotic rehabilitation device for post stroke patients, in *2009 Annual International Conference of the IEEE Engineering in Medicine and Biology Society* (IEEE, Piscataway, 2009), pp. 5026–5029
130. C. Qian, W. Li, T. Jia, C. Li, P.J. Lin, Y. Yang, L. Ji, Quantitative assessment of motor function by an end-effector upper limb rehabilitation robot based on admittance control. Appl. Sci. **11**(15), 6854 (2021)
131. C.A. Avizzano, M. Satler, G. Cappiello, A. Scoglio, E. Ruffaldi, M. Bergamasco, Motore: a mobile haptic interface for neuro-rehabilitation, in *2011 RO-MAN* (IEEE, Piscataway, 2011), pp. 383–388
132. T. Paolucci, F. Agostini, M. Mangone, A. Bernetti, L. Pezzi, V. Liotti, E. Recubini, C. Cantarella, R.G. Bellomo, C. D'Aurizio, et al., Robotic rehabilitation for end-effector device and botulinum toxin in upper limb rehabilitation in chronic post-stroke patients: an integrated rehabilitative approach. Neurol. Sci. **42**(12), 5219–5229 (2021)
133. S. Ghafoori, A. Rabiee, M. Norouzi, M.K. Jouaneh, R. Abiri, A novel seamless magnetic-based actuating mechanism for end-effector-based robotic rehabilitation platforms. IEEE Robot. Autom. Lett. **10**(5), 4516–4523 (2025)
134. R.C. Loureiro, B. Lamperd, C. Collin, W.S. Harwin, Reach & grasp therapy: effects of the gentle/g system assessing sub-acute stroke whole-arm rehabilitation, in *2009 IEEE International Conference on Rehabilitation Robotics* (2009), 755–760
135. H.A. Rahman, A.L.T. Narayanan, K.K. Xiang, E.S.L. Ming, Y.C. Fai, Q.I. Khan, IREST: interactive rehabilitation and assessment tool, in *2015 10th Asian Control Conference (ASCC)* (IEEE, Piscataway, 2015), pp. 1–6
136. A. Basteris, S.M. Nijenhuis, A.H. Stienen, J.H. Buurke, G.B. Prange, F. Amirabdollahian, Training modalities in robot-mediated upper limb rehabilitation in stroke: a framework for classification based on a systematic review. J. Neuroeng. Rehab. **11**, 1–15 (2014)
137. A. Van Delden, C.L.E. Peper, G. Kwakkel, P.J. Beek, et al., A systematic review of bilateral upper limb training devices for poststroke rehabilitation. Stroke Res. Treat. **2012**, 972069 (2012)
138. S. Dalla Gasperina, L. Roveda, A. Pedrocchi, F. Braghin, M. Gandolla, Review on patient-cooperative control strategies for upper-limb rehabilitation exoskeletons. Front. Robot. AI **8**, 745018 (2021)
139. T. Proietti, V. Crocher, A. Roby-Brami, N. Jarrasse, Upper-limb robotic exoskeletons for neurorehabilitation: a review on control strategies. IEEE Rev. Biomed. Eng. **9**, 4–14 (2016)
140. T. Desplenter, Y. Zhou, B.P. Edmonds, M. Lidka, A. Goldman, A.L. Trejos, Rehabilitative and assistive wearable mechatronic upper-limb devices: a review. J. Rehab. Assist. Technol. Eng. **7**, 2055668320917870 (2020)
141. I. O. for Standardization, *Robots and Robotic Devices: Safety Requirements for Personal Care Robots* (2014)
142. H. Jiang, T. Zhang, J.P. Wachs, B.S. Duerstock, Enhanced control of a wheelchair-mounted robotic manipulator using 3-d vision and multimodal interaction. Comput. Vis. Image Understand. **149**, 21–31 (2016)
143. R.M. Chapman, K.B. Taylor, E. Kaczynski, S. Khodabakhsh, S. Richards, J.B. Hutchinson, R.C. Marchand, Accuracy amidst errors: evaluating a commercially available wearable sensor system and its associated calibration procedures for monitoring sagittal knee motion in patients undergoing total knee arthroplasty. Knee **54**, 316–328 (2025)
144. M.P. De Looze, T. Bosch, F. Krause, K.S. Stadler, L.W. O'sullivan, Exoskeletons for industrial application and their potential effects on physical work load. Ergonomics **59**(5), 671–681 (2016)
145. R.A. Cooper, Intelligent control of power wheelchairs. IEEE Eng. Med. Biol. Mag. **14**(4), 423–431 (1995)

146. S. Udupa, V.R. Kamat, C.C. Menassa, Shared autonomy in assistive mobile robots: a review. Disabil. Rehab. Assist. Technol. **18**(6), 827–848 (2023)
147. A. Rabiee, S. Ghafoori, X. Bai, S. Ostadabbas, R. Abiri, Streams: an assistive multimodal AI framework for empowering biosignal based robotic controls (2024). arXiv:2410.03486
148. D. Park, Y. Hoshi, H.P. Mahajan, H.K. Kim, Z. Erickson, W.A. Rogers, C.C. Kemp, Active robot-assisted feeding with a general-purpose mobile manipulator: design, evaluation, and lessons learned. Robot. Autono. Syst. **124**, 103344 (2020)
149. E. Ackerman, JACO is a low-power robot arm that hooks to your wheelchair. *R* (2019)
150. A. Blom, H. Stuyt, Assistive robotic manipulators, in *Robotic Assistive Technologies* (CRC Press, Boca Raton, 2017), pp. 71–98
151. M.D. Paskett, M.R. Brinton, T.C. Hansen, J.A. George, T.S. Davis, C.C. Duncan, G.A. Clark, Activities of daily living with bionic arm improved by combination training and latching filter in prosthesis control comparison. J. NeuroEng. Rehabil. **18**(1), 45 (2021). https://doi.org/10.1186/s12984-021-00839-x
152. L.V. Herlant, R.M. Holladay, S.S. Srinivasa, Assistive teleoperation of robot arms via automatic time-optimal mode switching, in *2016 11th ACM/IEEE International Conference on Human-Robot Interaction (HRI)* (IEEE, Piscataway, 2016), pp. 35–42
153. N. Rudigkeit, M. Gebhard, A. Gräser, Towards a user-friendly AHRS-based human-machine interface for a semi-autonomous robot, in *Proceedings of the 2014 IEEE/RSJ International Conference on Intelligent Robots and Systems (IROS2014), Chicago* (2014), pp. 14–18
154. H.B. Surale, F. Matulic, D. Vogel, Experimental analysis of mode switching techniques in touch-based user interfaces, in *Proceedings of the 2017 CHI Conference on Human Factors in Computing Systems* (2017), pp. 3267–3280
155. M. Gao, J. Oberländer, T. Schamm, J.M. Zöllner, Contextual task-aware shared autonomy for assistive mobile robot teleoperation, in *2014 IEEE/RSJ International Conference on Intelligent Robots and Systems* (IEEE, Piscataway, 2014), pp. 3311–3318
156. D. Gopinath, S. Jain, B.D. Argall, Human-in-the-loop optimization of shared autonomy in assistive robotics. IEEE Robot. Autom. Lett. **2**(1), 247–254 (2016)
157. Y. Oh, M. Toussaint, J. Mainprice, Learning arbitration for shared autonomy by hindsight data aggregation (2019). arXiv:1906.12280
158. B. Shneiderman, Human-centered artificial intelligence: reliable, safe & trustworthy. Int. J. Hum. Comput. Interact. **36**(6), 495–504 (2020)
159. J.A. Yow, N.P. Garg, W.T. Ang, Shared autonomy of a robotic manipulator for grasping under human intent uncertainty using pomdps. IEEE Trans. Robot. **40**, 332–350 (2023)
160. S. Javdani, H. Admoni, S. Pellegrinelli, S.S. Srinivasa, J.A. Bagnell, Shared autonomy via hindsight optimization for teleoperation and teaming. Int. J. Robot. Res. **37**(7), 717–742 (2018)
161. A. Shafti, P. Orlov, A.A. Faisal, Gaze-based, context-aware robotic system for assisted reaching and grasping. in *2019 International Conference on Robotics and Automation (ICRA)* (IEEE, Piscataway, 2019), pp. 863–869
162. J.E. Downey, J.M. Weiss, K. Muelling, A. Venkatraman, J.S. Valois, M. Hebert, J.A. Bagnell, A.B. Schwartz, J.L. Collinger, Blending of brain-machine interface and vision-guided autonomous robotics improves neuroprosthetic arm performance during grasping. J. Neuroeng. Rehab. **13**, 1–12 (2016)
163. W. Xu, M.J. Dainoff, L. Ge, Z. Gao, Transitioning to human interaction with AI systems: new challenges and opportunities for HCI professionals to enable human-centered AI. Int. J. Hum. Comput. Interact. **39**(3), 494–518 (2023)
164. G. Beraldo, L. Tonin, J.D.R. Millán, E. Menegatti, Shared intelligence for robot teleoperation via BMI. IEEE Trans. Hum. Mach. Syst. **52**(3), 400–409 (2022)
165. A. Rabiee, S. Ghafoori, A. Cetera, W. Besio, R. Abiri, A comparative study of conventional and tripolar EEG for high-performance reach-to-grasp BCI systems (2024). arXiv:2402.09448

166. A. Rabiee, S. Ghafoori, A. Cetera, Y. Shahriari, R. Abiri, Wavelet analysis of noninvasive EEG signals discriminates complex and natural grasp types, in *2024 46th Annual International Conference of the IEEE Engineering in Medicine and Biology Society (EMBC)* (2024), pp. 1–5
167. S. Ghafoori, A. Rabiee, A. Cetera, Y. Shahriari, R. Abiri, Bispectrum analysis of noninvasive EEG signals discriminates complex and natural grasp types, in *2024 46th Annual International Conference of the IEEE Engineering in Medicine and Biology Society (EMBC)* (IEEE, Piscataway, 2024), pp. 1–5
168. A. Cetera, A. Rabiee, S. Ghafoori, Y. Shahriari, R. Abiri, Classification of emerging neural activity from planning to grasp execution using a novel EEG-based BCI platform, in *2024 46th Annual International Conference of the IEEE Engineering in Medicine and Biology Society (EMBC)* (IEEE, Piscataway, 2024), pp. 1–5
169. K. Tai, S. Blain, T. Chau, A review of emerging access technologies for individuals with severe motor impairments. Assist. Technol. **20**(4), 204–221 (2008)
170. E.M. Mugler, C.A. Ruf, S. Halder, M. Bensch, A. Kubler, Design and implementation of a p300-based brain-computer interface for controlling an internet browser. IEEE Trans. Neural Syst. Rehab. Eng. **18**(6), 599–609 (2010)
171. S. Stepputtis, J. Campbell, M. Phielipp, S. Lee, C. Baral, H. Ben Amor, Language-conditioned imitation learning for robot manipulation tasks. Adv. Neural Inf. Process. Syst. **33**, 13139–13150 (2020)
172. B.D. Ziebart, N. Ratliff, G. Gallagher, C. Mertz, K. Peterson, J.A. Bagnell, M. Hebert, A.K. Dey, S. Srinivasa, Planning-based prediction for pedestrians, in *2009 IEEE/RSJ International Conference on Intelligent Robots and Systems* (IEEE, Piscataway, 2009), pp. 3931–3936
173. G. Best, R. Fitch, Bayesian intention inference for trajectory prediction with an unknown goal destination, in *2015 IEEE/RSJ International Conference on Intelligent Robots and Systems (IROS)* (IEEE, Piscataway, 2015), pp. 5817–5823
174. A. Javernik, B. Buchmeister, R. Ojsteršek, The NASA-TLX approach to understand workers workload in humanrobot collaboration. Int. J. Simul. Model. **22**(4), 574–585 (2023)
175. Z. Erickson, V. Gangaram, A. Kapusta, C.K. Liu, C.C. Kemp, Assistive gym: a physics simulation framework for assistive robotics, in *2020 IEEE International Conference on Robotics and Automation (ICRA)* (IEEE, Piscataway, 2020), pp. 10169–10176
176. M. Ahn, A. Brohan, N. Brown, Y. Chebotar, O. Cortes, B. David, C. Finn, C. Fu, K. Gopalakrishnan, K. Hausman, et al., Do as i can, not as i say: grounding language in robotic affordances (2022). arXiv:2204.01691
177. D. Driess, F. Xia, M.S. Sajjadi, C. Lynch, A. Chowdhery, A. Wahid, J. Tompson, Q. Vuong, T. Yu, W. Huang, et al., PaLM-E: An Embodied Multimodal Language Model, in *Proceedings of the 40th International Conference on Machine Learning (ICML)* (ICML'23), vol. 340(20), (JMLR.org, Honolulu, Hawaii, USA, 2023). https://palm-e.github.io/. http://doi.org/10.5555/3618408.3618748
178. W. Huang, F. Xia, T. Xiao, H. Chan, J. Liang, P. Florence, A. Zeng, J. Tompson, I. Mordatch, Y. Chebotar, et al., Inner monologue: embodied reasoning through planning with language models (2022). arXiv:2207.05608
179. Y. Jin, D. Li, J. Shi, P. Hao, F. Sun, J. Zhang, B. Fang, et al. Robotgpt: robot manipulation learning from chatgpt. IEEE Robot. Autom. Lett. **9**(3), 2543–2550 (2024)
180. A.Z. Ren, A. Dixit, A. Bodrova, S. Singh, S. Tu, N. Brown, P. Xu, L. Takayama, F. Xia, J. Varley, et al., Robots that ask for help: uncertainty alignment for large language model planners. arXiv:2307.01928
181. S. Liu, L. Tor-Díez, M. Celorrio, S. Kambhampati, LLM-P: a framework for combining large language models and classical planners, in *Proceedings of the International Conference on Automated Planning and Scheduling*, vol. 34 (2024), pp. 568–576
182. M. Kabra, A. Soni, G. Singh, T. Chakraborti, Adaptbot: a knowledge-driven framework for adaptive task decomposition and action refinement (2024). arXiv:2402.02067
183. R. Paul, S. Chitta, C. Paxton, D. Fox, F. Zaidi, From language to action with object-level planning, in *2025 IEEE International Conference on Robotics and Automation (ICRA) Workshop on Long-Horizon Task and Motion Planning* (2025)

184. S. Chen, Z. Zhang, Y. Zhao, B. Li, J. Yin, Y. Liu, H. Zhang, An embodied LLM agent and text-based topological map for robotic guidance of people with visual impairments (2025). arXiv:2410.20666
185. Y. Wu, J. Zhang, Z. Yuan, J. Zhang, Y. Shi, Q. Zhu, Savor: skill affordance learning from visuo-haptic perception for robot-assisted bite acquisition (2025). arXiv:2506.02353
186. A. Brohan, N. Brown, J. Carbajal, Y. Chebotar, J. Dabis, C. Finn, K. Gong, R. Hadsell, K. Herzog, J. Ho, et al., Rt-2: vision-language-action models transfer web knowledge to robotic control. arXiv:2307.15818
187. W. Zhao, J. Chen, Z. Meng, D. Mao, R. Song, W. Zhang, VLMPC: vision-language model predictive control for robotic manipulation, *in Proceedings of Robotics: Science and Systems (RSS)* (Robotics: Science and Systems Foundation, Delft, Netherlands, 2024). https://github.com/PPjmchen/VLMPC. https://doi.org/10.48550/arXiv.2407.09829
188. Y. Fu, J. Liu, X. Chen, H. Mei, RH20T-P: a primitive-level robotic manipulation dataset towards composable generalization agents in real-world scenarios, in *Advances in Neural Information Processing Systems* (2024)
189. M. Hafner, M. Kirtay, G. Chalvatzaki, B. Bäuml, Safe and socially-aware robot navigation via scene-graph-based explanations, in *2024 IEEE/RSJ International Conference on Intelligent Robots and Systems (IROS)* (IEEE, Piscataway, 2024)
190. Y.W. Lai, C.C. Shen, S.C. Wang, Y.H. Liu, H.C. Chang, C.C. Lin, LaMI: a language-based multi-modal interaction framework for human-robot collaboration, in *2023 IEEE/RSJ International Conference on Intelligent Robots and Systems (IROS)* (IEEE, Piscataway, 2023), pp. 10149–10156
191. Z. Tan, Y. Zhu, S.C. Zhu, Reflect: summarizing robot experiences for failure explanation (2023). arXiv:2306.15724
192. J. van der Waa, J. van Diggelen, K. van den Bosch, M. Neerincx, Contrastive explanations for reinforcement learning in terms of expected consequences. arXiv preprint arXiv:1807.08706 (2018). https://arxiv.org/abs/1807.08706
193. A. Singh, S. Shah, Y. Patel, A. Gupta, P. Shah, Explainable social navigation for autonomous mobile robots via vision-language models (2025). arXiv:2504.05477
194. R. Parasuraman, V. Riley, Humans and automation: use, misuse, disuse, abuse. Hum. Factors **39**(2), 230–253 (1997)
195. M.E.P. Seligman, Learned helplessness. Annu. Rev. Med. **23**(1), 407–412 (1972)
196. S.F. Maier, M.E.P. Seligman, Learned helplessness at fifty: insights from neuroscience. Psychol. Rev. **123**(4), 349–367 (2016)
197. N. Carr, The high price of automation. MIT Technology Review (2019). Published online May 30, 2019
198. L. Bainbridge, Ironies of automation. Automatica **19**(6), 775–779 (1983)
199. A. Howard, J. Borenstein, The ugly truth about robots: they are tools of digital colonialism. The American Scholar, Winter 2019, published online December 2, 2019

Chapter 15
Enhancing Surgeon Feedback via LSTM Driven Prediction of Tissue Puncturing Events

Bulbul Behera, M. Felix Orlando, and R. S. Anand

Abstract Accurate recognition of puncturing events in heterogeneous tissue is pivotal for improving the interaction between humans and robotic systems in biomedical applications, such as clinical process and surgical analysis. This study presents a novel Human-Robot Interface (HRI) framework that leverages Long Short-Term Memory (LSTM) networks to detect tissue penetration events in real time. The system captures time-series force feedback data via sensors embedded in a 1-DOF prismatic robot during tissue insertion. This feedback serves as a communication channel between the user and the robotic interface, enabling the LSTM model to interpret subtle force patterns indicative of transitions between tissue layers. By translating complex physical interactions into interpretable feedback for the operator, the proposed HRI system enhances situational awareness and control during procedures. Experimental results confirm the model effectiveness in providing accurate, real-time feedback, marking a significant advancement in intuitive and safe human-robot interaction for medical environments.

Keywords Human robot interface · Long short term memory · multilayer tissue · Human robot interaction · Force pattern

M. Felix Orlando and R. S. Anand authors contributed equally to this work.

B. Behera · M. Felix Orlando (✉)
Advanced Robotics Laboratory, Department of Electrical Engineering, Indian Institute of Technology Roorkee, Roorkee, Uttarakhand, India
e-mail: bulbul_b@ee.iitr.ac.in; m.orlando@ee.iitr.ac.in

R. S. Anand
Virtual Laboratory, Department of Electrical Engineering, Indian Institute of Technology Roorkee, Roorkee, Uttarakhand, India
e-mail: r.anand@ee.iitr.ac.in; m.orlando@ee.iitr.ac.in

R. Vinjamuri (ed.), *Bridging the Gap between Mind and Machine*,
https://doi.org/10.1007/978-3-032-06713-5_15

15.1 Introduction

Brachytherapy, commonly used to treat cancers such as those of the lung and liver, often relies on precise insertion of clinical instruments like needles. The effectiveness of this treatment hinges on the clinicians ability to accurately perceive and respond to penetration events during tissue traversal. In percutaneous therapy, challenges such as limited target visibility, restricted access, complex tool maneuverability, and physiological variations introduce significant cognitive and operational demands on the user. Human-Robot Interface (HRI) technologies play a pivotal role in addressing these challenges by integrating robotic systems and real-time imaging to enhance user perception and control. These interfaces bridge the gap between the clinician and the robotic system, facilitating intuitive interaction, reducing patient trauma, and improving procedural outcomes. Despite these advancements, HRI systems must still contend with dynamic factors such as variable needle-tissue interactions across multiple layers, emphasizing the need for interfaces that can adapt to and interpret such changes to support informed decision-making during procedures.

15.1.1 Literature Review

Robotic systems have significantly enhanced the precision and safety of medical procedures involving tissue penetration by providing greater maneuverability, stability, and control. These capabilities directly improve the human operators ability to interact with the surgical environment, reducing patient trauma and improving outcomes. From a Human-Robot Interface (HRI) perspective, such systems act as intermediaries that translate user intent into precise robotic actions. Research by Fisher et al. and Arbabtafti et al. demonstrates the benefits of robotic-assisted interventions while highlighting a critical HRI challenge: the absence of real-time feedback mechanisms for accurately detecting penetration events [1–3].

Imaging technologies such as ultrasound, MRI, and CT scans serve as essential components of the HRI loop in percutaneous procedures, offering real-time spatial context to the user. Studies have shown that ultrasound imaging, in particular, enhances needle placement accuracy when effectively integrated into the interface layer [4, 5]. Despite these advantages, fusing real-time imaging with robotic control and sensory data remains a substantial systems integration challenge that directly impacts the interface design and user experience [6].

Machine learning especially neural networks has emerged as a powerful tool in strengthening the feedback loop between human operators and robotic systems. Long Short-Term Memory (LSTM) networks, a type of recurrent neural network (RNN), are especially suited for analyzing sequential biomedical data. Since their introduction by Hochreiter and Schmidhuber (1997), LSTMs have proven effective in a wide range of time-series prediction tasks [7]. In biomedical contexts,

Mirshekarian et al. demonstrated that LSTMs could predict physiological events from sensor data, outperforming conventional approaches [8]. These models serve as a crucial HRI component by interpreting sensor feedback in real time and translating it into meaningful cues for the user.

Current brachytherapy practice is limited to treating the entire prostate gland uniformly due to insufficient seed placement accuracy, even in patients with only localized prostate cancer. During the low dose rate brachytherapy process, multiple radioactive seeds are permanently implanted into the prostate and surrounding tissue. The location of seed placement is evaluated using computer-assisted presurgical outlining with the help of clinical image of the gland. After planning surgeon manually deposited the seeds with the help of transrectal ultrasound image but does not provide accurate location.

Hence by analyzing data on the mechanical properties of a patients tissue, researchers can develop needle design and surgical techniques that are optimized for that specific patient. Medical professionals can identify and mitigates potential risk and complications before they occur. However, accuracy is determined by the size and diversity of the experimental database.

Recent developments have begun to converge robotics, imaging, and machine learning to create cohesive HRI systems that improve the detection of tissue penetration events. For example, integrating LSTM-based force analysis into robotic platforms enhances the operator's awareness of needle-tissue interaction dynamics during soft-tissue insertions. These hybrid systems bridge the communication gap between machine and user, delivering more intuitive, context-aware feedback [9–12].

LSTM networks remain a popular choice for such interfaces due to their ability to capture long-term dependencies in sequential data essential for interpreting the gradual force buildup that typically precedes a penetration event. This temporal sensitivity aligns closely with the nature of percutaneous interventions, where real-time decisions must be based on evolving input streams. Although alternative architectures like GRU, CNN, transformers, and diffusion models exist, LSTM networks offer a balance between complexity, interpretability, and performance. Their temporal modeling capabilities make them particularly well-suited for supporting intelligent HRI systems that require real-time adaptation to dynamic surgical environments.

15.2 Research Design

This section introduces the proposed frame work for detecting penetration events in multilayered tissue. A time-series force dataset was generated during needle insertions to capture puncture events across different tissue layers. This dataset provides detailed temporal information aligned with insertion depth, enabling the identification of key transition points. The machine learning pipeline involves preprocessing, segmenting, and normalizing the time-series data to enhance model performance and ensure accurate event prediction.

15.2.1 Memory Model for Penetration Event Detection

Accurately identifying penetration events is essential for characterizing needle-tissue interaction forces during insertion procedures. In this study, a machine learning-based LSTM network is implemented to model puncture forces across various tissue layers and detect sharp force fluctuations indicative of needle puncture events. This task is framed as a classification problem, where the goal is to identify the occurrence of such events from sequential force data.

The LSTM model is selected due to its suitability for handling sequential, time-series data ideal for modeling temporal patterns in needle insertion procedures. LSTM networks are capable of learning long-term dependencies, which is crucial for recognizing the gradual force build-up leading to tissue puncture. This predictive capability contributes to better control over insertion forces, helping to avoid excessive force application and the formation of large cracks. Furthermore, LSTM is robust to noisy input, which is common in medical settings due to factors like patient motion, breathing, and the heterogeneous nature of biological tissues. They also enable real time feedback during procedures, enhancing safety and operator awareness.

The proposed LSTM-based neural network architecture, illustrated in Fig. 15.1, processes three-dimensional force data (in the x, y, and z directions) collected during needle insertion. The network includes an input layer, a hidden LSTM layer with five neurons, and an output layer. Each LSTM unit contains three gates forget, input, and output that control the flow of information through the cell [13].

These gates use sigmoidal activation functions to determine how much information is retained or discarded. The forget gate is defined as:

$$a_t = \sigma(b_f[c_{t-1}, s_t] + d_f) \tag{15.1}$$

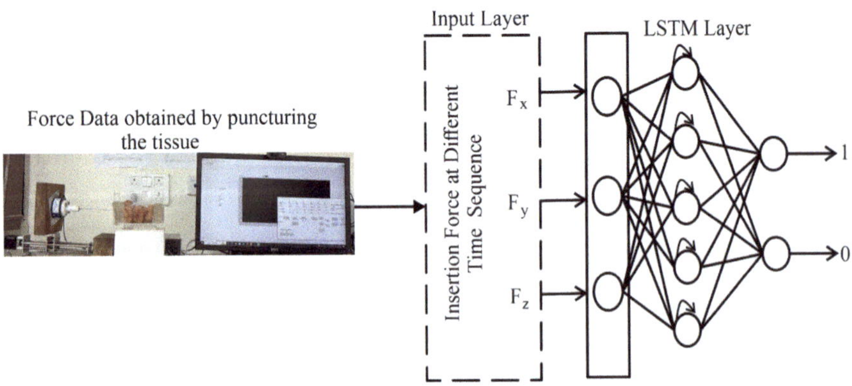

Fig. 15.1 Schematic of time dependent data captured and fed to the LSTM network

Here, a_t represents the forget gate's output at time step t, which determines the amount of information from the previous hidden state c_{t-1} and the current input s_t to retain. Weight matrix is denoted by b_f and d_f defined as bias term to the network. The input gate I_t, which decides what new information to store in the cell state, is defined as:

$$I_t = \sigma(b_i[c_{t-1}, s_t] + d_i) \tag{15.2}$$

The candidate values for the new cell state are generated using a hyperbolic tangent activation function:

$$\tilde{D}_t = \tanh(b_d[c_{t-1}, s_t] + d_d) \tag{15.3}$$

The updated cell state D_t is calculated by combining the old cell state and the new candidate values, modulated by the forget and input gates:

$$D_t = a_t \cdot D_{t-1} + I_t \cdot \tilde{D}_t \tag{15.4}$$

The hidden state h_t, which serves as the LSTM cell's output at each time step, is computed from the updated cell state c_t and the output gate O_t as follows:

$$c_t = O_t \cdot \tanh(C_t) \tag{15.5}$$

where the output gate is defined as:

$$O_t = \sigma(b_O[c_{t-1}, s_t] + d_O) \tag{15.6}$$

In these equations, b denotes the weight matrices and d the biases associated with each gate, all of which are learned during training using gradient descent. Since this is a classification task, the model is trained using a cross-entropy loss function:

$$L(\theta) = -\frac{1}{n} \sum_{t=1}^{n} \sum_{c=1}^{S} Y_{t,c} \log(\hat{Y}_{t,c}) \tag{15.7}$$

Here, n is the number of time steps, S is the number of classes, θ represents the model parameters (weights and biases), $Y_{t,c}$ is the true class distribution, and $\hat{Y}_{t,c}$ is the predicted class probability at time t.

15.2.2 Information Acquisition

The LSTM model utilizes three input signals corresponding to the force components in the x, y, and z directions. Data is collected using a multilayered tissue assembly composed of alternating layers of chicken breast and polyvinyl alcohol (PVA) phantom material, housed in a transparent container, as shown in Fig. 15.1. The structure consists of five distinct layers: a base layer of PVA, three intermediate layers of chicken tissue, and a top layer of PVA. Chicken breast is specifically used to reduce the risk of needle fracture during insertions.

A 1-DOF prismatic robot, powered by a stepper motor, is employed to perform controlled needle insertions. A force/torque (F/T) sensor is mounted at the rear end of the needle to record insertion forces along all three axes. This multidimensional force data is captured in real time and stored in .csv format for subsequent analysis. The tissue layers are arranged to simulate varying stiffness properties commonly encountered in real biological tissues, providing a realistic test environment.

To ensure data variability and minimize positional bias, each needle insertion is performed at a different location within the tissue stack, thereby avoiding previously punctured regions. All experiments are carried out under controlled environmental conditions, with the ambient temperature maintained at 25 °C. The insertion speed of the needle is uniformly set to 10 mm/s, in accordance with clinically established standards [11].

The dataset consists of 60 insertion trials, each capturing force data at 50 ms intervals over a period of 120 seconds. Each trial yields a complete time-series dataset with synchronized x, y, and z force measurements. The resulting .csv files are structured such that each row corresponds to a time-stamped sample, and each column represents one of the three force components. Before being fed into the LSTM network, the raw data undergoes min-max normalization to scale values to the [0, 1] range, ensuring uniform feature scaling across all inputs.

To develop a robust machine learning dataset, 100 needle insertion trials are performed using two different types of needles. Insertions are carried out at different spatial locations to further enrich the dataset. Force measurements are taken at every 0.25mm increment of needle depth, resulting in 3,840 data points per depth category (30, 40, 60, 80, and 100 mm). This comprehensive dataset supports reliable training, validation, and testing of the proposed LSTM-based model for penetration event detection.

15.2.3 Experimental Setup

A 1-DOF prismatic surgical platform (Huizhou Bachin Electronic Technology Co., LTD: T8-z60/T8-Z150) is employed to assist in human-robot collaborative needle insertion and extraction tasks. The linear motion of the system, essential for precise interaction, is actuated by a stepper motor controlled via an L298 H-bridge motor driver. The robotic platform is securely mounted on a wooden base, as shown in Fig. 15.2, ensuring stability during user-involved procedures. Motor control signals are generated through a user-programmable interface developed in the LabVIEW 2021 environment and transmitted via a National Instruments USB-

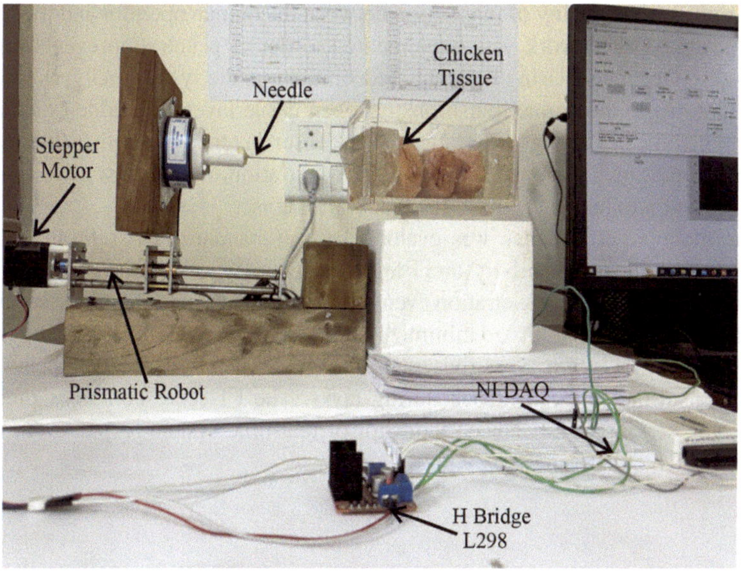

Fig. 15.2 Laboratory test bed for insertion episode identification

6008 data acquisition (DAQ) module, enabling intuitive real-time control by the human operator.

To facilitate interactive feedback during needle-tissue interaction, a JR3 force/-torque sensor (model 67M25A3-I40-DH-100N; JR3 Inc., Woodland, CA, USA) is integrated at the needles proximal end. This sensor captures multi-axis force and moment data (x, y, z, and corresponding torques), supporting high-fidelity force feedback mechanisms essential for safe and responsive HRI. With a force measurement range of $\pm$200 N and a moment range of $\pm$6 Nm, the sensor's integrated data logger enables real-time acquisition and seamless data export (.xlsx), aiding both autonomous and operator-in-the-loop decision-making processes.

15.3 Results and Discussion

A prismatic robotic system integrated with a multilayer tissue model comprising chicken tissue and PVA phantom is employed to investigate human-robot interaction during penetration event identification. The robot's motion is governed by a stepper motor operating at a constant speed, simulating a controlled surgical environment. This study focuses on how variations in tissue stiffness across different layers influence the interaction forces experienced during needle insertion, providing insight into how users might perceive and respond to haptic cues in real-time systems.

To enhance the quality of interaction between the human operator and the robotic system, a LSTM network was implemented to detect penetration events through force-based feedback. Using Python based simulation, the system analyzes insertion force patterns to determine transitions between tissue layers. A sudden increase in force typically indicating penetration of a denser layer is flagged by the LSTM model, outputting a binary response (1 for penetration, 0 otherwise), which could be translated into haptic or visual feedback for the user.

The models performance was evaluated using standard HRI relevant metrics: accuracy, specificity, sensitivity, and F1 score. Accuracy reflects the systems overall correctness in detecting penetration events, supporting system reliability. Specificity indicates the systems ability to minimize false alarms, essential for avoiding unnecessary user intervention. Sensitivity assesses the ability to detect true penetration events, ensuring timely feedback during tasks. The F1 score combines precision and recall, serving as a balanced indicator of real-time system responsiveness and user feedback effectiveness.

15.3.1 Results

This section outlines the proposed LSTM model developed to predict penetration events as the needle passes through new tissue layers. The model performance is assessed by comparing its predicted events with the actual recorded penetration events during needle insertions in a controlled laboratory setting. The results are illustrated in Fig. 15.3, which presents a time series plot of insertion force over the duration of the procedure.

In the figure, the *x*-axis represents the progression of time during the insertion, while the *y*-axis denotes the occurrence of penetration events. The blue line indicates the ground truth actual penetration points observed during the experiment while the orange line shows the corresponding predictions generated by the LSTM model.

Visual comparison reveals that the LSTM model successfully identifies most of the actual penetration events with high temporal precision. The close alignment between predicted and actual events highlights the models strong classification accuracy and its potential for real-time surgical assistance. To evaluate the models performance, key classification metrics were computed. The LSTM model achieved a recall of 0.89, a precision of 0.90, and an F1 score of 0.85. These values reflect the models strong reliability and robustness in accurately detecting needle penetration events, highlighting its potential to support clinical decision-making during minimally invasive procedures, as summarized in Table 15.1.

Beyond statistical performance, a temporal analysis was conducted to examine the latency between actual and predicted penetration events. As illustrated in Fig. 15.4, the average delay between an observed penetration and its corresponding model prediction was approximately 90.23 milliseconds. This minimal lag demonstrates the models capacity to deliver near real-time feedback an essential feature for enhancing procedural precision and patient safety. Overall, the results

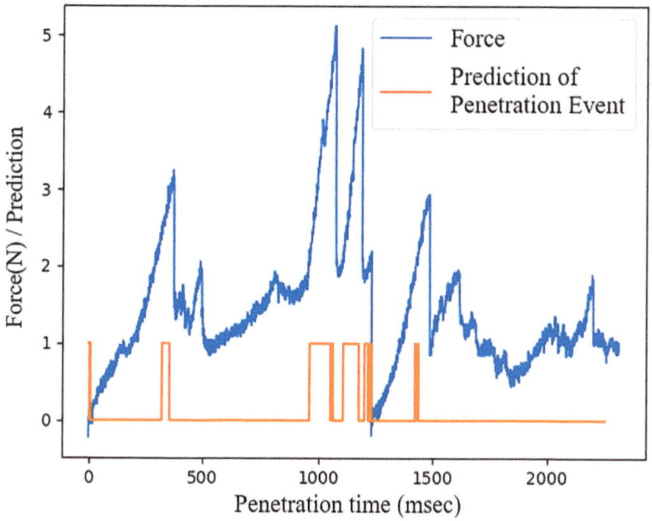

Fig. 15.3 Prediction of penetration episode through insertion force

Table 15.1 Performance evaluation of proposed LSTM network for training, validation and testing

Performance	Training	Validation	Testing
Metrics Values			
Accuracy	0.98	0.99	0.84
Recall	0.89	0.92	0.8
Precision	0.90	0.90	0.76
F1 score	0.85	0.95	0.73

affirm the LSTM models effectiveness in classifying needle-tissue interactions and suggest its promise as a real-time assistive tool for improving surgical outcomes in percutaneous interventions. Figure 15.5 illustrates the models throughput, defined as the number of data samples the LSTM network can process per second. Throughput, typically measured in samples per second, reflects the models ability to handle and analyze incoming data in real time. In this study, the LSTM achieved a throughput of 212.41 samples/sec, indicating its high processing efficiency.

A high throughput is particularly important in time-sensitive medical applications, where rapid decision-making is critical. The models ability to process sequences quickly enables faster surgical operations, which can significantly reduce the duration a patient remains under anesthesia. Shorter anesthesia times are associated with a lower risk of intraoperative complications and improved postoperative recovery, thereby enhancing overall surgical outcomes. Figure 15.6 shows the training and validation loss curves of the model over 500 epochs. Loss serves as a key indicator of model performance, with lower values signifying better predictive accuracy. At the beginning of training, both losses are relatively high due to the model initializing with random weights. As training progresses, the model learns from the data, resulting in a steady decline in training loss.

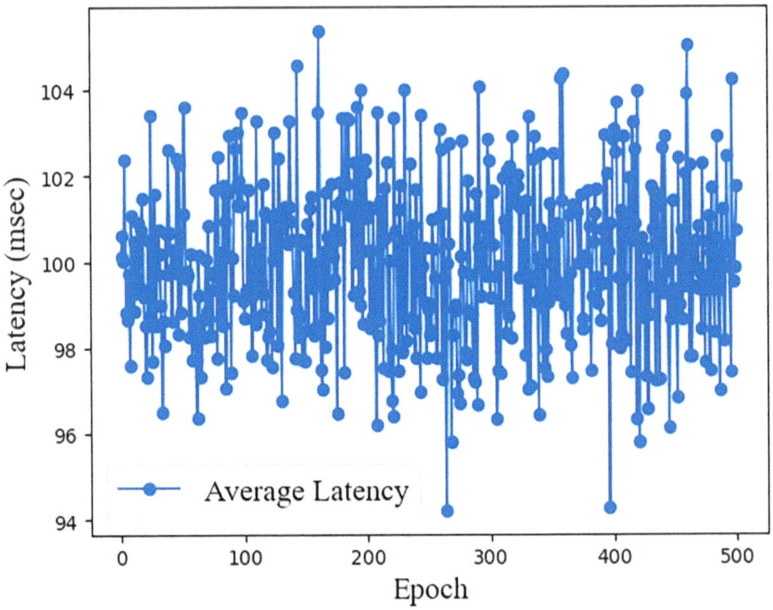

Fig. 15.4 Latency plot showing data condition over time

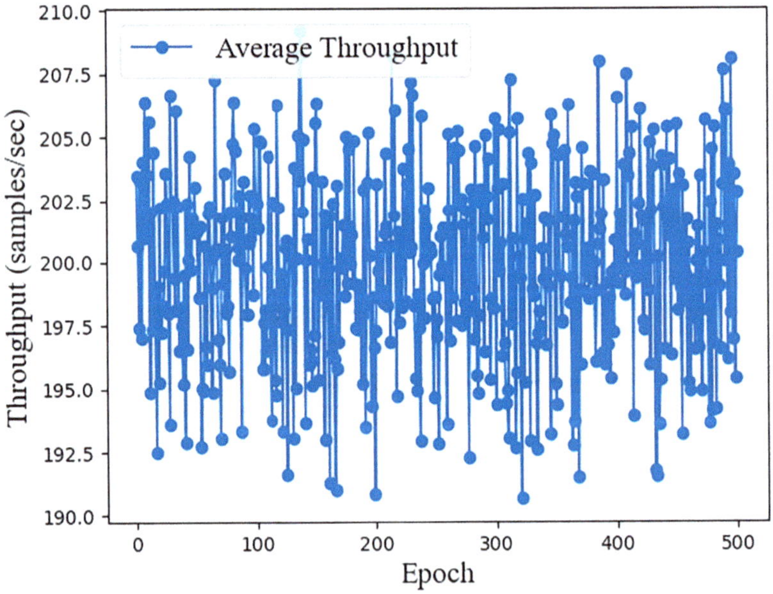

Fig. 15.5 Throughput of acquired data

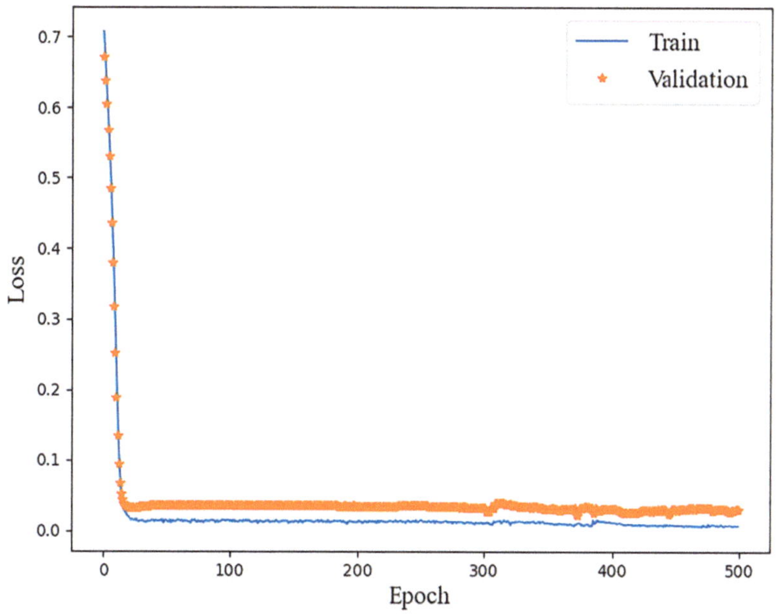

Fig. 15.6 Model loss propagation

Toward the end of the training process, the training loss converges to a minimal value, indicating successful learning. The validation loss also decreases and stabilizes, with only a slight gap between it and the training loss. This narrow gap suggests that the model maintains good generalization performance and is not overfitting to the training data, demonstrating robustness in handling unseen data. Figure 15.7 presents the models accuracy on both the training and validation datasets over the course of 500 epochs. The data was divided in a 70:30 ratio for training and validation, respectively. Throughout training, the model achieved a training accuracy of 0.998 and a validation accuracy of 0.995. The close alignment between these values suggests strong generalization capabilities and effective learning from the data.

The minimal difference between training and validation accuracy highlights the models robustness and resistance to overfitting. Furthermore, accuracy values stabilized as training progressed, and extending the training beyond 500 epochs showed no meaningful improvement, indicating that the model had fully converged.

Figure 15.8 displays the Autocorrelation Function (ACF) and Partial Autocorrelation Function (PACF), along with their respective confidence intervals two essential tools for analyzing time series data. The ACF illustrates the relationship between the current force measurement and its previous values over various time lags, revealing overall patterns of correlation. In contrast, the PACF isolates the direct relationship between the current value and its lagged values by removing the influence of intermediate lags. This provides a clearer understanding of the true lag-specific dependencies within the force data.

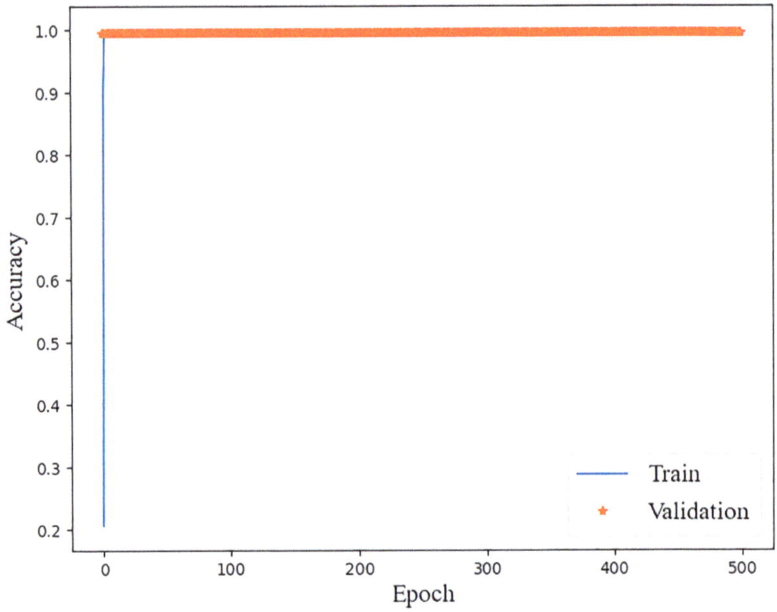

Fig. 15.7 Model accuracy

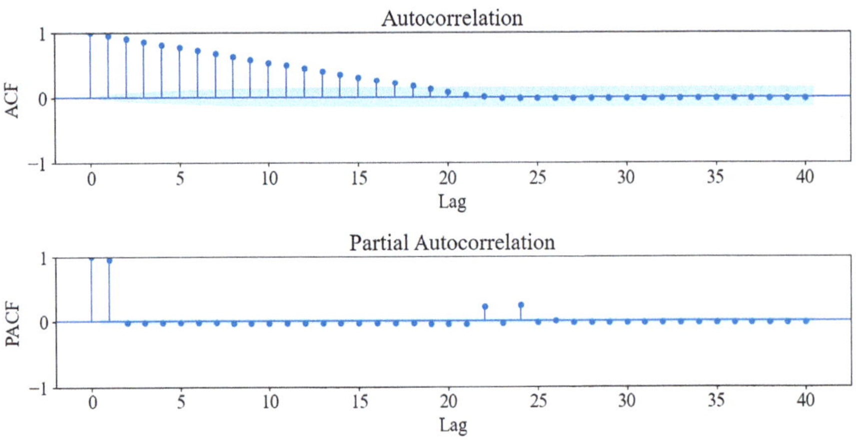

Fig. 15.8 Auto correlation function and particle auto correlation function for temporal data

15.3.2 Discussion

MIS critically depends on instantaneous monitoring and regulation systems [14, 15], where data is typically acquired as sequential streams such as sensor outputs from surgical tools, positional coordinates, and force feedback. LSTM networks are particularly well suited for handling this type of sequential data, as they are specifically designed to model temporal dependencies, making them ideal for applications involving dynamic, time sensitive tasks [10–13, 15].

Recent studies have explored multiple facets of needle tissue interaction, including control strategies, trajectory planning, and needle tracking. For instance, the study in [10] employed a Classification and Regression Tree (CART) algorithm to predict needle deflection. While decision trees can offer simplicity and interpretability, they often struggle to model complex nonlinear relationships. Their reliance on rigid node splitting rules leads to brittle models where minor changes in input features can cause large fluctuations in output, limiting their applicability in nuanced surgical environments.

In our previous work [11, 16], we have utilized a MLP model with an adaptive learning rate to predict tissue stiffness and needle deflection using preprocessed distributed load data. While this approach showed promise, it underscored the limitations of feedforward networks in capturing temporal patterns, highlighting the need for models like LSTM that can better leverage sequential dependencies inherent in MIS data.

Advanced and hybrid machine learning models are increasingly being integrated into MIS technologies. For example, the authors in [17] proposed an attention-guided CNN for real-time detection of surgical instruments in video frames. Similarly, [18] introduced a CNN-based classifier to identify different surgical tools. Although these vision-based techniques can enhance visual understanding in surgical environments, they often demand significant computational resources and involve extensive preprocessing pipelines. Moreover, their reliability may decrease with deeper tissue insertion, where visibility is reduced, making real-time analysis more challenging.

Needle penetration event detection using LSTM networks represents a user-centric approach to enhancing real-time surgical feedback in human-robot collaborative environments. LSTM networks, a type of RNN, are well-suited for interpreting time-series data generated during needle-tissue interaction. By learning from sequential force data, LSTM can detect subtle changes that indicate transitions between different tissue layers. This capability is crucial in designing intelligent systems that can inform or alert human operators such as surgeons or clinicians during minimally invasive procedures.

The proposed system offers precise depth estimation, achieving a resolution of just 0.25 mm, enabled by the integration of a stepper motor and prismatic robotic architecture. Each 1.8° motor step translates into a controlled 0.25 mm linear movement, ensuring highly accurate feedback to the user interface and minimizing unintended motion. This precision supports better alignment with the surgeons intent and improves user trust in the automation system.

From an HRI perspective, LSTM enhance the systems responsiveness by continuously analyzing sensor input and predicting needle penetration events in real time. This facilitates timely visual or haptic feedback, which can reduce cognitive load and support informed decision-making under complex conditions. By maintaining awareness of past interactions, the model improves continuity and coherence in the system's behavior essential features for user-friendly robotic assistance tools.

Moreover, LSTM networks are inherently adaptive, allowing them to handle input sequences of varying lengths. This is particularly useful in surgical procedures where complexity and duration differ from case to case. The adaptability of LSTM reduces the need for manual reconfiguration, promoting greater workflow flexibility and operational efficiency for healthcare professionals.

The integration of ACF and PACF analysis further supports system design by identifying how past penetration behavior influences current interaction dynamics. For instance, ACF reveals repetitive force patterns, while PACF isolates the most relevant previous inputs for modeling. This temporal insight informs more intuitive and effective HRI designs, such as predictive alerts or adaptive control mechanisms that respond proactively to user actions.

Finally, the proposed LSTM-based system was benchmarked against alternative architectures GRU, CNN, Transformers, and Diffusion Models to assess its suitability for medical HRI applications. LSTM consistently demonstrated strong performance in managing sequential dependencies, a critical requirement in MIS environments. These comparisons highlight the strengths of LSTMs in real-time interaction, low-latency prediction, and minimal user input demands, while also identifying areas where emerging models may offer future enhancements.

In summary, this study emphasizes how advanced machine learning models like LSTM can enhance HRI in surgical robotics by providing accurate, real-time, and adaptive support to medical practitioners thereby improving safety, precision, and user experience in high-stakes environments.

In HRI the unification of smart sensing and data-driven learning are pivotal for aiding safe, intrinsic, and conjoint process. This chapter presents the application of a LSTM framework to predict needle penetration events across multilayer tissue during clinical interventions. Precise identification of transitions between tissue layers is vital for enhancing clinician awareness and enhancing the safety and precision of automatic robotic assisted needle insertions. The proposed model utilizes real time force signals along the x, y, and z axes as input to the LSTM model, allowing it to learn temporal patterns associated with layer transitions. Unlike conventional threshold based methods, the learning based approach captures subtle variations in force profiles that correspond to different tissue properties, even under inter-patient variability.

Within an HRI framework, the output of the LSTM model can be used to provide context aware feedback to the human operator, either through haptic cues, visual indicators, or adaptive robot behavior. This enhances the users ability to make informed decisions during insertion, while also reducing the cognitive load and the risk of unintentional tissue damage. The system preserves human control while embedding autonomy in perception and decision support, aligning with the principles of shared autonomy. Experimental validation demonstrates a prediction accuracy of 95%, confirming the models reliability in identifying critical events during the procedure. Overall, this research highlights the potential of LSTM-based models to bridge the gap between low-level sensor data and high level human understanding, thereby fostering more intelligent and collaborative robotic systems. Such capabilities are particularly valuable in complex medical scenarios, where precision and safety are paramount, and where human expertise and machine intelligence must work seamlessly together.

15.4 Conclusion

This chapter introduces an LSTM-based neural network framework developed to predict penetration events during human-robot collaborative clinical procedures. By leveraging force measurements along the x, y, and z axes, the model enables real-time detection of transitions between different tissue layers information that is essential for providing meaningful feedback to the human operator. Accurate identification of these events enhances haptic perception, supports adaptive robotic responses, and contributes to safer, more precise interactions in variable tissue environments. Experimental validation indicates a high prediction accuracy of 95%, confirming the models effectiveness in dynamic HRI scenarios. These findings underscore the potential of LSTM-based learning systems to enrich intelligent control, support shared autonomy, and advance the next generation of machine learning-enhanced surgical assistance platforms.

References

1. T. Fisher, A. Hamed, P. Vartholomeos, K. Masamune, G. Tang, H. Ren, Z.T.H. Tse, Intraoperative magnetic resonance imaging conditional robotic devices for therapy and diagnosi, in *Proceedings of the Institution of Mechanical Engineers, Part H: Journal of Engineering in Medicine*, vol. 228, no. 3 (2014), pp. 303–318
2. M. Arbabtafti, M. Moghaddam, A. Nahvi, M. Mahvash, B. Richardson, B. Shirinzadeh,Physics-based haptic simulation of bone machining. IEEE Trans. Haptics **4**(1), 39–50 (2011)
3. M. Abayazid, R.J. Roesthuis, R. Reilink, S. Misra, Integrating deflection models and image feedback for real-time flexible needle steering. IEEE Trans. Robot. **29**(2), 542–553 (2013)
4. S. Xu, J. Kruecker, P. Guion, N. Glossop, Z. Neeman, P. Choyke, A.K. Singh, B.J. Wood, Closed-loop control in fused MR-TRUS image-guided prostate biopsy. Med. Image Comput. Assist. Interv. **10**, 128–135 (2007)
5. S. Kruecker, J. Guion, P. Glossop, N. Neeman, Z. Choyke, P. Singh, A.K. Wood, B.J., Closed-loop control in fused MR-TRUS image-guided prostate biopsy. Med. Image Comput. Assist. Interv. **10**, 128–135 (2007)
6. A.M. Okamura, C. Simone, M.D. O'Leary, Force modeling for needle insertion into soft tissue. IEEE Trans. Biomed. Eng. **51**(10), 1707-16 (2004)
7. S. Hochreiter, J. Schmidhuber, Long short-term memory. Neural Comput. **9**(8), 1735–1780 (1997)
8. S. Mirshekarian, R. Bunescu, C. Marling, F. Schwartz, Using LSTMs to learn physiological models of blood glucose behavior. Annu. Int. Conf. IEEE Eng. Med. Biol Soc. 2887-2891 (2017). https://doi.org/10.1109/EMBC.2017.8037460. PMID: 29060501; PMCID: PMC7888239
9. N. Agarwal, A.K. Yadav, A. Gupta, M. Felix Orlando, Real-time needle tip localization in 2D ultrasound images using Kalman filter, in *IEEE/ASME International Conference on Advanced Intelligent Mechatronics (AIM), Hong Kong, China* (2019), pp. 1008–1012

10. B. Behera, M. Felix Orlando, R.S. Anand, Prediction of bevel-tip needle deflection using CART based decision tree, in *Third International Conference on Power, Control and Computing Technologies (ICPC2T), Raipur, India* (2024), pp. 769–774
11. B. Behera, M. Felix Orlando, R.S. Anand, Prognosis of tissue stiffness through multilayer perceptron technique with adaptive learning rate in minimal invasive surgical procedures. IEEE Trans. Med. Robot. Bionics **6**(2), 769–781 (2024)
12. Y. Li, H. Li, Penetration event identification based on neural network for needle tip location in robot assisted lumbar puncture surgery. IEEE Robot. Autom. Lett. **9**(1), 343–350 (2024)
13. I. Goodfellow, Y. Bengio, A. Courville, *Deep Learning* (MIT Press, Cambridge, 2016)
14. S. Hans, M. Felix Orlando, Robust control of a bevel-tip needle for medical interventional procedures. IEEE/CAA J. Autom. Sin. **7**(1), 244–256 (2020)
15. S. Hans, M. Felix Orlando, Control of a flexible bevel-tipped needle using super-twisting controller based sliding mode observer. ISA Trans. **109**, 186–198 (2021)
16. B. Behera, V. Babu, S. Kumar, M. Felix Orlando, R.S. Anand, Neural network based predictive model for bevel-tip needle deflection in percutaneous interventions, in *28th International Conference on System Theory, Control and Computing (ICSTCC), Romania* (2024)
17. P. Shi, Z. Zhao, S. Hu, F. Chang, Real-time surgical tool detection in minimally invasive surgery based on attention-guided convolutional neural network. IEEE Access **8**, 228853–228862 (2020)
18. J. Jaafari, S. Douzi, K. Douzi, et al., Towards more efficient CNN-based surgical tools classification using transfer learning. J. Big Data **8**, 115 (2021)

Chapter 16
Reinforcement Learning-Driven Brain-Computer Interfaces for Neurorehabilitation: Trends, Methods, and Challenges

Parthan Olikkal, Sruthi Sundharram, Sina Makhdoomi Kaviri, Shivansh Sharma, and Ramana Vinjamuri

Abstract Neurological injuries such as stroke, spinal cord injury, and neurodegenerative diseases often result in long-term motor impairments that challenge conventional rehabilitation methods. Reinforcement learning (RL), when integrated with biosignal-driven brain-computer interfaces (BCIs), offers a promising framework for personalized, adaptive neurorehabilitation. This chapter presents a comprehensive review of RL applications in BCIs across three domains: electroencephalography (EEG), electromyography (EMG), and multimodal systems. In EEG-based rehabilitation, RL enables closed-loop systems that adapt therapy in real-time based on neural oscillatory patterns, promoting neuroplasticity and motor recovery. EMG-based systems leverage RL for precise gesture recognition, adaptive exoskeleton control, and immersive virtual reality therapy. Multimodal approaches—such as EEG-fMRI integration—enhance the spatial and temporal resolution of brain state estimation and support novel neurofeedback paradigms. Through thematic categorization, case studies, and architectural insights, this chapter outlines the current state of RL-based neurorehabilitation, identifies key challenges and highlights future directions for scalable, intelligent, and patient specific intervention strategies.

Keywords Electroencephalography · Electromyography · Reinforcement learning · Virtual reality · Neurorehabilitation · Clinical therapy · Closed-loop system

P. Olikkal · S. Sundharram · S. M. Kaviri · S. Sharma · R. Vinjamuri (✉)
Vinjamuri Lab, Department of Computer Science and Electrical Engineering, University of Maryland Baltimore County, Baltimore, MD, USA
e-mail: polikka1@umbc.edu; ssundha1@umbc.edu; sinam1@umbc.edu; shivans1@umbc.edu; rvinjam1@umbc.edu

R. Vinjamuri (ed.), *Bridging the Gap between Mind and Machine*,
https://doi.org/10.1007/978-3-032-06713-5_16

16.1 Introduction

Restoring motor functions after neurological injury remains a central challenge in neurorehabilitation. Conditions such as stroke, spinal cord injury, traumatic brain injury, and neurodegenerative diseases often leave patients with significant impairments that disrupt daily living. Rehabilitation traditionally relies on therapist-guided, repetitive motor training, which effectively is often limited. The diverse nature of patients differing in terms of motivation, cognitive capacity, and neuroplasticity makes it difficult to achieve consistent outcomes using rule-based protocols.

In recent years, advances in brain-computer interfaces (BCIs) and artificial intelligence (AI) have created new opportunities to develop a more individualized rehabilitation strategy. One particularly promising area involves integrating reinforcement learning (RL), a type of machine learning where an agent learns to take actions by interacting with their environment. By combining RL with BCIs that leverage biosignals to capture real-time activity, closed-loop rehabilitation systems can be designed to adapt interventions dynamically based on patients' response. Such systems hold considerable promise in delivering therapy that is responsive to the patient's moment-to-moment state, potentially enhancing engagement and promoting more effective motor recovery [1–3].

This chapter reviews the state-of-the-art use of reinforcement learning in BCI-driven neurorehabilitation across three domains: (i) EEG-based systems, which enable adaptive closed-loop feedback through neural patterns; (ii) EMG-based systems, which decode motor intent for prosthetic control, gesture recognition, and VR rehabilitation; and (iii) multimodal frameworks, such as EEG-fMRI integration, which provide rich spatial-temporal context for RL-guided neurofeedback and cognitive control as shown in Fig. 16.1. We examine the theoretical, practical and clinical potential of these approaches. In addition to motor rehabilitation, emerging applications of RL extend to immersive VR-based training, cognitive control, and emotional regulation through neurofeedback, enabling holistic rehabilitation paradigms. Figure 16.2 shows the growth of publication data of research on biosignals reinforcement learning from 2005 to 2024 on PubMed, National Library of Medicine. The remainder of this chapter is organized as follows: Sect. 16.2 introduces the formal RL problem and its biological parallels. Section 16.3 discusses EEG-RL systems, Sect. 16.4 covers EMG-RL applications, and Sect. 16.5 explores multimodal integration. Finally, Sect. 16.6 highlights key challenges, open problems, and future directions.

16.2 Reinforcement Learning Problem

Reinforcement Learning is an area of machine learning focused on training agents to make optimal decisions through interactions within dynamic environments. The foundational concept of RL revolves around an agent continuously interacting with

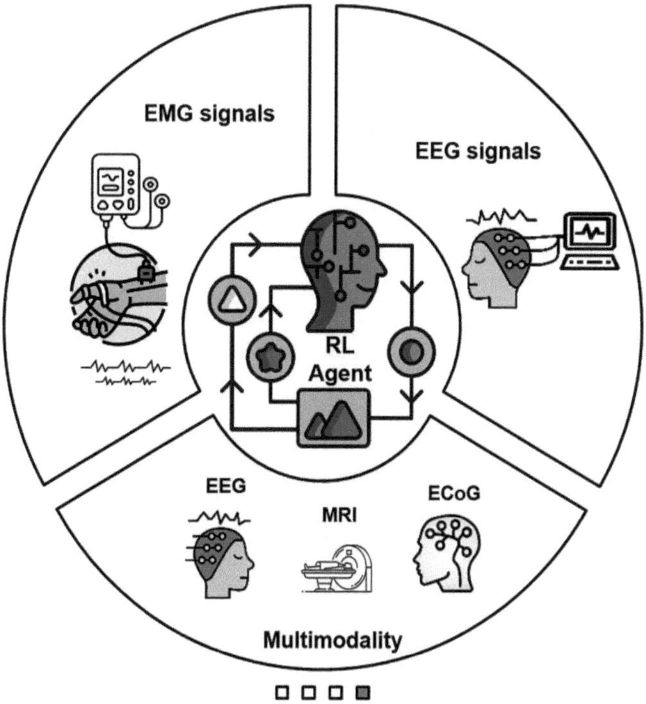

Fig. 16.1 Systematic organization of literature on electromyography (EMG), electroencephalography (EEG) and multimodal approaches highlighting the role of reinforcement learning in advancing neurorehabilitation

an environment, where it selects actions and receives feedback in the form of rewards or penalties. This trial-and-error method enables the agent to learn a policy aimed at maximizing cumulative rewards over time [4, 5]. The RL problem can be formally described using a Markov Decision Process (MDP). An MDP consists of a set of states S, a set of actions A, state transition probabilities $T(s,a,s')$, a reward function $\mathcal{R}(s,a)$, and discount factor $\gamma \in [0, 1]$, which determines the importance of future rewards.

$$MDP = (S, A, T, R, \gamma) \tag{16.1}$$

The agent's goal is to find an optimal policy π^*, mapping states to actions, which maximizes the expected discounted cumulative reward:

$$\pi^* = argmax_\pi \mathbb{E}_\pi \left[\sum_{t=0}^{\infty} \gamma^t \mathcal{R}(s_t, a_t) \right] \tag{16.2}$$

Central to solving RL problems are the value functions, namely the state-value function $V^\pi(s)$ and the action-value function $Q^\pi(s, a)$ defined as follows:

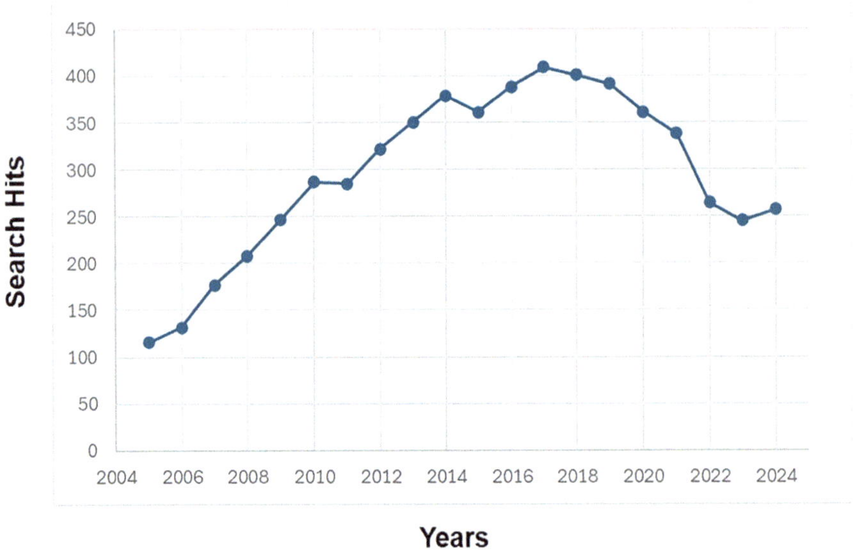

Fig. 16.2 A search for "biosignals reinforcement learning" on PubMed, National Library of Medicine, reveals a rapidly growing body of research, indicating increasing interest and activity in this interdisciplinary field

$$V^{\pi}(s) = \mathbb{E}_{\pi}\left[\sum_{t=0}^{\infty} \gamma^{t}\mathcal{R}(s_t, a_t) \mid s_0 = s\right] \tag{16.3}$$

$$Q^{\pi}(s, a) = \mathbb{E}_{\pi}\left[\sum_{t=0}^{\infty} \gamma^{t}\mathcal{R}(s_t, a_t) \mid s_0 = s, a_0 = a\right] \tag{16.4}$$

RL algorithms broadly divide into two categories: model-free and model-based methods [6]. Model-free approaches learn directly from interaction with the environment without explicitly modeling transition probabilities. A notable model-free algorithm is Q-learning [7], which iteratively updates the action-value function based on temporal-difference errors. The updating rule of Q-learning is formulated as follows:

$$Q_{t+1}(S_t, A_t) \leftarrow Q_t(S_t, A_t) + \alpha\left[\mathcal{R}_{t+1} + \gamma \max_a Q_t(S_{t+1}, a) - Q_t(S_t, A_t)\right] \tag{16.5}$$

In contrast, model-based RL explicitly models the environment's dynamics, including state transitions and reward functions, providing superior sample efficiency and generalization. This makes model-based methods highly suitable for applications in robotics and dynamic environments [8, 9].

16.2.1 Reinforcement Learning in the Context of Human Motor Learning

Reinforcement learning process resembles how humans acquire new motor skills through repeated practice, feedback, and gradual adjustment. In neurorehabilitation, RL has been applied in several areas, such as optimizing control strategies for robotic exoskeletons, adjusting therapy difficulty levels, and modifying training protocols based on a patient's ongoing performance and progress. Unlike rule-based systems that operate under fixed protocols, RL-based approaches are capable of adapting to the changing needs of each patient in real-time. For example, RL algorithm can take into account variations in a patient's fatigue level, motivation, or functional ability during therapy session, and modify the intervention accordingly. This dynamic capability makes RL particularly valuable for rehabilitation, where patients often present with diverse and evolving needs. By tailoring therapy to the individual, RL has the potential to improve outcomes and enhance patient engagement throughout the rehabilitation process.

16.2.2 Biological Inspiration: The Cerebellum as a Reinforcement Learning System

The cerebellum has long been recognized for its role in motor coordination and sensorimotor integration, but emerging evidence highlights its integral involvement in RL processes, positioning it as a biologically inspired model for RL systems. RL fundamentally involves an agent acquiring sensory information, selecting actions, receiving evaluative feedback, and modifying behavior to maximize rewards through trial-and-error learning [10]. The cerebellum's precise anatomical organization, as proposed by Eccles et al. [11] supports its ability to store and integrate information essential for such adaptive learning. Classical conditioning paradigms, such as the rabbit eye-blink response, have provided foundational evidence of cerebellar-mediated RL mechanisms [10]. Beyond motor learning, recent research has demonstrated the cerebellum's critical role in encoding reward prediction errors (RPEs), a central component of RL. For instance, studies using EEG have shown that disruption of cerebellar output, through stroke or transcranial magnetic stimulation, blunts RPE processing in the forebrain regions traditionally associated with RL, such as the anterior cingulate cortex and striatum [12]. Moreover, theta and delta oscillations in EEG have been found to reflect anticipatory RPE-like signals during decision-making processes, further implicating cerebellar involvement in both motor and cognitive domains of RL [13]. These findings emphasize that the cerebellum, through its predictive coding, error correction, and evaluative feedback functions, embodies many principles central to RL algorithms, offering rich inspiration for biologically informed AI models.

16.3 EEG Based Reinforcement Learning

16.3.1 Role of EEG in Monitoring Neural States

Electroencephalography (EEG) is a widely used, non-invasive technique that records electrical activity from the brain with high temporal resolution. This makes EEG especially well-suited for real-time applications in neurorehabilitation, where continuous monitoring of brain activity can provide valuable insights into a patients' neural state. In motor rehabilitation, EEG recordings often focus on detecting patterns such as event-related desynchronization (ERD) and event-related synchronization (ERS), which reflect changes in neural oscillations linked to motor-related brain activity. These oscillations typically occur within the *mu* (8–12 Hz) and *beta* (13–30 Hz) frequency bands over sensorimotor regions and are associated with processes such as motor planning, movement execution, and motor imagery. Importantly, even in patients with limited or no physical movement, EEG can capture these neural markers, allowing clinicians or researchers to infer motor intent and engagement. This capability enables the development of BCIs that can translate brain activity into control signals for assistive devices or provide neurofeedback to the user. Because EEG is non-invasive, relatively affordable, and portable, it has become a practical tool for long-term use in clinical neurorehabilitation, especially in settings where frequent and adaptive monitoring of brain activity is needed [14, 15].

16.3.2 Integrating EEG and RL for Adaptive Therapy

Integrating EEG with RL offers a promising approach to building adaptive neurorehabilitation systems that are highly responsive to individual patient needs. Continuous EEG monitoring provides real-time information about the patient's level of cognition and motor engagement. This information can be fed into an RL algorithm, which in turn can adjust various aspects of the therapy such as task difficulty, feedback timing, or the degree of robotic assistance in order to maintain an optimal level of challenge and engagement [16, 17].

In addition to tailoring therapy, the use of reward-based feedback mechanisms within RL frameworks can help sustain patient motivation, which is a critical factor for long-term participation in rehabilitation. For example, gamified rehabilitation platforms that provide positive reinforcement when patients produce desired neural or behavioral responses have been shown to increase engagement and support neuroplastic changes that are essential for motor recovery. Several studies combining EEG-based BCIs with RL have reported not only improvements in motor function but also enhanced cognitive engagement during therapy, both of which are important for achieving meaningful functional gains [18].

Overall, the combination of real-time brain monitoring with adaptive, reward-driven learning algorithms represents an innovative direction for developing the next generation of neurorehabilitation systems. These systems have the potential to provide highly personalized, flexible interventions that can better address the complex and variable nature of motor recovery in patients with neurological injuries.

16.3.3 Closed-Loop BCI-RL Framework

Closed-loop BCI systems have become a cornerstone of modern neurorehabilitation, offering adaptive, real-time feedback that directly responds to a user's ongoing brain activity. These systems rely on EEG to continuously capture neural signals, which are then processed to extract meaningful features such as ERD/ERS or specific frequency band powers [19, 20]. These features serve as inputs to RL algorithms, which dynamically select and adjust therapeutic interventions such as modulating robotic assistance, delivering visual or haptic feedback, or administering functional electrical stimulation as indicated in Fig. 16.3.

What makes closed-loop BCIs especially powerful is their ability to adapt in real time, shaping interventions based on immediate neural feedback. This approach reinforces desirable brain activity patterns and promotes neuroplastic changes essential for recovery [19].

In stroke rehabilitation, closed-loop BCIs are particularly valuable when traditional therapies have limited effectiveness. By detecting volitional brain activity often through motor imagery, these systems engage patients in goal-directed tasks even when overt movement is not possible. Over repeated sessions, RL agents refine their control policies to maximize therapeutic benefit, guided by EEG-derived performance metrics such as increased beta-band modulation or enhanced detection of motor intent. Studies have demonstrated that these adaptive interventions can help overcome maladaptive behaviors like learned non-use and reactive motor cortex areas in chronic stroke patients [21, 22].

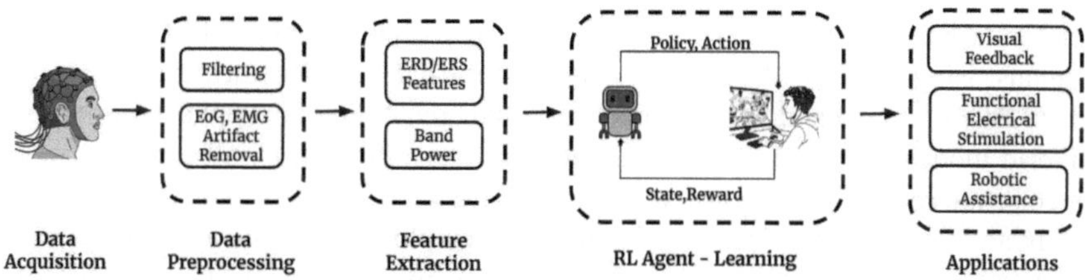

Fig. 16.3 Closed-loop BCI-RL framework for neurorehabilitation is illustrated here. EEG signals are acquired, preprocessed, and transformed into features. A RL agent selects therapeutic actions like visual feedback, FES, or robotic assistance based on task performance, with policy updates occurring over repeated sessions to enhance recovery

A critical feature distinguishing restorative BCIs from purely assistive ones is their ability to induce lasting functional brain changes. This involves operant learning mechanisms where users progressively modulate specific neural oscillations, especially sensorimotor beta rhythms which correlate to improved motor outcomes. For instance, volitional increase in beta-band activity has been linked to heightened corticospinal excitability and motor learning in both healthy individuals and stroke survivors [20, 23, 24].

Beyond motor recovery, closed-loop BCIs are expanding into cognitive rehabilitation. A notable example involves a system developed for individuals with autism spectrum disorder (ASD), which used EEG-detected error-related potentials (ErrPs) during observation of a digital agent's actions. The agent adapted its behavior via RL informed by these neural error signals, while users implicitly trained their error-monitoring processes. This method shows promise for enhancing cognitive function in populations with impaired feedback processing, such as ASD [25].

Overall, closed-loop BCI-RL frameworks offer versatile, personalized rehabilitation approaches that dynamically adjust to the user's internal brain states, fostering engagement and promoting functional recovery across a range of neurological conditions.

16.3.4 EEG and RL for BCI Optimization and Feature Selection

RL has gained increased attention as a method to improve the performance of EEG-based BCIs. Its ability to adaptively learn from interaction and improve decoding accuracy, providing personalized feedback, handling the variability inherent in EEG data makes it suited for BCI. One area of active research has involved combining RL with graph neural networks (GNNs), which are capable of modeling spatial relationships across EEG channels. Aung et al. [26] demonstrated that incorporating GNNs into RL frameworks led to substantial improvements in motor imagery classification accuracy by better capturing the spatial dependencies between electrodes.

Hybrid approaches have also shown promise in enhancing decoding for complex motor tasks. Fu et al. [27] developed a system that combined EEG and EMG signals using a GCN-LSTM architecture optimized with RL, achieving effective decoding of fine hand movements for stroke rehabilitation applications. These hybrid models leverage both neural and muscular information, allowing for more precise interpretation of motor intent.

In addition to motor decoding, RL has been employed to improve neurofeedback protocols. Xu et al. [28], applied RL to adaptively modulate EEG alpha activity during mindfulness training, enabling a feedback system that adjusts in real-time without relying on manually defined rules. Another direction involves using ErrPs as implicit feedback signals. Xu et al. [29], demonstrated that incorporating ErrPs

Traditional EMG-based control systems have predominantly utilized supervised learning algorithms to classify discrete gestures or predict continuous movements. While these methods have demonstrated efficacy in controlled environments, they often lack adaptability and robustness in dynamic, real-world situations. Challenges such as inter-subject variability, electrode displacement, and muscle fatigue can significantly affect performance. To address these limitations, reinforcement learning (RL) has been introduced as a promising alternative, enabling systems to learn optimal control policies through interaction with the environment and feedback mechanisms [6, 39].

The integration of deep learning with reinforcement learning, known as Deep Reinforcement Learning (DRL), has further enhanced the capability to manage high-dimensional and temporally complex data. In the context of EMG-based applications, DRL facilitates adaptive control strategies that can personalize assistance levels in real-time. Moreover, the convergence of EMG signal processing with immersive technologies like Virtual Reality (VR) has opened new avenues for upper limb rehabilitation [40]. VR environments, when combined with EMG-based feedback, have been shown to enhance motor imagery training and promote neuroplasticity in post-stroke rehabilitation scenarios [41]. These systems provide engaging and interactive platforms that can adapt to the user's performance, thereby improving the efficacy of rehabilitation protocols.

This section delves into the current state-of-the-art research at the intersection of using EMG signal, reinforcement learning and virtual reality. As shown in Fig. 16.4 we categorize literature into three main domains– Gesture Recognition and Control, Rehabilitation and Assistive Systems, Immersive Virtual Reality Environments.

16.4.1 Gesture Recognition Using EMG and RL

EMG based hand gesture recognition is a foundational technique in human-robot interaction, enabling intuitive control of prosthetic limbs, robotic systems, and virtual interfaces. By decoding muscle activity into distinct gesture classes, EMG systems offer a non-invasive and responsive means to interpret user intent in real-time. In this section, we explore how RL algorithms enhance the decoding of EMG signals for precise gesture recognition and control in prosthetic and robotic applications.

(a) Gesture Recognition

Recent advancements in EMG-based hand gesture recognition have explored the integration of RL techniques to enhance classification accuracy and adaptability. A foundational experiment [42] applied dueling DQN atop convolutional neural network for classifying six discrete hand movements, significantly improving robustness over class supervised methods across ten subjects. Extending this, Caraguay et al. [43] evaluated both DQN and Double DQN—with and without an

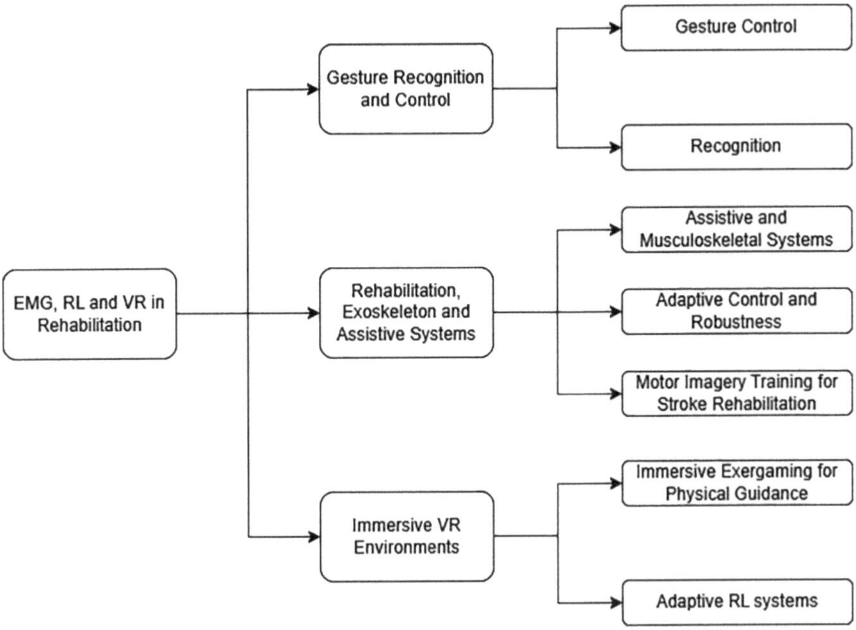

Fig. 16.4 Thematic organization of research domains at the intersection of Electromyography (EMG), Reinforcement Learning (RL) and Virtual Reality (VR) in rehabilitation

added LSTM layer—on the EMG-EPN-612 dataset, achieving peak classification accuracy of 90% and recognition accuracy of 82.5% demonstrating the potential of RL in capturing the temporal dynamics of EMG signals. Similarly, another work [44] developed an RL-based agent capable of learning to classify and recognize diverse hand gestures using DQN algorithm incorporating both EMG and IMU information, highlighting the versatility of RL in handling complex gesture pattern. Comparative analyses between RL and traditional supervised learning approaches have provided insights into their respective strengths. Berman et al. [45] compared RL agents and supervised learning neural networks for EMG-based decoding of continuous movements, revealing that RL-based kinematic decoders were more robust to changes in movement speeds between training and testing data, suggesting their potential for applications requiring adaptability to dynamic conditions.

(b) Gesture Control

Recent advances have demonstrated the effectiveness of deep RL agents—particularly DQN—in translating EMG signals into robust control commands for robotic platforms. For example, Cruz et al. [46] implemented a DQN-based hand gesture recognition system using EMG-IMU inputs from the Myo armband, achieving classification accuracy of 97.45% and recognition accuracy of 88.05%. This system was successfully tested in real-world control tasks, including maneuvering a 3-DOF tandem helicopter and a 6-DOF UR5 manipulator under

PID control, demonstrating both high responsiveness and robustness in control applications.

Beyond gesture-based control, RL has been effectively applied to nuanced motion predictions and multitasking finger control. Gardner et al. [47] integrated a CNN with an A2C actor-critic algorithm to simultaneously predict wrist flexion, extension, supination, and pronation from EMG signals. In parallel, Freitag et al., fine-tuned a supervised myoelectric classifier with RL during human-in-the-loop gameplay, doubling simultaneous finger-movement decoding accuracy and achieving a 39% improvement in human-machine task performance [48]. Moreover, a neural RL classifier [49] was able to identify forearm EMG-driven finger movements for typing, attaining up to 99% accuracy per channel and 92.7% across ten users, highlighting RL's adaptability for fine-grained motor intent decoding.

16.4.2 Rehabilitation and Assistive Systems

Rehabilitation and assistive systems, such as exoskeletons, benefit significantly from control strategies that can personalize assistance based on individual user needs. RL has emerged as a powerful tool in this domain, enabling systems to learn optimal assistance strategies through interaction with the user. In this section, we investigate the role of RL in personalizing rehabilitation strategies and controlling assistive devices like exoskeletons based on EMG inputs.

(a) Assistive Exoskeleton and Musculoskeletal Systems

RL has shown strong potential in developing EMG-driven, personalized assistance for musculoskeletal systems and exoskeletons. Oghogho et al. [50] introduced a real-time Twin Delayed Deep Deterministic Policy Gradient (TD3) deep-RL framework that dynamically adjusts exoskeleton assistance gains based on EMG and motion error, significantly reducing user muscle effort in point-to-point reaching movements. Similarly, Berman et al. [51] proposed a rapid RL-based personalized framework that tunes a generic musculoskeletal model to individual users—model optimization took under one second, delivering more accurate joint-angle estimation than traditional methods. Pelosi et al. [52] extended this to interactive rehabilitation using Q-learning to adapt VR-based reaching exercises to patient-specific capabilities. Additionally, Wu et al. [53] demonstrated that RL algorithms—particularly PPO—could estimate wrist and finger joint moments from EMG and kinematic data with accuracy comparable to inverse dynamics models, laying the foundation for RL-based biomechanical inference.

(b) Adaptive Control and Robustness

Beyond assistance, RL has been instrumental in creating robust and adaptable control frameworks for noisy or variable EMG environments. An actor-critic RL algorithm was employed to reduce contamination in EMG signal classification, improving reliability in real-word conditions [54]. Notably, Pilarski et al. [55] introduced an online human-in-the-loop training framework where users co-adapt with actor-critic-controlled myoelectric prosthesis, enabling real-time policy refinement and personalized control performance through direct user interaction. Campbell et al. [56] further advanced this concept with context-informed incremental learning, which adapts myoelectric control policies on the fly, ensuring sustained performance amidst EMG signal drift or fatigue. Complementing this, Garenfeld et al. [57] developed a compact closed-loop system integrating simulation, recording, and RL-driven adaptation, showcasing a complete hardware-in-the-loop pipeline for stable and resilient myoelectric interfaces.

These advancements demonstrate the potential of RL in enhancing the adaptability and personalization of rehabilitation and assistive systems, leading to more effective and patient-tailored interventions.

16.4.3 Immersive Virtual Reality Environments

The integration of EMG signals and RL with immersive virtual reality (VR) environments has opened new avenues for rehabilitation, offering engaging and adaptive training experiences. By leveraging real-time EMG feedback, VR systems can tailor rehabilitation exercises to individual users, enhancing motivation and effectiveness. In this section, we examine the integration of EMG-driven control within VR settings to create adaptive and engaging rehabilitation experiences.

(a) EMG-enhanced Motor Imagery Training for Stroke Rehabilitation

Several studies have explored the integration of EMG-based real-time feedback within immersive VR to enhance motor imagery (MI) training for post-stroke rehabilitation [40]. Lin et al. [58] developed a VR system featuring basketball shooting exercises, where real-time EMG metrics informed users' performance, notably improving muscle engagement and promoting stronger event-related desynchronization in both alpha and beta EEG bands when MI was paired with action observation. Huang et al. [59] further validated this approach, demonstrating that EMG-informed VR MI systems led to significantly enhanced muscle strength and exercise engagement across multiple patient groups. Complementing these efforts, Yang et al. [60] designed a virtual reality rehabilitation system where EMG signals controlled a hand avatar, offering gesture-based feedback during interactive tasks. Likewise, a modular VR platform proposed by Al-Jumaily et al. [61] used EMG-driven control to simulate prosthetic interactions and therapeutic exercises, supporting diverse upper-limb motion tasks such forearm pronation and supination, wrist extension and flexion, radial and ulnar deviation, ball grip and release, hand open and rest states in real-time.

Collectively, these works suggest EMG-backed VR MI systems improve patient motivation and enable precise quantification of training participation.

(b) Immersive Exergaming with Deep RL for Physical Guidance

Building on VR's motivational framework, Elor et al. [62], introduced a DRL—powered exergame that employed a virtual "ghost arm" to guide players during multi-joint movements in an immersive environment. Using Proximal Policy Optimization (PPO) and Generative Adversarial.

Imitation Learning (GAIL), the system learned to provide both cooperative and competitive guidance, showing early evidence of RL agents effectively supporting selective motor exercise in VR. Expanding on this concept, multiple studies have demonstrated how gamification can enhance motor engagement and compliance in upper-limb rehabilitation [63]. For example, mixed-reality systems have been designed to gamify upper limb rehabilitation therapeutic routines, increasing user motivation through reward-driven tasks and interactive virtual environments [64]. In patients with C7 spinal cord injury [65], myoelectric gaming approaches have shown promise as a motivational and complementary rehabilitation tool, enabling real-time biofeedback and quick adaptation to control interfaces. Spinal cord injured patients were able to demonstrate comparable scores to control subjects, supporting its use in regular therapy routines. Additionally, RL agents have been employed as intelligent movement guides within VR, nudging users to specific spatial targets in rehabilitation scenarios [66]. In post-stroke context through "Project Star Catcher", an immersive VR game that uses psychological constraint and reward mechanics to encourage use of the impaired limb without physical restraints. The system demonstrated high user engagement, increased weak-arm usage suggesting strong potential for enhancing compliance and personalizing therapy protocols in neurorehabilitation [67]. These works emphasize the potential of combining reinforcement learning and gamified VR to deliver personalized, engaging, and adaptive rehabilitation experiences.

(c) Adaptive RL systems

Recent progress has focused on continuous adaptation of EMG-driven controllers within immersive VR environments, leveraging employing RL to personalize user-machine interaction in real-time. Brenneis et al. [68] conducted a pilot study using continually learning RL agents in to predict and respond dynamically to user behavior, showing improved interaction fluency and enhanced user intent and system feedback alignment. Building on this, Yu et al. proposed a [69] user-centric, heterogeneous-action deep RL model tailored for VR applications in wireless metaverse contexts, aiming to support diverse rehabilitation needs through scalable and latency-aware learning mechanisms. Complementing these adaptive control approaches, Lee et al. [70] introduced an evidence-based intelligent VR training system for assessing upper-limb motor function. Their framework combined sensor-driven motion tracking with adaptive scoring metrics to provide quantifiable, personalized evaluations of stroke rehabilitation progress, reinforcing the role of intelligent feedback in next-generation VR therapy systems.

16.5 Multimodal BCIs and Reinforcement Learning

16.5.1 Enhancing EEG with Multimodal Integration

Conventional scalp EEG provides millisecond-scale tracking of neural dynamics but is hampered by volume conduction and limited sensor density [71, 72]. To overcome these constraints, contemporary datasets increasingly combine EEG with complementary biosignals:

(a) Structural and diffusion MRI co-registration
    High-resolution T1/T2 volumes plus diffusion-weighted MRI supply subject-specific conductivity priors and white-matter orientation, improving forward models and source localization [73–76].
(b) Peripheral physiology (EOG, EMG, ECG, GSR)
    Simultaneous ocular, muscular, cardiac and electrodermal recordings permit regression of artifacts and yield behavioral state labels for sleep, workload or affective studies [77–81].
(c) Optical and ultrasonic hemodynamics
    Combine EEG with near-infrared spectroscopy (EEG-NIRS) or transcranial function ultrasound to link fast electrical events to slower vascular responses at the cortical surface [82–84].
(d) Motion capture and inertial sensing (MoBI)
    Head-mounted inertial units and infrared marker systems track full-body kinematics, enabling artifact-aware analysis of locomotion and natural interaction in mobile brain/body imaging [85–88].
(e) High-density sensor arrays and hybrid caps
    Arrays of 256–512 electrodes, sometimes integrated with dry or water-based sensors, increase spatial sampling and improve conductivity model fitting when paired with individual MR head meshes [89–91].

Each additional data stream contributes orthogonal priors—tissue geometry, artifact markers, hemodynamic context or biomechanical constraints—that ameliorate the ill-posed EEG inverse problem and boost signal-to-noise ratios across cognitive, clinical and mobile paradigms.

16.5.2 Why Combine EEG and fMRI?

EEG and fMRI are widely used non-invasive neuroimaging techniques that offer complementary features: EEG provides high temporal resolution, while fMRI yields high spatial resolution. This complementarity has motivated their combined use in various domains, including neurofeedback, BCI and cognitive neuroscience [92]. In the context of neurofeedback, simultaneous EEG-fMRI acquisition allows for the investigation of the temporal dynamics of neuronal activity alongside

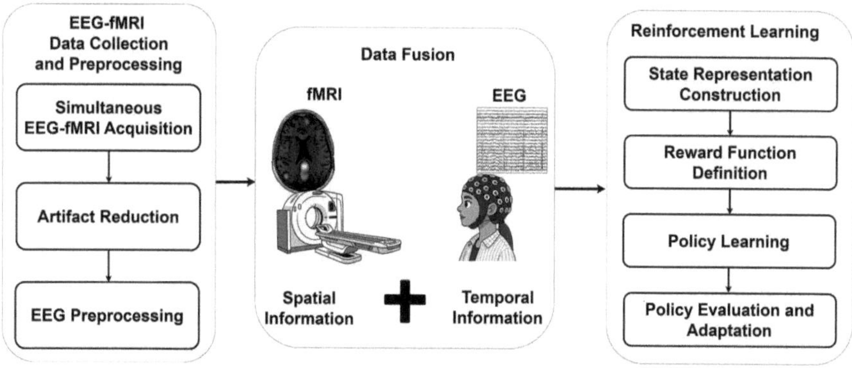

Fig. 16.5 A unified pipeline for multimodal EEG-fMRI neuroimaging and reinforcement learning integration is shown here. The left panel outlines the preprocessing stages, beginning with simultaneous EEG (electrophysiological signals) and fMRI (hemodynamic responses) acquisition, followed by artifact reduction and EEG signal refinement. In the center, a fusion module combines spatially rich fMRI data with temporally precise EEG recordings, enabling comprehensive neuroimaging. The right panel demonstrates the RL-based decision modeling framework that follows fused data, including state representation construction, reward function definition, policy learning, and adaptive evaluation. This integrated framework allows temporally aligned, spatially resolved, and policy-driven interpretation of brain function for applications in closed-loop neurofeedback, BCI, and neurorehabilitation

spatially resolved hemodynamic responses. This combination has been explored to enhance the precision of neurofeedback protocols and improve clinical outcomes by providing more detailed information about brain states.

However, the integration of EEG and fMRI is associated with several technical and methodological challenges. Blood Oxygenation Level Dependent (BOLD) signals measured by fMRI represent delayed and indirect hemodynamic responses to neuronal events, which limits their temporal accuracy [93]. In contrast, EEG captures direct electrophysiological activity but is spatially smeared due to volume conduction. Early studies of simultaneous EEG-fMRI acquisition [94] demonstrated the feasibility of this approach for localizing epileptic foci but also identified several limitations, including gradient-induced artifacts, RF-related heating of electrode leads, and field-related distortions. Subsequent developments have focused on artifact correction techniques and hardware adaptations to enable reliable data acquisition in simultaneous settings [95]. Together, the joint application of EEG and fMRI continues to support improved characterization of spatiotemporal patterns of brain activity in both basic and translational research.

As illustrated in Fig. 16.5, simultaneous EEG-fMRI enables the integration of electrophysiological and hemodynamic measurements within the context of brain activity. This multimodal setup leverages the distinct advantages of each modality: EEG detects rapid neuronal oscillations with millisecond resolution, while fMRI captures spatially precise hemodynamic responses. Their overlap corresponds to brain processes that exhibit both electrical and vascular dynamics, offering a more comprehensive representation of functional activity. Simultaneous acquisition, as shown by the illustration, ensures temporal synchronization and facilitates joint analysis of spatiotemporal brain dynamics [96].

Based on this, three principal analysis strategies have matured:

(a) EEG-informed fMRI: Trial-wise EEG features e.g., alpha power, ERP amplitudes are convoluted with a canonical hemodynamic response and entered as regressors in the fMRI general-linear model, revealing voxels whose BOLD signal covaries with fast electrophysiological markers [97, 98].
(b) fMRI-constrained EEG source imaging: fMRI activation maps serve as spatial priors to guide EEG source localization, improving spatial precision by constraining dipole estimates to BOLD-active regions. The Dynamic Brain Transition Network (DBTN) method advances this by incorporating spatiotemporal fMRI priors within a Bayesian framework, achieving higher spatial and temporal accuracy than traditional approaches like fMRIMNE [99].
(c) Joint data-fusion: Joint data-fusion approaches integrate EEG and fMRI signals simultaneously within unified mathematical frameworks to exploit both spatial and temporal information. Techniques such as joint independent component analysis (jICA), canonical correlation analysis (CCA), and tensor decomposition methods like parallel factor analysis (PARAFAC) allow the identification of latent sources that drive neural dynamics across modalities. These methods enable the exploration of coupled neural processes, providing improved sensitivity to dynamic brain states and connectivity patterns [100–102].

16.5.3 Multimodal Neurofeedback and Applications in Cognitive Control, Reward, and Learning

The integration of EEG and biosignals offers a powerful framework to investigate the spatiotemporal dynamics of RL in the human brain. In this regard, a foundational extension of multimodal learning involves combining EEG with EMG to investigate the interaction between motor effort and reward valuation [103]. In a notable study, participants were asked to generate precise quadriceps muscle contractions while EEG and EMG data were simultaneously recorded. Tasks varied in effort level and were reinforced with binary feedback based on the accuracy of EMG-target achievement. While early ERP components remained stable, high effort trials elicited enhanced late ERP amplitudes—specifically in the P300 window—during reward outcome processing. Source decomposition attributed early components to general reward response and late components to retrospective effort valuation, suggesting a delayed integration of physical cost into reward circuits [104].

On the other hand, EEG-fMRI multimodality comes up with electrical brain activity in correlation with hemodynamic changes in brain during the electrical activity. EEG provides high temporal resolution necessary to capture the rapid neural computations underlying value updates and decision-making processes, while fMRI contributes precise spatial localization of the involved brain structures. Their combined use allows for the disentanglement of the chronological and anatomical cascade of RL-related activity, particularly relevant in both basic neuroscience and translational neurofeedback research. Advancing this line, a drift diffusion modeling (DDM) framework paired with EEG-fMRI recordings has illustrated how brain networks dynamically adjust decision thresholds during RL tasks. Participants made trial-by-trial decisions based on learned reward contingencies. BOLD signals in the subthalamic nucleus (STN) and EEG-recorded mediofrontal theta oscillations co-varied with trial-level conflict, modulating decision thresholds adaptively. Increased STN activation and theta power corresponded to slower, more deliberate choices under uncertainty, implicating pre-SMA and STN in cognitive control during value-guided decisions [105].

EEG-fMRI integration offers critical insights into spatiotemporal dynamics of reward processing during binary decisions. In one study, a computational model-based analysis of EEG-fMRI data revealed temporally distinct neural activations. Approximately 200 ms post-stimulus, value-related EEG components were localized to posterior regions, particularly the intraparietal sulcus. This activity then shifted anteriorly around 850 ms to involve the ventromedial and lateral prefrontal cortices. By 1050 ms, dorsomedial prefrontal regions exhibited signals reflecting value comparisons between chosen and unchosen potions. These findings suggest a posterior-to-anterior progression of decision-related processing, highlighting distinct temporal phases underlying choice evaluation and post-decision action selection [106].

In reinforcement-related neurofeedback, multimodal techniques have yielded clinically actionable biomarkers. A ventral-striatum EEG fingerprint (VS-EFP) model was trained using music-evoked BOLD activity and cross-validated using a monetary reward paradigm. The resulting EEG model accurately predicted VS BOLD responses, providing a non-invasive method to track deep reward system activity in real-time [107].

Neurofeedback paradigms based on positive emotion regulation have similarly benefited from EEG-fMRI co-registration. Participants recalling autobiographical memories underwent alpha asymmetry-based neurofeedback while fMRI measured BOLD changes. Results showed increased activation in prefrontal, parietal, limbic, and insular regions, with new inter-network connectivity emerging during upregulation tasks. These changes correlated with psychometric mood improvements, confirming the efficacy of EEG-guided neuromodulation [108].

Simultaneous EEG-fMRI has advanced our understanding of perceptual learning by dissociating bottom-up sensory plasticity from top-down attentional modulation. During auditory frequency discrimination training, participants exhibited behavioral improvements that correlated with increased gamma-band EEG activity localized to the right superior temporal gyrus (STG) and left inferior frontal gyrus (IFG). fMRI data further supported the involvement of these regions, indicating their role in learning-related plasticity. Variational Bayesian source modeling of EEG data confirmed coactivation of STG and IFG, suggesting a dynamic interplay between sensory and higher-order cortical areas during perpetual learning [109].

Simultaneous EEG-fMRI during Stroop tasks has enabled fine-grained mapping of the spatiotemporal dynamics of cognitive control. Using jICA, conflict-related EEG components—such as the N200—were linked to anterior cingulate cortex (ACC) activation, while slower conflict potential were associated with broader engagement of the frontoparietal control network. Connectivity analyses, including dynamic causal modeling, indicated that the ACC mediates adaptive control by modulating executive regions following incongruent trials, supporting performance improvements through top-down regulatory mechanisms [110].

Multimodal meta-analyses across EEG and fMRI have consolidated reward-related abnormalities in psychiatric conditions. In depression, particularly in adolescents, studies report blunted feedback-related negativity in EEG and reduced striatal BOLD responses during reward anticipation and outcome. These signals were predictive of depressive symptom onset, offering multimodal biomarkers for early intervention [111].

Collectively, these studies illustrate the utility of EEG-fMRI and EEG-EMG integration in elucidating the neural substrates of reinforcement learning. By capturing both fast neural dynamics and deep structural activations, these approaches enable a more complete understanding of cognitive and affective processes central to adaptive behavior and psychiatric conditions.

16.6 Conclusion and Future Directions

RL integrated with biosignal-driven BCIs represents a paradigm shift in neurorehabilitation, offering dynamic, personalized interventions responsive to individual neural states and motor capabilities. This chapter reviews the applications of RL in EEG-based, EMG-based, and multimodal BCI systems, highlighting their capacity to enhance motor decoding, tailor assistive feedback, and promote cortical plasticity across diverse rehabilitation scenarios. From stroke recovery and prosthetic control to immersive virtual reality training, RL-driven systems have demonstrated promising gains in motor function, user engagement, and adaptability.

In EEG-based frameworks, RL enables closed-loop modulation of therapy based on neural oscillatory patterns such as ERD/ERS and beta rhythms, enhancing operant conditioning of motor-related brain activity. EMG-RL systems, particularly when combined with VR, have further enriched user interaction by decoding fine motor intent and adapting robotic or virtual assistance accordingly. The fusion of EEG with other modalities—such as EMG and fMRI—has deepened our understanding of the spatiotemporal mechanisms underlying decision-making, cognitive control, and learning, reinforcing the value of multimodal approaches in shaping intelligent neurorehabilitation systems.

Despite substantial progress, several key challenges and opportunities remain:

- Data variability and non-stationarity: EEG and EMG signals are inherently noisy and subject to variability across sessions and individuals [112, 113]. Future

work should focus on robust RL architecture (e.g., meta-RL, continual learning) capable of handling such fluctuations over time.
- Reward signal design: The efficacy of RL hinges on accurate and biologically meaningful reward feedback. Developing hybrid reward models that integrate behavioral, physiological, and neurophysiological markers will be critical for effective learning.
- Clinical generalizability: While promising in experimental settings, RL-BCI systems must be validated across larger, diverse patient populations. Real-world deployment will require scalable systems that balance personalization with standardization.
- Ethics and transparency: Adaptive systems that modulate neural activity raise important concerns about safety, interpretability, and agency. Transparent models and explainable RL policies should be prioritized to build user trust and meet regulatory standards.
- Next-generation interfaces: Future systems may incorporate neuromorphic hardware, wireless multimodal sensing, and cloud-based RL agents to facilitate continuous, home-based rehabilitation. Integration with digital twins and simulation environments can also accelerate training and personalization.

Ultimately, the convergence of reinforcement learning, biosignal processing, and intelligent feedback mechanisms hold immense potential for redefining neurorehabilitation. By enabling responsive, patient-centered therapy, these systems can drive not only motor recovery but also cognitive and emotional resilience—paving the way for more holistic and adaptive human-machine interaction.

References

1. F. Safavi et al., Emerging frontiers in human–robot interaction. J. Intell. Robot. Syst. Theory Appl. **110**(2) (2024). https://doi.org/10.1007/s10846-024-02074-7
2. F. Safavi, D. Pei, P. Olikkal, R. Vinjamuri, New horizons in human–robot interaction: Synergy, cognition, and emotion, in *Discovering the Frontiers of Human-Robot Interaction*, (Springer, Cham, 2024), pp. 103–133. https://doi.org/10.1007/978-3-031-66656-8_5
3. H. Meyerson, P. Olikkal, D. Pei, R. Vinjamuri, Introductory chapter: Human-robot interaction—advances and applications, in *Human-Robot Interaction - Perspectives and Applications*, (IntechOpen, 2023). https://doi.org/10.5772/intechopen.109343
4. R.S. Sutton, A.G. Barto, Reinforcement learning: An introduction. IEEE Trans. Neural Netw. **9**(5), 1054–1054 (1998). https://doi.org/10.1109/TNN.1998.712192
5. L. P. Kaelbling, M. L. Littman, A. W. Moore, "Reinforcement Learning: A Survey", 1996
6. A.K. Shakya, G. Pillai, S. Chakrabarty, Reinforcement learning algorithms: A brief survey. Expert Syst. Appl. **231**, 120495 (2023). https://doi.org/10.1016/j.eswa.2023.120495
7. C.J.C.H. Watkins, P. Dayan, Q-learning. Mach. Learn **8**(3), 279–292 (1992). https://doi.org/10.1007/BF00992698
8. A.S. Polydoros, L. Nalpantidis, Survey of model-based reinforcement learning: Applications on robotics. J. Intell. Robot. Syst. **86**(2), 153–173 (2017). https://doi.org/10.1007/s10846-017-0468-y
9. B. Singh, R. Kumar, V.P. Singh, Reinforcement learning in robotic applications: A comprehensive survey. Artif. Intell. Rev. **55**(2), 945–990 (2022). https://doi.org/10.1007/s10462-021-09997-9

10. R.A. Swain, A.L. Kerr, R.F. Thompson, The cerebellum: A neural system for the study of reinforcement learning. Front. Behav. Neurosci. **5** (2011). https://doi.org/10.3389/fnbeh.2011.00008
11. J.C. Eccles, M. Ito, J. Szentágothai, *The Cerebellum as a Neuronal Machine* (Springer, Berlin/Heidelberg, 1967). https://doi.org/10.1007/978-3-662-13147-3
12. D.M. Huvermann et al., The cerebellum contributes to prediction error coding in reinforcement learning in humans. J. Neurosci. **45**(19), e1972242025 (2025). https://doi.org/10.1523/JNEUROSCI.1972-24.2025
13. W. Zhang, X. Tang, M. Wang, Attention model of EEG signals based on reinforcement learning. Front. Hum. Neurosci. **18** (2024). https://doi.org/10.3389/fnhum.2024.1442398
14. S.R. Soekadar, M. Witkowski, J. Mellinger, A. Ramos, N. Birbaumer, L.G. Cohen, ERD-based online brain–machine interfaces (BMI) in the context of neurorehabilitation: Optimizing BMI learning and performance. IEEE Trans. Neural Syst. Rehabil. Eng. **19**(5), 542–549 (2011). https://doi.org/10.1109/TNSRE.2011.2166809
15. S.A. Roset, H.F. Gonzalez, J.C. Sanchez, Development of an EEG based reinforcement learning Brain-Computer Interface system for rehabilitation, in *Conference proceedings - IEEE engineering in medicine and biology society*, (2013), pp. 1563–1566. https://doi.org/10.1109/EMBC.2013.6609812
16. S. Song et al., Deep reinforcement learning for modeling human locomotion control in neuromechanical simulation. J. Neuroeng. Rehabil. **18**(1), 126 (2021). https://doi.org/10.1186/s12984-021-00919-y
17. S. Liu, J. Luo, C. He, Optimizing the dynamic treatment regime of outpatient rehabilitation in patients with knee osteoarthritis using reinforcement learning. J. Neuroeng. Rehabil. **22**(1), 107 (2025). https://doi.org/10.1186/s12984-025-01609-9
18. J. Zhao et al., Improving rehabilitation motivation and motor learning ability of stroke patients using different reward strategies: Study protocol for a single-center, randomized controlled trial. Front. Neurol. **15** (2024). https://doi.org/10.3389/fneur.2024.1418247
19. W. Jin et al., Electroencephalogram-based adaptive closed-loop brain-computer interface in neurorehabilitation: A review. Front. Comput. Neurosci. **18** (2024). https://doi.org/10.3389/fncom.2024.1431815
20. G. Naros, A. Gharabaghi, Reinforcement learning of self-regulated β-oscillations for motor restoration in chronic stroke. Front. Hum. Neurosci. **9** (2015). https://doi.org/10.3389/fnhum.2015.00391
21. B.R. Ballester, M. Maier, R.M.S.S. Mozo, V. Castañeda, A. Duff, P.F.M.J. Verschure, Counteracting learned non-use in chronic stroke patients with reinforcement-induced movement therapy. J. Neuroeng. Rehabil. **13**(1), 74 (2016). https://doi.org/10.1186/s12984-016-0178-x
22. H. Jeong, M. Song, S.-H. Jang, J. Kim, Investigating the cortical effect of false positive feedback on motor learning in motor imagery based rehabilitative BCI training. J. Neuroeng. Rehabil. **22**(1), 61 (2025). https://doi.org/10.1186/s12984-025-01597-w
23. D. Kraus, G. Naros, R. Bauer, M.T. Leão, U. Ziemann, A. Gharabaghi, Brain–robot interface driven plasticity: Distributed modulation of corticospinal excitability. NeuroImage **125**, 522–532 (2016). https://doi.org/10.1016/j.neuroimage.2015.09.074
24. M. Vukelić, A. Gharabaghi, Self-regulation of circumscribed brain activity modulates spatially selective and frequency specific connectivity of distributed resting state networks. Front. Behav. Neurosci. **9** (2015). https://doi.org/10.3389/fnbeh.2015.00181
25. G. Pires et al., A new error-monitoring brain–computer interface based on reinforcement learning for people with autism spectrum disorders. J. Neural Eng. **19**(6), 066032 (2022). https://doi.org/10.1088/1741-2552/aca798
26. H.W. Aung, J.J. Li, Y. An, S.W. Su, *EEG_RL-Net: Enhancing EEG MI Classification through Reinforcement Learning-Optimised Graph Neural Networks* (2024). https://doi.org/10.48550/arXiv.2405.00723
27. L. Fu, H. Li, H. Ji, J. Li, EEG-EMG analysis method in hybrid brain computer Interface for hand rehabilitation training. Comput. Inform. **42**(3), 741–761 (2023). https://doi.org/10.31577/cai_2023_3_741

28. T. Xu, D. Wang, X. You, Mindgame, in *Adjunct Proceedings of the 31st Annual ACM Symposium on User Interface Software and Technology*, (2018), pp. 5–6. https://doi.org/10.1145/3266037.3266083
29. D. Xu, M. Agarwal, E. Gupta, F. Fekri, R. Sivakumar, Accelerating reinforcement learning using EEG-based implicit human feedback. Neurocomputing **460**, 139–153 (2021). https://doi.org/10.1016/j.neucom.2021.06.064
30. Y. An, Y. Tong, W. Wang, S. W. Su, "Enhancing EEG Signal Generation Through a Hybrid Approach Integrating Reinforcement Learning and Diffusion Models", 2024
31. X. Liu, X. Ding, J. Liu, W. Nie, Q. Yuan, Automatic focal EEG identification based on deep reinforcement learning. Biomed. Signal Process. Control **83**, 104693 (2023). https://doi.org/10.1016/j.bspc.2023.104693
32. M. Ren, R. Xu, T. Zhu, Double deep Q-network decoder based on EEG brain-computer interface. ZTE Commun. **21**(3), 3–10 (2023). https://doi.org/10.12142/ZTECOM.202303002
33. S. Pareek, H.J. Nisar, T. Kesavadas, AR3n: A reinforcement learning-based assist-as-needed controller for robotic rehabilitation. IEEE Robot. Autom. Mag. **31**(3), 74–82 (2024). https://doi.org/10.1109/MRA.2023.3282434
34. J.G.A. Cashaback et al., NSF DARE—Transforming modeling in neurorehabilitation: A patient-in-the-loop framework. J. Neuroeng. Rehabil. **21**(1), 23 (2024). https://doi.org/10.1186/s12984-024-01318-9
35. Y. Huo, X. Wang, W. Zhao, H. Hu, L. Li, Effects of EMG-based robot for upper extremity rehabilitation on post-stroke patients: A systematic review and meta-analysis. Front. Physiol. **14** (2023). https://doi.org/10.3389/fphys.2023.1172958
36. L. Bi, A.G. Feleke, C. Guan, A review on EMG-based motor intention prediction of continuous human upper limb motion for human-robot collaboration. Biomed. Signal Process. Control **51**, 113–127 (2019). https://doi.org/10.1016/j.bspc.2019.02.011
37. C. Fang, B. He, Y. Wang, J. Cao, S. Gao, EMG-centered multisensory based technologies for pattern recognition in rehabilitation: State of the art and challenges. Biosensors **10**(8), 85 (2020). https://doi.org/10.3390/bios10080085
38. T. Teramae, T. Noda, J. Morimoto, EMG-based model predictive control for physical human–robot interaction: Application for assist-as-needed control. IEEE Robot. Autom. Lett. **3**(1), 210–217 (2018). https://doi.org/10.1109/LRA.2017.2737478
39. A. Coronato, M. Naeem, G. De Pietro, G. Paragliola, Reinforcement learning for intelligent healthcare applications: A survey. Artif. Intell. Med. **109**, 101964 (2020). https://doi.org/10.1016/j.artmed.2020.101964
40. C.S. Choy, S.L. Cloherty, E. Pirogova, Q. Fang, Virtual reality assisted motor imagery for early post-stroke recovery: A review. IEEE Rev. Biomed. Eng. **16**, 487–498 (2023). https://doi.org/10.1109/RBME.2022.3165062
41. J. Bui, J. Luauté, A. Farnè, Enhancing upper limb rehabilitation of stroke patients with virtual reality: A mini review. Front. Virtual Reality **2** (2021). https://doi.org/10.3389/frvir.2021.595771
42. C. Song, C. Chen, Y. Li, X. Wu, Deep reinforcement learning apply in electromyography data classification, in *2018 IEEE International Conference on Cyborg and Bionic Systems (CBS)*, (2018), pp. 505–510. https://doi.org/10.1109/CBS.2018.8612213
43. Á.L. Valdivieso Caraguay, J.P. Vásconez, L.I. Barona López, M.E. Benalcázar, Recognition of hand gestures based on EMG signals with deep and double-deep Q-networks. Sensors **23**(8), 3905 (2023). https://doi.org/10.3390/s23083905
44. J.P. Vásconez, L.I. Barona López, Á.L. Valdivieso Caraguay, M.E. Benalcázar, Hand gesture recognition using EMG-IMU signals and deep Q-networks. Sensors **22**(24), 9613 (2022). https://doi.org/10.3390/s22249613
45. J. Berman, R. Hinson, H. Huang, Comparing reinforcement learning agents and supervised learning neural networks for EMG-based decoding of continuous movements, in *2021 43rd Annual International Conference of the IEEE Engineering in Medicine & Biology Society (EMBC)*, (2021), pp. 6297–6300. https://doi.org/10.1109/EMBC46164.2021.9630744

46. P.J. Cruz et al., A deep Q-network based hand gesture recognition system for control of robotic platforms. Sci. Rep. **13**(1), 7956 (2023). https://doi.org/10.1038/s41598-023-34540-x
47. N. Gardner et al., EMG based simultaneous wrist motion prediction using reinforcement learning, in *2020 IEEE 20th International Conference on Bioinformatics and Bioengineering (BIBE)*, (2020), pp. 1016–1021. https://doi.org/10.1109/BIBE50027.2020.00172
48. K. Freitag, Y. Karayiannidis, J. Zbinden, R. Laezza, "Fine-tuning Myoelectric Control through Reinforcement Learning in a Game Environment", 2024., doi: arXiv:2411.13327
49. R. Sharma, A. Kukker, Neural reinforcement learning based identifier for typing keys using forearm EMG signals, in *Proceedings of the 9th International Conference on Signal Processing Systems*, (2017), pp. 225–229. https://doi.org/10.1145/3163080.3163117
50. M. Oghogho, M. Sharifi, M. Vukadin, C. Chin, V.K. Mushahwar, M. Tavakoli, Deep Reinforcement Learning for EMG-based Control of Assistance Level in Upper-limb Exoskeletons, in *2022 International Symposium on Medical Robotics (ISMR)*, (2022), pp. 1–7. https://doi.org/10.1109/ISMR48347.2022.9807562
51. J. Berman, I.-C. Lee, J. Yin, H. Huang, An efficient framework for personalizing EMG-driven musculoskeletal models based on reinforcement learning. IEEE Trans. Neural Syst. Rehabil. Eng. **32**, 4174–4185 (2024). https://doi.org/10.1109/TNSRE.2024.3483150
52. A.D. Pelosi, N. Roth, T. Yehoshua, D. Itah, O. Braun Benyamin, A. Dahan, Personalized rehabilitation approach for reaching movement using reinforcement learning. Sci. Rep. **14**(1), 17675 (2024). https://doi.org/10.1038/s41598-024-64514-6
53. W. Wu, K.R. Saul, H.(H.) Huang, Using reinforcement learning to estimate human joint moments from electromyography or joint kinematics: An alternative solution to musculoskeletal-based biomechanics. J. Biomech. Eng **143**(4) (2021). https://doi.org/10.1115/1.4049333
54. M.C. Tosin, L.B. Bagesteiro, A. Balbinot, Actor-critic reinforcement learning based algorithm for contaminant minimization in sEMG movement recognition, in *2022 44th Annual International Conference of the IEEE Engineering in Medicine & Biology Society (EMBC)*, (2022), pp. 3636–3639. https://doi.org/10.1109/EMBC48229.2022.9871412
55. P.M. Pilarski, M.R. Dawson, T. Degris, F. Fahimi, J.P. Carey, R.S. Sutton, Online human training of a myoelectric prosthesis controller via actor-critic reinforcement learning, in *2011 IEEE International Conference on Rehabilitation Robotics*, (2011), pp. 1–7. https://doi.org/10.1109/ICORR.2011.5975338
56. E. Campbell, E. Eddy, S. Bateman, U. Côté-Allard, E. Scheme, Context-informed incremental learning improves both the performance and resilience of myoelectric control. J. Neuroeng. Rehabil. **21**(1), 70 (2024). https://doi.org/10.1186/s12984-024-01355-4
57. M.A. Garenfeld et al., A compact system for simultaneous stimulation and recording for closed-loop myoelectric control. J. Neuroeng. Rehabil. **18**(1), 87 (2021). https://doi.org/10.1186/s12984-021-00877-5
58. M. Lin, J. Huang, J. Fu, Y. Sun, Q. Fang, A VR-based motor imagery training system with EMG-based real-time feedback for post-stroke rehabilitation. IEEE Trans. Neural Syst. Rehabil. Eng. **31**, 1–10 (2023). https://doi.org/10.1109/TNSRE.2022.3210258
59. J. Huang, M. Lin, J. Fu, Y. Sun, Q. Fang, An immersive motor imagery training system for post-stroke rehabilitation combining VR and EMG-based real-time feedback, in *2021 43rd Annual International Conference of the IEEE Engineering in Medicine & Biology Society (EMBC)*, (2021), pp. 7590–7593. https://doi.org/10.1109/EMBC46164.2021.9629767
60. X. Yang, S.-C. Yeh, J. Niu, Y. Gong, G. Yang, Hand rehabilitation using virtual reality electromyography signals, in *2017 5th International Conference on Enterprise Systems (ES)*, (2017), pp. 125–131. https://doi.org/10.1109/ES.2017.27
61. A. Al-Jumaily, R.A. Olivares, Electromyogram (EMG) driven system based virtual reality for prosthetic and rehabilitation devices, in *Proceedings of the 11th International Conference on Information Integration and Web-based Applications & Services*, (2009), pp. 582–586. https://doi.org/10.1145/1806338.1806448

62. A. Elor, S. Kurniawan, Deep Reinforcement Learning in Immersive Virtual Reality Exergame for Agent Movement Guidance, in *2020 IEEE 8th International Conference on Serious Games and Applications for Health (SeGAH)*, (2020), pp. 1–7. https://doi.org/10.1109/SeGAH49190.2020.9201901
63. W.-S. Kim et al., Clinical application of virtual reality for upper limb motor rehabilitation in stroke: Review of technologies and clinical evidence. J. Clin. Med. **9**(10), 3369 (2020). https://doi.org/10.3390/jcm9103369
64. A. Pillai, M.S.H. Sunny, M.T. Shahria, N. Banik, M.H. Rahman, Gamification of upper limb rehabilitation in mixed-reality environment. Appl. Sci. **12**(23), 12260 (2022). https://doi.org/10.3390/app122312260
65. R. de la Rosa, A. Carrera, A. Alonso, B. Peñasco-Martín, A. Gil-Agudo, E.J. Abril, Myoelectric gaming in the rehabilitation of patients with C7 spinal cord injury. Appl. Sci. **9**(9), 1912 (2019). https://doi.org/10.3390/app9091912
66. A. Rovira, M. Slater, Reinforcement learning as a tool to make people move to a specific location in immersive virtual reality. Int. J. Hum. Comput. Stud. **98**, 89–94 (2017). https://doi.org/10.1016/j.ijhcs.2016.10.007
67. A. Elor, S. Kurniawan, M. Teodorescu, Towards an Immersive Virtual Reality Game for Smarter Post-Stroke Rehabilitation, in *2018 IEEE International Conference on Smart Computing (SMARTCOMP)*, (2018), pp. 219–225. https://doi.org/10.1109/SMARTCOMP.2018.00094
68. D. J. A. Brenneis *et al.*, "Assessing Human Interaction in Virtual Reality With Continually Learning Prediction Agents Based on Reinforcement Learning Algorithms: A Pilot Study", 2021
69. W. Yu, T.J. Chua, J. Zhao, User-centric heterogeneous-action deep reinforcement learning for virtual reality in the Metaverse over wireless networks. IEEE Trans. Wirel. Commun., 1–1 (2024). https://doi.org/10.1109/TWC.2023.3277226
70. S.-H. Lee, J. Cui, L. Liu, M.-C. Su, L. Zheng, S.-C. Yeh, An evidence-based intelligent method for upper-limb motor assessment via a VR training system on stroke rehabilitation. IEEE Access **9**, 65871–65881 (2021). https://doi.org/10.1109/ACCESS.2021.3075778
71. P.L. Nunez, R.B. Silberstein, On the relationship of synaptic activity to macroscopic measurements: Does co-registration of EEG with fMRI make sense? Brain Topogr. **13**(2), 79–96 (2000). https://doi.org/10.1023/A:1026683200895
72. C.M. Michel, D. Brunet, EEG source imaging: A practical review of the analysis steps. Front. Neurol. **10** (2019). https://doi.org/10.3389/fneur.2019.00325
73. A.M. Dale et al., Dynamic statistical parametric mapping. Neuron **26**(1), 55–67 (2000). https://doi.org/10.1016/S0896-6273(00)81138-1
74. S. Baillet, J.C. Mosher, R.M. Leahy, Electromagnetic brain mapping. IEEE Signal Process. Mag. **18**(6), 14–30 (2001). https://doi.org/10.1109/79.962275
75. A. Gramfort, T. Papadopoulo, E. Olivi, M. Clerc, OpenMEEG: Opensource software for quasistatic bioelectromagnetics. Biomed. Eng. Online **9**(1), 45 (2010). https://doi.org/10.1186/1475-925X-9-45
76. R. Oostenveld, P. Praamstra, The five percent electrode system for high-resolution EEG and ERP measurements. Clin. Neurophysiol. **112**(4), 713–719 (2001). https://doi.org/10.1016/S1388-2457(00)00527-7
77. A. Delorme, T. Sejnowski, S. Makeig, Enhanced detection of artifacts in EEG data using higher-order statistics and independent component analysis. NeuroImage **34**(4), 1443–1449 (2007). https://doi.org/10.1016/j.neuroimage.2006.11.004
78. R.J. Croft, R.J. Barry, Removal of ocular artifact from the EEG: A review. Neurophysiol. Clin. Neurophysiol. **30**(1), 5–19 (2000). https://doi.org/10.1016/S0987-7053(00)00055-1
79. E.S. Kappenman, S.J. Luck, The effects of electrode impedance on data quality and statistical significance in ERP recordings. Psychophysiology (2010). https://doi.org/10.1111/j.1469-8986.2010.01009.x

80. G.N. Tripathi, H. Wagatsuma, PCA-based algorithms to find synergies for humanoid robot motion behavior. Int. J. Humanoid Rob. **13**(2) (2016). https://doi.org/10.1142/S0219843615500371
81. W. Boucsein, *Electrodermal Activity* (Springer, Boston, MA, 2012). https://doi.org/10.1007/978-1-4614-1126-0
82. H. Obrig, A. Villringer, Beyond the visible—Imaging the human brain with light. J. Cereb. Blood Flow Metab. **23**(1), 1–18 (2003). https://doi.org/10.1097/01.WCB.0000043472.45775.29
83. X. Cui, S. Bray, A.L. Reiss, Functional near infrared spectroscopy (NIRS) signal improvement based on negative correlation between oxygenated and deoxygenated hemoglobin dynamics. NeuroImage **49**(4), 3039–3046 (2010). https://doi.org/10.1016/j.neuroimage.2009.11.050
84. F. Wallois, M. Mahmoudzadeh, A. Patil, R. Grebe, Usefulness of simultaneous EEG–NIRS recording in language studies. Brain Lang. **121**(2), 110–123 (2012). https://doi.org/10.1016/j.bandl.2011.03.010
85. J.T. Gwin, K. Gramann, S. Makeig, D.P. Ferris, Removal of movement artifact from high-density EEG recorded during walking and running. J. Neurophysiol. **103**(6), 3526–3534 (2010). https://doi.org/10.1152/jn.00105.2010
86. K. Gramann et al., Cognition in action: Imaging brain/body dynamics in mobile humans. Rev. Neurosci. **22**(6) (2011). https://doi.org/10.1515/RNS.2011.047
87. S. Makeig et al., Dynamic brain sources of visual evoked responses. Science (80-.) **295**(5555), 690–694 (2002). https://doi.org/10.1126/science.1066168
88. S. Ladouce, D.I. Donaldson, P.A. Dudchenko, M. Ietswaart, Mobile EEG identifies the re-allocation of attention during real-world activity. Sci. Rep. **9**(1), 15851 (2019). https://doi.org/10.1038/s41598-019-51996-y
89. G. Lantz, R. Grave de Peralta, L. Spinelli, M. Seeck, C. Michel, Epileptic source localization with high density EEG: How many electrodes are needed? Clin. Neurophysiol. **114**(1), 63–69 (2003). https://doi.org/10.1016/S1388-2457(02)00337-1
90. D. Brunet, M.M. Murray, C.M. Michel, Spatiotemporal analysis of multichannel EEG: CARTOOL. Comput. Intell. Neurosci. **2011**, 1–15 (2011). https://doi.org/10.1155/2011/813870
91. M.M. Murray, D. Brunet, C.M. Michel, Topographic ERP analyses: A step-by-step tutorial review. Brain Topogr. **20**(4), 249–264 (2008). https://doi.org/10.1007/s10548-008-0054-5
92. M. Fleury, P. Figueiredo, A. Vourvopoulos, A. Lécuyer, Two is better? Combining EEG and fMRI for BCI and neurofeedback: A systematic review. J. Neural Eng. **20**(5), 051003 (2023). https://doi.org/10.1088/1741-2552/ad06e1
93. N.K. Logothetis, What we can do and what we cannot do with fMRI. Nature **453**(7197), 869–878 (2008). https://doi.org/10.1038/nature06976
94. J.R. Ives, S. Warach, F. Schmitt, R.R. Edelman, D.L. Schomer, Monitoring the patient's EEG during echo planar MRI. Electroencephalogr. Clin. Neurophysiol. **87**(6), 417–420 (1993). https://doi.org/10.1016/0013-4694(93)90156-P
95. A.C. Coan et al., EEG-fMRI in the presurgical evaluation of temporal lobe epilepsy. J. Neurol. Neurosurg. Psychiatry **87**(6), 642–649 (2016). https://doi.org/10.1136/jnnp-2015-310401
96. F. Biessmann, S. Plis, F.C. Meinecke, T. Eichele, K.-R. Muller, Analysis of multimodal neuroimaging data. IEEE Rev. Biomed. Eng. **4**, 26–58 (2011). https://doi.org/10.1109/RBME.2011.2170675
97. S. Debener, M. Ullsperger, M. Siegel, A.K. Engel, Single-trial EEG–fMRI reveals the dynamics of cognitive function. Trends Cogn. Sci. **10**(12), 558–563 (2006). https://doi.org/10.1016/j.tics.2006.09.010
98. R.J. Huster, S. Debener, T. Eichele, C.S. Herrmann, Methods for simultaneous EEG-fMRI: An introductory review. J. Neurosci. **32**(18), 6053–6060 (2012). https://doi.org/10.1523/JNEUROSCI.0447-12.2012

99. T. Nguyen, T. Potter, R. Grossman, Y. Zhang, Characterization of dynamic changes of current source localization based on spatiotemporal fMRI constrained EEG source imaging. J. Neural Eng. **15**(3), 036017 (2018). https://doi.org/10.1088/1741-2552/aa9fb2
100. V.D. Calhoun, J. Liu, T. Adalı, A review of group ICA for fMRI data and ICA for joint inference of imaging, genetic, and ERP data. NeuroImage **45**(1), S163–S172 (2009). https://doi.org/10.1016/j.neuroimage.2008.10.057
101. P.A. Valdés-Sosa et al., Estimating brain functional connectivity with sparse multivariate autoregression. Philos. Trans. R. Soc. B Biol. Sci. **360**(1457), 969–981 (2005). https://doi.org/10.1098/rstb.2005.1654
102. E. Martínez-Montes, P.A. Valdés-Sosa, F. Miwakeichi, R.I. Goldman, M.S. Cohen, Concurrent EEG/fMRI analysis by multiway partial least squares. NeuroImage **22**(3), 1023–1034 (2004). https://doi.org/10.1016/j.neuroimage.2004.03.038
103. P. Olikkal, B.M. Pillai, J. Suthakorn, H. Ali, R. Vinjamuri, A hybrid EEG-EMG framework for humanoid control using deep learning transformers, in *2024 IEEE International Conference on Robotics and Biomimetics (ROBIO)*, (2024), pp. 899–904. https://doi.org/10.1109/ROBIO64047.2024.10907308
104. D.J. Palidis, P.L. Gribble, EEG correlates of physical effort and reward processing during reinforcement learning. J. Neurophysiol. **124**(2), 610–622 (2020). https://doi.org/10.1152/jn.00370.2020
105. M.J. Frank et al., fMRI and EEG predictors of dynamic decision parameters during human reinforcement learning. J. Neurosci. **35**(2), 485–494 (2015). https://doi.org/10.1523/JNEUROSCI.2036-14.2015
106. T. Larsen, J.P. O'Doherty, Uncovering the spatio-temporal dynamics of value-based decision-making in the human brain: A combined fMRI–EEG study. Philos. Trans. R. Soc. B Biol. Sci. **369**(1655), 20130473 (2014). https://doi.org/10.1098/rstb.2013.0473
107. N. Singer et al., Development and validation of an fMRI-informed EEG model of reward-related ventral striatum activation. NeuroImage **276**, 120183 (2023). https://doi.org/10.1016/j.neuroimage.2023.120183
108. A. Dehghani, H. Soltanian-Zadeh, G.-A. Hossein-Zadeh, Probing fMRI brain connectivity and activity changes during emotion regulation by EEG neurofeedback. Front. Hum. Neurosci. **16** (2023). https://doi.org/10.3389/fnhum.2022.988890
109. A.C.S. de Souza, H.C. Yehia, M. Sato, D. Callan, Brain activity underlying auditory perceptual learning during short period training: Simultaneous fMRI and EEG recording. BMC Neurosci. **14**(1), 8 (2013). https://doi.org/10.1186/1471-2202-14-8
110. T. Hinault, K. Larcher, N. Zazubovits, J. Gotman, A. Dagher, Spatio–temporal patterns of cognitive control revealed with simultaneous electroencephalography and functional magnetic resonance imaging. Hum. Brain Mapp. **40, no 1**, 80–97 (2019). https://doi.org/10.1002/hbm.24356
111. H. Keren et al., Reward processing in depression: A conceptual and meta-analytic review across fMRI and EEG studies. Am. J. Psychiatry **175**(11), 1111–1120 (2018). https://doi.org/10.1176/appi.ajp.2018.17101124
112. P. Olikkal, D. Pei, B.K. Karri, A. Satyanarayana, N.M. Kakoty, R. Vinjamuri, Biomimetic learning of hand gestures in a humanoid robot. Front. Hum. Neurosci. **18**, 1391531 (2024). https://doi.org/10.3389/fnhum.2024.1391531
113. P. Olikkal, D. Pei, B.K. Karri, A. Satyanarayana, N.M. Kakoty, R. Vinjamuri, Learning hand gestures using synergies in a humanoid robot, in *2023 IEEE International Conference on Robotics and Biomimetics ROBIO 2023*, (2023), pp. 1–6. https://doi.org/10.1109/ROBIO58561.2023.10354698

Chapter 17
SIVAM: Synergy-Based Intuitive Virtual and Augmented Therapy for Mental Health

Parthan Olikkal, Oritsejolomisan Mebaghanje, Viraj Janeja, Golnaz Moharrer, Akshara Ajendla, Sruthi Sundharram, Andrea Kleinsmith, Ann Sofie Clemmensen, and Ramana Vinjamuri

Abstract With growing mental health challenges among populations and limited access to in-person therapy, there is a critical need for accessible, personalized, and non-pharmacological interventions. To address this, we developed the Synergy-based Intuitive Virtual and Augmented Therapy for Mental Health (SIVAM) platform, a real-time home-deployable system that delivers emotionally responsive dance movement therapy (DMT). SIVAM integrates full-body and hand motion capture, avatar-based interaction, and humanoid robot mirroring, while simultaneously recording multimodal physiological signals (EEG, EMG, ECG, GSR, temperature). By extracting motor synergies and affective biomarkers, the system aims to adapt choreography and feedback to the user's emotional and physical state. The platform demonstrates low-latency communication, high-fidelity mapping of body landmarks to avatars and robots, and seamless synchronization between movement and biofeedback. A pilot study confirmed SIVAM's effectiveness across user profiles, supporting its potential as an emotion-aware, scalable therapeutic solution tailored to the need of individuals.

Keywords Human-robot interaction · Virtual reality · Synergies · Affective computing · Mental health therapy · Dance

P. Olikkal · O. Mebaghanje · V. Janeja · A. Ajendla · S. Sundharram · R. Vinjamuri (✉)
Vinjamuri Lab, Department of Computer Science and Electrical Engineering, University of Maryland Baltimore County, Baltimore, MD, USA
e-mail: rvinjam1@umbc.edu

G. Moharrer · A. Kleinsmith
Affective Behavior Interaction Lab, Department of Information Systems, University of Maryland Baltimore County, Baltimore, MD, USA

A. S. Clemmensen
Department of Dance, University of Maryland Baltimore County, Baltimore, MD, USA

R. Vinjamuri (ed.), *Bridging the Gap between Mind and Machine*,
https://doi.org/10.1007/978-3-032-06713-5_17

17.1 Introduction

The global demographic shift towards an aging population presents a growing public health challenge, with mental health disorders significantly impacting the quality of life for older adults. Affecting approximately 1 in 4 U.S. adults annually and accounting for 1 in 6 years lived with disability globally, these conditions often go untreated or are inadequately managed due to persistent barriers [1, 2]. Traditional mental health interventions frequently face challenges such as pervasive stigma, high costs, and limited access to specialized care, leading to intervention delays of up to 11 years. Such delays exacerbate symptoms and diminish overall well-being for this vulnerable demographic.

In response to these pervasive challenges, dance movement interventions (DMI), including dance/movement therapy (DMT), emerge as a highly promising non-pharmacological solution. DMT uniquely integrates physical movement, sensory stimulation, cognitive engagement, and social interaction, offering a holistic approach to mental well-being. A systematic review and meta-analysis of 14 randomized controlled trials (RCTs) with 983 older adults without dementia rigorously demonstrated that DMI improves overall psychological health and enhances general cognitive function compared to controls [3]. As emphasized by Goodill, DMT harnesses these multimodal benefits to alleviate stress, anxiety, and depression, while simultaneously enhancing mood, body awareness, and social connectedness, particularly within elderly populations [4–6]. Furthermore, DMI has been shown to improve postural stability, sensorimotor function, and cognitive performance in older individuals, underpinning its potential role in healthy aging and Alzheimer's disease (AD) prevention [3, 7].

The profound therapeutic potential of DMT is fundamentally rooted in its reliance on interlimb coordination and dynamic muscle synergies, which not only facilitate rhythmic movements but also may serve as robust biomarkers for mental health, reflecting underlying neural and emotional states [8]. Affective neuroscience supports this through technique like multivariate pattern analysis (MVPA), offering real-time insights into DMT's immediate effects on neural networks and emotional valence [9]. Although direct neuroplasticity measures were absent in the RCTs reviewed by Podolski et al. [3], Goodill strongly emphasizes DMT's capacity to influence brain function through movement. This aligns with a growing body of evidence demonstrating increased functional connectivity in critical cognitive and emotional networks (e.g., default mode and fronto-parietal networks) and positive changes in white matter integrity in regions such as the fornix following dance interventions [10, 11]. These findings collectively suggest that DMT actively promotes neural plasticity and enhances emotional regulation, particularly benefiting older adults at heightened risk of cognitive decline, including those with subject cognitive decline (SCD) or mild cognitive impairment (MCI).

Despite the clear promise of DMT, its widespread deployment, particularly for older adults, faces significant limitations. Traditional in-person DMT sessions are often constrained by issues of scalability, the availability of trained therapists, high

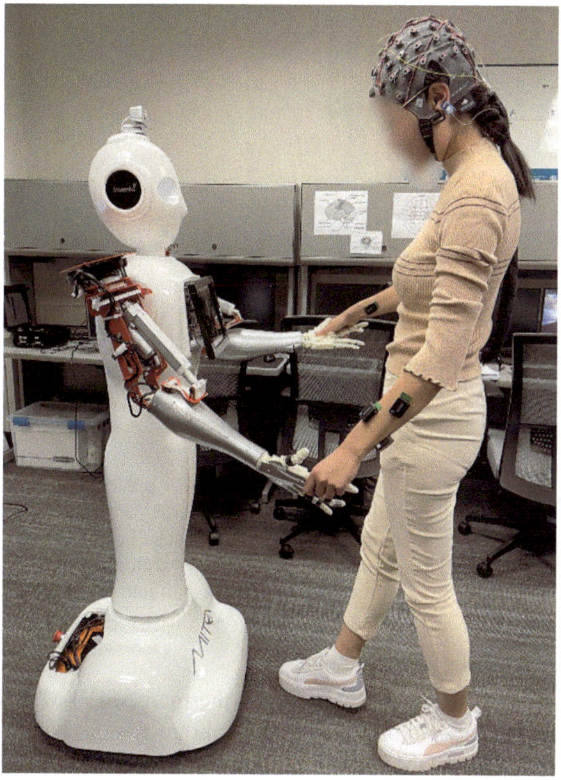

Fig. 17.1 SIVAM Platform where the user can interact with a humanoid robot as a dance partner for mental stress support

individual costs, and logistical barriers such as transportation or mobility limitations for home-bound individuals. These practical challenges hinder consistent access and adherence, thereby limiting the full realization of DMT's therapeutic potential on a broader scale. Furthermore, while the social aspect of DMT is crucial, often conducted in group settings which amplify its therapeutic impact (Podolski et al., reported a large but non-significant effect on social integration), the current deployment models do not fully address the need for personalized, accessible, and destigmatized solutions that can reach individuals in their own environments.

To overcome these deployment challenges and enhance the accessibility and efficacy of DMT, particularly for aging populations, this research introduces the Synergy-based Intuitive Virtual and Augmented Therapy for Mental Health (SIVAM) platform. SIVAM represents a novel technology-enabled solution that pairs the principles of DMT with a humanoid robot to deliver personalized, home-based therapeutic interventions as shown in Fig. 17.1. Drawing on the benefits of socially assistive robotics, which offer non-judgmental companionship and motivation [12], SIVAM mirrors DMT's inherent social advantages while significantly expanding reach and reducing logistical barriers and stigma. This platform leverages interlimb coordination as a potential biomarker for mental health, utilizing advanced motion capture and neurophysiological measures to objectively assess progress and

Fig. 17.2 Users engaging with the SIVAM platform during dance-based interaction. As illustrated in (**a**), swivel-mounted RGB cameras capture full-body landmarks in real-time. These landmarks are then mapped onto virtual avatars in Unity showing in the background (**b**), enabling synchronized avatar motion. Simultaneously, physiological signals such as EEG, HRV, and skin temperature are recorded (**c**), offering insight into the user's emotional and cognitive state during the session

provide real-time insights. Furthermore, SIVAM uniquely analyzes connectivity in frontal, motor, and premotor brain regions to explore the neural mechanisms underlying DMT's effect, allowing for a comparative analysis of human and robot dance partners to optimize both social and therapeutic outcomes.

SIVAM's interdisciplinary approach bridges cutting-edge affective neuroscience, precise motor coordination, and innovative human-robot symbiosis. This synergy aims not only to advance mental health interventions but also to contribute significantly to neuromotor rehabilitation and 'aging in place' strategies. Goodill [4] advocates for the seamless integration of DMT into broader healthcare systems—a vision SIVAM helps realize through its scalable, home-based delivery model (Fig. 17.2). This directly aligns with the urgent call by Podolski et al. [3], for accessible and equitable mental health interventions for older adults. For individuals susceptible to cognitive decline, such as those with SCD or MCI, DMT's multimodal stimulation of physical, psychological, and social well-being builds crucial resilience against age-related conditions. SIVAM further enhances this by expertly tailoring therapy to individual needs via robotics, thereby addressing critical disparities in care access for underserved populations, including rural residents or those with significant mobility limitations.

This chapter aims to present a comprehensive overview of SIVAM as a pioneering approach to delivering Dance Movement Therapy for mental health and cognitive well-being in aging populations. The rest of the chapter follows with Related Works presented in Sect. 17.2; System Architecture of SIVAM explained in Sect. 17.3. Section 17.4 deals with system testing and validation and ends with conclusion and future work in Sect. 17.5.

17.2 Related Works

This section provides key literature and technological foundations relevant to the development of the SIVAM platform.

17.2.1 Dance Movement Interventions and Technological Integration

DMI, including formal DMT, are well-documented for their efficacy in supporting mental and emotional well-being. Recent research has increasingly incorporated technology into the delivery of DMI. Radanliev in the study [13], emphasized the importance of embodied experience, wearable biofeedback, and ethical considerations in XR-based therapeutic environments. In a subsequent study Radanliev [14] demonstrated the efficacy of Virtual Dance Movement Therapy (XR-DMT) in reducing anxiety through the integration of biometric signals—such as electroencephalography (EEG), electrodermal activity (EDA), heart rate variability (HRV) and skin conductance—and AI to personalize therapeutic dance experiences, demonstrating the power of real-time biofeedback in emotionally adaptive interventions. While these systems show promise in leveraging affective computing and immersive environments, they often lack real-time physiological feedback integration with robotic agents or interactive avatars.

17.2.2 Synergy-Based Encoding and Robotics

Motor synergies—coordinated activations of muscles or joints—reflect the nervous system's strategy for simplifying complex movement control, making them ideal for mapping expressive human motion onto avatars or robots. Recent work by Olikkal et al. [15], introduces a MediaPipe-based gesture recognition system capable of encoding 33 distinct hand gestures into low-dimensional kinematic synergies [16, 17]. These synergies preserved over 98% of the original motion variance and were effectively used to reconstruct gestures with high fidelity on a humanoid robot platform. Building on this concept, a related study [18] demonstrated how learned movement synergies could be used to control a humanoid robot's upper limbs in a biomimetic manner. The authors showed that synergy-driven gestures encoding simplifies the control space and enables real-time adaptation of robot movements to human demonstrations. This biologically inspired approach underpins SIVAM's architecture, enabling emotion-sensitive motion analysis and personalized therapeutic feedback through interpretable, low-dimensional movement patterns.

17.2.3 Robotics in Therapeutic Movement Systems

Recent advancements in technology-enhanced dance therapy underscore a growing interest in combining movement, robotics, affective sensing, and immersive environments to support mental and physical well-being. The study [19] piloted a robot-enabled exergaming system for Parkinson's patients, combining Irish dance sessions with real-time robotic guidance and posture monitoring, resulting in improved balance, gait, and motor symptoms. While effective for physical rehabilitation, it lacks emotion sensing and avatar-based interaction—key features that distinguish SIVAM's more holistic design. Other studies have explored robotic companions in therapeutic dance settings [20]. highlights how older adults accept and engage positively with robot-facilitated dance, reinforcing the importance of social presence and enjoyment. Chen et al., expand on this with a study [21] on robot partner dance exercise, where a mobile manipulator robot was used for physical interaction and admittance control—an important step toward real-time, embodied co-creation with non-human agents [22]. further adds that real-time adaptation and leader/follower roles can significantly enhance user engagement, especially among vulnerable populations such as children or the elderly. These studies support the feasibility of robotic companions in movement therapy but often address isolated components.

17.2.4 Emotion Detection Through Movement and Multimodal Signals

Motion-based emotion classification is another growing area. Early systems like [23] used motion capture for mirrored avatar training, a precursor to SIVAM's use of Unity avatars and real-time mapping [24]. introduces a hybrid machine learning approach to classify motions from dance movement, validating the concept of motion-driven emotion inference—central to SIVAM's use of interlimb coordination as a mental health biomarker. However, these systems typically focus on either physical training, emotion classification, or robotic interaction. In contrast, SIVAM integrates all these elements—emotion sensing, robotic and avatar feedback, multiplayer tracking, and home-based XR delivery—into one cohesive platform for mental health and aging support.

17.2.5 Embodied Emotion Regulation and Affective Computing

Emotional awareness plays a critical role in enhancing performance and reducing harmful rumination, ultimately supporting more effective navigation of personal and professional challenges [25, 26]. Poor emotional regulation can deplete psychological, emotional, and social well-being, particularly in high-stress environments [27]. Movement-based activities such as dance have been shown to positively impact emotional and mental well-being [28, 29], although existing studies often focus on individuals already experienced in physical expression [30]. To broaden access and impact, researchers recommend technologies that support embodied, tangible, and reflective practices to help regulate emotions and enhance mood [31, 32]. Advances in affective computing further support this direction, as physiological signals—such as EEG, ECG and galvanic skin response (GSR)—can be used to detect emotional states. While general classification rates remain moderate, significantly better accuracy is achieved when analyzing extreme valence and arousal levels, reaching up to 81% for arousal detection [33]. These findings underscore the value of integrating emotion-aware, body-based, and self-reflective tools into interactive systems to promote emotional well-being and resilience.

Despite significant advancements in DMT, affective computing, and robot-assisted interventions, current systems still exhibit notable limitations. Most platforms do not offer real-time integration of physiological signals such as EEG and EDA will full-body motion tracking, which restricts their ability to provide synchronized, adaptive emotional feedback. Additionally, the absence of emotionally responsive avatars and real-time social feedback mechanisms—such as multiplayer or partner-based engagement—limits the therapeutic depth and interactivity of these systems. Accessibility also remains a major barrier, as many solutions depend on clinic-based equipment, technical supervision, or high-bandwidth environments, making them unsuitable for older adults or underserved populations. The SIVAM platform directly addresses these gaps by integrating markerless full-body motion tracking through MediaPipe and YOLO, real-time physiological monitoring (EEG, EDA), adaptive avatar feedback, virtual reality interfaces, and multiplayer functionality, all within a low-cost, home-based deployment model. By leveraging movement synergies as emotional biomarkers, SIVAM offers a scalable, emotion-aware, and deeply personalized therapeutic experience for aging individuals, positioning itself as the first of its kind in this domain.

17.3 System Architecture

SIVAM is an immersive, real-time platform that integrates body motion capture, gesture recognition, and avatar-based feedback to support therapeutic interventions

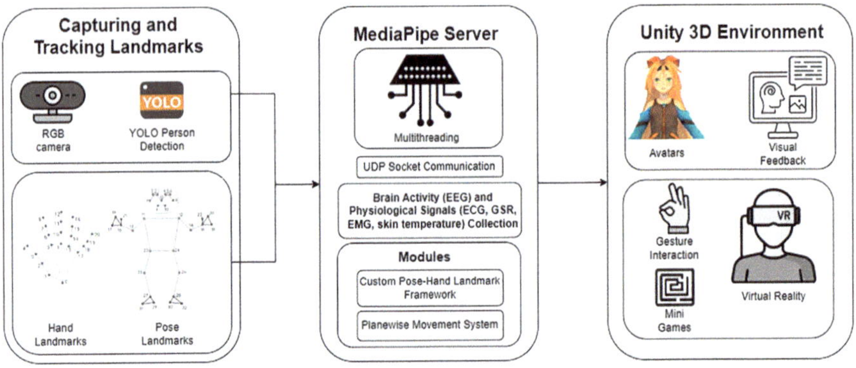

Fig. 17.3 Architecture Diagram of SIVAM. The main components are a MediaPipe Server that detects and overlays key body and hand landmarks from an RGB camera stream. These landmarks are transmitted via socket communication to a Unity environment, where they are mapped onto virtual avatars in real time. Unity provides visual feedback, supports gesture-based interaction, and enables mini-games for therapeutic engagement. The platform is also compatible with virtual reality to enhance user embodiment. Simultaneously, SIVAM records multimodal physiological signals—such as EEG, ECG, GSR, EMG, skin temperature—enhancing synchronized analysis of emotional and cognitive states during interaction

through dance as shown in Fig. 17.3. The system is designed with modularity, low latency, and high adaptability in mind.

17.3.1 Core Components

The platform comprises two major components that operate simultaneously: A backend MediaPipe Server and a Unity 3D environment. The backend server is built in Python and utilizes Google's MediaPipe framework for real-time pose and hand tracking. The MediaPipe Pose landmark model [34] captures full-body skeletal landmarks (33 landmarks) whereas the MediaPipe Hand landmark model [35] details hand gestures landmarks (21 landmarks per hand) using a single RGB camera. These two streams are fused into a unified 67-point landmark representation uniquely for the platform. To ensure smooth data flow and minimal latency, the backend employs multithreading and transmits the landmark data to the Unity environment through UDP sockets. The Unity environment receives the landmark data and animates avatars in real-time. This is achieved using custom C# scripts that map the incoming data to the corresponding avatar joints and apply inverse kinematics or interpolation to simulate smooth and natural motion. The Unity also handles visual feedback, gesture-based interaction, mini-games and virtual reality integration.

17.3.2 Landmark Capture and Tracking

The system captures full-body and hand movements simultaneously by stitching a custom framework that integrates MediaPipe Pose and MediaPipe Hand landmark model. This fusion enables both gross motor (arm and leg movements) and fine motor (finger gesture) analysis making a combined 67 key landmark points for tracking. Additionally, YOLO is used for multi-person detection, allowing the system to track multiple individuals in the camera frame. Although currently only one user actively controls the avatar, the architecture is scalable to support future multiplayer interactions, such as cooperative dance therapy, social engagement games or comparative performance analysis.

17.3.3 Avatar Models, Mapping and Motion

Each set of captured body and hand landmarks is mapped to corresponding joints in a 3D avatar within the Unity environment. The SIVAM system supports two primary avatars, each optimized for different tracking and therapeutic functions: Unity-Chan [36] for fully-body tracking and immersive Virtual Reality (VR) interaction and Kyle Robot [37] for high fidelity hand tracking and gesture-specific tracking.

Unity-Chan is a humanoid anime-style character developed by Unity Technologies Japan, is used for full-body motion tracking. This avatar supports naturalistic limb articulation and expressive movement across the spine, head, and extremities, making it ideal for immersive VR sessions and therapeutic interventions that involve dance choreography, coordination tasks, and embodied interaction. Unity-Chan is primarily deployed in gameplay modes and rhythm-based gesture matching exercises, where full-body expressiveness enhances engagement and emotional resonance.

Kyle Robot, in contrast, is a simplified robotic avatar designed for precise hand and finger movement representation. With a streamlined joint hierarchy, Kyle allows for accurate mapping of the 21-point MediaPipe hand landmarks, enabling expressive finger articulation critical for fine motor rehabilitation and gesture-based emotion recognition. It is particularly valuable in therapy modes that emphasize hand gestures as emotional indications or functional tasks, such as mindfulness-based motion therapy or symbolic expression exercises.

This avatar-mapping system allows SIVAM to flexibly switch between avatars based on therapeutic needs, ensuring both gross and fine motor gestures are rendered accurately and responsively in real-time.

MediaPipe Pose landmark, MediaPipe Hand landmark and custom MediaPipe landmark frameworks were utilized to detect the key points from the RGB camera. The tracked movements were sent through UDP from the server side to Unity. These tracking models extract positions and orientations and map them to the avatars in

Unity. This enables nuanced expressive capabilities, such as sign language gestures or intricate hand-based choreographies.

A key innovation in this platform is the planewise movement system that allows users to move in three dimensions (left-right, up-down, forward-backward) based on their relative position and scale within the camera frame as shown in Fig. 17.5a. This significantly enhances the realism and engagement of the virtual environment.

17.3.4 Dance Combo, Gesture Matching and Playback

To support complex choreographies and enhance the interactive experience, the system features a combo playback mechanism that enables sequential dance moves to be executed and replicated. For example, the "Dance" move, which consists of a coordinated sequence: Raise Left Arm ➔ Raise Right Arm ➔ Raise Both Arms. This entire sequence is stored as a CSV file which captures the joint positions of each gestures frame by frame. These CSV files are manually recorded in real-time using the Python based MediaPipe capture system, which exports 3D landmark positions for all 33 tracked joints. During playback, the Unity component reads the CSV file and animates the avatar accordingly by interpolating between the recorded poses. Once the playback is completed, control is handed back to the user to mirror the sequence. The system then checks if the user performs the full combo in the correct order, using real-time pose recognition and temporal alignment. This allows for more dynamic and expressive user interaction, moving beyond single static gestures to fluid, choregraphed movement sequences.

A key feature of the platform is a gesture matching mini game designed to assess coordination between the user and the virtual avatar as shown in Fig. 17.4. In each round, the avatar performs a predefined dance move, and the user is prompted to replicate the gesture within a ten second window. During this period, the system continuously monitors the users body pose in real-time using landmark streamed from MediaPipe. Each gesture has a predefined detection logic for example, raising the left arm is recognized when the left wrist is positioned higher than the left shoulder. While hands on the hips pose are detected if both wrists are within a short distance of the hips. These pose conditions are evaluated live by comparing the relevant 3D joint positions. If the user correctly performs the gesture within the allotted time, a green "Great!" message appears on screen and the user's score increases. If the user fails to match the pose in time, a red "Missed!" message is shown instead. All instructions and responses are color coded providing clear intuitive feedback throughout the interaction. Music is also played in the background to enhance immersion and create a rhythmic environment, helping users stay engaged and perform gestures in tempo.

In order to allow pre-recorded dance combos and gesture movements, a gesture recording and playback strategy was developed with Unity-side CSV logging and server-side socket logging. In the Unity-side logging method, Unity logs avatar joint rotations over time into CSV files. These files can be replayed to reanimate the

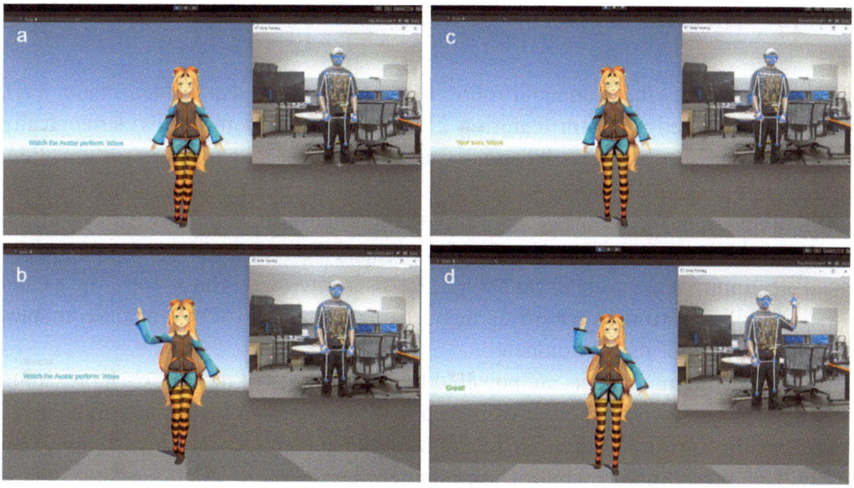

Fig. 17.4 Gesture Matching Mini-Game in SIVAM. The sequence illustrates the interactive workflow of the gesture matching module. In (**a**), the avatar prepares to demonstrate the target gestures, "Wave". In (**b**), the avatar performs the gesture as a reference for the user. In (**c**), the platform prompts the user to replicate the movement within a set time window. In (**d**), upon successful execution above a predefined similarity threshold, the user receives positive feedback and a score, enhancing engagement and motivation

avatar's past movements, enabling a lead-follow paradigm. This is especially useful for mirrored practice sessions, movement imitation, and rehabilitation progress tracking. Whereas the server-side can record raw landmark data when users press a designated key ('r' for record, 'p' for playback). This data is stored in a session stack, preserving multiple gesture recordings. During playback, the stored data is re-injected into the MediaPipe pipeline, allowing the avatar to reproduce previously recorded gestures seamlessly.

17.3.5 Virtual Reality Integration

As the instructions for the mini game can be quite difficult to read from a distance, this platform is ensured to be fully compatible with virtual reality headsets (Meta Quest 3, Meta Quest 2), offering users a deeply immersive therapeutic experience. In VR mode, users can either inhabit their avatar from a first-person perspective or interact with them in a third-person view. All gesture recognition and feedback mechanisms remain functional in this mode, allowing seamless transition from 2D to 3D environments. This immersive mode greatly enhances presence, embodiment, and engagement, particularly valuable for affective therapy.

17.3.6 Humanoid Robot—Mitra

This work utilizes the humanoid robot, Mitra (Invento Research Inc., Plano, TX), a custom-designed platform featuring 21 degrees of freedom (DoFs) strategically designed across its body. Each hand is equipped with 5 DoFs for finger articulation, along with additional joints providing 1 DoF for the wrist, 1 for the elbow, 2 for each shoulder, 1 for the head, and 2 at the mobile base. An RGB camera (1280 × 720 resolution) is mounted on the robot's head for real-time visual input, while an iPad (ninth Gen) on its chest serves as a user interface. The robot's right hand is structurally reinforced for lifting heavier objects, and a LiDAR module enables spatial mapping of its environment.

Mitra's hand movements are powered by three classes of motors, each providing different torque levels suited to their respective roles. High-torque motors manage shoulder flexion and extension, medium-torque motors actuate the elbows, and individual servo motors control each finger. These servo motors operate via a microcontroller network using three input lines—ground, power, and control. Finger movements are achieved by adjusting control voltage, which modulates the servo's rotation to pull internal cables for flexion or extension; higher voltages result in stronger, more pronounced movements.

The robot's motion commands are handled via asynchronous socket communication between Mitra and an external control system. Joint movement instructions are issued from the controlling interface, processed onboard, and relayed to the appropriate microcontroller. These microcontrollers then regulate motor voltages to achieve the intended joint actions, enabling precise and responsive physical interaction.

17.3.7 Multimodal Data Collection

The SIVAM platform is equipped with the capability to capture a rich array of physiological signals in real-time, enabling a deeper understanding of the user's affective and neuromotor states during therapeutic interaction. Specifically, SIVAM integrates with Wearable Sensing DSI-24 dry electrode head mount system (San Diego, USA) that can simultaneously record EEG, electromyography (EMG), GSR, ECG, and skin temperature signals during dance combos or gesture matching mini-games. This multimodal signal acquisition allows SIVAM to go beyond mere motion tracking by capturing the emotional and cognitive correlations of embodied movement. For instance, EEG provides insights into attentional engagement, affective arousal, and cognitive load [38]; EMG reflects muscle activation patterns underlying motor execution and effort [39]; GSR and skin temperature serve as reliable indicators of autonomic arousal linked to emotional valence [40]; and ECG can reveal variations in heart rate variability (HRV), often associated with stress, anxiety, or relaxation [41]. This simultaneous availability of these data streams enables the extraction of movement synergies—low-dimensional coordination patterns across muscles, joints—that may serve as biomarkers. By aligning real-time avatar behavior and feedback mechanisms with multimodal affective responses, SIVAM delivers a closed-loop therapeutic framework. This enables the system to intelligently adjust exercises, challenge levels, and emotional responses in real time based on the user's physiological and emotional state, delivering a highly personalized and emotionally responsive therapeutic experience.

17.4 Testing and Validation

To evaluate the performance and usability of the SIVAM platform, we conducted a pilot study involving five participants—three individuals with no formal dance training and two trained dancers. This diverse sample allowed us to access both accessibility for general users and performance robustness under complex motion conditions.

Motion data were captured using the Obsbot Tiny 2 AI swivel camera system, which automatically tracked user movement with a resolution of 1920 × 1080 pixels and a frame rate of 60 fps. This enabled consistent tracking even as users moved across physical space. MediaPipe's real-time landmark models for pose and hand tracking were overlaid onto the live camera feed to extract precise skeletal and hand joint coordinates as shown in Figs. 17.2a and 17.4. These landmark points were then transmitted to the Unity-based frontend, where they were mapped onto virtual avatars (Unity-Chan or Kyle Robot) depending on the selected therapy mode—either full-body tracking for dance sequences or fine-motor gesture tracking for hand-based interactions as shown in Fig. 17.5a, b.

To accommodate user skill levels, the system presented different levels of task complexity. Non-dancers engaged with simple, rhythmic dance combos and gesture-matching mini-games designed to promote motor coordination and emotional engagement without requiring choreography expertise as shown in Fig. 17.3b. In contrast, the dancer group was encouraged to explore more expressive, high-mobility sequences realized through planewise movement component (refer Fig. 17.3a), testing the system's response time, fidelity of motion replication, and real-time feedback accuracy.

In parallel with motion tracking, physiological data were collected simultaneously through the integrated SIVAM biosensing pipeline, using the DSI-24 EEG head mount cap equipped with additional modules for ECG and skin temperature monitoring. Data acquisition was performed at a sampling rate of 600 Hz, allowing for high-resolution capture of rapid changes in neural and autonomic activity during task performance as shown in Fig. 17.2c. This multimodal data collection facilitated synchronized analysis of motor execution, affective response, and neural engagement, contributing to the development of personalized feedback algorithms and synergy-based emotional biomarkers.

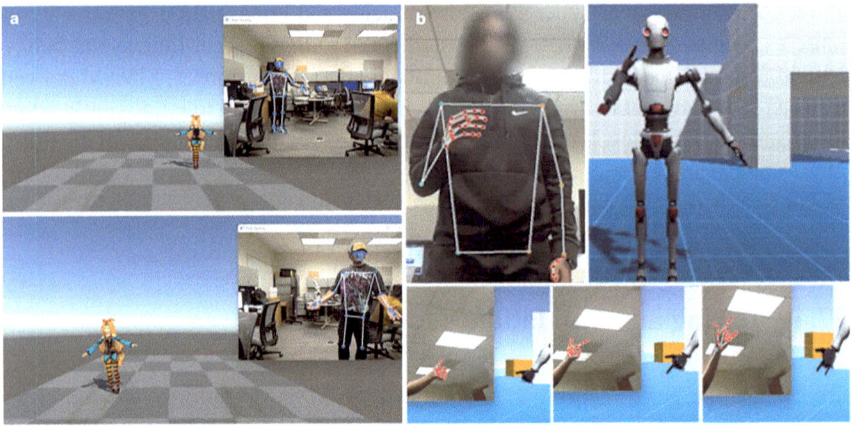

Fig. 17.5 (**a**) The Planewise movement system enables user navigation along vertical, horizontal, and depth planes. The top figure shows the user at a greater distance from the camera, resulting in a smaller virtual avatar, while the bottom figure shows the user closer to the camera, with the avatar dynamically adjusted to reflect this proximity. (**b**) The Kyle Robot supports fine-grained hand gesture mirroring; various user hand gestures are captured and instantly replicated by the robot, as illustrated in the lower figures

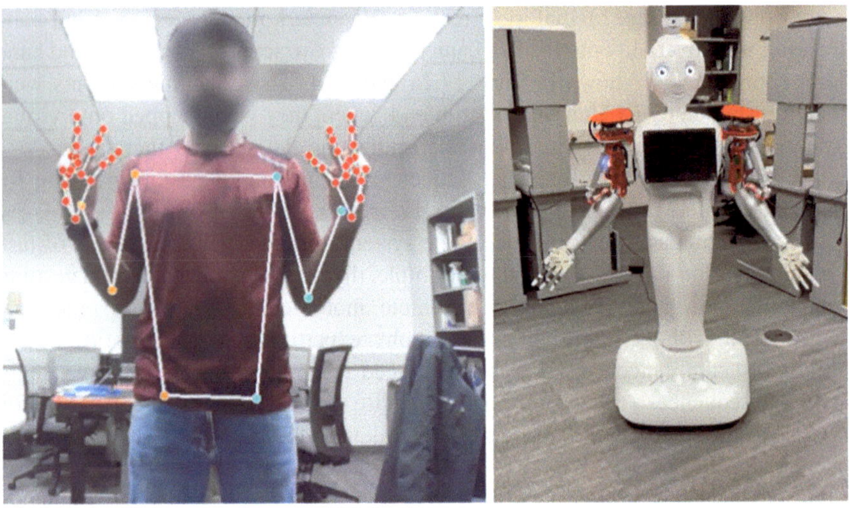

Fig. 17.6 This figure demonstrates the integration of the custom pose-hand landmark framework with Mitra. The user's movements, captured via RGB camera and processed through MediaPipe, were mapped to Mitra's joints. The robot successfully mirrors the user's gestures in real-time, validating the feasibility of human-robot interaction for therapeutic applications

Moreover, the custom pose-hand integration framework was employed to map the key landmarks captured from the RGB camera to the corresponding DoFs of Mitra through a non-linear joint-joint mapping method as shown in Fig. 17.6. This allowed Mitra to successfully mirror user-performed gesture-based tasks in real-time, within the mechanical and kinematic constraints of its design. Mitra was able to replicate upper-limb movements successfully, demonstrating the feasibility of human-robot mirroring for therapeutic interaction.

This pilot validation demonstrated the SIVAM platform's technical reliability, multimodal integration capabilities, and flexibility across different user profiles. It also provided valuable insights into how motion complexity, emotional state, and physiological responses co-vary during interactive dance therapy, informing future iterations of adaptive feedback and personalized mechanisms.

During testing, we observed specific limitations in the MediaPipe-based tracking system that affected avatar consistency. Since MediaPipe provides 2D landmark coordinates, the system occasionally loses accurate avatar mapping when users rotate or turn away from the camera. This occlusion leads to brief lapses in full-body tracking, especially during dynamic sequences that involve significant changes in body orientation. While evaluating the gesture matching module, we noted inconsistencies in hand tracking when the user's fingers came into close proximity or overlapped. In such cases, the MediaPipe Hand landmark model struggled to distinguish individual digits accurately, resulting in misclassification or temporary loss of hand pose fidelity. These challenges likely stem from the inherent limitations of the model's reliance on 2D projections and its sensitivity to occlusions in fine-motor tasks.

A noticeable delay in Mitra's elbow flexion and extension responses was observed compared to other joint movements. While the systems successfully mapped human gestures to Mitra's kinematic structure, the response latency at the elbow joint was consistently higher. This lag is likely attributed to the type and torque characteristics of the actuator motor responsible for controlling elbow motion. Unlike the high-torque motors used at the shoulder joint—which provide faster and more powerful actuation—the elbow relies on a medium-torque motor, which may prioritize smoothness and load management over speed. This trade-off, although beneficial for ensuring motor longevity and safety, introduces the delay in real-time mirroring, especially during rapid or high-frequency elbow movements.

17.5 Conclusion and Future Works

This chapter presented the SIVAM platform, a novel system integrating DMT, immersive virtual environments, multimodal neurophysiological sensing, and humanoid robotic mirroring to deliver emotionally intelligent, personalized mental health interventions for aging populations. By integrating full-body and fine-motor motion tracking with simultaneous EEG, EMG, ECG, GSR, and skin temperature acquisition, SIVAM creates a rich, closed-loop feedback environment. Movement synergies serve as interpretable biomarkers to assess motor coordination and emotional state in real-time, enabling dynamic adjustments to the therapy process. Avatar-based feedback, VR immersion, and real-time robotic mirroring promote embodied engagement, social connectedness, and accessibility. In the future, we aim to enable dual-avatar interaction from a single RGB camera, enhancing multiplayer co-therapy experiences and real-time social mirroring. Additionally, we will expand our movement library to include diverse and emotionally intelligent choreographed dance sequences, designed to elicit, reflect, and regulate affective states, through curated movement patterns. Also, we are developing custom pose-hand landmark detection to replace the default hand tracking in the Kyle Robot avatar to enable more expressive gestures and symbolic communication. Through these advancements, SIVAM aspires to become a robust, adaptive, and scalable platform for home-based mental health care—merging the science of embodied emotion with the precision of real-time computing and robotics.

References

1. Mental Health Statistics (2025)—Forbes Health. https://www.forbes.com/health/mind/mental-health-statistics/. Toegang verkry 01 Julie 2025
2. Mental health. https://www.who.int/health-topics/mental-health#tab=tab_1. Toegang verkry 01 Julie 2025
3. O.S. Podolski et al., The impact of dance movement interventions on psychological health in older adults without dementia: A systematic review and meta-analysis. Brain Sci. **13**(7), 981 (2023). https://doi.org/10.3390/brainsci13070981
4. S. Goodill, *An Introduction to Medical Dance/Movement Therapy: Health Care in Motion* (2005), p. 240
5. S.C. Koch, R.F.F. Riege, K. Tisborn, J. Biondo, L. Martin, A. Beelmann, Effects of dance movement therapy and dance on health-related psychological outcomes. A meta-analysis update. Front. Psychol. **10** (2019). https://doi.org/10.3389/fpsyg.2019.01806
6. C. Quiroga Murcia, G. Kreutz, S. Clift, S. Bongard, Shall we dance? An exploration of the perceived benefits of dancing on well-being. Arts Health **2**(2), 149–163 (2010). https://doi.org/10.1080/17533010903488582
7. J.-C. Kattenstroth, T. Kalisch, S. Holt, M. Tegenthoff, H.R. Dinse, Six months of dance intervention enhances postural, sensorimotor, and cognitive performance in elderly without affecting cardio-respiratory functions. Front. Aging Neurosci. **5** (2013). https://doi.org/10.3389/fnagi.2013.00005
8. A. d'Avella, F. Lacquaniti, Control of reaching movements by muscle synergy combinations. Front. Comput. Neurosci. **7** (2013). https://doi.org/10.3389/fncom.2013.00042
9. P.A. Kragel, K.S. LaBar, Decoding the nature of emotion in the brain. Trends Cogn. Sci. **20**(6), 444–455 (2016). https://doi.org/10.1016/j.tics.2016.03.011
10. Z. Balazova et al., Dance intervention impact on brain plasticity: A randomized 6-month fMRI study in non-expert older adults. Front. Aging Neurosci. **13** (2021). https://doi.org/10.3389/fnagi.2021.724064
11. A.Z. Burzynska et al., White matter integrity declined over 6-months, but dance intervention improved integrity of the fornix of older adults. Front. Aging Neurosci. **9** (2017). https://doi.org/10.3389/fnagi.2017.00059
12. A. Tapus, M. Mataric, B. Scassellati, Socially assistive robotics [grand challenges of robotics]. IEEE Robot. Autom. Mag. **14**(1), 35–42 (2007). https://doi.org/10.1109/MRA.2007.339605
13. P. Radanliev, Dance sport movement therapy in the metaverse: A new frontier for alternative mental health therapies. Digital Health **10** (2024). https://doi.org/10.1177/20552076241258273

14. P. Radanliev, Virtual dance movement therapy for reducing anxiety, and artificial intelligence for monitoring the body and mind during therapy. Res. Dance Educ., 1–15 (2025). https://doi.org/10.1080/14647893.2025.2486256
15. P. Olikkal, D. Pei, B.K. Karri, A. Satyanarayana, N.M. Kakoty, R. Vinjamuri, Biomimetic learning of hand gestures in a humanoid robot. Front Hum Neurosci **18**, 1391531 (2024). https://doi.org/10.3389/fnhum.2024.1391531
16. P. Olikkal, D. Pei, T. Adali, N. Banerjee, R. Vinjamuri, Data fusion-based musculoskeletal synergies in the grasping hand. Sensors (Basel) **22**(19) (2022). https://doi.org/10.3390/S22197417
17. P. Olikkal, D. Pei, T. Adali, N. Banerjee, R. Vinjamuri, Musculoskeletal synergies in the grasping hand. Proc. Annu. Int. Conf. IEEE Eng. Med. Biol. Soc. EMBS **2022**(July), 3649–3652 (2022). https://doi.org/10.1109/EMBC48229.2022.9871023
18. P. Olikkal, D. Pei, B.K. Karri, A. Satyanarayana, N.M. Kakoty, R. Vinjamuri, Learning Hand Gestures using Synergies in a Humanoid Robot, in *2023 IEEE International Conference on Robotics and Biomimetics ROBIO 2023*, (2023), pp. 1–6. https://doi.org/10.1109/ROBIO58561.2023.10354698
19. R. Bevilacqua et al., Exploring dance as a therapeutic approach for Parkinson disease through the social robotics for active and healthy ageing (SI-robotics): Results from a technical feasibility study. JMIR Aging **8**, e62930–e62930 (2025). https://doi.org/10.2196/62930
20. Y. Li, N. Liang, M. Effati, G. Nejat, Dances with social robots: A pilot study at long-term care. Robotics **11**(5), 96 (2022). https://doi.org/10.3390/robotics11050096
21. T.L. Chen et al., Evaluation by expert dancers of a robot that performs partnered stepping via haptic interaction. *PLoS One***10**(5), e0125179 (2015). https://doi.org/10.1371/journal.pone.0125179
22. H. Javed, C.H. Park, Promoting social engagement with a multi-role dancing robot for in-home autism care. Front. Robot. AI **9** (2022). https://doi.org/10.3389/frobt.2022.880691
23. J.C.P. Chan, H. Leung, J.K.T. Tang, T. Komura, A virtual reality dance training system using motion capture technology. IEEE Trans. Learn. Technol. **4**(2), 187–195 (2011). https://doi.org/10.1109/TLT.2010.27
24. J. Bai, R. Dai, J. Dai, J. Pan, EmoDescriptor: A hybrid feature for emotional classification in dance movements. Comput. Animat. Virtual Worlds **32**(6) (2021). https://doi.org/10.1002/cav.1996
25. C.L. Ferguson, J.A. Lockman, Increasing PhD student self-awareness and self-confidence through strengths-based professional development. Front. Educ. **9** (2024). https://doi.org/10.3389/feduc.2024.1379859
26. A. Sutton, Measuring the effects of self-awareness: Construction of the self-awareness outcomes questionnaire. Eur. J. Psychol. **12**(4), 645–658 (2016). https://doi.org/10.5964/ejop.v12i4.1178
27. S. Akiri, V. Joshi, S. Taherzadeh, G. Williams, H.M. Mentis, A. Kleinsmith, Design and preliminary evaluation of a stress reflection system for high-stress training environments, in *Companion Proceedings of the 26th International Conference on Multimodal Interaction*, (2024), pp. 11–15. https://doi.org/10.1145/3686215.3690148
28. S.A. Mackie, G.W. Bates, Contribution of the doctoral education environment to PhD candidates' mental health problems: A scoping review. High. Educ. Res. Dev. **38**(3), 565–578 (2019). https://doi.org/10.1080/07294360.2018.1556620
29. A. Ajendla, M. Patel, P. Olikkal, R.K. Vinjamuri, Mental health management through wearables and AI innovation, in *Smart Healthcare, Clinical Diagnostics, and Bioprinting Solutions for Modern Medicine*, (IGI Global, 2025), pp. 193–212. https://doi.org/10.4018/979-8-3373-0659-9.ch010
30. M. Mateu, S. Garcías, L. Spadafora, A. Andrés, E. Febrer, Student moods before and after body expression and dance assessments. Gender perspective. Front. Psychol. **11** (2021). https://doi.org/10.3389/fpsyg.2020.612811
31. C. Daudén Roquet et al., Exploring situated & embodied support for youth's mental health: design opportunities for interactive tangible device, in *CHI Conference on Human Factors in Computing Systems*, (2022), pp. 1–16. https://doi.org/10.1145/3491102.3502135

32. V. Hollis et al., What does all this data mean for my future mood? Actionable analytics and targeted reflection for emotional well-being. Human–Computer Interact **32**(5–6), 208–267 (2017). https://doi.org/10.1080/07370024.2016.1277724
33. R. Horlings, D. Datcu, Emotion recognition using brain activity, in *Proceedings of the 9th International Conference on Computer Systems and Technologies and Workshop for PhD Students in Computing – CompSysTech '08*, ed. by L.J.M. Rothkrantz, (2008), p. II.1. https://doi.org/10.1145/1500879.1500888
34. V. Bazarevsky, I. Grishchenko, K. Raveendran, T. Zhu, F. Zhang, M. Grundmann, BlazePose: On-device Real-time Body Pose tracking (2020)
35. F. Zhang, et al., MediaPipe Hands: On-Device Real-Time Hand Tracking (2020)
36. Unity-Chan! Model | 3D Characters | Unity Asset Store. https://assetstore.unity.com/packages/3d/characters/unity-chan-model-18705?srsltid=AfmBOoqTb-EIr5ogCW-XdgmXlwGF_8NRuT8kcZUJrhAvNTFKEqSkiPaI. Toegang verkry 03 Julie 2025
37. Robot Kyle | URP | 3D Robots | Unity Asset Store. https://assetstore.unity.com/packages/3d/characters/robots/robot-kyle-urp-4696. Toegang verkry 03 Julie 2025
38. S.M. Alarcao, M.J. Fonseca, Emotions recognition using EEG signals: A survey. IEEE Trans. Affect. Comput. **10**(3), 374–393 (2019). https://doi.org/10.1109/TAFFC.2017.2714671
39. P. Olikkal, B.M. Pillai, J. Suthakorn, H. Ali, R. Vinjamuri, A hybrid EEG-EMG framework for humanoid control using deep learning transformers, in *2024 IEEE International Conference on Robotics and Biomimetics (ROBIO)*, (2024), pp. 899–904. https://doi.org/10.1109/ROBIO64047.2024.10907308
40. W. Boucsein, *Electrodermal Activity* (Springer, Boston, MA, 2012). https://doi.org/10.1007/978-1-4614-1126-0
41. F. Shaffer, J.P. Ginsberg, An overview of heart rate variability metrics and norms. Front. Public Heal. **5** (2017). https://doi.org/10.3389/fpubh.2017.00258

Chapter 18
NEUROiD: Building Neuro-musculoskeletal Models of Upper and Lower Limb *In-Silico* by Bridging Neural Circuitry, Biomechanics, and Intelligent Control

Avinash Kumar Singh, Kapardi Mallampalli, Yashaswini Mandayam Rangayyan, Raghu Sesha Iyengar, and Mohan Raghavan

Abstract This chapter introduces NEUROiD, an integrated co-simulation framework designed to model human movement by coupling neural circuitry with biomechanical systems. NEUROiD combines NEURON-based simulations of spinal and cortical networks with OpenSim musculoskeletal models, enabling the simulation of closed-loop motor behaviour driven by realistic neural dynamics and proprioceptive feedback. The framework supports subject-specific personalization, allowing detailed exploration of how anatomical variability and neural impairments interact to shape motor outcomes. Case studies demonstrate its versatility: spinal microstimulation experiments replicate segmental movement patterns; virtual tendon transfer surgeries reveal the interplay between altered musculotendon geometry and reflex modulation; simulations of spasticity using the Modified Tardieu Test reproduce velocity-dependent resistance and clasp-knife effects; and reinforcement learning agents trained within NEUROiD learn biologically plausible control strategies through interaction with the virtual body. Together, these studies underscore NEUROiD's utility in bridging brain, body, and environment for research in motor control, rehabilitation, and intelligent neuro-prosthetic design.

Keywords Neuro-musculoskeletal modelling · NEUROiD · Biomechanics · Neurorehabilitation · Motor control · Spinal circuits · Proprioception · Virtual

Authors Avinash Kumar Singh and Kapardi Mallampalli have equally contributed to this chapter.

A. K. Singh · K. Mallampalli · Y. M. Rangayyan · R. S. Iyengar · M. Raghavan (✉)
Indian Institute of Technology Hyderabad, Hyderabad, Telangana, India
e-mail: bm18resch11012@iith.ac.in; bm17resch11002@iith.ac.in; bm20resch02003@iith.ac.in; mohanr@bme.iith.ac.in

R. Vinjamuri (ed.), *Bridging the Gap between Mind and Machine*,
https://doi.org/10.1007/978-3-032-06713-5_18

patients · Reinforcement learning · Spasticity · OpenSim · Closed-loop · Sensorimotor integration · Motor learning · Surgical simulation · In-silico trials

18.1 Introduction

In the pursuit of emulating human movement in machines and understanding the biological roots of motion, researchers have long stood at the interface of two powerful but distinct domains: neuroscience and biomechanics. The former aims to uncover the internal processes of the mind, including computations, reflexes, and feedback loops that coordinate movement, while the latter offers the mechanical foundation through which these commands are carried out. Bridging these domains is no longer just a theoretical exercise; it is the basis of a new era of medical simulation, rehabilitation, and intelligent neuro-prosthetic design.

In this chapter, we explore how advanced in silico neuromusculoskeletal models can serve as a bridge between the mind, represented by neural circuitry and intelligent control, and the machine, which includes the musculoskeletal system, actuators, and prosthetic interfaces. This convergence is embodied in our work through NEUROiD [1], a co-simulation platform that unifies NEURON [2] based spinal models with OpenSim based [3] biomechanical representations. Through NEUROiD, we have built highly detailed, modular models of upper and lower limb systems that allow us to simulate movement as an emergent phenomenon of neural activity, proprioceptive feedback, and mechanical dynamics.

Virtual patients have become a compelling concept in biomedical engineering, driven by the US FDA's growing emphasis on in-silico trials to de-risk and expedite therapeutic pipelines [4]. Most virtual patients, however, focus on cardiac, respiratory, or orthopaedic simulations, with limited neural representation [5–7]. Our work significantly extends this space by taking a step towards virtual patients with detailed neural controllers that reproduce sensorimotor behaviour in both health and disease. This is not merely a computational exercise, but a means to embody the mind within mechanical systems in a clinically and physiologically meaningful way.

The need for such systems is manifold. Traditional rehabilitation and surgical planning rely heavily on empirical outcomes and heuristic-based predictions. Conditions like spasticity, paralysis, or post-surgical neuromotor recovery are not only difficult to model but also exhibit high inter-individual variability [8, 9]. Our work addresses this complexity by enabling simulations of neuromotor responses under varying conditions of spinal cord integrity, proprioceptive disruption, and motor learning. For instance, in simulating the Modified Tardieu Test [10], we recreate clasp-knife effects and other velocity-dependent responses that are hallmarks of spasticity [11]. This level of realism opens the door to predictive simulation-based triage, where treatment paths can be tested virtually before implementation.

An essential part of bridging the gap between mind and machine is understanding how feedback mechanisms shape behaviour. In our studies involving virtual tendon transfer and pulley-based surgery [12] models [13], we examine how proprioceptive afferents (Ia, Ib, II) alter motor neuron firing patterns and thereby reshape force profiles during pincer grasp tasks. These simulations, conducted on wrist models with different configurations such as healthy, pathological, and surgically repaired, emphasize the importance of internal sensing in restoring motor function. Our virtual patients show that afferents play a crucial role in both improving and constraining performance, a detail that is difficult to isolate in living subjects.

But perhaps most critically, bridging this gap is not just about reproducing movement, it is more about enabling learning. The human motor system is adaptive, continuously reshaping itself in response to experience and environment. To mimic this property, we have explored reinforcement learning (RL) as a model of cortical adaptation. By wrapping our spino-musculoskeletal model into an OpenAI Gym environment, we allowed artificial agents to learn control policies for elbow flexion tasks. These agents, trained with sparse feedback, develop biologically plausible firing patterns that activate spinal circuits in coordination with biomechanical constraints. This fusion of machine learning with mechanistic neurophysiology not only aids in understanding cortical control but also provides a testbed for designing intelligent rehabilitation protocols.

Our approach highlights a central idea: movement arises from the interaction of nested systems including the neural, muscular, and skeletal components, which are naturally connected through feedback. By integrating these systems, each modelled with suitable detail and scale, we create a virtual platform where hypotheses about motor control, disease progression, and surgical recovery can be tested in a safe, consistent, and scalable manner.

This chapter brings together several strands of our research. The first involves creating biologically realistic spinal circuits and integrating them with limb mechanics - both upper and lower. This is exemplified by our upper limb movement simulations that replicate spinal micro-stimulation studies. The second is our investigation into surgical interventions through virtual pincer grasp models that simulate tendon transfer and analyse proprioceptive roles. The third is our work on lower limb spasticity, where we explore the onset of pathology through loss of descending modulation and conduct in-silico clinical assessments. The fourth and emerging thread is motor learning, where RL frameworks simulate cortical plasticity and its interactions with spinal feedback.

Together, these efforts form the basis for virtual patients that not only move but also learn, adapt, and respond. They establish a foundation for subject-specific simulations that consider differences in body structure, unique patterns of neural control, and the impact of treatments. These models evolve from abstract representations into functional stand-ins for human experience, capable of producing motion while capturing the complexity of the connection between mind and body.

18.2 Steps in Neuro-musculoskeletal Modelling

The process of neuromusculoskeletal modelling involves curating extensive data across scales and modalities [1]. The integration of multi-scale data requires careful selection of methodologies, as each assumption introduces trade-offs that can influence model accuracy, interpretability, and practical applicability. A model built on overly simplistic assumptions may be computationally efficient but lack physiological realism, while an overly detailed model may be intractable for practical use. The consequences of these choices manifest in the model's ability to generalize across different populations, predict novel motor behaviours, and guide clinical and technological applications. In the coming sections we extensively discuss the steps involved in a typical neuro-musculoskeletal modelling exercise.

18.2.1 Data Collection and Anatomical Reconstruction

Human subject experiments, including motion capture, MRI, and EMG recordings, provide the anatomical and functional data needed for model construction. The process of neuromusculoskeletal modelling begins with an extensive phase of data collection and anatomical reconstruction, where various experimental techniques are employed to obtain accurate representations of musculoskeletal geometry, and functional biomechanics.

Motion capture (MoCap) technology, which involves high-speed cameras and reflective markers placed on anatomical landmarks, is widely used to record human movement and joint kinematics. Optical motion capture systems, such as Vicon and OptiTrack, are among the most commonly used due to their high precision in tracking three-dimensional movements. These systems can be categorized into passive and active marker-based technologies. Passive marker systems use retroreflective markers illuminated by infrared cameras, while active marker-based systems utilize infrared markers that emit light signals, improving marker identification and reducing mislabelling errors [14].

However, marker-based systems suffer from soft tissue artifacts, where skin movement introduces inaccuracies in joint position estimations, leading researchers to explore marker-less motion capture alternatives. Recent advancements in deep-learning-based computer vision methods have shown promise in extracting joint kinematics from standard video footage without markers [15]. Additionally, inertial measurement units (IMUs), which use accelerometers and gyroscopes, provide an alternative for capturing movement outside laboratory settings. While IMUs offer portability and usability in diverse environments, they require extensive calibration and suffer from drift errors, affecting long-term accuracy [16].

In addition to motion capture, medical imaging techniques such as Magnetic Resonance Imaging (MRI) and Computed Tomography (CT) scans are used to reconstruct anatomical structures with high spatial resolution. MRI, particularly diffusion tensor imaging (DTI), provides detailed insights into muscle architecture, fibre orientations, and neural pathways, making it a preferred choice for soft tissue visualization [17, 18]. CT imaging, on the other hand, is more effective for capturing high-contrast skeletal structures, facilitating precise bone geometry reconstructions [19]. While MRI offers superior soft-tissue contrast, it is more expensive, time-consuming, and requires specialized post-processing algorithms to extract anatomical features. CT scans, though more accessible, expose subjects to ionizing radiation, making their repeated use for dynamic studies less desirable. The selection between these imaging modalities depends on the specific requirements of the study, where musculoskeletal models requiring fine details of bone structure benefit from CT, whereas those focusing on muscle dynamics and fibre architecture are best served by MRI [20].

Electromyography (EMG) is an essential technique for measuring muscle activity and neural drive to muscles. EMG recordings, obtained through surface electrodes or fine-wire intramuscular electrodes, provide crucial data for validating neuromuscular models by capturing activation patterns during movement [21].

Surface EMG is non-invasive and widely used in clinical and research settings. However, it has limitations in detecting deep muscle activity and differentiating signals from adjacent muscles due to cross-talk [22]. Intramuscular EMG, which involves inserting fine-wire electrodes into the muscle tissue, offers greater specificity but is invasive, making it less feasible for large-scale studies [23].

Processing EMG signals involves several key steps, including rectification, filtering, and normalization, to ensure that the extracted activation signals accurately reflect neuromuscular function [24]. Advances in signal processing techniques, such as high-density EMG and decomposition algorithms, have improved the resolution and interpretability of EMG data, enabling more precise assessments of motor unit activity [25].

Despite its limitations, EMG remains indispensable for validating neuromuscular control strategies within computational models. It has been particularly useful in understanding motor unit recruitment, muscle coordination, and neuromechanical interactions [26]. As neuromusculoskeletal models become increasingly sophisticated, integrating EMG data with simulations offers a robust method for enhancing predictions of muscle function and motor control [27]. The ultimate validation of these models will be achieved when simulated EMG data from tasks closely match the data collected in laboratory experiments, thereby ensuring the accuracy and reliability of the simulations.

18.2.2 Musculoskeletal Model Development

Once the necessary anatomical and functional data have been collected, the next step involves the construction of a musculoskeletal model that accurately represents the mechanical and physiological properties of the human body. The skeletal system is typically modelled as a series of rigid bodies connected by joints, where kinematic constraints define the degrees of freedom (DoF) allowed at each joint. These models are often developed in biomechanical simulation software such as OpenSim [3] AnyBody [28], or SIMM [29], which provide standardized musculoskeletal representations. Rigid-body models assume that bones do not deform under loading, which simplifies computations and is sufficient for most movement analyses, though finite element (FE) methods may be used for applications requiring stress-strain analysis of bone deformation [30].

OpenSim, in particular, provides multiple levels of muscle model abstraction, ranging from simple torque actuators to lumped fibre Hill-type muscle models [31, 32] as well as custom user defined models. This flexibility allows users to select an appropriate model based on computational demands and the required physiological detail. Furthermore, it is possible to develop highly detailed models of the musculoskeletal system by incorporating contact forces, wrap geometry, and ligaments, joint constraints etc., in the model. These features enable the realistic representation of muscle paths, complex joints with significant multi-degree-of-freedom movement, such as the shoulder and hip, thereby enhancing the accuracy of muscle force estimations and joint moment calculations.

The representation of muscles in models is crucial for accurately predicting forces and movement in neuromusculoskeletal simulations. Lumped fibre models, simplify complex internal structure of muscle systems while maintaining essential physiological features, enhancing computational efficiency. While these models are widely used due to their efficiency, and have been validated in specific contexts, they are not universally the standard; other approaches like finite element models based on continuum mechanics, are employed when detailed muscle properties are needed [33]. The choice between Hill-type and FE models depends on the trade-off between computational cost and physiological realism, with Hill-type models being preferable for whole-body movement simulations and FE models being used in localized studies requiring detailed tissue mechanics.

Biomechanical simulations can be broadly categorized based on how they handle movement and force estimation. Two common approaches are forward dynamic simulations, where muscle activations drive the system and the resulting motion is computed, and inverse dynamic simulations, where recorded motion data is used to estimate muscle forces and joint loads. Forward simulations are particularly useful for predictive modelling and understanding neuromuscular control, while inverse simulations help in analysing existing movement patterns [34].

Several validated musculoskeletal models have been developed using OpenSim to provide anatomically and functionally realistic representations of human movement. For example, Holzbaur's upper extremity model was developed based on detailed cadaveric data, including muscle architecture and joint kinematics, to accurately capture shoulder and elbow function [35]. Similarly, Rajagopal lower limb model incorporated extensive anatomical data from cadaver studies and imaging techniques to refine muscle attachment points, joint structures, and physiological cross-sectional areas, making it one of the most detailed lower limb models available [36]. These studies exemplify how researchers integrate high-fidelity anatomical data into their models to improve realism and predictive accuracy.

18.2.3 Neural Control Integration

Integrating neural control into a musculoskeletal model involves incorporating representations of spinal reflex pathways, central pattern generators (CPGs), and cortical control mechanisms to simulate movement generation and motor adaptation. Neural dynamics in neuromusculoskeletal models vary in complexity, from simple empirical functions mapping muscle activations to forces, to detailed spiking neuron models that simulate action potentials and synaptic transmission [37–45].

Reflex pathways are commonly modelled using delayed feedback control, where sensory inputs from muscle spindles and Golgi tendon organs modulate muscle activation in response to external perturbations. These models can be implemented using proportional-integral-derivative (PID) controllers or more advanced control-theoretic approaches such as optimal control and reinforcement learning [46]. While PID controllers are computationally simple and effective for basic movements, they lack adaptability and fail to capture the complexity of biological motor learning. Optimal control methods, which minimize cost functions representing energy expenditure or movement smoothness, provide more realistic motor behaviours but require extensive optimization and may not fully capture the variability observed in human movement [47, 48].

Central pattern generators (CPGs), which are neural circuits in the spinal cord responsible for generating rhythmic movements such as walking, can be modelled using coupled oscillators or artificial neural networks. These models replicate locomotor patterns observed in both animals and humans and are particularly useful in studies on gait disorders and rehabilitation strategies [49]. However, CPG models often require extensive parameter tuning to match experimental data, and their applicability to voluntary movements beyond locomotion is limited. In animal models such as cats, CPGs have been well characterized and shown to operate in the absence of cortical input. In contrast, the extent to which CPGs contribute to human locomotion remains uncertain due to greater reliance on descending control [50, 51]. Recent developments incorporate neuromodulatory effects and afferent feedback for improved adaptability [52].

To enhance our understanding and simulation of motor control complexities and movement disorders, it is crucial to focus on developing more detailed and physiologically relevant models of spinal circuitry. This approach will help move beyond abstracted mathematical controllers that lack the nuance of biological systems. Spinal circuits can be modelled based on functional spinal circuit literature, incorporating known pathways such as the corticospinal, reticulospinal, and propriospinal systems to simulate movement disorders in an integrated fashion [53–57]. These models can help explore mechanisms underlying conditions such as spasticity, dystonia, and spinal cord injury, offering insights into potential therapeutic interventions. Advances in computational frameworks allow for the integration of multiple spinal pathways, enabling more physiologically realistic simulations of movement impairments and recovery mechanisms [58].

Various tools exist for modelling neural dynamics and their integration with musculoskeletal models. Simulators such as NEURON [2], NEST [59], and Brian [60] allow for the development of detailed spiking neural network models, while OpenSim (Delp et al., 2007) and Mujoco [61] provide platforms for musculoskeletal simulations. The simulation engines for neural dynamics and musculoskeletal dynamics used together can enable researchers to study the interaction between neural control and biomechanics in a unified framework [1].

Supraspinal control mechanisms, which include cortical and cerebellar contributions, are more difficult to model due to their complex connectivity and task-dependent behaviour. Some models use simplified representations of cortical activation derived from electroencephalography (EEG) or functional MRI (fMRI) data, while others rely on detailed spiking neural networks [62–65].

18.2.4 Parameter Tuning and Validation

After developing musculoskeletal and neural models, parameter tuning, and validation are essential to ensure simulations accurately reflect experimental data. Optimization techniques such as genetic algorithms, simulated annealing, and Bayesian inference are employed to refine model parameters, including muscle force-length properties, neural activation patterns, and joint stiffness.

Genetic algorithms are favoured for their efficiency in exploring large parameter spaces, despite requiring numerous iterations to converge [66]. These algorithms simulate natural selection processes to iteratively improve parameter estimates. Simulated annealing, inspired by the annealing process in metallurgy, is another optimization method used in this context. It probabilistically explores the parameter space, allowing occasional acceptance of worse solutions to escape local minima, which can be beneficial in complex model tuning [67, 68]. Bayesian inference offers a probabilistic framework for parameter estimation, incorporating prior knowledge of parameter distributions. While it provides a comprehensive uncertainty quantification, it is computationally intensive and necessitates well-defined prior information [69].

A critical step in model evaluation is validation, which ensures that the simulated outputs align with experimental data. This process involves systematically comparing model predictions with empirical measurements obtained from motion capture systems, electromyography (EMG), and force plate data [70]. Discrepancies between simulated and real-world data highlight areas requiring refinement, allowing for iterative improvements to the model's predictive accuracy. Motion capture provides detailed kinematic data, EMG records neural activation patterns, and force plates measure ground reaction forces, all contributing to a holistic validation framework [71]. Additionally, joint torque measurements have been increasingly used to cross-validate muscle force estimations, adding further depth to the validation process [72, 73].

Sensitivity analysis plays a crucial role in assessing how variations in model parameters affect predictions. By systematically perturbing input values and analysing their impact on outcomes, researchers can determine which parameters have the greatest influence on model behaviour, ensuring robustness against measurement uncertainties [74]. This approach enhances confidence in model predictions by identifying key determinants of variability and guiding targeted refinements [70].

In summary, the integration of advanced optimization techniques with robust validation protocols is crucial for developing reliable musculoskeletal and neural models. By leveraging experimental data, sensitivity analysis, and innovative computational techniques, researchers can enhance model accuracy and applicability in both clinical and research settings.

18.2.5 Simulation and Analysis

Once validated, the neuromusculoskeletal model serves as a powerful tool for simulating and analysing human movement, enabling researchers to test hypotheses related to biomechanics, pathology, and rehabilitation. These simulations can be classified into two main approaches: forward dynamics and inverse dynamics.

In forward dynamics simulations, movement is generated by applying known neural inputs, such as muscle activation patterns, to predict resulting kinematics and kinetics. This approach is particularly valuable for investigating novel movement strategies, optimizing motor control theories, and assessing the effects of neuromuscular impairments [75–77]. However, forward dynamics requires precise input parameters, including realistic neural excitations and musculoskeletal properties, which can be challenging to obtain.

Conversely, inverse dynamics relies on experimental motion capture data to estimate underlying joint torques, muscle forces, and activations. This method is widely used in clinical gait analysis and rehabilitation research, as it provides insight into musculoskeletal loading patterns and compensatory mechanisms in patients with movement disorders [78]. However, inverse dynamics is constrained by experimental limitations, such as marker placement errors and soft tissue artifacts, which can introduce inaccuracies in force estimations.

Applications of neuromusculoskeletal modelling span a wide range of fields. In the study of neuromuscular disorders, such as cerebral palsy or stroke, these models help assess motor deficits and evaluate the effectiveness of surgical or rehabilitative interventions [79]. In prosthetics research, simulations aid in the design and optimization of lower-limb prostheses, ensuring improved stability and energy efficiency for users [80]. Additionally, in the development of assistive exoskeletons, neuromusculoskeletal models facilitate the tuning of control algorithms to enhance user comfort and mobility [81].

By enabling virtual experiments that would be difficult or impossible to conduct in vivo, neuromusculoskeletal modelling has emerged as a critical tool in neuroscience, biomechanics, rehabilitation, and clinical decision-making.

18.3 NEUROiD: Platform for Simulating Neuromusculoskeletal Systems

NEUROiD (NEUROmotor Integration and Design) is an integrated co-simulation platform that bridges detailed neural modelling with biomechanical simulation. It enables real-time, bidirectional communication between the nervous system and the musculoskeletal system by coupling the NEURON simulator, which handles electrophysiologically detailed neural circuits and OpenSim, which provides anatomically accurate musculoskeletal dynamics. This coupling is essential for modelling closed-loop movement generation, where descending neural commands activate muscles, produce motion, and generate proprioceptive feedback that modulates ongoing neural activity. By replicating this biological feedback cycle, NEUROiD supports the simulation of both healthy and pathological human movement with a level of precision and adaptability not previously possible.

18.3.1 Design Goals

NEUROiD was built with three primary goals: scalability, subject specificity, and biological fidelity. The platform enables modelling across spatial and temporal scales, from single neurons to full-body systems, and supports large-scale simulations using high-performance computing infrastructure. Subject-specific details such as body dimensions, muscle architecture, reflex patterns, and surgical modifications can be incorporated to create personalized virtual patients. NEUROiD achieves high biological realism through biophysically accurate neuron models, validated muscle dynamics, and integrated reflex loops. A more comprehensive and in-depth description of the platform's design, simulation framework, and capabilities is provided elsewhere [1, 82].

18.3.2 System Architecture

NEUROiD is designed as a modular and extensible platform, where neural and musculoskeletal components are built, tested, and reused independently. The architecture supports tight coupling between NEURON and OpenSim through a Python-based middleware, allowing simulations to proceed in synchronized time steps.

NEURON generates motor commands in the form of spike trains, which are translated into muscle activations, while OpenSim updates the biomechanical state of the system. Feedback such as muscle stretch, velocity, and force are computed and routed back to NEURON to influence ongoing neural activity. The communication interface is configurable for latency and update frequency, enabling either high-fidelity simulation or fast prototyping depending on the user's goals.

18.3.3 Neural Module

The neural modelling component (based on NEURON) supports detailed simulation of spinal circuits, including alpha motor neurons, Ia, Ib, and group II interneurons (both excitatory and inhibitory), and Renshaw cells. Descending tracts such as corticospinal, vestibulospinal, reticulospinal, and rubrospinal pathways can be included as modulatory inputs or explicit spiking populations. Sensory afferents (Ia, Ib, II) are modelled based on muscle states including stretch, velocity, and force. The neural elements can be built using either biophysically detailed Hodgkin-Huxley-type neurons or simplified spiking units, offering flexibility in balancing detail and computational load. Connectivity rules are defined using structured templates, allowing replication of known reflex pathways or exploration of pathological scenarios such as spinal injury or hyperreflexia.

18.3.4 Musculoskeletal Module

OpenSim provides the biomechanical foundation of NEUROiD. It simulates body dynamics using anatomically accurate models that include joints with degrees of freedom, muscle-tendon units, ligaments, and soft-tissue contacts. Muscles are represented using Hill-type models with customizable force-length-velocity properties. The platform supports subject-specific scaling using motion capture data and morphometric measurements. Outputs from this module—such as joint angles, muscle lengths, moment arms, and generated forces—are used to calculate afferent feedback that informs the neural module.

18.3.5 Model Development Workflow

Model development in NEUROiD follows a structured pipeline that supports subject-specific neuromusculoskeletal simulation (refer Fig. 18.1). The process begins by selecting a joint or motion of interest, identifying the muscles and associated motor neurons, and gathering circuit-level parameters from literature. These parameters are organized into standardized spreadsheet templates, which are parsed

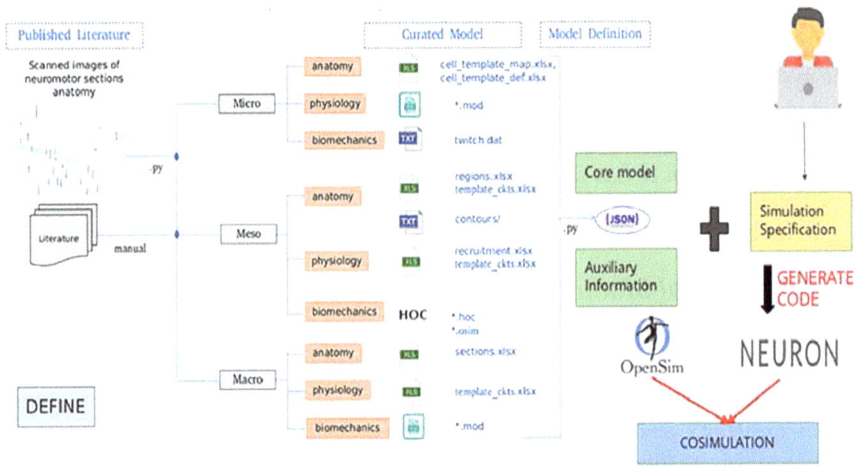

Fig. 18.1 The workflow for model development proceeds from left to right. Relevant parameters are manually extracted from published literature based on the model of interest. These parameters, spanning multiple scales and domains, are organized into structured Excel tables. Python scripts then convert these curated tables into composite model definitions formatted as JSON files. Simulation settings provided by the user are used to run the model using the NEURON simulation engine. (Adapted from Iyengar et al. [1])

using Python scripts to generate JSON-based composite model definitions. These definitions are then used to instantiate models within NEURON and OpenSim, guided by the specific objectives of the simulation, whether it be to observe spinal reflex dynamics or simulate full-limb movement following a surgical intervention.

Models are built across multiple scales. At the microscopic level, neuron morphology, ion channel properties, and muscle fibre twitch dynamics are defined. Mesoscale definitions include laminar organization of spinal segments, synaptic connectivity rules, and muscle architecture. At the macroscale, full-body dynamics, local field potentials, and joint kinematics are simulated. Model components are sourced from publicly available repositories such as ModelDB [83] and Channelpedia [84], or custom-built as needed. Through this scalable and modular workflow, NEUROiD enables rapid development and testing of both generic and patient-specific models.

18.4 Modelling Movement and Movement Disorders in NEUROiD

18.4.1 Spinal Microstimulation and Feedback-Driven Arm Movement

One of the fundamental demonstrations of embodied neural control is the ability to generate naturalistic movement by stimulating the spinal cord. In experimental neuroscience, intraspinal microstimulation (ISMS) has shown that stimulating specific spinal segments can evoke coordinated limb movements [85]. Our work replicates this finding in-silico using NEUROiD's upper limb model [86]. We conducted several simulation experiments to assess the neuro-musculoskeletal behaviour of upper limb models under spinal cord stimulation.

Three OpenSim models were to study upper limb muscle dynamics. The Arms26 model [35] includes six muscles—biceps (two heads), triceps (three heads), and brachialis—with shoulder constraints. The MOBL-ARMS model [35] has 50 muscles across the shoulder, elbow, and wrist with seven degrees of freedom but lacks digit and neck muscles. The wrist-only model [87] has 15 muscles and supports wrist flexion-extension. Gravity was disabled in all simulations. Arms26 was used to test spinal-motor interfaces and afferent pathways; MOBL-ARMS for multi-joint coordination; and the wrist model to assess digit manipulation. Initial postures ensured equilibrium with no residual forces or stimulation. Sanity tests confirmed the integrity of muscle-motor neuron connections and the model's capacity to exercise its available degrees of freedom.

One of the key trends in neuromusculoskeletal modelling is the use of abstraction to simplify biological systems while retaining essential functional characteristics. Rather than modelling every neuron and muscle fibre individually, researchers use reduced-order models that capture the critical dynamics of movement control. For example, simplified models of spinal circuits can accurately replicate reflexive responses without needing to simulate the full CNS [39]. Similarly, lumped-parameter muscle models, such as Hill-type actuators, are widely used to approximate muscle contraction dynamics based on empirical data [88]. These abstractions make computational modelling more feasible while preserving biological realism.

The ability to generate, control, and adapt movement in-silico requires more than reproducing biomechanical motion—it demands the embedding of neural intent, proprioceptive feedback, and plasticity. In this section, we highlight three major case studies from our research that demonstrate how NEUROiD-based models can function as intelligent, embodied systems that connect neural control with physical embodiment, offering a biologically grounded approach to understanding sensorimotor integration.

18.4.1.1 Individual Segment Test

This test evaluated how stimulation of different spinal cord segments influenced limb movement. It was inspired by in-vivo spinal micro-stimulation studies [85], though specific motor neuron targeting data were unavailable. Therefore, all motor neurons within each spinal segment (C4–T1) were stimulated simultaneously using intracellular current injection. Parameters such as frequency, duration, and target neurons were compiled and translated into NEURON's netstim format. A square pulse of 380 nA and 200 ms duration was used to evoke approximately 20 Hz firing in alpha motor neurons. For the wrist model, stimulation lasted 100 ms.

Simulations showed that each spinal segment generated characteristic joint motions, with joint displacements mapped for all degrees of freedom. The movement types generally aligned with those reported in segmental movement maps [85, 89], though some differences were expected due to the coarse stimulation approach. Notably, flexors elicited a larger range of motion than extensors in OL, particularly from a 60° elbow flexion starting posture. This reflected the mechanical advantage of flexors and their physiological dominance in that configuration.

In CL mode, the inclusion of afferents consistently reduced the range of motion, confirming the expected modulatory effect of sensory feedback. This afferent-mediated resistance was highly dependent on the amount of muscle stretch from optimal lengths, consistent with known physiological behaviour [90, 91].

18.4.1.2 Varying Initial Postures Test

This test examined how different joint configurations at the start of stimulation affect motor output. Identical stimulation inputs were applied with the elbow initially flexed at 90°, 60°, or 30°. This setup allowed analysis of how muscle length-tension properties and mechanical advantage interact with spinal feedback to produce movement.

Results revealed that initial posture had a significant effect on movement outcomes. For example, starting from a highly flexed position led to greater extension, while lower flexion postures favoured larger flexion responses. These effects were attributed to proprioceptive afferents attempting to return muscles to their optimal length [55]. The results confirmed the model's capacity to reflect realistic, posture-dependent variations in movement due to both mechanical and neural factors.

18.4.1.3 Conduction Delay Test

This experiment assessed the role of afferent conduction delays on motor output. Different afferent types have distinct peripheral latencies, influencing the timing of their feedback. The model was tested with three latency configurations (Table 4), altering the arrival times of Ia, Ib, and type II inputs to motor neurons and interneurons.

Simulation results showed that afferent timing critically shaped the movement response. In configurations where Ia arrived much earlier than Ib and II, elbow extension movements closely followed Ib dynamics. As the latency differences narrowed, flexion behaviour shifted to reflect more balanced input from Ia and II afferents. These outcomes validated the model's sensitivity to afferent latency and confirmed that movement is governed by the relative timing of converging sensory feedback signals.

18.4.1.4 Connection Strength Test

Since physiological measurements of afferent synaptic weights are limited, this test explored how different afferent-to-motor neuron weight configurations affect behaviour. Weights were tuned using inhibitory-to-excitatory (I/E) ratios for Ia, Ib, and II afferents, along with a global scale parameter. Firing rates in the absence of gamma drive were matched to [92, 93].

Three levels of synaptic strengths were tested: nominal (NW), high (HW = NW × 5), and low (LW = NW ÷ 5). In OL conditions, movements were robust, but in CL mode, increasing afferent weight led to greater resistance and reduced range of motion. At high weights, antagonistic muscles sometimes overpowered agonists, inverting the expected motion (e.g., flexion during extensor stimulation). These results underscore the importance of balanced afferent connectivity to maintain stable, physiological motor control.

18.4.1.5 All Segment Test

To evaluate the model's ability to handle dynamic, sequential motor commands, all spinal segments from C4 to T1 were stimulated in ascending and descending order. Each segment received a 50 ms pulse at 20 Hz, with a total simulation time of 600 ms. The MOBL-ARMS model was used for this test.

The limb exhibited smooth, continuous motion across multiple joints, including coordinated flexion/extension and supination/pronation. Transitions between antagonistic movements were fluid, reflecting realistic motor coordination even under simplified, untargeted segmental stimulation. These results matched prior findings [94] and demonstrated that the model could generate realistic, temporally coherent responses to spatiotemporal motor inputs.

18.4.1.6 Summary

We used virtual arms, integrating a realistic musculoskeletal model of the shoulder, elbow, and wrist with a spinal neural circuit representing cervical segments C4 to T1. The spinal model included alpha motor neurons, Ia/Ib/II afferents, interneurons, and Renshaw cells, organized according to established segmental anatomy.

By delivering spatiotemporal current pulses to specific spinal segments (such as C5 for elbow flexion), we observed emergent arm movements spanning multiple degrees of freedom. These responses closely resembled experimental movement maps reported in primates and rodents undergoing intraspinal micro-stimulation (ISMS), validating the physiological accuracy of the model.

We further explored closed-loop control by activating proprioceptive afferent pathways. For instance, changes in muscle stretch velocity sensed by Ia fibres led to real-time modulation of motor neuron output, dynamically adjusting joint trajectories. This demonstrated self-regulating control, where motor commands induced movement, movement altered sensory feedback, and the feedback refined ongoing motor output.

Each experiment captured a critical element of motor behaviour, the segmental mapping, posture sensitivity, conduction latency, synaptic strength, and sequential activation, collectively advancing our understanding of spinal control in upper limb function and its implications for neuro-prosthetics and neurorehabilitation.

18.4.2 Modelling Tendon Transfer: Surgery, Recovery, and Feedback Control

Surgical interventions that rewire musculotendon architecture such as tendon transfers pose a unique challenge to the nervous system. In particular, the BRD-to-FPL transfer, used in upper limb reconstruction after spinal cord injury or radial nerve palsy, involves rerouting the tendon of the brachioradialis (BRD) muscle to replace the damaged flexor pollicis longus (FPL), restoring thumb function [95].

Using NEUROiD, we simulated a pincer grasp task, an essential indicator of fine motor ability, to compare two surgical approaches: tendon transfer and novel pulley-based reconstruction [13]. These approaches are commonly applied in patients with damaged or non-functional thumb muscles. We employed 4 four wrist models for our study, representing healthy hand, pathological case (damaged FPL), conventional tendon transfer and novel pulley-based surgery. Simulating the task across these wrist models allowed us to study not only task performance but also the underlying neuromechanical dynamics in a reproducible and quantifiable way. This setup provides a safe and flexible means of analysing how structural changes in the musculoskeletal system affect force generation, motion trajectories, and neural control.

18.4.2.1 Pincer Grasp Task: Motion Analysis

Figures 18.2, 18.3 and 18.4 show the frame-by-frame motion of the thumb and index finger during the simulated pincer grasp task across three wrist models: healthy, tendon-transfer, and pulley-based. In the healthy model, the Flexor Pollicis Longus (FPL) motor neurons were stimulated at 50 Hz in spinal cord segments C7 and C8, resulting in thumb flexion and abduction. This enabled the contact pad on the thumb to move toward the index finger, ultimately achieving a coordinated and stable pincer grip. The initial contact between the thumb and touchpad was recorded at 289 ms and was maintained until the end of the simulation at 1000 ms (Fig. 18.2).

In the tendon-transfer model, the Brachioradialis (BR) tendon was surgically connected to the FPL muscle. Upon stimulation of BR-related motor neurons at 50 Hz in segments C5 and C6, the thumb underwent a combination of flexion and adduction. This resulted in the thumb moving toward the index finger and

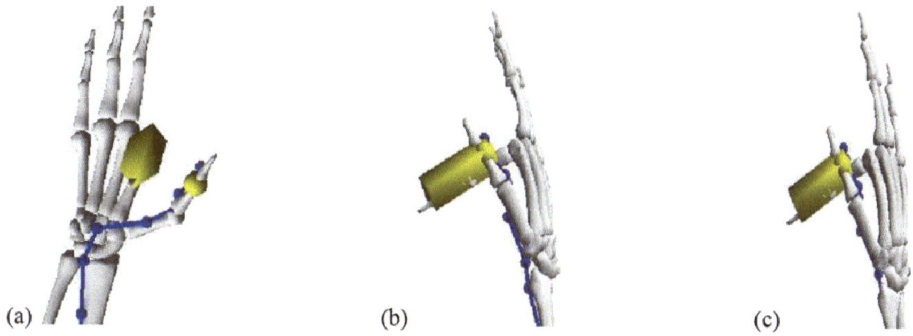

Fig. 18.2 Demonstrates pincer grasp task occurring from frame to frame, with the thumb touching the touchpad on index digit by healthy wrist model. The first frame (**a**) shows the initial position of the model at 0 ms. (**b**) represents the instance where thumb made a contact on the touchpad for first time at 289 ms, and third frame (**c**) represents the task at the end of the simulation at 1000 ms. The pincer grasp task was achieved by stimulating FPL related motor neurons in C7 and C8 spinal cord segments at 50 Hz respectively. The above frames are recorded during CL condition. The model is rotated within frames in (b) and (c) for better visualization of contact of thumb and touchpad

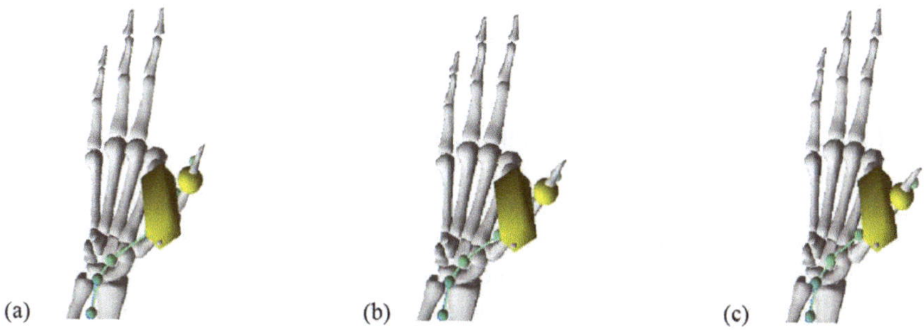

Fig. 18.3 The motion captured in these frames depicts a pincer grasp task performed by tendon-transfer model with the thumb touching the touchpad on the index digit. In the first frame (a), the thumb is initially positioned away from the index digit. As the simulation progresses, (b) represents the contact made by the thumb on the touchpad for the first time at 925 ms and the third frame (c) represents the task at the end of the simulation at 1000 ms. The motion is characterized by smooth and extended movement by thumb. The pincer grasp task was achieved by stimulating BRD related motor neurons in the C5 and C6 spinal cord segments at 50 Hz respectively. This simulation was performed under CL condition

achieving contact at 925 ms (Fig. 18.3). The pulley-based model involved the insertion of a mechanical pulley between BR and FPL tendons to introduce a mechanical advantage. With similar BR motor neuron stimulation parameters, the thumb reached the touchpad earlier than in the tendon-transfer model, achieving contact at 812 ms (Fig. 18.4). The pathological model, lacking sufficient motor control, was unable to complete the task and was excluded from detailed kinematic analysis.

The motion sequences provide a clear view of the task execution timeline across models. The healthy wrist completed the task in the shortest time, followed by the

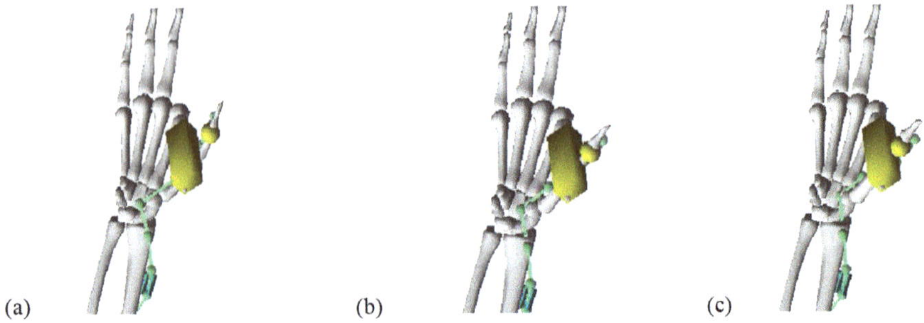

Fig. 18.4 Represents the frames capturing pincer grasp task by pulley-based surgical model. The first frame (a) shows the initial position of the model at 0 ms. The second frame (b) represents the first instance where the thumb touched the touchpad at 812 ms and the third frame (c) represents the task at the end of the simulation at 1000 ms. Throughout the simulation, BRD related motor neurons in the C5 and C6 spinal cord segments were stimulated at 50 Hz to achieve precise movement and control. The model was simulated under CL condition.

pulley-based and tendon-transfer models. These results are consistent with timing data reported in prior literature [96, 97].

18.4.2.2 Kinematic and Joint Angle Analysis

To further evaluate motion quality, we analysed the changes in the carpometacarpal joint angles of the thumb during the grasp task (Fig. 18.5). The healthy model exhibited rapid flexion followed by minor adjustments after initial contact, maintaining a stable and controlled grasp throughout the simulation. Both surgical models required a longer time to achieve contact and exhibited greater thumb flexion angles compared to the healthy model. After initial contact, the thumb in both surgical cases continued to slide along the touchpad surface to stabilize the grasp.

The pathological model did not demonstrate significant thumb movement or flexion, confirming its inability to perform the task. These joint angle trajectories illustrate the compensatory kinematics introduced by surgical modifications. The greater flexion angles in the tendon-transfer and pulley-based models suggest that although the task is achieved, it is done so through altered mechanical pathways that require increased joint displacement.

18.4.2.3 Muscle Force and Afferent Activity

Muscle force data for the BR and FPL muscles were recorded for the tendon-transfer and pulley-based models to evaluate the effectiveness of each surgical configuration in transmitting force during grasp. Figure 18.6a shows that in the tendon-transfer model, the direct latching of BR to FPL allowed reasonable force transfer, with FPL and BR muscles generating similar force levels. In contrast, the pulley-based model

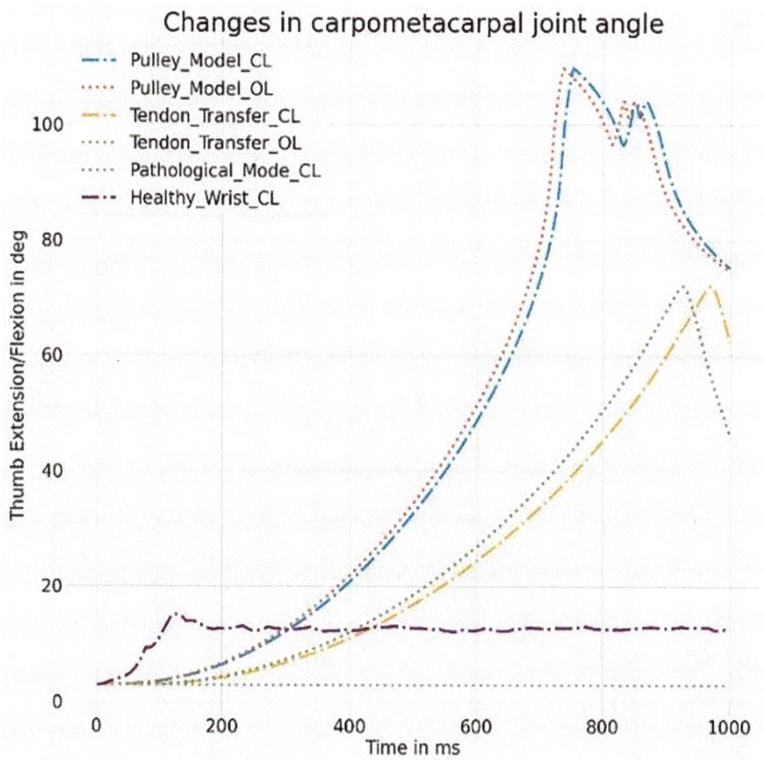

Fig. 18.5 Changes in carpometacarpal joint angles for different wrist models during OL and CL conditions. This figure represents the flexion of thumb during the simulation of pincer grasp task using the considered wrist models. The stimulation of BRD/FPL muscle had incited thumb flexion in the healthy wrist model. After initial contact with the touchpad, it made minute adjustments to achieve pincer grasp task and remained steady until the end of simulation. The position of the thumb was unchanged during the entire simulation for the pathological model. However, both the surgical models took longer duration to achieve pincer grasp task. After the thumb made initial contact with the index digit in both the surgical models, it started to slide on the touchpad, to achieve pincer grasp task. The thumb in healthy wrist model flexed less compared to surgical models

demonstrated a marked mechanical advantage; the pulley amplified FPL output such that it exerted approximately 1.75 times the force of BR. This result aligns with prior biomechanical observations [12].

To understand the neural implications of these configurations, we recorded firing rates of Ia and Ib afferent neurons associated with the BR and FPL muscles (Fig. 18.6b). Both models showed higher Ib firing rates compared to Ia, consistent with the known sensitivity of Ib afferents to tendon tension. FPL-related afferents exhibited higher firing rates than BR-related ones across both models. Notably, in the pulley-based configuration, the amplified FPL activity resulted in significantly increased Ib firing, which in turn led to greater inhibition of BR motor neurons. This

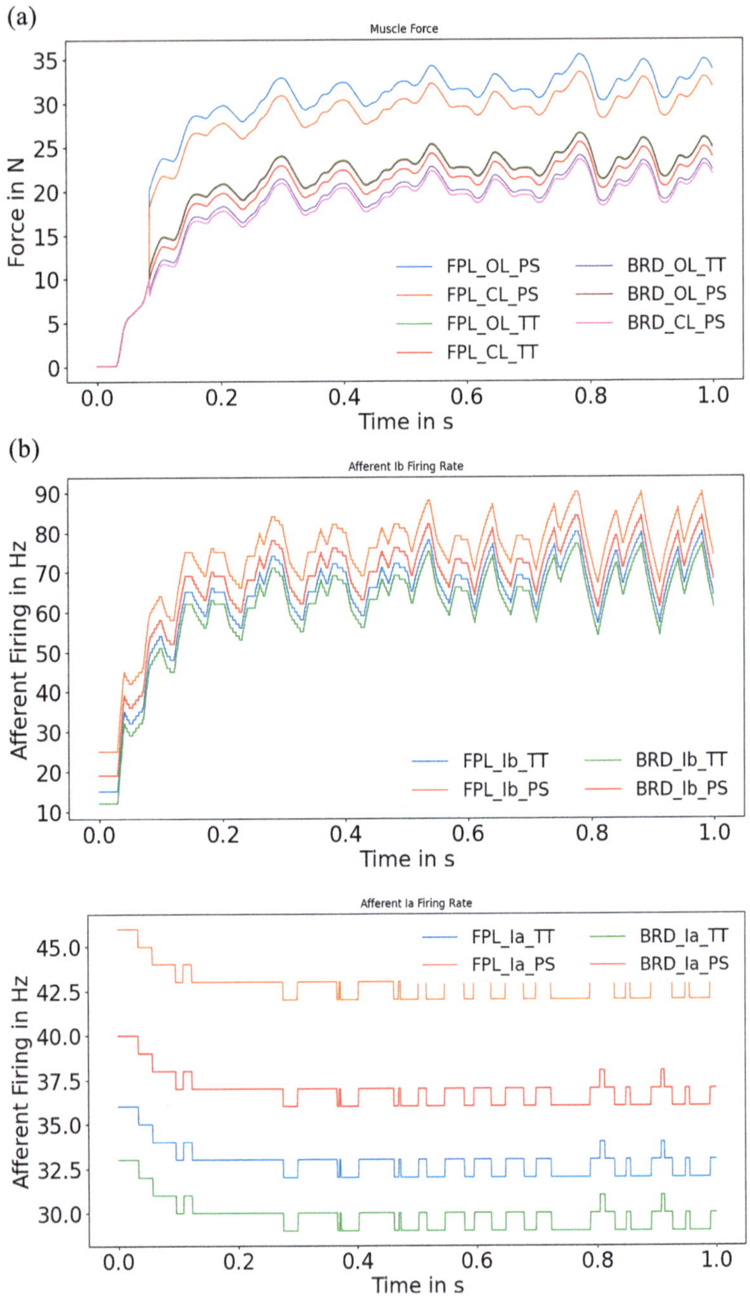

Fig. 18.6 (a) and (b), TT represent Tendon-transfer and PS refers Pulley-based surgery model. Fig. 18.6 (b) represents Ib and Ia firing rates of FPL and BRD based afferent cells in the motor circuit. From Fig. 18.6 (b), it is observed that due to facilitation of FPL tendon in both the models, the Ib afferent firing rate has been higher when compared to Ia afferent firing. Among both models,

Table 18.1 Force exerted by thumb on the touchpad

Wrist model	Force exerted during CL condition (in N)	Force exerted during OL condition (in N)
Pulley-based model	11	14
Tendon-transfer model	7	9
Healthy model	20	22

inhibitory effect was reflected in reduced BR force output under closed-loop (CL) control.

In the tendon-transfer model, the balance of afferent feedback led to more equal contributions from both muscles. Table 18.1 presents the net force exerted by the thumb on the touchpad under both CL and open-loop (OL) conditions. The healthy model produced the highest force (20 N CL, 22 N OL), followed by the pulley-based model (11 N CL, 14 N OL) and the tendon-transfer model (7 N CL, 9 N OL). These values indicate that while the surgical models enable task completion, the mechanical and neural differences between them have a significant impact on task strength and timing.

18.4.2.4 Neural Modulation and Sensory Feedback

The afferent analysis illustrates the complex neural dynamics that emerge following structural surgical changes. In the pulley-based model, the heightened Ib afferent activity due to amplified FPL force induced a stronger inhibitory signal on BR motor neurons, thereby reducing BR's contribution to the task. This mechanism reflects an adaptation in the motor control strategy driven by changes in proprioceptive input. The modulation of thumb motion through afferent signalling is also visible in Fig. 18.6, where afferent input appeared to constrain thumb joint angles during task execution, leading to smoother and more stable grasping.

These findings underscore the integral role of proprioception and feedback loops in functional task recovery following surgery. Even when mechanical structures are appropriately altered, the success of the surgical intervention is partially dependent on how the neural system adapts and redistributes control.

Fig. 18.6 (continued) the FPL afferent firing rate of Ib and Ia cells was higher when compared to BRD related afferents. In the Pulley-based model, the higher firing rate of FPL afferents had increased the inhibition of BRD motor neurons. Hence, muscle force of BRD in pulley-based model under CL condition has been significantly lower compared to other conditions and models (Fig. 18.6a). In tendon-transfer model, muscle force of FPL and BRD was in similar range, but in pulley-based surgery model FPL muscle force was ~1.75 times BRD muscle. The wave pattern in muscle force results from the interaction of biomechanical factors, including tendon compliance post-surgery, and joint mechanics as the thumb adjusts during contact with the index finger. (**a**) Muscle force of BRD and FPL in tendon-transfer and pulley-based models. (**b**) Afferent firing rates of Ib and Ia cells

18.4.2.5 Summary

Using NEUROiD, we simulated four wrist configurations: a healthy model with normal FPL, a pathological model with disabled FPL, a tendon transfer model connecting BRD to FPL, and a pulley enhanced transfer model with the tendon looped around bone. The pincer grasp task, requiring thumb and index coordination, was performed under open loop and closed loop conditions to assess the role of proprioceptive afferents.

The tendon transfer model restored basic thumb flexion but produced limited force, while the pulley model significantly improved force output. Afferent feedback enabled finer control but also led to inhibition of BRD activation in some cases, revealing reflex driven effects from altered tendon routing. These results show that changes in musculotendon geometry affect both mechanics and neural feedback, with spinal reflexes adapting in ways that are not always beneficial.

This framework allows testing of alternate configurations and control schemes before surgery, supporting patient specific virtual planning to predict functional outcomes based on surgical choices and neural integrity.

18.4.3 Learning Movement Using Reinforcement Learning in NEUROiD

Understanding the neural basis of movement control, particularly the role of the cortex in processing sensory feedback and issuing motor commands is a major focus in neuroscience. Spinal networks have been widely studied in various contexts: undulatory motion [98, 99], flexor-extensor coordination in mammals [100], left-right coordination [101], and limb activation patterns [102]. Movement control fundamentally depends on sensory feedback from muscle spindles, joints, and cutaneous receptors. Specifically, Ia afferents respond to changes in muscle length and velocity, while II afferents primarily respond to muscle stretch [6, 55]. The complex role of Ia, II, and Ib afferents in motor feedback during locomotion has been extensively investigated and modelled [90, 103].

Despite these insights, the exact mechanisms by which the motor cortex selects and controls movements, especially in the vertebrate upper limb, remain unresolved [104]. It is generally accepted that cortical activity encodes more than just muscle activation, yet there is no consensus on what additional parameters are encoded or which neuronal subsets are responsible [105]. Cortico-motoneurons (CM), located in the caudal region of the primary motor cortex (M1), have been shown to correlate with various movement features, such as position [106], velocity [12, 13, 107, 108], force [109], torque dynamics [110], and EMG activity [111]. However, their activity is specific to certain movement types and does not directly mirror alpha motor neuron activation.

An alternative theory suggests that M1 neurons encode temporally extended movement trajectories [112, 113], though how such neurons acquire these properties and how different neuron types in M1 interact to generate cohesive movement is still poorly understood.

Further, the human motor system is not a static controller; it is adaptive, learning to adjust its commands based on experience, success, and failure. This capacity for motor learning is a cornerstone of cortical control, allowing humans to fine-tune their movements through trial and error, repetition, and goal-oriented feedback. To bridge the gap between machine and mind, it is essential not just to simulate movement, but to enable learning over neuromusculoskeletal substrates.

In this section, we present our integration of Reinforcement Learning (RL) [114] into the NEUROiD platform, where artificial agents act as virtual cortical controllers that learn to move a realistic arm model by discovering optimal patterns of stimulation to spinal motor circuits. This approach not only allows movement to be generated from neural input, but also enables the emergence of cortical representations that mimic biological systems.

18.4.3.1 Reinforcement Learning Over Neuromusculoskeletal Substrates

To simulate cortical-level motor learning, we cast the movement generation problem as a reinforcement learning task [115]. In this paradigm, an agent interacts with the environment in this case, a virtual upper limb driven by a spinal-musculoskeletal system and learns a control policy by maximizing a reward signal tied to task success.

Simulation Environment

We wrapped our NEUROiD model into an OpenAI Gym [116]-compatible interface, allowing direct integration with modern RL algorithms. The environment consists of: A spinal cord segment (e.g., C5–C6), with motor neurons and interneurons simulated in NEURON. A biomechanical model of the elbow joint (e.g., Arms26), with biceps, brachialis, and triceps modelled in OpenSim (refer Fig. 18.7).

Proprioceptive feedback in the form of Ia, Ib, and II afferent firing rates derived from muscle length, force, and velocity. An action space where the RL agent modulates descending stimulation patterns mimicking cortical output from supraspinal centres. A reward function based on task objectives, such as achieving a target joint angle, minimizing energy, or maximizing movement smoothness (penalty for biologically unrealistic configurations).

Muscles involved include the biceps, brachialis, and triceps, grouped into flexors and extensors. The spinal network encompasses C5 and C6 segments and includes motoneurons, interneurons, and sensory afferents. Movement planning and intention are simulated as descending inputs to this spinal network.

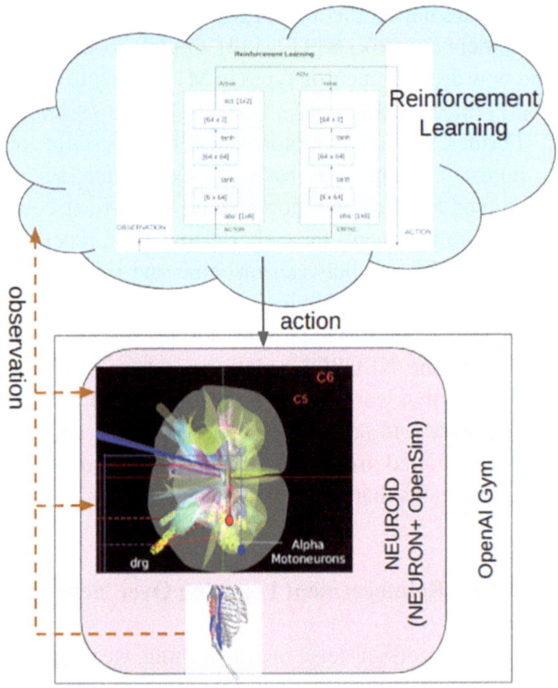

Fig. 18.7 Spinal and musculoskeletal models were built on NEUROiD. Spinal networks were simulated on NEURON and musculoskeletal models were simulated on OpenSim. The NEUROiD (NEURON+OpenSim) model was encapsulated as an OpenAI Gym environment. Reinforcement Learning was used to learn an optimal policy to approximate the Central Motor Program for a specific movement task

The RL controller provides a time series of activations $\{a_{flex}, a_{ext}\}$, which stimulate flexor and extensor networks. The muscle efferent signal (spike trains) is modelled as a sum of delayed delta functions derived from axonal potentials [37].

$$\mathrm{s}(\mathrm{t}) = \sum \delta\ (\mathrm{t} - t_{AP}) \tag{18.1}$$

s(t) is the superposition of all the action potentials occurring at time t_{AP}. A normalized efferent activation (in range [0,1]) is passed to OpenSim musculoskeletal model. The musculoskeletal model returns joint angles, muscle lengths, velocities, and forces at each simulation step. Proprioceptive afferent rates (Ia, II, Ib) were computed using equations adapted from [25].

$$I_a(t) = pIa_m + pIa_p * x_m(t) + pIa_{v1} * x'^{v2}{}_m(t) \tag{18.2}$$

$$II(t) = pII_m + pII_p * x_m(t) \tag{18.3}$$

$$I_b(t) = pI_b \frac{f * a(t)\left(F_{max} + K_m x_m(t) + B_m x'_m(t)\right)}{F_{max}} \tag{18.4}$$

Here, I_a(t), $II(t)$ and $I_b(t)$ are the instantaneous values of the Ia, II and Ib afferent firing rates (in PPS). pIa_m (=80PPS) is the firing rate constant for Ia. $x_m(t)$ is muscle stretch in mm, $x_m^{'}$ (t) is muscle stretch velocity in mm/s. pIa_p (=2) is length change constant, pIa_{v1} (=4.3) and $v2$ (=0.6) are velocity constants. pII_m(=5PPS) is the firing rate constant and pII_p (=8) is length change constant for II. pI_bf (=200) is the force constant, K_m (=56.3 kN/m) is muscle stiffness, B_m (=2.81 kNs/m) is damping. F_{max} is the maximum isometric force of the muscle. a(t) is the muscle activation (normalized to be in range [0,1]).

Motoneuron properties were adapted from [93], with soma diameter set at 80 ± 2.5 μm, consistent with S-type neurons [37]. Each spinal segment modelled includes 169 alpha motoneurons, 60 Ia afferents, 60 II afferents, and 400 interneurons.

Cortical Representations and Emerging Control Policies

The learning task was simple but representative: move the elbow from a neutral position (0.17 rad) to a target position (0.43 ± 0.05 rad) using only indirect stimulation of spinal networks. Over episodes, the agent adjusted its stimulation inputs (in amplitude, timing, and target segment) based on observed proprioceptive feedback and resulting joint movement.

Generalization is a hallmark of intelligent motor control and serves as a test of the robustness of learned cortical representations. The trained model was evaluated with a randomized starting elbow angle between 0.17 rad and 0.7 rad, to test whether it could consistently reach the target angle (0.43 rad). The learned policy succeeded across this range. This supports the idea that the simulated M1 module developed a flexible motor strategy, akin to state-dependent muscle activation patterns observed in cortical recordings [117]. Figure 18.8 shows elbow angle trajectories from four different starting positions.

While performance-based success indicates functional learning, we further investigate whether the learned control network exhibits neural properties similar to those observed in biological M1. Specifically, we explore whether internal representations encode movement variables like velocity or trajectory goals. We analysed neuron activations in the trained policy network, which consists of two hidden layers (each with 64 neurons) and a two-dimensional output. By testing the network on different initial joint angles, we examined the tuning characteristics of neurons. We found that the output neurons strongly correlate with joint velocity mirroring velocity-tuned cortical neurons [108, 118].

To identify neurons encoding position or trajectory information, we introduced two metrics: λn, which captures the difference in neural activity for above-target vs. below-target initial angles, and rn, which measures correlation of neural activity with rate of elbow angle change. Neuron #13 consistently ranked among the top candidates for both metrics, suggesting its role in encoding movement direction or trajectory [119]. Figures 18.9 and 18.10 illustrate neural activations and their relationships with movement variables.

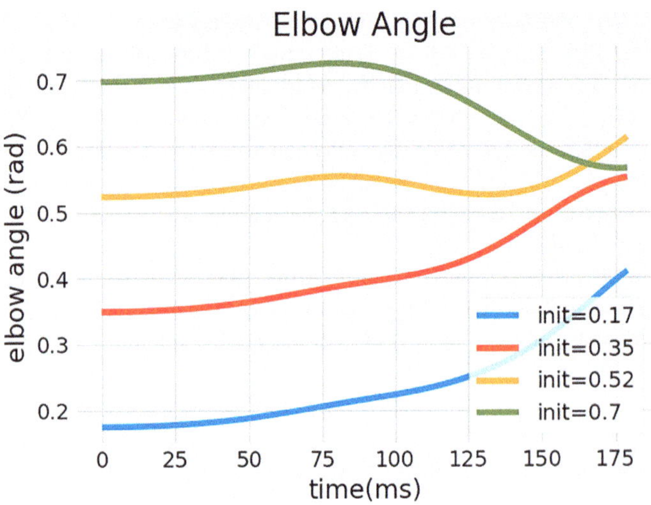

Fig. 18.8 Plot of elbow angle for four different initial angles in range 0.17 rad - 0.7 rad

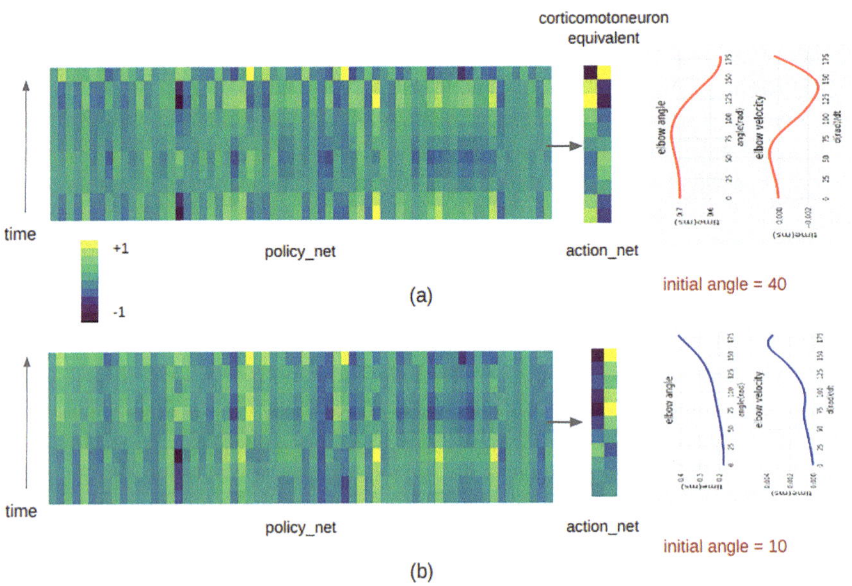

Fig. 18.9 Neuron activation pattern of the 64 neurons in the outer layer of policy net and 2 neurons in the outer layer of action net. The plot of elbow angle and rate of change of elbow angle is also shown. (**a**) initial elbow angle = 0.7 rad (**b**) initial elbow angle = 0.17 rad. The activation pattern of the 2 neurons in the outer layer of action net is correlated with rate of change of elbow angle

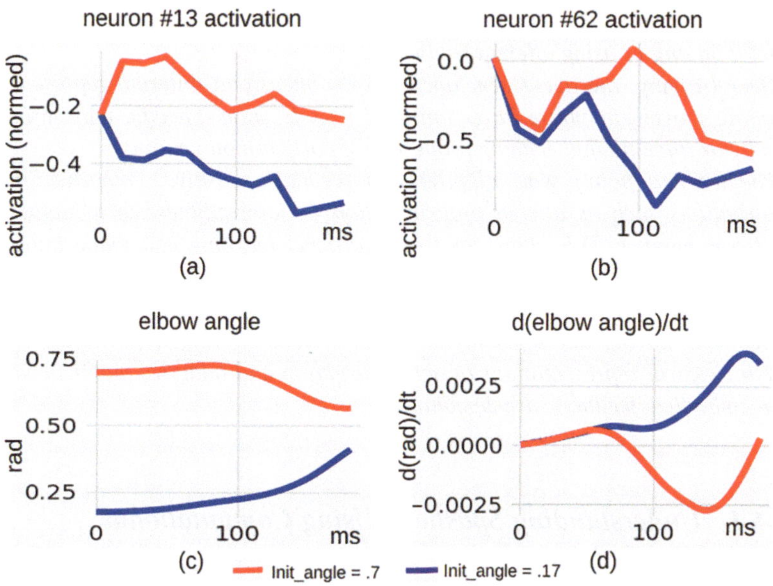

Fig. 18.10 (**a**) Activation pattern of neuron (#13) in policy net. This neuron consistently showed high value for dm v in all experiments indicating that it could potentially be responsible for trajectory selection. (**b**) Activation pattern of neuron (#13) in policy net. The activation pattern of this neuron showed high correlation with rate of change of elbow angle. (**c**) Plot of elbow angle for two initial angles (0.17 and 0.7 rad) (**d**) Plot of rate of change of elbow angle for two initial angles (0.17 and 0.7 rad)

The policy obtained represents one among many possible solutions. This aligns with evidence that biological systems exhibit degeneracy multiple strategies to achieve the same goal [120–122]. Extending this framework, we propose that complex behaviours could be generated by a compositional structure of CMP sub-modules (C1, C2, ..., Cn), modulated by descending cortical drives (D1, D2, ...) to produce a repertoire of goal-directed movements. Such compositional organization has been observed in motor control studies [123, 124].

18.4.3.2 Summary

To investigate, the mechanisms by which the motor cortex selects and coordinates actions, we used NEUROiD–Reinforcement Learning (RL) environment that integrates a biologically detailed spinal-musculoskeletal system with a cortical-level learning model.

We framed movement generation as a reinforcement learning task, an agent learned to control a virtual upper limb by maximizing task-based rewards.

The environment included a NEURON-simulated spinal segment (C5–C6), an OpenSim elbow model with relevant muscles, and proprioceptive feedback through

Ia, Ib, and II afferents. The agent modulated descending stimulation to achieve target joint angles using afferent feedbacks.

Over training, the agent developed a low-dimensional internal representation encoding movement phase, direction, and muscle state. Specific units exhibited velocity or force tuning, echoing properties of cortico-motoneurons.

The learned policy was adaptive and feedback-sensitive, compensating for perturbations such as altered proprioception or joint stiffness in a biologically consistent manner. This shows that cortical-level learning can guide movement through fixed spinal circuits.

The model reflects how the brain might learn to interface with spinal control mechanisms during development or rehabilitation, offering a powerful tool for studying neural representations of action selection and adaptive motor behaviour in an embodied, feedback-driven context.

18.4.4 Understanding Spasticity Using Computational Simulations

Movement disorders, such as spasticity, result from complex interactions between the nervous system, muscles, and skeletal structures [125]. Traditional methods of studying these disorders often rely on animal models or limited human experiments due to ethical constraints [126]. However, advances in computational neuroscience and biomechanics have led to the development of neuromusculoskeletal models for simulating movement dynamics under various conditions [27, 41, 127–129].

By leveraging in-silico models, researchers can gain deeper insights into the neural mechanisms of motor disorders like spasticity and can potentially assist them in development of targeted rehabilitation strategies. Incorporating patient-specific modelling approach in model development, the computational simulations could also be used to design individualized treatment plans [130]. By adjusting model parameters to match a patient's neuromuscular characteristics, clinicians could predict the effects of various interventions and optimize therapy. As these models mature, they could potentially become indispensable tools for understanding movement and its associated disorders. The goal is to create comprehensive models simplifying complex biological systems into manageable models while retaining essential features necessary for simulating specific phenomena, such that both healthy physiological states and pathological conditions such as spasticity, can be replicated in-silico.

18.4.4.1 Challenges in Spasticity Management

Clinically, spasticity manifests as increased muscle tone [131], exaggerated tendon reflexes, and abnormal resistance to passive movement, particularly at high velocities [132]. One of the characteristic signs of spasticity is the clasp-knife phenomenon, where an initial resistance to passive movement is suddenly released, mimicking the action of a closing knife [11]. Understanding the precise neurophysiological mechanisms responsible for spasticity is crucial for developing effective rehabilitation strategies and improving patient outcomes.

Spasticity is conventionally managed through a combination of physiotherapy, pharmacological interventions, and surgery. While these approaches offer symptomatic relief, they suffer from several drawbacks [9, 133–135]. Current treatment methods are largely generalized and do not account for patient-specific neurophysiological variations. Most clinical assessments rely on subjective rating scales, such as the Modified Ashworth Scale and the Modified Tardieu Scale etc. [136, 137]. Pharmacological interventions often have side effects and do not permanently restore normal motor function. Studying spasticity in human subjects presents ethical concerns, and hypothesis tested in animal models do not always translate well to human physiology [138]. Given these limitations, computational modelling provides an alternative avenue for studying spasticity by enabling controlled and repeatable simulations of neuromuscular behaviour under different conditions.

18.4.4.2 Understanding Mechanism Underlying Spasticity: An -In-Silico Case Study

Computational simulations offer several advantages over traditional experimental and clinical methods. Unlike human or animal studies, in-silico modelling allows for testing hypotheses without ethical constraints. Neural and biomechanical parameters can be precisely adjusted to analyse their individual contributions to spasticity.

A neuro-musculoskeletal simulation framework was employed to model spasticity in the lower limb, particularly the Modified Tardieu Test, using OpenSim and NEURON simulations by [10]. By integrating α-motor neurons, Ia afferents, and flexor reflex afferents, the model replicates the velocity-dependent resistance seen in spastic individuals, offering a mechanistic explanation for the clasp-knife effect. Variations in neural reflex transmission efficacy and testing velocities were examined to assess their influence on spastic responses, shedding light on the interplay between excitatory and inhibitory spinal pathways in spasticity generation.

A modified **gait2392** (see Fig. 18.11) musculoskeletal model [139] was employed, incorporating 92 muscle-tendon units modelled as Hill-type actuators [140]. Passive knee flexion was simulated using external forces applied at the tibia, reproducing conditions of the Modified Tardieu Test [141] in a prone position. The study modelled basic spinal reflex circuits, including the Ia-α motor reflex loop responsible for stretch-induced reflex contraction in knee extensors and flexor reflex afferents mediating inhibition of extensors and excitation of flexors, influencing the release phase in the clasp-knife response [142, 143]. Neural simulations were conducted using the NEURON environment [2], and estimated muscle activation patterns were fed into OpenSim in a co-simulation loop. To evaluate the robustness of neural control mechanisms, the study introduced temporal perturbations in

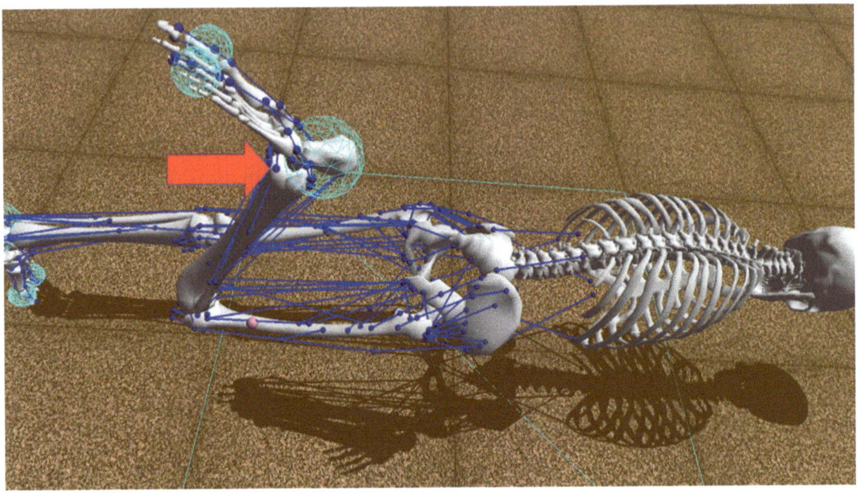

Fig. 18.11 The musculoskeletal model used in this study, depicts the position for administering the Modified Tardiue Test in-silico, the red arrows indicate the site of the external force application

extensor and flexor activation signals. The resulting joint profiles were analysed using k-means clustering to assess variability and stability in simulated spastic responses.

The simulations successfully replicated the catch-and-release pattern characteristic of spasticity (refer Fig. 18.12). At high velocity, knee flexion was initially resisted due to strong Ia-mediated extensor activation, followed by a sharp release due to FRA-mediated inhibition. This profile closely matched human spasticity data from previous experimental studies [132]. At the heart of the investigation lies the successful emulation of the clasp-knife effect. The simulation produced a knee joint angle profile that closely mirrors the characteristic "catch-and-release" behaviour observed in clinical settings. In the spastic condition, the model demonstrated an initial abrupt resistance during knee flexion- the "catch"—followed by a rapid release where the joint resumes its movement. This behaviour was not only evident in the joint angle curves (as illustrated in Fig. 18.13) but was further corroborated by the corresponding joint velocity plots. The authors note that, in contrast to the smooth, uninterrupted flexion seen in normal physiology, the spastic case displays a distinct segmentation. This segmentation arises from an early, forceful activation of the knee extensors, which temporarily impedes motion before the subsequent inhibitory mechanisms take over.

Another remarkable facet of the study is the systematic exploration of the impact of Ia-α synaptic weights and the speed at which the test is administered. The simulations reveal a direct correlation between increased synaptic efficacy and the intensity of the initial catch. Increasing the Ia-α reflex transmission strength resulted in a more pronounced catch phase, whereas reducing it led to smoother motion (refer Fig. 18.14 top). When the weight of the Ia-α reflex loop was elevated, the

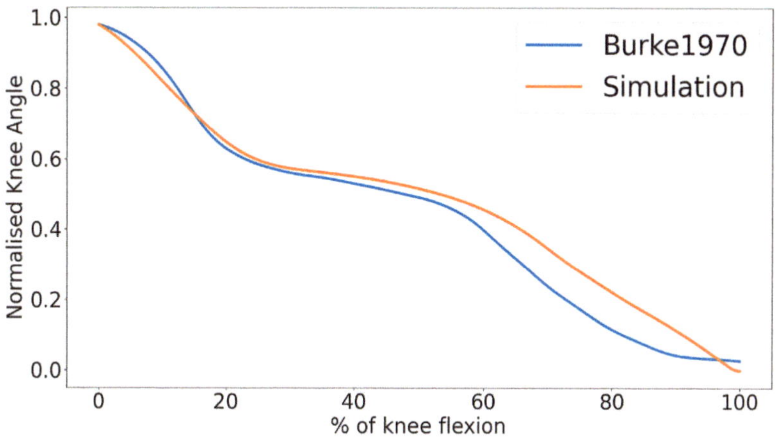

Fig. 18.12 Joint profile comparison between Burke 1970 and simulation. The joint profile from [132] was digitally reconstructed and normalized to compare with simulation results

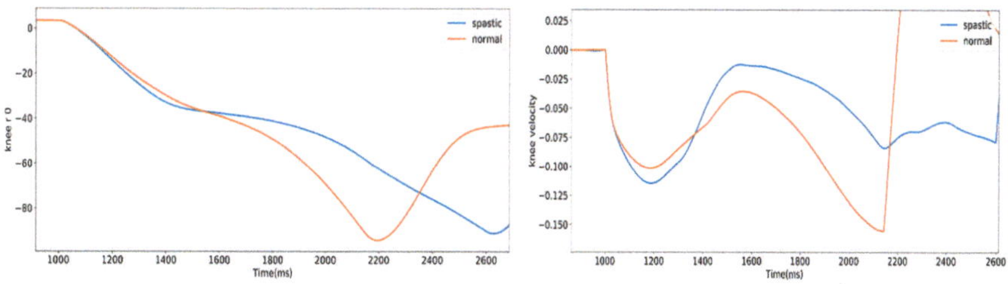

Fig. 18.13 Spastic (blue curve) and normal knee profile (orange curve) when the test is administered in the prone position, left plot shows knee-position (in degrees), right shows corresponding knee-velocities (degree/ms)

initial resistance during knee flexion was markedly pronounced. This heightened excitation simulates the loss of descending inhibitory control, a common scenario in spasticity due to spinal injury, thus reinforcing the exaggerated stretch reflex.

Higher test velocities exacerbated the spastic response (refer Fig. 18.14 bottom), reinforcing the velocity-dependence of hyperreflexia [144]. As the velocity of the passive limb movement increased, the simulation recorded an amplified excitation of the knee extensors, culminating in a more prominent catch. This was subsequently followed by an even more robust inhibition (or “release”) phase, as the system endeavoured to counterbalance the initial reflex contraction. The interplay between these factors is captured eloquently in the joint profiles, where both the initial excitation and subsequent inhibitory phases are delineated with clarity in higher testing velocity(force) cases.

Robustness testing formed a significant pillar of the study. By introducing controlled perturbations—specifically, temporal delays and advancements (of ±50 ms and ± 100 ms) in the neural control signals—the authors could scrutinize the

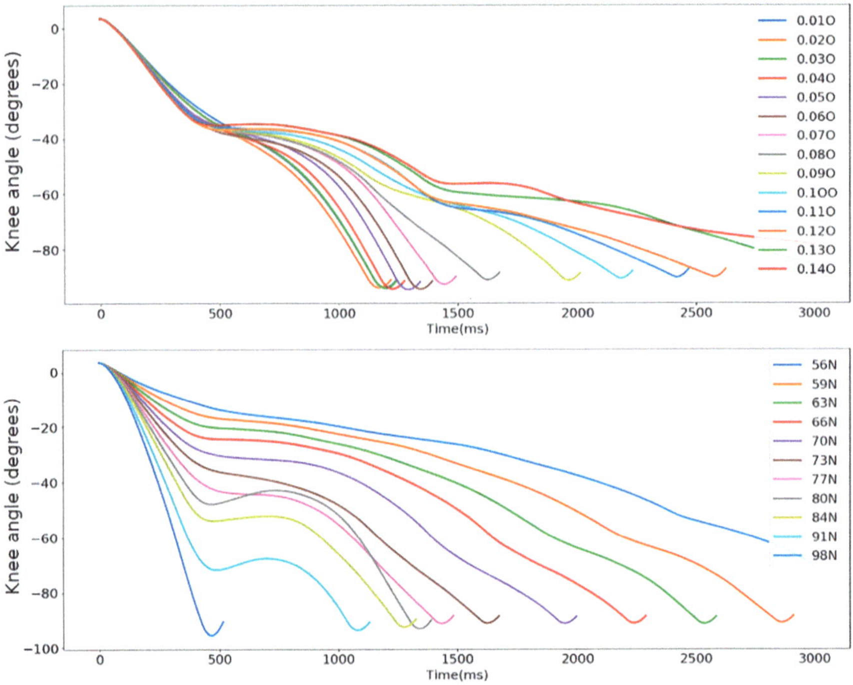

Fig. 18.14 **Top:** Joint profiles comparing varying Ia-α synaptic transmission efficacy. Higher weights represent greater loss of supra-spinal drive (spastic case), while lower weights indicate minimal or no loss. **Bottom:** Knee profiles at different testing speeds (force)

resilience of the simulated spastic response. Their extensive simulation campaign, encompassing over 13,000 distinct combinations, yielded insightful results. The temporal perturbations in extensor activation significantly altered the timing of the clasp-knife effect, whereas flexor activation delays had minimal impact (refer Fig. 18.15). The joint profiles remained largely stable when temporal jitter was applied solely to the knee flexor inputs. In stark contrast, perturbations to the knee extensor signals led to discernible variations in the onset and progression of the catch and release phases.

To systematically interpret these deviations, a k-means clustering algorithm was also employed by authors. The resulting clusters (displayed in Fig. 18.16) underscored a clear demarcation—profiles close to the ideal spastic case emerged predominantly when only the flexor signals were perturbed, while the furthest deviations were associated with extensor delays. This analysis reinforces the pivotal role of extensor control in the manifestation of the clasp-knife phenomenon. These findings suggest that extensor reflex circuits play a dominant role in spasticity manifestation, while flexor pathways contribute more to fine-tuning post-catch relaxation.

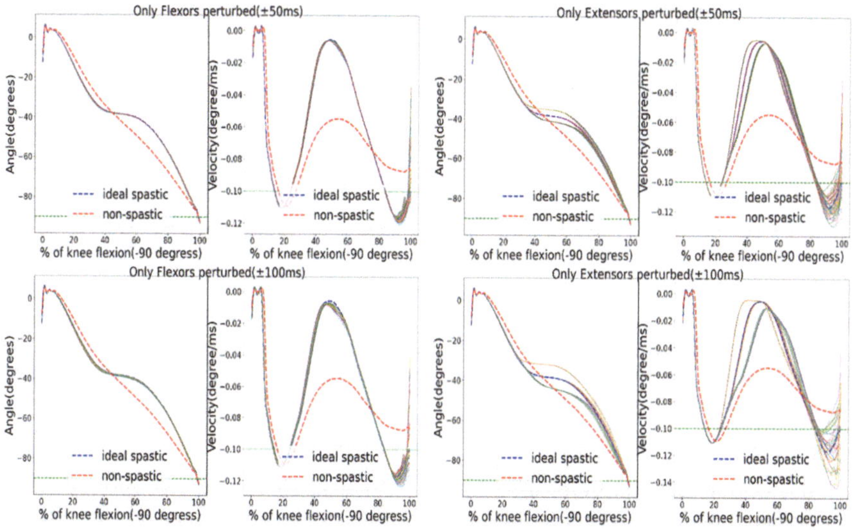

Fig. 18.15 Knee joint angle and velocity variations with ±50 ms (top plots) and ± 100 ms (bottom plots): Left pair shows spastic cases with delays introduced in knee flexors muscles' neural control input, and the right pair shows cases with delays in extensors muscles control inputs. The green fine dotted horizontal line indicates the 90-degree mark for angle profiles on the left and the 0.10 degree/ms (100 degree/s) mark

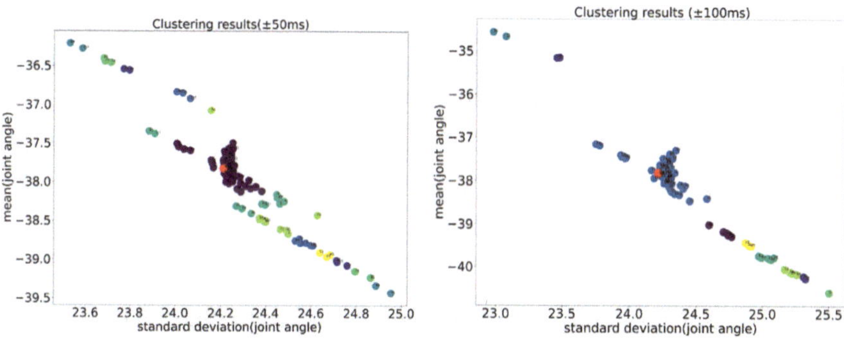

Fig. 18.16 Clustering output of selected joint angle profiles from jitter experiment,left plot is the case of ±50 ms and right represents cases of ±100 ms, the points clustering around red dot (ideal spastic case) are the cases where delays were added only to knee flexors, the points that are away from the ideal spastic case are the cases where delays were introduced into extensors

18.4.4.3 Subject-Specific In-Silico Assessment of Spasticity

A central goal of digital patient modelling is the ability to replicate and study motor disorders as they manifest in specific individuals, accounting for anatomical and physiological variability. To move beyond generic simulations, we used motion capture data to personalize the musculoskeletal model for individual subjects and

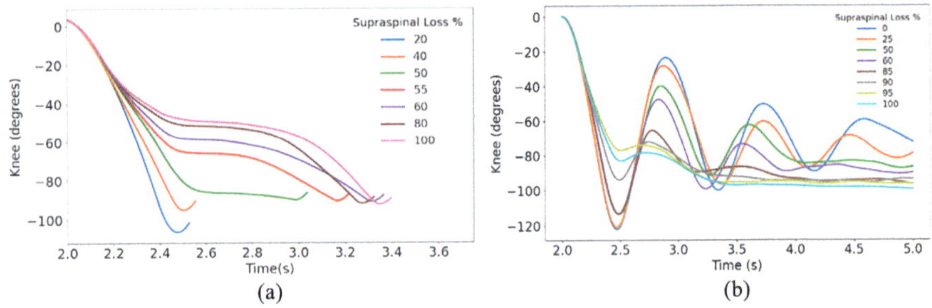

Fig. 18.17 Knee profile profiles with varying supraspinal injury levels for (a) Modified Tardieu Test conducted and (**b**) Pendulum Test conducted on the generic musculoskeletal model. The increase in knee resistance in (**a**) and the decrement in first swing amplitudes with increasing loss of supraspinal influence is clear

conducted clinical spasticity assessments virtually. This approach allowed us to investigate how morphometric differences influence the emergence and severity of spasticity under controlled neural conditions.

Using the NEUROiD framework, we integrated detailed spinal circuitry with subject-specific musculoskeletal geometries to simulate how reflexes and tone evolve with spinal cord injury across different body types. We focused on two clinical assessments widely used to evaluate spasticity, the Modified Tardieu Test (MTT) [136] and the Pendulum Test (PT) [145]. For each test, we conducted a baseline simulation (intact supraspinal input) and multiple injury scenarios (graded supraspinal loss) to observe the onset and progression of spastic behaviours across subjects (generic and personalised virtual subjects) [146].

Personalised Modified Tardieu Test (MTT) Simulation

The MTT is designed to assess velocity-dependent muscle resistance and is sensitive to hyperreflexia and clasp-knife phenomena, the hallmarks of spasticity. Our aim was to evaluate how with injury spasticity emerges (refer Fig. 18.17) and how subject-specific musculoskeletal properties influence the timing and nature of these responses, even when the underlying spinal circuitry is held constant.

We scaled a generic OpenSim musculoskeletal model using motion capture data from four healthy subjects. While neural parameters (e.g., motor pool size, connection weights) remained unchanged, each model reflected the unique bone lengths, segment masses, and muscle paths of the individual subject. The test involved applying a passive, fast stretch to the knee joint, simulating increasing levels of supraspinal disruption.

The simulations revealed marked variability across subjects in both the threshold and expression of spasticity (refer Fig. 18.18). One virtual subject (VS2), with the highest body mass index, exhibited early onset of the clasp-knife response at moderate injury levels (~40% supraspinal loss), while others only showed similar

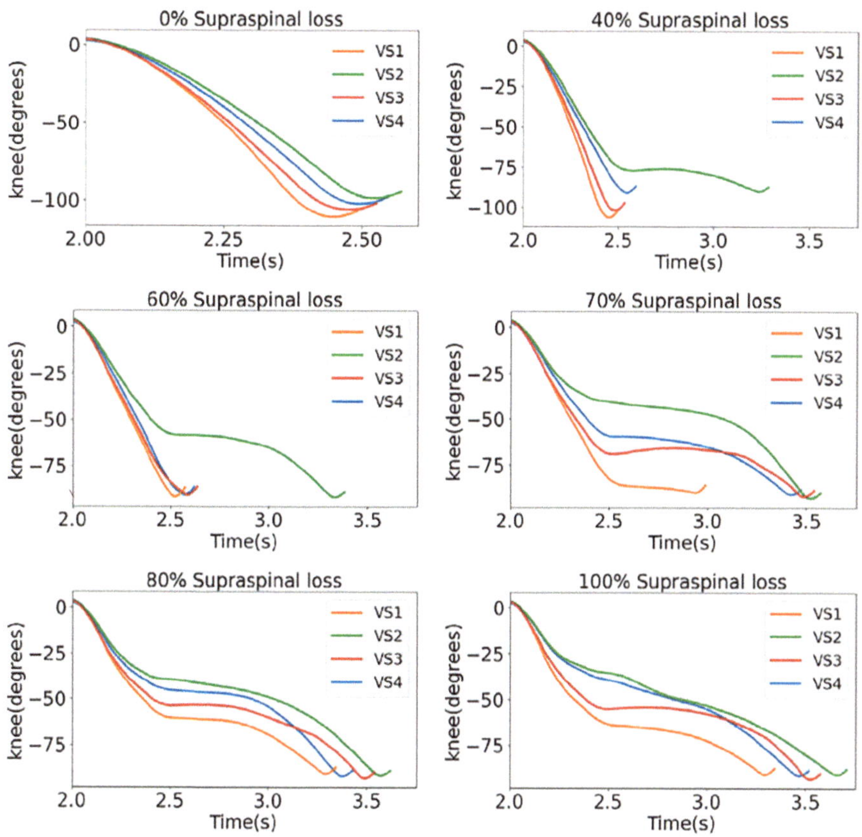

Fig. 18.18 Virtual Subject Experiments for MTT simulations: Knee profiles with subjectspecific variations for virtual subjects with varying supraspinal inputs, the emergence of spastic behavior with spinal injury is also shown

behaviour at higher injury levels. Another subject (VS1), with lower BMI, showed delayed onset and less pronounced resistance throughout.

These differences emerged despite identical neural drive and reflex circuitry, highlighting how mechanical properties, such as tendon slack, segment inertia, and muscle paths shape reflex expression. Importantly, repeated trials for each subject showed minimal intra-subject variation, confirming that the observed differences were due to model personalization and not simulation noise.

Personalised Pendulum Test (PT) Simulations

The PT evaluates passive resistance to movement and is widely used as a quantitative measure of lower limb spasticity. Unlike the MTT, it does not rely on externally applied stretch velocities but rather reveals stiffness through damped oscillations

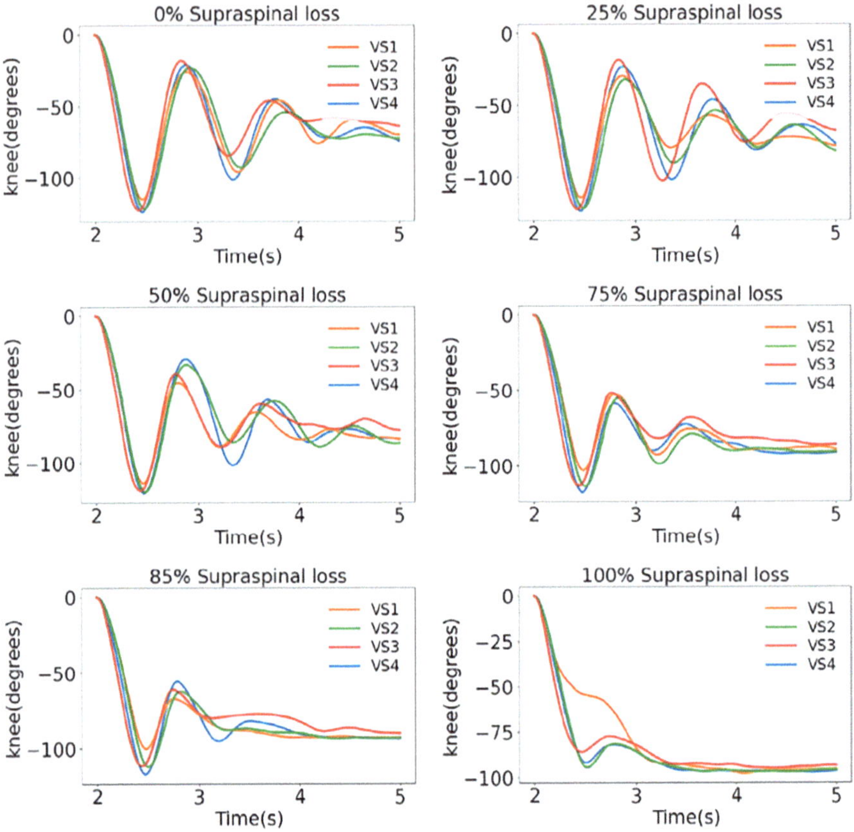

Fig. 18.19 Virtual Subject Experiments for PT simulations: Knee profiles with subjectspecific variations for virtual subject with varying supraspinal inputs, the emergence of spastic behavior with spinal injury is also shown

under gravity. We aimed to assess whether subject-specific musculoskeletal differences influence PT outcomes under similar spinal injury levels.

Each virtual subject's model was placed in a seated posture with the knee fully extended. The leg was then released and allowed to swing freely. We simulated multiple supraspinal injury levels by progressively reducing descending inputs to the spinal cord, observing how knee kinematics changed with injury severity.

The PT simulations showed less inter-subject variability compared to the MTT. Across all virtual subjects, increased injury consistently led to reduced swing amplitude, fewer oscillations, and a more extended resting angle, characteristics typical of spasticity. However, distinct subject-specific deviations only became apparent at the most severe injury level (100% supraspinal loss), with one subject (VS1) displaying a notably stiffer response (refer Fig. 18.19).

The consistency of PT responses across trials and subjects suggests that this test is more robust to individual morphometric variation and may be better suited for generalized spasticity quantification. Nonetheless, the slight variations observed at higher injury levels reaffirm the influence of subject-specific inertial and muscle-tendon properties on passive limb dynamics.

18.4.4.4 Cracking the Code of Spasticity: Why Virtual Experiments Beat Real-World Limits

The computational model of spasticity discussed above, provided a unique opportunity to validate the contributions of Ia afferents and flexor reflex afferent (FRA) pathways to the clasp-knife phenomenon. By simulating the Modified Tardieu Test, this study has successfully replicated velocity-dependent resistance, reinforcing the critical role of extensor reflex control in spastic responses. The findings not only advance our understanding of the neurophysiological mechanisms underlying spasticity but also emphasize the necessity of computational approaches in studying complex neuromuscular disorders.

The insights derived from this model were only made possible through in-silico simulations, which offer a level of experimental control and reproducibility that is unattainable in purely empirical studies. The ability to manipulate neural circuit parameters, isolate specific reflex pathways, and systematically test hypotheses in a controlled virtual environment underscores the advantages of computational modelling. In contrast to clinical and in vivo studies, which are often constrained by ethical considerations, inter-individual variability, and limited intervention capabilities, in-silico models allow for precise dissection of underlying mechanisms without the confounding effects present in human subjects or animal models.

Furthermore, computational models provide a framework for understanding movement disorders beyond traditional clinical observation. Spasticity is a multifaceted condition influenced by spinal and supraspinal pathways, biomechanical properties, and individual physiological variations. By integrating musculoskeletal dynamics with detailed neural circuitry, models could bridge the gap between theoretical neuroscience and applied rehabilitation. This approach facilitates hypothesis testing that would be impractical or impossible in human subjects, such as selectively ablating specific pathways or systematically varying muscle stiffness and reflex gain. As a result, in-silico models enable a more mechanistic understanding of pathological movement patterns, guiding the development of targeted therapeutic interventions.

18.5 Conclusions

Movement arises from a coordinated hierarchy of neural and biomechanical processes involving intention, planning, reflexes, muscle dynamics, and sensory feedback. This integration, spanning from synapses to joint motion, forms the basis of movement intelligence. To replicate such intelligence, it is essential to build systems that combine neural computation with physical embodiment, enabling learning and adaptation rather than predefined motion.

This chapter described the development of in-silico neuromusculoskeletal models using NEUROiD, a platform that integrates anatomically realistic musculoskeletal systems with biologically grounded spinal and cortical control. By embedding reflex loops, afferent feedback, and reinforcement learning, we demonstrated how motor behaviour can emerge through interaction with the environment.

Case studies involving spinal micro-stimulation, tendon transfer, spasticity modelling, and cortical reinforcement learning showed that realistic motor behaviour can be learned within these models. Reflexes and proprioceptive pathways play active roles in adaptation, while artificial neural controllers develop features resembling cortical representations. Personalized models allow exploration of compensatory strategies following injury, aiding clinical planning and therapy design.

A particularly compelling demonstration of this capability was provided through simulations of spasticity progression using subject-specific neuro-musculoskeletal models. By varying individual anatomy while keeping neural circuitry constant, we showed that identical neural impairments can lead to distinct motor outcomes, highlighting the importance of biomechanics in shaping clinical expression. The Modified Tardieu Test was found to be more sensitive to such variations, supporting its use in personalized assessment, while the Pendulum Test offered consistent results across individuals, making it suitable for general screening. These findings underscore the need to interpret motor dysfunction through an integrated neural-mechanical lens and suggest that virtual patients can serve as non-invasive platforms for diagnosis, therapy design, and recovery simulation.

The integration of learning with spino-musculoskeletal dynamics has significant implications for brain-machine interfaces and rehabilitation. In prosthetic control, learned neural policies can drive spinal or robotic actuators in response to cortical intent. In neurorehabilitation, virtual agents can simulate recovery strategies and identify effective stimulation targets. The framework also enables modelling of impaired systems, such as spastic limbs or partial motor loss, offering a platform for studying recovery and planning interventions.

By uniting biomechanical realism with adaptive neural control, the NEUROiD-RL framework moves beyond simulation to model how the nervous system learns to act through the body. It provides a testbed for understanding motor learning, control variability, and plasticity. These models do not merely reproduce movement but capture how it is shaped by feedback and experience. As the line between biological and artificial systems continues to shift, platforms like NEUROiD will play a central role in bridging neural intent with physical execution. They offer a path toward understanding and designing systems that learn to move as living beings do, i.e., through continuous interaction between brain, body, and environment.

References

1. R.S. Iyengar, M.V. Pithapuram, A.K. Singh, Curated model development using NEUROiD: A web-based NEUROmotor integration and design platform. Front. Neuroinf. **13**(August), 56 (2019). https://doi.org/10.3389/fninf.2019.00056
2. M.L. Hines, N.T. Carnevale, The NEURON simulation environment. Neural Comput. **9**(6), 1179–1209 (1997). https://doi.org/10.1162/neco.1997.9.6.1179
3. S.L. Delp et al., OpenSim: Open-source software to create and analyze dynamic simulations of movement. I.E.E.E. Trans. Biomed. Eng. **54**(11), 1940–1950 (2007). https://doi.org/10.1109/TBME.2007.901024
4. A. Simalatsar, Synthetic biomedical data generation in support of in silico clinical trials. Front. Big Data **6**, 1085571 (2023). https://doi.org/10.3389/fdata.2023.1085571
5. D.M. Hannon et al., Modeling mechanical ventilation in silico-potential and pitfalls. Semin. Respir. Crit. Care Med. **43**(3), 335–345 (2022). https://doi.org/10.1055/s-0042-1744446
6. B.A. Killen, A. Falisse, F. De Groote, I. Jonkers, In silico-enhanced treatment and rehabilitation planning for patients with musculoskeletal disorders: Can musculoskeletal modelling and dynamic simulations really impact current clinical practice? Appl. Sci. **10**(20), 7255 (2020). https://doi.org/10.3390/app10207255
7. K. Zieliński, T. Gólczewski, M. Kozarski, M. Darowski, Virtual and artificial cardiorespiratory patients in medicine and biomedical engineering. Membranes **12**(6), 548 (2022). https://doi.org/10.3390/membranes12060548
8. M. Banky, R.A. Clark, Y.H. Pua, B.F. Mentiplay, J.H. Olver, G. Williams, Inter- and intra-rater variability of testing velocity when assessing lower limb spasticity. J. Rehabil. Med. **51**(1), 54–60 (2019). https://doi.org/10.2340/16501977-2496
9. T. Rekand, Clinical assessment and management of spasticity: A review. Acta Neurol. Scand. **122**(SUPPL. 190), 62–66 (2010). https://doi.org/10.1111/j.1600-0404.2010.01378.x
10. A.K. Singh, K.S. Sridharan, S. Wu, O. Tirosh, M. Raghavan, Simulated clinical triage of modified Tardieu test for lower limb spasticity, in *2024 IEEE International Symposium on Medical Measurements and Applications (MeMeA)*, (IEEE, Eindhoven, 2024), pp. 1–6. https://doi.org/10.1109/MeMeA60663.2024.10596866. Available: https://ieeexplore.ieee.org/document/10596866/. Accessed 28 Jan 2025]
11. D. Burke, L. Knowles, C. Andrews, P. Ashby, Spasticity, decerebrate rigidity and the clasp-knife phenomenon: An experimental study in the cat. Brain **95**(1), 31–48 (1972). https://doi.org/10.1093/brain/95.1.31
12. H. Ling, K.L. Roberts, D. Kao, R. Balasubramanian, Force-amplifying implant to improve key pinch strength in tendon transfer surgery: Cadaver model proof-of-concept. J. Orthop. Res. **41**(8), 1653–1660 (2023). https://doi.org/10.1002/jor.25511
13. K. Mallampalli, K. Sarathy, R. Balasubramanian, M. Raghavan, Studying spinal feedback loops post tendon transfer surgery using the pincer grasp task: A neuro-musculoskeletal study using NEUROiD, in *2024 46th Annual International Conference of the IEEE Engineering in Medicine and Biology Society (EMBC)*, (IEEE, Orlando, FL, 2024), pp. 1–4. https://doi.org/10.1109/EMBC53108.2024.10782641. Available: https://ieeexplore.ieee.org/document/10782641/. Accessed 04 July 2025
14. T.B. Moeslund, A. Hilton, V. Krüger, A survey of advances in vision-based human motion capture and analysis. Comput. Vis. Image Underst. **104**(2–3), 90–126 (2006)
15. Z. Cao, G. Hidalgo, T. Simon, S.-E. Wei, Y. Sheikh, OpenPose: Realtime multi-person 2D pose estimation using part affinity fields. IEEE Trans. Pattern Anal. Mach. Intell. **43**(1), 172–186 (2019)
16. D. Roetenberg, H.J. Luinge, P.J. Slycke, Xsens MVN: Full 6DOF human motion tracking using miniature inertial sensors. Xsens Technol. BV **1**(1), 1–7 (2009)
17. M.A. Dresner, G.H. Rose, P.J. Rossman, R. Muthupillai, A. Manduca, R.L. Ehman, Magnetic resonance elastography of skeletal muscle. J. Magn. Reson. Imaging **13**(2), 269–276 (2001). https://doi.org/10.1002/1522-2586(200102)13:2<269::AID-JMRI1039>3.0.CO;2-1

18. J. Oudeman, A.J. Nederveen, G.J. Strijkers, M. Maas, P.R. Luijten, M. Froeling, Techniques and applications of skeletal muscle diffusion tensor imaging: A review. J. Magn. Reson. Imaging **43**(4), 773–788 (2016). https://doi.org/10.1002/jmri.25016
19. E.A. Lalone, R.T. Willing, H.L. Shannon, G.J.W. King, J.A. Johnson, Accuracy assessment of 3D bone reconstructions using CT: An intro comparison. Med. Eng. Phys. **37**(8), 729–738 (2015). https://doi.org/10.1016/j.medengphy.2015.04.010
20. M.C. Florkow, K. Willemsen, V.V. Mascarenhas, E.H.G. Oei, M. Van Stralen, P.R. Seevinck, Magnetic resonance imaging versus computed tomography for three-dimensional bone imaging of musculoskeletal pathologies: A review. J. Magn. Reson. Imaging **56**(1), 11–34 (2022). https://doi.org/10.1002/jmri.28067
21. D. Farina, R. Merletti, R.M. Enoka, The extraction of neural strategies from the surface EMG. J. Appl. Physiol. **96**, 1486–1495 (2004)
22. C.J. De Luca, The use of surface electromyography in biomechanics. J. Appl. Biomech. **13**(2), 135–163 (1997). https://doi.org/10.1123/jab.13.2.135
23. R. Merletti, D. Farina, *Surface Electromyography: Physiology, Engineering, and Applications* (Wiley, 2016). https://doi.org/10.1002/9781119061971
24. P. Konrad, *The ABC of EMG: A Practical Introduction to Kinesiological Electromyography* (Noraxon USA, Inc, 2005)
25. D. Farina, A. Holobar, R. Merletti, R.M. Enoka, Decoding the neural drive to muscles from the surface electromyogram. Clin. Neurophysiol. **121**(10), 1616–1623 (2010). https://doi.org/10.1016/j.clinph.2009.10.040
26. J.L. Dideriksen, A. Del Vecchio, D. Farina, Neural and muscular determinants of maximal rate of force development. J. Neurophysiol. **123**(1), 149–157 (2020). https://doi.org/10.1152/jn.00330.2019
27. T.S. Buchanan, D.G. Lloyd, K. Manal, T.F. Besier, Neuromusculoskeletal modeling: Estimation of muscle forces and joint moments and movements from measurements of neural command. J. Appl. Biomech. **20**(4), 367–395 (2004). https://doi.org/10.1123/jab.20.4.367
28. M. Damsgaard, J. Rasmussen, S.T. Christensen, E. Surma, M. de Zee, Analysis of musculoskeletal systems in the AnyBody Modeling System. Simul. Model. Pract. Theory **14**(8), 1100–1111 (2006). https://doi.org/10.1016/j.simpat.2006.09.001
29. S.L. Delp, J.P. Loan, A graphics-based software system to develop and analyze models of musculoskeletal structures. Comput. Biol. Med. **25**(1), 21–34 (1995). https://doi.org/10.1016/0010-4825(95)98882-E
30. L. Cristofolini, E. Schileo, M. Juszczyk, F. Taddei, S. Martelli, M. Viceconti, Mechanical testing of bones: The positive synergy of finite–element models and *in vitro* experiments. Philos. Trans. R. Soc. Math. Phys. Eng. Sci. **368**(1920), 2725–2763 (2010). https://doi.org/10.1098/rsta.2010.0046
31. M. Millard, T. Uchida, A. Seth, S.L. Delp, Flexing computational muscle: Modeling and simulation of musculotendon dynamics. J. Biomech. Eng. **135**(2), 1–11 (2013). https://doi.org/10.1115/1.4023390
32. D.G. Thelen, Adjustment of muscle mechanics model parameters to simulate dynamic contractions in older adults. J. Biomech. Eng. **125**(1), 70–77 (2003). https://doi.org/10.1115/1.1531112
33. S.S. Blemker, P.M. Pinsky, S.L. Delp, A 3D model of muscle reveals the causes of nonuniform strains in the biceps brachii. J. Biomech. **38**(4), 657–665 (2005). https://doi.org/10.1016/j.jbiomech.2004.04.009
34. F.E. Zajac, J.M. Winters, Modeling musculoskeletal movement systems: Joint and body segmental dynamics, musculoskeletal actuation, and neuromuscular control, in *Multiple Muscle Systems*, (1990), pp. 121–148. https://doi.org/10.1007/978-1-4613-9030-5_8
35. K.R.S. Holzbaur, W.M. Murray, S.L. Delp, A model of the upper extremity for simulating musculoskeletal surgery and analyzing neuromuscular control. Ann. Biomed. Eng. **33**(6), 829–840 (2005). https://doi.org/10.1007/s10439-005-3320-7

36. A. Rajagopal, C.L. Dembia, M.S. DeMers, D.D. Delp, J.L. Hicks, S.L. Delp, Full-body musculoskeletal model for muscle-driven simulation of human gait. I.E.E.E. Trans. Biomed. Eng. **63**(10), 2068–2079 (2016). https://doi.org/10.1109/TBME.2016.2586891
37. R.R.L. Cisi, A.F. Kohn, Simulation system of spinal cord motor nuclei and associated nerves and muscles, in a web-based architecture. J. Comput. Neurosci. **25**(3), 520–542 (2008). https://doi.org/10.1007/s10827-008-0092-8
38. S.M. Danner, N.A. Shevtsova, A. Frigon, I.A. Rybak, Computational modeling of spinal circuits controlling limb coordination and gaits in quadrupeds. eLife **6**, e31050 (2017). https://doi.org/10.7554/eLife.31050
39. H. Geyer, H. Herr, A muscle-reflex model that encodes principles of legged mechanics produces human walking dynamics and muscle activities. IEEE Trans. Neural Syst. Rehabil. Eng. (2010). https://doi.org/10.1109/TNSRE.2010.2047592
40. S.N. Markin, A.N. Klishko, N.A. Shevtsova, M.A. Lemay, B.I. Prilutsky, I.A. Rybak, Afferent control of locomotor CPG: Insights from a simple neuromechanical model. Ann. N. Y. Acad. Sci. (2010). https://doi.org/10.1111/j.1749-6632.2010.05435.x
41. M. Sartori, U.Ş. Yavuz, D. Farina, In vivo neuromechanics: Decoding causal motor neuron behavior with resulting musculoskeletal function. Sci. Rep. **7**(1), 13465 (2017). https://doi.org/10.1038/s41598-017-13766-6
42. D. Song, N. Lan, G.E. Loeb, J. Gordon, Model-based sensorimotor integration for multi-joint control: Development of a virtual arm model. Ann. Biomed. Eng. **36**(6), 1033–1048 (2008). https://doi.org/10.1007/s10439-008-9461-8
43. S. Song, H. Geyer, A neural circuitry that emphasizes spinal feedback generates diverse behaviours of human locomotion. J. Physiol. **593**(16), 3493–3511 (2015). https://doi.org/10.1113/JP270228
44. S. Song, H. Geyer, Predictive neuromechanical simulations indicate why walking performance declines with ageing. J. Physiol. **596**(7), 1199–1210 (2018). https://doi.org/10.1113/JP275166
45. G. Taga, Y. Yamaguchi, H. Shimizu, Self-organized control of bipedal locomotion by neural oscillators in unpredictable environment. Biol. Cybern. **65**(3), 147–159 (1991). https://doi.org/10.1007/BF00198086
46. D.W. Franklin, D.M. Wolpert, Computational mechanisms of sensorimotor control. Neuron **72**(3), 425–442 (2011). https://doi.org/10.1016/j.neuron.2011.10.006
47. J. Diedrichsen, R. Shadmehr, R.B. Ivry, The coordination of movement: Optimal feedback control and beyond. Trends Cogn. Sci. **14**(1), 31–39 (2010). https://doi.org/10.1016/j.tics.2009.11.004
48. E. Todorov, Optimality principles in sensorimotor control. Nat. Neurosci. **7**(9), 907–915 (2004). https://doi.org/10.1038/nn1309
49. A.J. Ijspeert, Central pattern generators for locomotion control in animals and robots: A review. Neural Netw. **21**(4), 642–653 (2008). https://doi.org/10.1016/j.neunet.2008.03.014
50. K. Minassian, U.S. Hofstoetter, F. Dzeladini, P.A. Guertin, A. Ijspeert, The human central pattern generator for locomotion: Does it exist and contribute to walking? Neuroscientist **23**(6), 649–663 (2017). https://doi.org/10.1177/1073858417699790
51. J.B. Nielsen, How we walk: Central control of muscle activity during human walking. Neuroscientist **9**(3), 195–204 (2003). https://doi.org/10.1177/1073858403009003012
52. D.A. McCrea, I.A. Rybak, Organization of mammalian locomotor rhythm and pattern generation. Brain Res. Rev. **57**(1), 134–146 (2008). https://doi.org/10.1016/j.brainresrev.2007.08.006
53. T.G. Deliagina, I.N. Beloozerova, G.N. Orlovsky, P.V. Zelenin, Contribution of supraspinal systems to generation of automatic postural responses. Front. Integr. Neurosci. **8** (2014). https://doi.org/10.3389/fnint.2014.00076. Available: http://journal.frontiersin.org/article/10.3389/fnint.2014.00076/abstract. Accessed 09 Feb 2025
54. J.C. Houk, W.Z. Rymer, Neural control of muscle length and tension. Compr. Physiol., 257–323 (1981). https://doi.org/10.1002/cphy.cp010208

55. E. Pierrot-Deseilligny, D. Burke, *The Circuitry of the Human Spinal Cord: Spinal and Corticospinal Mechanisms of Movement*, 1st edn. (Cambridge University Press, 2012). https://doi.org/10.1017/CBO9781139026727. Available: https://www.cambridge.org/core/product/identifier/9781139026727/type/book. Accessed 05 July 2025
56. E.D. Schomburg, Spinal sensorimotor systems and their supraspinal control. Neurosci Res **7**(4), 265–340 (1990)
57. K. Takakusaki, M. Takahashi, K. Obara, R. Chiba, Neural substrates involved in the control of posture. Adv. Robot. **31**(1–2), 2–23 (2017). https://doi.org/10.1080/01691864.2016.1252690
58. G. Courtine et al., Transformation of nonfunctional spinal circuits into functional states after the loss of brain input. Nat. Neurosci. **12**(10), 1333–1342 (2009). https://doi.org/10.1038/nn.2401
59. M.-O. Gewaltig, M. Diesmann, NEST (NEural simulation tool). Scholarpedia **2**(4), 1430 (2007). https://doi.org/10.4249/scholarpedia.1430
60. D. Goodman, Brian: A simulator for spiking neural networks in python. Front. Neuroinform. **2** (2008). https://doi.org/10.3389/neuro.11.005.2008. Available: http://journal.frontiersin.org/article/10.3389/neuro.11.005.2008/abstract. Accessed 25 Feb 2025
61. E. Todorov, T. Erez, Y. Tassa, MuJoCo: A physics engine for model-based control, in *2012 IEEE/RSJ International Conference on Intelligent Robots and Systems*, (IEEE, Vilamoura-Algarve, 2012), pp. 5026–5033. https://doi.org/10.1109/IROS.2012.6386109. Available: http://ieeexplore.ieee.org/document/6386109/. Accessed 25 Feb 2025
62. C. Eliasmith et al., A large-scale model of the functioning brain. Science **338**(6111), 1202–1205 (2012). https://doi.org/10.1126/science.1225466
63. H. Markram et al., Reconstruction and simulation of neocortical microcircuitry. Cell (2015). https://doi.org/10.1016/j.cell.2015.09.029
64. K.E. Stephan, W.D. Penny, R.J. Moran, H.E.M. Den Ouden, J. Daunizeau, K.J. Friston, Ten simple rules for dynamic causal modeling. NeuroImage **49**(4), 3099–3109 (2010). https://doi.org/10.1016/j.neuroimage.2009.11.015
65. R.D. Traub, Single-column thalamocortical network model exhibiting gamma oscillations, sleep spindles, and epileptogenic bursts. J. Neurophysiol. (2004). https://doi.org/10.1152/jn.00983.2004
66. K.A. De Jong, *Evolutionary Computation: A Unified Approach* (MIT Press, Cambridge, MA, 2006)
67. E.H.L. Aarts, J. Korst, *Simulated Annealing and Boltzmann Machines: A Stochastic Approach to Combinatorial Optimization and Neural Computing*, Reprinted in Wiley-Interscience series in discrete mathematics and optimization (Wiley, Chichester, 1997)
68. S. Kirkpatrick, C.D. Gelatt, M.P. Vecchi, *Optimization by Simulated Annealing*, vol 220 (1983)
69. A. Gelman, J.B. Carlin, H.S. Stern, D.B. Dunson, A. Vehtari, D.B. Rubin, *Bayesian Data Analysis*, Chapman & Hall/CRC Texts in Statistical Science, 3rd edn. (CRC Press, Hoboken, 2013)
70. J.L. Hicks, T.K. Uchida, A. Seth, A. Rajagopal, S.L. Delp, Is my model good enough? Best practices for verification and validation of musculoskeletal models and simulations of movement. J. Biomech. Eng. **137**(2), 020905 (2015). https://doi.org/10.1115/1.4029304
71. D.A. Winter, *Biomechanics and Motor Control of Human Movement*, 4th edn. (Wiley, Hoboken, 2009)
72. B.J. Fregly et al., Grand challenge competition to predict in vivo knee loads. J. Orthop. Res. **30**(4), 503–513 (2012). https://doi.org/10.1002/jor.22023
73. D.G. Lloyd, T.F. Besier, An EMG-driven musculoskeletal model to estimate muscle forces and knee joint moments in vivo. J. Biomech. **36**(6), 765–776 (2003). https://doi.org/10.1016/S0021-9290(03)00010-1
74. A. Saltelli et al., *Global Sensitivity Analysis. The Primer*, 1st edn. (Wiley, 2007). https://doi.org/10.1002/9780470725184. Available: https://onlinelibrary.wiley.com/doi/book/10.1002/9780470725184. Accessed 26 Feb 2025

75. A. Seth et al., OpenSim: Simulating musculoskeletal dynamics and neuromuscular control to study human and animal movement. PLoS Comput. Biol. **14**(7), 1–20 (2018). https://doi.org/10.1371/journal.pcbi.1006223
76. A. Seth et al., OpenSim: Simulating musculoskeletal dynamics and neuromuscular control. Adv. Exp. Med. Biol. **1152**, 1–12 (2019). https://doi.org/10.1007/978-3-030-16034-1_1
77. F.E. Zajac, R.R. Neptune, S.A. Kautz, *Biomechanics and muscle coordination of human walking Part I: Introduction to concepts, power transfer, dynamics and simulations* (2002). https://doi.org/10.1016/S0966-6362(02)00068-1
78. J.A. Reinbolt, R.T. Haftka, T.L. Chmielewski, B.J. Fregly, Are patient-specific joint and inertial parameters necessary for accurate inverse dynamics analyses of gait? I.E.E.E. Trans. Biomed. Eng. **54**(5), 782–793 (2007). https://doi.org/10.1109/TBME.2006.889187
79. A. Falisse et al., Physics-based simulations to predict the differential effects of motor control and musculoskeletal deficits on gait dysfunction in cerebral palsy: A retrospective case study. Front. Hum. Neurosci. **14**(February), 1–17 (2020). https://doi.org/10.3389/fnhum.2020.00040
80. E. Russell Esposito, R.H. Miller, Maintenance of muscle strength retains a normal metabolic cost in simulated walking after transtibial limb loss. PLoS One **13**(1), e0191310 (2018). https://doi.org/10.1371/journal.pone.0191310
81. D. Zanotto, Y. Akiyama, P. Stegall, S.K. Agrawal, Knee joint misalignment in exoskeletons for the lower extremities: Effects on user's gait. IEEE Trans. Robot. **31**(4), 978–987 (2015). https://doi.org/10.1109/TRO.2015.2450414
82. R.S. Iyengar, K. Mallampalli, A.K. Singh, A. Koppula, K.S. Sridharan, M. Raghavan, The NEUROiD neuromusculoskeletal movement simulation platform, in *Digital Human Modeling and Medicine*, (Elsevier, 2023), pp. 161–197. https://doi.org/10.1016/B978-0-12-823913-1.00015-4. Available: https://linkinghub.elsevier.com/retrieve/pii/B9780128239131000154. Accessed 05 July 2025
83. SenseLab. ModelDB: A database for computational neuroscience models. 1996. Available: https://senselab.med.yale.edu/ModelDB/
84. Blue Brain Project. Channelpedia: Ion channel information resource. 2011. Available: https://channelpedia.epfl.ch/
85. M.D. Sunshine, F.S. Cho, D.R. Lockwood, A.S. Fechko, M.R. Kasten, C.T. Moritz, Cervical intraspinal microstimulation evokes robust forelimb movements before and after injury. J. Neural Eng. **10**(3), 036001 (2013). https://doi.org/10.1088/1741-2560/10/3/036001
86. M. Kapardi et al., In-silico neuro musculoskeletal model reproduces the movement types obtained by spinal micro stimulation. Comput. Methods Prog. Biomed. **220**, 106804 (2022). https://doi.org/10.1016/j.cmpb.2022.106804
87. R.V. Gonzalez, T.S. Buchanan, S.L. Delp, How muscle architecture and moment arms affect wrist flexion-extension moments. J. Biomech. **30**(7), 705–712 (1997). https://doi.org/10.1016/S0021-9290(97)00015-8
88. F.E. Zajac, Muscle and tendon: Properties, models, scaling, and application to biomechanics and motor control. Rev. Biomed. Eng. **17**(4) (1989) Available: https://hal.science/hal-04849267
89. D. Avers, M. Brown, *Daniels and Worthingham's Muscle Testing: Techniques of Manual Examination and Performance Testing*, 10th edn. (Elsevier, St. Louis, MO, 2019)
90. A. Prochazka, Chapter 11 Quantifying proprioception, in *Progress in Brain Research*, (Elsevier, 1999), pp. 133–142. https://doi.org/10.1016/S0079-6123(08)62850-2. Available: https://linkinghub.elsevier.com/retrieve/pii/S0079612308628502. Accessed 05 July 2025
91. W.W. Teka et al., From the motor cortex to the movement and back again. PLoS One **12**(6), e0179288 (2017). https://doi.org/10.1371/journal.pone.0179288
92. K. Jalaleddini et al., Neuromorphic meets neuromechanics, part II: The role of fusimotor drive. J. Neural Eng. **14**(2), 025002 (2017). https://doi.org/10.1088/1741-2552/aa59bd
93. E.M. Moraud et al., Mechanisms underlying the neuromodulation of spinal circuits for correcting gait and balance deficits after spinal cord injury. Neuron **89**(4), 814–828 (2016). https://doi.org/10.1016/j.neuron.2016.01.009

94. L.H. Salmond, A.D. Davidson, S.K. Charles, Proximal-distal differences in movement smoothness reflect differences in biomechanics. J. Neurophysiol. **117**(3), 1239–1257 (2017). https://doi.org/10.1152/jn.00712.2015
95. J. Gardenier, R. Garg, C. Mudgal, Upper extremity tendon transfers: A brief review of history, common applications, and technical tips. Indian J. Plast. Surg. **53**(02), 177–190 (2020). https://doi.org/10.1055/s-0040-1716456
96. D.S. Asakawa, G.H. Crocker, A. Schmaltz, D.L. Jindrich, Fingertip forces and completion time for index finger and thumb touchscreen gestures. J. Electromyogr. Kinesiol. **34**, 6–13 (2017). https://doi.org/10.1016/j.jelekin.2017.02.007
97. I.M. Comsa, T. Fischbacher, K. Potempa, A. Gesmundo, L. Versari, J. Alakuijala, Temporal coding in spiking neural networks with alpha synaptic function. ICASSP IEEE Int. Conf. Acoust. Speech Signal Process. - Proc. **2020**(May), 8529–8533 (2020). https://doi.org/10.1109/ICASSP40776.2020.9053856
98. A. Kozlov, M. Huss, A. Lansner, J.H. Kotaleski, S. Grillner, Simple cellular and network control principles govern complex patterns of motor behavior. Proc. Natl. Acad. Sci. USA **106**(47), 20027–20032 (2009). https://doi.org/10.1073/pnas.0906722106
99. A.K. Kozlov, A.A. Kardamakis, J. Hellgren Kotaleski, S. Grillner, Gating of steering signals through phasic modulation of reticulospinal neurons during locomotion. Proc. Natl. Acad. Sci. **111**(9), 3591–3596 (2014). https://doi.org/10.1073/pnas.1401459111
100. A.E. Talpalar et al., Identification of minimal neuronal networks involved in flexor-extensor alternation in the mammalian spinal cord. Neuron **71**(6), 1071–1084 (2011). https://doi.org/10.1016/j.neuron.2011.07.011
101. S.J.B. Butt, O. Kiehn, Functional identification of interneurons responsible for left-right coordination of Hindlimbs in mammals. Neuron **38**(6), 953–963 (2003). https://doi.org/10.1016/S0896-6273(03)00353-2
102. O. Kiehn, Decoding the organization of spinal circuits that control locomotion. Nat. Rev. Neurosci. **17**(4), 224–238 (2016). https://doi.org/10.1038/nrn.2016.9
103. A.H.A. Stienen, A.C. Schouten, J. Schuurmans, F.C.T. Van Der Helm, Analysis of reflex modulation with a biologically realistic neural network. J. Comput. Neurosci. **23**(3), 333 (2007). https://doi.org/10.1007/s10827-007-0037-7
104. A. Georgopoulos, J. Kalaska, R. Caminiti, J. Massey, On the relations between the direction of two-dimensional arm movements and cell discharge in primate motor cortex. J. Neurosci. **2**(11), 1527–1537 (1982). https://doi.org/10.1523/JNEUROSCI.02-11-01527.1982
105. M. Omrani, M.T. Kaufman, N.G. Hatsopoulos, P.D. Cheney, Perspectives on classical controversies about the motor cortex. J. Neurophysiol. **118**(3), 1828–1848 (2017). https://doi.org/10.1152/jn.00795.2016
106. T.N. Aflalo, M.S.A. Graziano, Partial tuning of motor cortex neurons to final posture in a free-moving paradigm. Proc. Natl. Acad. Sci. **103**(8), 2909–2914 (2006). https://doi.org/10.1073/pnas.0511139103
107. J. Ashe, A.P. Georgopoulos, Movement parameters and neural activity in motor cortex and area 5. Cereb. Cortex **4**(6), 590–600 (1994). https://doi.org/10.1093/cercor/4.6.590
108. D.W. Moran, A.B. Schwartz, Motor cortical representation of speed and direction during reaching. J. Neurophysiol. **82**(5), 2676–2692 (1999). https://doi.org/10.1152/jn.1999.82.5.2676
109. A.M. Smith, M.-C. Hepp-Reymond, U.R. Wyss, Relation of activity in precentral cortical neurons to force and rate of force change during isometric contractions of finger muscles. Exp. Brain Res **23**(3) (1975). https://doi.org/10.1007/BF00239743. Available: http://link.springer.com/10.1007/BF00239743. Accessed 06 July 2025
110. P.D. Cheney, E.E. Fetz, Functional classes of primate corticomotoneuronal cells and their relation to active force. J. Neurophysiol. **44**(4), 773–791 (1980). https://doi.org/10.1152/jn.1980.44.4.773
111. D.M. Griffin, H.M. Hudson, A. Belhaj-Saïf, B.J. McKiernan, P.D. Cheney, Do Corticomotoneuronal cells predict target muscle EMG activity? J. Neurophysiol. **99**(3), 1169–1986 (2008). https://doi.org/10.1152/jn.00906.2007

112. N.G. Hatsopoulos, Q. Xu, Y. Amit, Encoding of movement fragments in the motor cortex. J. Neurosci. **27**(19), 5105–5114 (2007). https://doi.org/10.1523/JNEUROSCI.3570-06.2007
113. N.G. Hatsopoulos, Y. Amit, Synthesizing complex movement fragment representations from motor cortical ensembles. J. Physiol.-Paris **106**(3–4), 112–119 (2012). https://doi.org/10.1016/j.jphysparis.2011.09.003
114. R.S. Sutton, A. Barto, Reinforcement learning: An introduction, in *Adaptive Computation and Machine Learning*, 2nd edn., (The MIT Press, Cambridge, MA/London, 2018)
115. R.S. Iyengar, K. Mallampalli, M. Raghavan, A Deep Reinforcement Learning based approach for movement training of neuro-musculoskeletal systems. Bioengineering (2021). https://doi.org/10.1101/2021.03.28.437396. Available: http://biorxiv.org/lookup/doi/10.1101/2021.03.28.437396. Accessed 06 July 2025
116. G. Brockman et al., OpenAI Gym. arXiv, June 05, 2016. https://doi.org/10.48550/arXiv.1606.01540. Available: http://arxiv.org/abs/1606.01540. Accessed 06 July 2025
117. M.S.A. Graziano, K.T. Patel, C.S.R. Taylor, Mapping from motor cortex to biceps and triceps altered by elbow angle. J. Neurophysiol. **92**(1), 395–407 (2004). https://doi.org/10.1152/jn.01241.2003
118. C. Genna et al., Spatiotemporal dynamics of the cortical responses induced by a prolonged tactile stimulation of the human fingertips. Brain Topogr. **30**(4), 473–485 (2017). https://doi.org/10.1007/s10548-017-0569-8
119. M. Omrani, C.D. Murnaghan, J.A. Pruszynski, S.H. Scott, Distributed task-specific processing of somatosensory feedback for voluntary motor control. eLife **5**, e13141 (2016). https://doi.org/10.7554/eLife.13141
120. G.M. Edelman, J.A. Gally, Degeneracy and complexity in biological systems. Proc. Natl. Acad. Sci. **98**(24), 13763–13768 (2001). https://doi.org/10.1073/pnas.231499798
121. A.A. Prinz, D. Bucher, E. Marder, Similar network activity from disparate circuit parameters. Nat. Neurosci. **7**(12), 1345–1352 (2004). https://doi.org/10.1038/nn1352
122. R.K. Rathour, R. Narayanan, Homeostasis of functional maps in active dendrites emerges in the absence of individual channelostasis. Proc. Natl. Acad. Sci **111**(17) (2014). https://doi.org/10.1073/pnas.1316599111. Available: https://pnas.org/doi/full/10.1073/pnas.1316599111. Accessed 05 July 2025
123. E. Bizzi, F.A. Mussa-Ivaldi, S. Giszter, Computations underlying the execution of movement: A biological perspective. Science **253**(5017), 287–291 (1991). https://doi.org/10.1126/science.1857964
124. M.S.A. Graziano, C.S.R. Taylor, T. Moore, Complex movements evoked by microstimulation of precentral cortex. Neuron **34**(5), 841–851 (2002). https://doi.org/10.1016/S0896-6273(02)00698-0
125. J.B. Nielsen, C. Crone, H. Hultborn, The spinal pathophysiology of spasticity - from a basic science point of view. Acta Physiol. **189**(2), 171–180 (2007). https://doi.org/10.1111/j.1748-1716.2006.01652.x
126. E.L. Lane, S.B. Dunnett (eds.), *Animal Models of Movement Disorders: Volume I*, Neuromethods, vol 61 (Humana Press, Totowa, NJ, 2012). https://doi.org/10.1007/978-1-61779-298-4. Available: https://link.springer.com/10.1007/978-1-61779-298-4. Accessed 26 Feb 2025
127. S. Aoi et al., Neuromusculoskeletal model that walks and runs across a speed range with a few motor control parameter changes based on the muscle synergy hypothesis. Sci. Rep. **9**(1), 1–5 (2019). https://doi.org/10.1038/s41598-018-37460-3
128. A. Bruel et al., Investigation of neural and biomechanical impairments leading to pathological toe and heel gaits using neuromusculoskeletal modelling. J. Physiol. **600**(11), 2691–2712 (2022). https://doi.org/10.1113/JP282609
129. L.A. Elias, R.N. Watanabe, A.F. Kohn, Spinal mechanisms may provide a combination of intermittent and continuous control of human posture: Predictions from a biologically based Neuromusculoskeletal model. PLoS Comput. Biol. **10**(11) (2014). https://doi.org/10.1371/journal.pcbi.1003944
130. B.J. Fregly, A conceptual blueprint for making Neuromusculoskeletal models clinically useful. Appl. Sci. **11**(5), 2037 (2021). https://doi.org/10.3390/app11052037

131. J.W. Lance, The control of muscle tone, reflexes, and movement: Robert Wartenbeg lecture. Neurology **30**(12), 1303–1303 (1980). https://doi.org/10.1212/WNL.30.12.1303
132. D. Burke, J.D. Gillies, J.W. Lance, The quadriceps stretch reflex in human spasticity. J. Neurol. Neurosurg. Psychiatry **33**(2), 216–223 (1970). https://doi.org/10.1136/jnnp.33.2.216
133. L.E. Krach, Pharmacotherapy of spasticity: Oral medications and intrathecal baclofen. J. Child Neurol. **16**(1), 31–36 (2001). https://doi.org/10.1177/088307380101600106
134. D.M. Simpson et al., Assessment: Botulinum neurotoxin for the treatment of spasticity (an evidence-based review): [RETIRED]: Report of the therapeutics and technology assessment Subcommittee of the American Academy of neurology. Neurology **70**(19), 1691–1698 (2008). https://doi.org/10.1212/01.wnl.0000311391.00944.c4
135. P. Steinbok, C. Schrag, Complications after selective posterior rhizotomy for spasticity in children with cerebral palsy. Pediatr. Neurosurg. **28**(6), 300–313 (1998). https://doi.org/10.1159/000028668
136. F. Biering-Sørensen, J.B. Nielsen, K. Klinge, Spasticity-assessment: A review. Spinal Cord **44**(12), 708–722 (2006). https://doi.org/10.1038/sj.sc.3101928
137. B. Petek Balci, Spasticty measurement. Arch. Neuropsychiatry (2018). https://doi.org/10.29399/npa.23339. Available: http://submission.noropsikiyatriarsivi.com/default.aspx?s=public~kabul&mId=23339
138. P. Perel, et al., Comparison of treatment effects between animal experiments and clinical trials: systematic review, 2006, https://doi.org/10.1136/bmj.39048.407928.BE
139. S.L. Delp, J.P. Loan, M.G. Hoy, F.E. Zajac, E.L. Topp, J.M. Rosen, An interactive graphics-based model of the lower extremity to study orthopaedic surgical procedures. I.E.E.E. Trans. Biomed. Eng. **37**(8), 757–767 (1990). https://doi.org/10.1109/10.102791
140. A.V. Hill, The heat of shortening and the dynamic constants of muscle. Proc. R. Soc. Lond. Ser. B - Biol. Sci. **126**(843), 136–195 (1938). https://doi.org/10.1098/rspb.1938.0050
141. J. Mehrholz et al., Reliability of the modified Tardieu scale and the modified Ashworth scale in adult patients with severe brain injury: A comparison study. Clin. Rehabil. **19**(7), 751–759 (2005). https://doi.org/10.1191/0269215505cr889oa
142. C.L. Cleland, L. Hayward, W.Z. Rymer, Neural mechanisms underlying the clasp-knife reflex in the cat. II. Stretch-sensitive muscular-free nerve endings. J. Neurophysiol. **64**(4), 1319–1330 (1990). https://doi.org/10.1152/jn.1990.64.4.1319
143. C.L. Cleland, W.Z. Rymer, Neural mechanisms underlying the clasp-knife reflex in the cat. I. Characteristics of the reflex. J. Neurophysiol. **64**(4), 1303–1318 (1990). https://doi.org/10.1152/jn.1990.64.4.1303
144. Y.N. Wu, H.S. Park, J.J. Chen, Y. Ren, E.J. Roth, L.Q. Zhang, Position as well as velocity dependence of spasticity-four-dimensional characterizations of catch angle. Front. Neurol. **9**(OCT), 1–10 (2018). https://doi.org/10.3389/fneur.2018.00863
145. E.G. Fowler, A.I. Nwigwe, T.W. Ho, Sensitivity of the pendulum test for assessing spasticity in persons with cerebral palsy. Dev. Med. Child Neurol. **42**(3), 182–189 (2000). https://doi.org/10.1111/j.1469-8749.2000.tb00067.x
146. A.K. Singh, K.S. Sridharan, S. Wu, O. Tirosh, M. Raghavan, In-silico neuro-musculoskeletal model demonstrates spasticity progression with descending motor tracts loss in simulated clinical triage. Comput. Biol. Med. **192**, 110270 (2025). https://doi.org/10.1016/j.compbiomed.2025.110270

Chapter 19
Soft Robotics for Neuroadaptive Assistance: Actuation, Sensing, and Brain-Computer Interface Integration

Abhay Raina, Arjun Raina, Parthan Olikkal, Viraj Janeja, Oritsejolomisan Mebaghanje, and Ramana Vinjamuri

Abstract Soft robotics has emerged as a transformative field aimed at creating machines with compliant, adaptable, and bioinspired machines for human-centered applications. This chapter presents an integrative overview of the key technologies that drive soft robotic systems, particularly in assistive and neuroadaptive contexts. We begin by examining diverse actuation strategies—such as pneumatic, hydraulic, shape memory alloys, dielectric elastomers, and magnetic actuation—highlighting their mechanical principles and suitability for flexible, human-interactive tasks. The role of sensing and control is explored through resistive, capacitive, fiber-optic, and proprioceptive sensor modalities, and we describe the use of open-loop, closed-loop, and autonomous control strategies. Emphasis is placed on the synergy between soft robotics and brain-computer interfaces (BCIs), detailing both invasive and non-invasive paradigms such as EEG, fNIRS, and MEG. Applications discussed include rehabilitation for motor impairments, lower-limb gait assistance, facial expression recognition, and emotion-aware interaction. Finally, recent trends over the last five years are covered, including advancements in AI integration, embodied intelligence, smart materials, and multi-robot systems. This multidisciplinary view provides a foundation for researchers developing soft robotic systems for real-world human interaction and neuroadaptive control.

Keywords Actuation mechanisms · Brain-computer Interface · Human-robot interaction · Neurorehabilitation · Open-loop and closed-loop control · Soft robotics

Authors Abhay Raina and Arjun Raina have equally contributed to this chapter.

A. Raina · A. Raina · P. Olikkal · V. Janeja · O. Mebaghanje · R. Vinjamuri (✉)
Vinjamuri Lab, Department of Computer Science and Electrical Engineering, University of Maryland Baltimore County, Baltimore, MD, USA
e-mail: polikka1@umbc.edu; vjaneja1@umbc.edu; xz94254@umbc.edu; rvinjam1@umbc.edu

R. Vinjamuri (ed.), *Bridging the Gap between Mind and Machine*,
https://doi.org/10.1007/978-3-032-06713-5_19

19.1 Introduction

The human brain, often described as the most complex and essential organ in the body, serves as the command center for all physiological, cognitive, and motor activities. It operates in close coordination with the central and peripheral nervous systems to regulate voluntary and involuntary functions. Among these functions, motor control is particularly crucial for performing everyday actions such as walking, grasping, speaking, and expressing emotion [1, 2]. This seamless interaction between brain and body is disrupted in conditions involving neural injury or degeneration, such as stroke, spinal cord injury, or neurodegenerative diseases. One of the most severe manifestations of these disruptions is paralysis, a condition that results in the partial or complete loss of motor function and affects approximately 5.4 million individuals in the United States alone [3]. Replicating these tasks has proved to be an extremely challenging task [4].

Traditional rehabilitation methods have had limited success in restoring motor capabilities for individuals with paralysis, particularly in cases involving chronic neurological damage. In recent years, emerging technologies—most notably soft robotics and brain-computer interfaces (BCIs)—have offered promising alternatives. Soft robotic systems provide flexible, adaptive mechanical support that can safely interface with the human body. Meanwhile, BCIs serve as a neural bridge by capturing brain signals and translating them into control commands, bypassing the damaged neuromuscular pathways.

Together, soft robotics and BCIs represent a paradigm shift in assistive and rehabilitative technologies. This chapter explores the interdisciplinary foundations of these technologies and how they coalesce to create human-centered solutions for regaining lost function. Key areas explored include the mechanics of soft robotic actuation, the design of biologically inspired materials, neural decoding through non-invasive brain monitoring, closed-loop control strategies for feedback integration, and real-world applications such as rehabilitation, facial expression recognition (FER), and intelligent human-robot interaction. The remainder of the chapter is structured as follows: Sect. 19.2 delves into various actuation strategies and material innovations that enable the flexibility and adaptability of soft robots. Section 19.3 focuses on Brain-Computer Interface (BCI) paradigms, highlighting both invasive and non-invasive techniques for neural decoding. Section 19.4 explores sensing technologies and control architecture, including open-loops, closed-loops, and autonomous systems. Section 19.5 examines real-world applications of soft robotics in areas such as rehabilitation, locomotion, and emotion recognition. Section 19.6 presents recent advancements in the field along with emerging trends and future directions. Finally, Sect. 19.7 concludes the chapter by summarizing key insights and discussing the broader implications of integrating soft robotics with BCI technologies.

19.2 Actuation and Materials in Soft Robotics

Soft robotic actuators are engineered through a variety of mechanisms, with some of the most widely adopted approaches including fluid-powered actuation, dielectric elastomer actuators, magnetic materials and shape memory alloys as shown in Fig. 19.1. Among these, fluidic actuation—where motion is generated through the controlled inflation and deflation of internal chambers—has proven particularly effective in applications requiring adaptability, compliance, and safe physical interaction [5]. This class of actuation is broadly divided into pneumatic and hydraulic systems [6], both of which operate based on pressure regulation to achieve precise deformation and motion control.

Pneumatic actuators, driven by compressed air, are especially valued in wearable and assistive technologies due to their inherent lightness, flexibility, and safety. These characteristics make them ideal for close human interaction. Common examples include soft robotic gloves for hand rehabilitation, soft elbow trainers, wearable devices for lower back support, and exoskeletal assistive systems [7–11]. The historical roots of pneumatic actuation trace back to the artificial pneumatic muscle developed by Richard Gaylord in 1958, later refined and popularized by Joseph McKibben. These "McKibben muscles" mimicked the behavior of human muscle tissue and laid the foundation for modern pneumatic soft actuators. Over the past decade, advances in materials and control systems have accelerated their adoption in rehabilitative and assistive robotics [12]. Compared to hydraulic systems, pneumatic actuators are significantly lighter, making them more suitable for wearable applications such as exosuits and mobile robotic devices [13].

By contrast, **hydraulic actuators** operate using pressurized fluids and offer considerably higher power density and bandwidth. This makes them well-suited for applications demanding strong, rapid actuation—especially in compact or confined spaces. Hydraulic systems are particularly effective when embedded in soft robotic devices that require a high degree of control, such as glove-like interfaces or

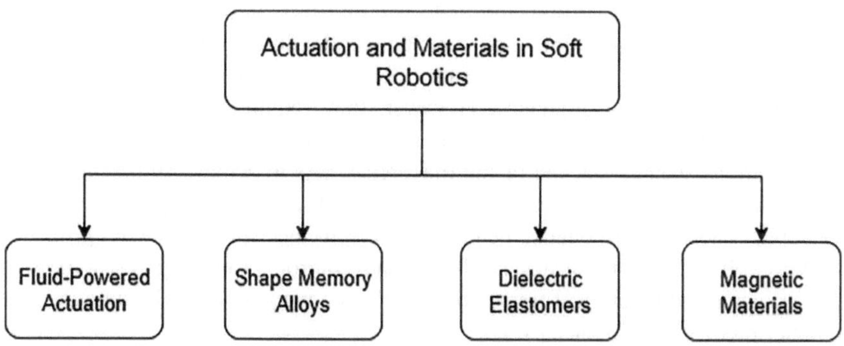

Fig. 19.1 Overview of the major actuation mechanisms and material types discussed, thematically categorized for clarity and application relevance in soft robotics

grippers. In hybrid configurations, the combination of pneumatic and hydraulic actuation has led to the development of highly versatile robotic platforms. A notable example includes a soft robot capable of underwater swimming and crawling, designed to perform exploration tasks in complex aquatic environments [12, 14]. This hybrid approach harnesses the strength of both modalities, enabling precise control, adaptability, and robust force generation.

Another widely used class of actuators in soft robotics is **Shape Memory Alloys (SMAs)**. These are metallic materials known for their remarkable ability to return to a pre-defined shape when subjected to external stimuli, such as heat or mechanical stress. These metallic alloys undergo a reversible phase transformation between a stiff, high-temperature austenite phase and a more flexible, low-temperature martensite phase, allowing them to generate large strains and strong recovery forces [15]. Their compactness, silent operation, and muscle-like motion make them ideal for wearable applications in rehabilitation and assistance. SMAs have been successfully incorporated into various devices, such as soft knee-ankle-foot orthoses for lower limb therapy [16], suit-type wearable robots for muscular support [17], and soft gloves for aiding hand movement and recovery [18]. However, one of their primary limitations is the relatively slow response time—typically between 10 to 30 seconds—due to the need for significant temperature changes to induce actuation, which complicates system design and limits real-time responsiveness [12].

Elastomeric materials play a crucial role in soft robotics, particularly in ensuring the mechanical stability of components like grippers during deformation. These materials often work in tandem with fluidic actuation—typically using air or liquid—to enable smooth and controlled motion across a wide range of applications [5]. A prominent example of elastomer-based actuation is the **Dielectric Elastomer Actuator (DEA)**, which consists of a dielectric polymer film sandwiched between two compliant electrodes. When voltage is applied, the electrodes generate electrostatic forces that compress the elastomer, causing lateral expansion and movement. DEAs have been successfully used in various assistive technologies, including soft orthotic gloves for hand rehabilitation, soft robotic grippers, and lower-extremity compression bands [19, 20]. The performance of DEAs strongly depends on the material properties of both the dielectric and the electrodes—requiring high conductivity, strong adhesion, and mechanical flexibility for optimal actuation [21]. Key advantages of DEAs include their high strain capacity, compliance, bandwidth, and energy efficiency. However, they require high-voltage inputs—often in the kilovolt range—and typically produce low force, necessitating complex configurations such as stacked or multilayered DEAs to achieve sufficient output. These multi-component designs can become bulky and limit practical use in wearable systems. Additionally, the nonlinear electromechanical behavior governed by Maxwell stress complicates modeling and control, prompting researchers to explore advanced theoretical frameworks such as nonequilibrium thermodynamics to better predict performance [5, 12].

Magnetic actuation offers a versatile and remote method for powering soft robots, eliminating the need for onboard power or physical connections. This approach utilizes magnetic fields to induce motion and deformation in soft robotic structures, enabling wireless and untethered operation [5]. Magnetic actuation can be categorized into two main types: direct and indirect activation, which corresponds to the use of ferromagnetic and paramagnetic materials, respectively. Ferromagnetic materials possess permanent magnetic dipoles that continuously interact with external magnetic fields. In contrast, paramagnetic materials only exhibit magnetic dipoles when exposed to a field, allowing for field-dependent control. When these materials are placed in magnetic field gradients, they experience localized forces that can drive motion or change in stiffness. Dipole-dipole interactions can also be harnessed to achieve complex deformations or mechanical responses. Soft robots can use these interactions to generate torque or force, enabling movement across various terrains or tasks. Alternating current (AC) magnetic fields are particularly effective, as they can induce mechanical oscillations at specific frequencies, allowing for precise and selective control. AC fields also tend to produce larger actuation forces compared to direct current (DC) fields. The combination of AC and DC inputs allows for a broader and more dynamic range of motion. **Magnetic Shape Memory Alloys (MSMAs)** further enhance this modality by deforming in response to magnetic stimuli, though they present challenges in fabrication and shaping [22].

The material composition of soft robots is tailored to their intended function, with specific components chosen to meet mechanical, sensory, and adaptive performance needs. A core element across most designs is the integration of sensors that allow the robot to detect and respond to environmental stimuli in real time [23]. In addition to sensing capabilities, soft robots often incorporate smart materials such as hydrogels, fibers, fabrics, rubber polymers, and shape memory materials, each contributing to flexibility, resilience, and functional responsiveness [24, 25]. Rubber polymers, in particular, are commonly used in biologically inspired robots due to their ability to enhance load-bearing capacity, endurance, and shape retention. Hydrogels, composed of three-dimensional networks of cross-linked polymer chains, are notable for their capacity to respond to diverse stimuli—such as light, temperature, pH, electric and magnetic fields—by undergoing significant and reversibly shape changes. Their flexibility, transparency, and compatibility with biological systems make them especially valuable in applications requiring fine-tuned responsiveness and adaptability.

Shape Memory Materials (SMMs) possess the unique ability to return to a pre-defined original shape after being deformed by external stimuli such as heat, stress, or light. This "memory" effect allows them to recover their initial form once the environment trigger is removed, making them valuable in various soft robotic systems [26].

Fibers, often incorporated using weaving or knitting techniques, are lightweight, highly flexible, and porous—properties that align well with the demands of soft robots requiring high degree of freedom and comfort during wear. Their performance can also be influenced by moisture levels, making them responsive to hydration-based actuation.

Another notable advancement is the development of **Variable Length Tendons (VLTs)**, which are designed to replicate the function of biological tendons [27]. VLT systems are typically divided into two main types: those that rely on SMAs for contraction, and those that use tension cables. While tension cable-driven systems were more common in early designs, they have largely been replaced by actuator-based methods due to improved precision and integration potential.

Fluidic Elastomer Actuators (FEAs) operate through internal pressure changes to induce deformation and motion [22]. These actuators inflate embedded chambers to generate localized forces that drive movement and are commonly powered by either pneumatic or hydraulic pressure. Pneumatic actuation, using compressed air, is preferred for its rapid response and availability, though it requires airtight sealing and controlled environments. Hydraulic systems, on the other hand, produce greater force, enabling stronger locomotion, but may add complexity due to fluid containment. Despite these differences, both systems rely on the same principle of pressure-driven actuation to achieve dynamic deformation.

Lastly, **Electroactive Polymers (EAPs)** are a class of smart materials that deform in response to electrical stimulation [28]. A prominent subtype is the **Dielectric Elastomer (DE)**, which is known for its fast actuation speed, customizability, silent operation, and low material cost [29, 30]. However, DEs require high driving voltages and may suffer from operational instability, posing challenges for long-term and scalable deployment in soft robotic platforms.

19.3 Brain Computer Interface and Paradigms

BCIs have undergone continuous innovation since their inception in the 1970s, achieving major milestones across biomedical, assistive, and rehabilitation domains. A BCI is a closed-loop system that measures brain signals, decodes user intentions, and provides input to external devices. BCIs are broadly classified into invasive and non-invasive modalities as presented in Fig. 19.2. Invasive BCIs involve surgically implanted electrodes, such as intracortical arrays or electrocorticography (ECoG), which capture high-resolution neural activity. These systems often record stronger signals because of their proximity to neuronal tissue. Non-invasive BCIs use external sensors like electroencephalography (EEG), functional near-infrared spectroscopy (fNIRS), or magnetoencephalography (MEG), eliminating the need for surgery but sacrificing signal fidelity and spatial resolution. Both invasive and non-invasive BCIs typically process data in real-time to generate actionable commands [1, 4, 31, 32].

Non-invasive BCIs have gained popularity due to their ease of use, safety, and accessibility across different user groups, despite offering lower fidelity than invasive systems. These were first introduced by Jonathan Wolpaw in the 1990s. They are especially beneficial for individuals with severe motor impairments, offering an alternative communication channel without requiring muscle control. Signal amplification is typically applied to these recordings to facilitate easier interpretation and command generation [4]. A major use case for BCIs is in motor recovery, where they are integrated with assistive technologies like soft robotic suits

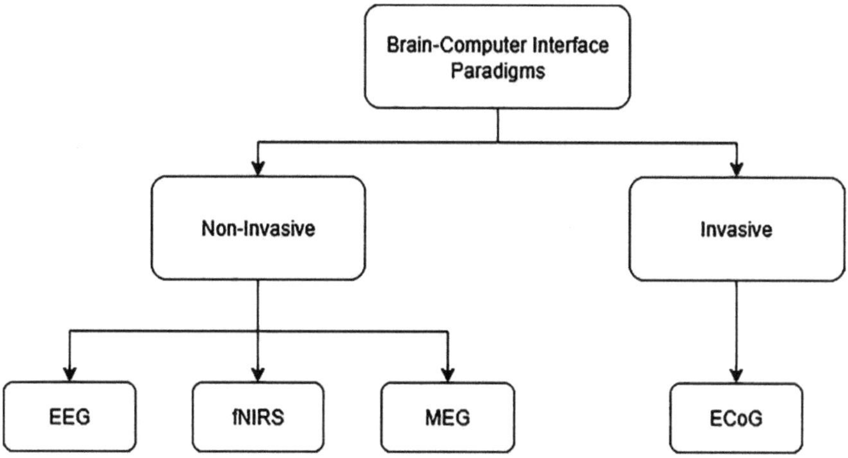

Fig. 19.2 Categorized overview of major Brain-Computer Interface (BCI) paradigms, distinguishing between invasive and non-invasive approaches and illustrating their relevance to neural decoding and control in soft robotics

(e.g., HEXOES [33]) or used in neurorehabilitation to promote neural reorganization [1].

Rehabilitation BCIs typically follow one of two paradigms: motor execution (ME) or motor imagery (MI). The ME approach decodes cortical activity associated with voluntary movement, while MI focuses on imagined actions, enabling control even for users who are physically unable to move [2].

EEG, which measures cortical electrical activity through non-invasive electrodes placed on the scalp, remains one of the most widely used modalities in BCI research, particularly for motor imagery (MI)-based systems. EEG captures electrical activity through electrodes placed on the scalp, often using conductive gels or pastes to improve signal quality. Depending on the mental condition of the patient, different frequency ranges and brainwave types are detected. Although limited in spatial resolution, EEG remains a preferred method due to its portability, affordability, strong temporal resolution, and ease of use. Recent research demonstrates the application of EEG in decoding not only the intent of the user's actions but also affective states such as eustress (positive stress) and distress (negative stress). Modern applications have expanded to include emotional state recognition, with EEG outperforming other biosignals in stress detection tasks. By combining EEG with machine learning and deep learning models, researchers have significantly improved cognitive state decoding accuracy and enabled the development of wearable neurotechnologies [34]. In fact, EEG has outperformed other biosignals in stress detection tasks, emphasizing its practical utility [1]. It also plays a critical role in diagnosing neurological disorders like Parkinson's disease, epilepsy, and ADHD [35].

Recent research indicates that low-frequency EEG can be used to reconstruct intended hand movements. The ultimate aim is to enable users to control robotic systems using thought alone, with neural signals serving as control inputs. Researchers have mapped directional intent by monitoring neuron activity patterns in studies involving primates [36]. A notable contribution by Mondini et al. demonstrated that real-time decoding of hand trajectories from EEG is feasible, moving beyond prior studies that relied solely on offline analysis. Their online closed-loop control framework integrated EEG signals into robotic actuation. Through partial least squares (PLS) regression, meaningful correlations between brain signals and movement parameters were identified. This work supports the trend of making BCIs more natural and intuitive in actuator control. Comparable success has also been achieved using MEG, highlighting its complementary role [37].

Other non-invasive BCI modalities include fNIRS and MEG. fNIRS uses light in the near-infrared range to monitor brain activity by detecting fluctuations in oxygenated and deoxygenated hemoglobin concentration across cortical regions. These hemodynamic responses serve as a proxy for neural activation and are particularly useful in cognitive neuroscience and rehabilitation. fNIRS can operate even in the presence of slight subject movement and provides outputs such as signal averages and variance across tasks. The spatial resolution of fNIRS depends on the configuration and density of its optodes. Too few optodes may lead to poor spatial fidelity, while an overly dense setup can cause light scattering and crosstalk, reducing signal clarity. Therefore, optimal optode placement is critical for accurate measurement [31].

MEG, on the other hand, visualizes brain activity by recording the tiny magnetic fields produced by electrical currents in neurons, especially in cortical pyramidal cells. These magnetic signals are typically characterized by temporal properties and spatial distribution. MEG provides high temporal resolution and excellent spatial accuracy, making it especially effective in auditory BCI applications. Recent efforts have focused on expanding MEG use to visual paradigms and multimodal setups that incorporate EEG and fNIRS for enhanced spatial-temporal resolution. Unlike fNIRS, MEG can accommodate increased sensor density without significant degradation in signal quality, enhancing its utility for high-resolution tasks. Nonetheless, the system's cost and physical complexity make it impractical for routine or wearable use [4]. Precision in positioning MEG sensors over target brain regions is essential to maintain signal quality and minimal noise [31].

Although this chapter focuses primarily on non-invasive BCIs, it is important to acknowledge a prominent invasive modality ECoG. Originally developed by W. Penfield and H. Jasper in the 1930s for seizure detection, ECoG involves the surgical placement of macroelectrodes directly on the surface of the cerebral cortex. This proximity allows for the recording of electrophysiological activity with extremely high temporal and spatial precision. Compared to non-invasive techniques, ECoG offers significantly improved signal fidelity, reduced noise, and enhanced signal-to-noise ratios due to the absence of skull-based signal attenuation. This direct cortical access allows ECoG to capture subtle, high-speed neuronal dynamics, making it ideal for applications requiring fine motor control and high-resolution decoding. While increasing the number of implanted electrodes generally enhances performance, studies suggest that accuracy gain tend to plateau after approximately eight electrodes. Beyond this threshold, additional electrodes contribute marginal improvements, making further surgical or computational investment difficult to justify [31].

Designing an effective BCI system also hinges on the underlying BCI paradigm, which serves as the bridge between the user's intention and neural signal interpretation. These paradigms translate brain activity into interpretable commands based on task-specific neural patterns. To ensure usability, paradigms must prioritize user comfort, task simplicity, and alignment with transparent, intuitive mappings between intention and control. User satisfaction plays a central role in paradigm success. A complete BCI framework consists of three core components: the signal acquisition modality, the neural encoding of brain activity, and the decoding algorithm that translates these signals into actionable outputs. Decoding relies heavily on both the paradigm structure and the selected neural coding model. While traditional BCI system often use classical machine learning approaches for decoding due to their interpretability and low computational demand, deep learning is gaining traction for its superior performance on large and complex datasets [38].

Importantly, each BCI paradigm may require a tailored neural coding model depending on the imaging modality and experimental design. The resolution, latency, and sensor configuration of each modality influence how neural activity is represented and encoded. For instance, EEG-based systems require different feature extraction strategies than fNIRS or MEG systems, even for similar motor imagery tasks. In the context of EEG, neural coding strategies may include time-domain analysis of signal amplitude, frequency-domain extraction of oscillatory rhythms, and spatial-domain mapping across motor cortex regions. In contrast, fNIRS paradigms typically adopt temporal coding models that analyze hemodynamic responses during cognitive or motor tasks. MEG paradigms may also apply time-domain models to detect transient peaks in the brain's magnetic field following specific stimuli. Standard BCI development begins with task design, followed by the construction of appropriate neural coding models and signal acquisition using the chosen imaging technique [38].

Among the most established BCI paradigms, the P300 event-related potential paradigm is widely recognized for its reliability and low training requirements. First observed 300 milliseconds after stimulus presentation, the P300 wave is used to detect a user's recognition of a target stimulus among non-targets. This method enables robust stimulus discrimination and is especially effective in visual tasks due to clearer signal separation. Another popular paradigm is the Steady-State Visual Evoked Potential (SSVEP), known for its rapid response time and minimal calibration requirements. However, SSEVPs can be visually fatiguing and uncomfortable for extended use, limiting their applicability in real-world daily settings. Despite this, SSVEPs offer a strong signal-to-noise ratio, making them effective in high-speed and high-accuracy BCI applications [4, 31].

19.4 Open-Loop, Closed-Loop, and Autonomous Control in Soft Robotics

The importance of sensing and control in soft robotics stems from the inherent flexible and deformable nature of these systems, which possess numerous degrees of freedom. As a result of these characteristics, it becomes challenging for centralized controllers to accurately determine the robot's configuration at a given time. Programmable materials are often employed to accommodate this variability, yet accurately modeling such dynamic systems remains complex. Therefore, sensors become essential for real-time tracking of environmental and internal variations [5].

However, these sensors must also be stretchable and flexible to align with the robot's mechanical properties. This requirement introduces further challenges, necessitating sophisticated modeling and high-speed data processing to interpret sensor inputs effectively. Given the diversity of tasks in soft robotics, selecting the most appropriate sensor type is crucial for achieving precise and efficient performance. Commonly used sensors include resistive, piezoresistive, capacitive, and fiber optic shape sensors [5].

Resistive and piezoresistive sensors rely on Ohm's law to detect changes in electrical resistance caused by external stimuli, which are then used to infer the robot's current state. These changes in resistance inform decisions about actuation or adjustment [5].

Fiber optic shape sensors (FOSS) operate using optical frequency domain reflectometry, in which a later sweep across a range of optical frequencies to produce interference fringes. These fringes are converted into reflection coefficients via Fourier transform, and strain is measured at multiple points along the sensor's core embedded within the robot. This method enables extremely high spatial resolution, with sensors capable of detecting changes every 0.8 mm. Galloway et al. demonstrated the effectiveness of FOSS when placed along the robot's neutral axis, enabling precise shape detection with minimal noise [39].

Capacitive sensors measure environmental interactions by detecting variations in capacitance, which correspond to deformation or external contact. Like other sensor types, they require flexible and stretchable materials to maintain compatibility with soft robotic substrates [5].

Soft robots employ several control strategies to process sensory data and regulate actuation, with open-loop, close-loop, and autonomous control being the primary approaches. Among these, open-loop control remains the most widely used due to its simplicity, although more sophisticated alternatives are gaining traction. Open-loop systems function without any feedback from the robot, operating on pre-programmed instructions similar to muscle memory [5]. These systems require comprehensive knowledge of the robot's shape, dimensions, and environment, as there is no dynamic adjustment based on feedback. Finite Element Modeling (FEM) is often used to simulate the robot's behavior under specific conditions, reducing the need for real-time feedback. FEM can model complex deformations and material properties, though it is computationally intensive and time-consuming [5].

In contrast, closed-loop control leverages sensory feedback to dynamically adjust the robot's movements in real time, making it ideal for applications requiring precision and adaptability. This form of control often incorporates reinforcement learning to optimize behavior through interaction with the environment. Feedback comes from proprioceptive sensors, which provide two key layers: one associated with actuator deformation, and the other related to geometric positioning and body dynamics [22]. Initial feedback includes physical stimuli like pressure or temperature, which influences actuator output. The secondary layer includes kinematic feedback, allowing the robot to modulate its shape and navigate varied environments. Closed-loop systems are especially promising for human-interactive technologies, such as prosthetics, although obtaining reliable signals from the human body remains a challenge. Depending on the actuation method, different sensor types—like EEG—may be required to detect human intent and drive the system [5]. Once data is collected and interpreted, necessary adjustments are made to achieve the desired behavior. A predefined model of robot is usually used in this control mode to simplify computation.

Autonomous control extends closed-loop capabilities by incorporating learning algorithms that adept in real time based on environmental conditions. This intelligent framework enables robots to self-adjust their parameters and configurations without predefined instructions, making them ideal for unfamiliar or dynamic terrains [22]. The core of autonomous control lies in machine learning algorithms trained on large datasets to recognize patterns and optimize responses. However, the effectiveness of these systems is contingent on the availability of sufficient training data. Additionally, the computational load required to run these algorithms in real time is substantial, presenting another limitation to widespread adoption [22].

19.5 Application of Soft Robotics

From restoring motor function in rehabilitation to decoding subtle muscle activity for facial expression recognition, soft robotic systems offer unmatched versatility. Moreover, their potential in assisting with locomotion and enhancing precision in minimally invasive surgery highlights their growing impact in clinical and assistive technologies. These key application areas (in Fig. 19.3) are discussed here.

Locomotion remains one of the most essential yet underexplored aspects of soft robotics, encompassing biomechanical patterns and physical principles that govern movement in both biological organisms and robotic constructs. Research in this domain has heavily drawn from biological systems, taking advantage of evolutionary refined strategies to develop efficient and adaptive robotic mobility. Various soft robotic locomotion strategies such as crawling, peristalsis, walking, leaping, and bio-inspired aerial motion have been examined. These biologically inspired movements offer significant benefits for navigating unstructured environments and enhancing interactions between humans and robots [40].

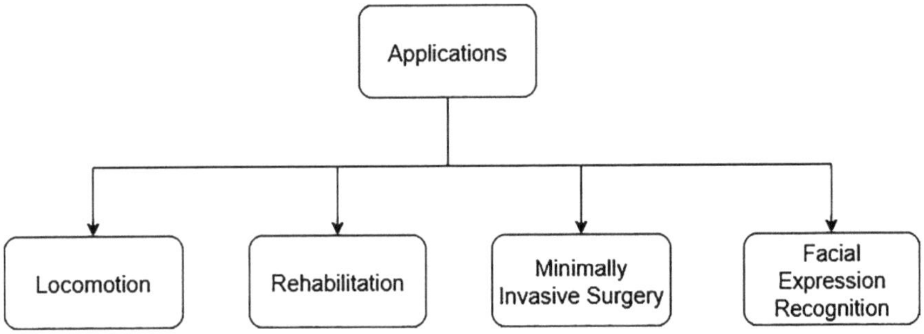

Fig. 19.3 Visual representation of key application domains of soft robotics and BCI integration, including rehabilitation, facial expression recognition (FER), locomotion, and minimally invasive surgery—each mapped to relevant human-centered use cases

One of the key areas of application is supporting Activities of Daily Living (ADLs), which often depend on fine motor skills involving hand movements. Tasks such as buttoning clothes, typing, or brushing teeth highlight how hand functionality underpins independence and quality of life [41, 42]. Stroke and Parkinson's disease are two major impairments that compromise the ability to perform ADLs. Recovery through physical rehabilitation hinges on neuroplasticity, the brain's capacity to reorganize itself by forming new neural pathways after damage [35].

BCI-assisted soft robotic systems for stroke rehabilitation show substantial promise in restoring upper-limb motor control and perceptual feedback [43]. In one example, Shi et al. conducted a study administering twenty-one-hour sessions of EMG-controlled soft robotic hand therapy to stroke survivors, reporting notable gains in upper limb performance. Statistically significant outcomes were observed across various metrics, including ARAT ($p = 0.032$), FMA-UE ($p = 0.003$), BBT ($p = 0.024$), and grip strength ($p < 0.001$) [44]. Kim et al. demonstrated a wearable robotic garment targeting hip flexion, which effectively mitigated gait freezing in individuals with Parkinson's disease over a six-month period, increasing indoor ambulation and travel distance. The same intervention also reduce outdoor gait freezing and enhanced overall walking distance [45]. Tang et al. validated motor improvements using the Timed Up and Go (TUG) test, driven by motor imagery BCI rehabilitation with visual feedback [46, 47].

Soft robotics also finds considerable use in medical domains, particularly in devices designed for surgery, diagnostics, drug administration, wearable aids, prosthetics, organ simulation, and biomechanical training tools [24].

In the realm of emotional intelligence, the ability of robots to interpret facial expressions is crucial for effective interpersonal communication. Recognizing universal emotions facilitates meaningful human-robot interaction across cultural and linguistic boundaries. To support this, FER technologies have been developed, using deep learning to classify facial cues into emotional states. These systems help robots identify and respond to human emotions accurately, enhancing communication. FER systems are trained in both static and dynamic visual inputs, using neural networks to categorize emotional content from human expressions. Convolutional Neural Networks (CNNs) are frequently used in such classification tasks.

Static input methods focus on still frames, allowing spatial features such as pose and facial geometry to inform emotion recognition. These typically analyze single-frame expressions like photographs. In contrast, dynamic media such as video captures allow models to monitor subtle changes in facial expressions over time. The goal is to detect transitional emotional cues to enable more responsive and empathetic robot behavior.

Several innovative models have emerged for training on static images. CAKE (Compact and Accurate K-dimensional Embedding) is a recent architecture designed for classifying emotions from static inputs. FaceBehaviorNet leverages detailed facial behavior analysis to determine emotional state, while Hybrid MultiTask Learning (HMTL) employs self-supervised learning to enrich facial feature representations. Distract Your Attention Network (DAN) isolates key facial regions and combines them to improve emotion interpretation, often in conjunction with CNNs.

For dynamic visual input, models such as EmoAffectNet concentrate on critical facial landmarks to optimize emotion recognition efficiency. These models often utilize ensembles of neural networks instead of a single one, thereby improving classification accuracy. This multi-network approach yields superior emotional understanding and response capabilities. Nonetheless, inherent biases related to race, age, gender, or lighting conditions can impair accuracy. These factors influence the model's effectiveness, as facial cues may vary across demographics and lighting can obscure essential expression features.

These models primarily rely on embedded cameras within robots to gather visual input. The captured data—ranging from facial movements to posture and gestures—is analyzed in real time to drive robot responses. Additional emotional indicators such as micro-expressions and body stance enhance the system's context awareness. This multimodal sensing improves the robot's emotional understanding, leading to richer human-robot engagement.

Once the data is captured, automated analysis is triggered using features like motion dynamics, fluidity, and gesture kinematics. These behavioral attributes, especially in facial and bodily expressions, are essential to inferring the emotional condition of the human user.

A commonly adopted framework for emotion recognition in human-robot interaction is Emotion Body Gesture Recognition (EGBR), which comprises three primary components: body detection, pose estimation, and representation learning. **Body detection** serves as the foundational step, focusing on isolating the human subject as the sole object of interest within the visual frame by removing the background to prevent interference in data interpretation [48, 49]. This process enhances emotion recognition accuracy by ensuring that only relevant features are passed to the model. Within body detection, two subcategories exist: geometric features, which capture specific facial or bodily elements such as hands, eyes, or the mouth; and appearance features, which analyze the entire body or face to infer emotional states. **Pose estimation** follows as the second stage, involving model training to track body configurations and the emotional context they convey, whether from static postures or dynamic movements, with emphasis on facial expressions, hand gestures, and full-body positioning [50]. The final stage, **representation learning**, applies machine learning techniques to classify emotional states by mapping observable features to categorical or dimensional emotion scales. Some approaches quantify emotion using regression or classification models across three- or five-point scales based on the prominence of features in the body or face. Others analyze the spatial positions of joints and body parts—such as the head or hands—viewed from different angles to infer emotional content. Convolutional Neural Networks (CNNs) have also been developed by combining spatial features and camera-derived data to enhance classification accuracy. In addition to single-modality inputs, multimodal models have been proposed that integrate facial expressions, speech cues, and full-body posture—including limb and head motion—to develop more reliable, context-aware systems for emotion interpretation. This structured approach to EGBR enables robots to better understand human emotional states, ultimately leading to more natural and emotionally intelligent human-robot communication.

19.5.1 Minimally Invasive Surgery

Minimally Invasive Surgery (MIS) enhanced by soft robotic systems is emerging as a transformative field with significant implications for both rehabilitative and assistive medical interventions. Unlike traditional rigid surgical manipulators, soft robots offer superior compliance and adaptability, making them more suitable for maneuvering within the complex and delicate environments of the human body. Their flexibility allows them to navigate around irregular anatomical structures and avoid damaging sensitive tissues such as blood vessels, muscles, and bones, thereby enabling safer and more precise surgical procedures [51]. This represents a notable improvement over existing rigid surgical tools, which lack such dexterity. Additionally, soft robotic systems, including serial and continuum robots, demonstrate the potential for sub-millimeter positioning accuracy while offering enhanced flexibility and a higher number of degrees of freedom than current rigid counterparts. The long-term goal of this achievement is to make these surgical tools more accessible and affordable for broader patient populations. However, a key challenge remains: soft actuators are typically unable to produce the high mechanical forces required for certain surgical tasks, limiting their standalone applicability. As a result, considerable research is focused on developing stiffness-tunable robotic systems that can switch between soft and rigid states, thus combining the navigational advantages of soft robotics with the strength of rigid systems. Such hybrid platforms promise to revolutionize the landscape of surgical robotics by unifying the benefits of compliance and precision in a single adaptable system [51].

19.5.2 Artificial Intelligence and Machine Learning

The integration of Artificial Intelligence (AI) in soft robotics marks a major advancement aimed at improving interactions among humans, robots, and their environments. AI enables soft robots to better perceive, interpret, and respond to human behavior in complex, changing settings. In these systems, robots must collect and process data effectively in order to determine context-appropriate responses. To achieve this, they are outfitted with diverse sensory modalities—such as tactile sensors, cameras, microphones, and environmental detectors—that continuously collect real-time information from their surroundings [52].

Two prominent AI techniques used in soft robotics are supervised learning and reinforcement learning [53]. Supervised learning trains robots on labeled datasets to detect patterns and make predictions, which are then validated against actual outcomes. In contrast, reinforcement learning enables robots to learn from experience through feedback acquired during interaction with the environment. By comparing expected versus actual outcomes, robots adaptively refine their behavior to maximize reward. Over time, this results in improved pattern recognition and decision-making, particularly in interpreting complex human communication signals like gestures, tone, and facial expressions.

Soft robots are particularly suitable for human-robot interaction due to their embodiment of intelligence, a concept where physical form and cognitive processes are deeply intertwined. These robots can utilize distributed processing through sensory skins, sometimes processing information locally without central computational units [54]. Unlike traditional rigid body that follow pre-defined instructions, soft robots leverage body flexibility to interact fluidly with their surroundings. This makes them more adept in unpredictable environments, enabling lifelike responsiveness. Their biomechanical similarity to living organisms also makes them more relatable and compatible in human-centered applications [55].

A promising development in this field is **Reservoir Computing (RC)**, a neural network model that specializes in processing temporal patterns through a dynamic system. RC is composed of an input layer, a nonlinear reservoir, and an output layer. The input layer receives environmental data, which passes through randomly connected nonlinear units in the reservoir, allowing for short-term memory and dynamic response. The processed patterns then move to the output layer, which generates the final system behavior. This structure allows soft robots to analyze and respond to stimuli in real time, enhancing adaptive capabilities.

Feedback mechanisms form a critical part of intelligent control in soft robots. These systems enable autonomous execution of complex motions based on internal and external sensory inputs. Feedback not only enhances autonomy but also aligns the robot's behavior with environmental cues, improving real-world task performance. This autonomy is closely tied to the principle of embodied cognition, which suggests that intelligent behavior emerges from the coupling of brain and body dynamics.

Recent efforts have focused on the dual evolution of cognitive and physical capabilities in soft robots. Cognitive development is largely driven by reinforcement learning algorithms that allow systems to evolve toward autonomy. This evolution is further enhanced through **Central Pattern Generators (CPGs)**—neural circuits that produce rhythmic movements without rhythmic stimuli. CPGs support smooth, coordinated motor behaviors and help accelerate the development of independent function in robotic agents.

Another emerging area involves the design of multi-robot organisms, where multiple soft robots work cooperatively as a single, modular entity. Each unit specializes in a particular function and communicates with others to complete complex tasks. This resource-sharing model increases scalability, promotes task distribution, and enhances the flexibility of robotic systems.

As AI technology matures, soft robots are becoming more adept at performing tasks that require adaptability, contextual understanding, and real-time learning. This broadens their utility across domains like healthcare, industry, exploration, and personal care.

One of the most widely used AI architectures in soft robotic control is CNN. CNNs are highly effective at visual analysis, taking input from onboard cameras to identify and classify objects. Due to their high accuracy, CNNs are integral to robotic perception systems. To enhance performance in low-light environments, some systems now incorporate infrared cameras. Robots also use specialized tactile sensors to interpret sensory data, helping them react appropriately to detected stimuli—such as choosing how to grasp an object based on visual recognition [56].

While CNNs are dominant, other types of neural networks are also employed for tasks like sensing, object recognition, and locomotion control [57]. These networks are used to direct the robot's movements through vector-based controls that guide limb position and orientation. As the robot gains experience through environmental interaction, it gradually improves its accuracy and responsiveness, reducing reliance on human supervision.

Robotic control is also often achieved using kinematic models. These models are generally trained using demonstrations by humans who perform the desired motion patterns. The resulting data is then used to train AI models—often neural networks—that learn to replicate these movements with precision. Over time, this leads to more refined and purposeful motor behavior, allowing the robot to operate efficiently and independently.

19.6 Recent Trends in Soft Robotics for BCI Systems

The integration of soft robotics into BCI systems has rapidly evolved over the last five years, driven by the growing demand for safer, adaptive, and biologically inspired human-machine interactions. Key considerations such as user comfort, biomechanical compatibility, and functional adaptability have led to the development of soft robots that mimic the compliant nature of human tissues and joints. These robots are designed to respond dynamically to the user's motion and physiological feedback, thereby improving usability over prolonged interaction. The structure of these robots absorbs mechanical stress and enhances safety, while materials such as electroactive polymers (EAPs) offer lightweight and scalable actuation. EAPs, broadly categorized into ionic and field-activated types, exhibit large deformation in response to electrical stimuli, making them ideal for compact assistive applications. While ionic EAPs operate at low voltages but yield limited force, field-activated EAPs produce higher actuation forces but require kilovolt-level inputs, presenting a design challenge for wearable systems [58, 59].

A recurring theme in recent research is the importance of clean, high-fidelity brain signal acquisition. Studies have consistently emphasized that the clarity and stability of EEG signals directly influence the accuracy of motor intent decoding and real-time control of robotic systems [60]. This has led to the exploration of multimodal and flexible EEG acquisition devices, aiming to overcome motion artifacts and electrode instability in real-world settings [61].

The progression of BCI technologies since 2020 has been marked by several pivotal developments. Early work identified the limitations of rigid-body robotics in HRI and advocated for malleable, user-adaptive platforms. Around the same time, visual stimulus-based P300 BCIs gained popularity due to their compatibility with wireless EEG setups [62]. In 2021, researchers began advocating for the synergistic potential of soft robotics with BCIs in rehabilitation and assistance, noting their ability to circumvent constraints imposed by rigid actuation [60, 63].

By 2022, significant advances were made in reducing false positives in MI-BCI systems through asynchronous classification strategies [64]. Reinforcement learning and deep learning architectures began to be integrated with EEG-based interfaces, improving decoding accuracy and enabling more complex motor control paradigms [65]. These improvements in decoding facilitated real-time closed-loop systems capable of adapting to a user's evolving intent and neuromotor state.

In recent years, hybrid models combining EEG and fNIRS signals have improved motor imagery classification performance, as evidenced by gains in clinical scores like Fugl-Meyer Motor Assessment (FMA) and Action Research Arm Test (ARAT) [66]. Studies such as Zhang et al. [67] have implemented MI and SSEVP-based systems to control soft robotic gloves, achieving control accuracies above 90% in healthy users and demonstrating promising functionality even in stroke patients. Similarly, the NeuroFlex system demonstrated the feasibility of translating MI EEG patterns into multi-gesture hand commands, achieving an AUC of 0.88 for key tasks like fist closure [68].

Furthermore, CNNs, and actor-critic reinforcement learning models have combined to enhance BCI command classification, achieving over 93% accuracy in real-time control scenarios [69]. Another novel approach involved alpha-band visual stimulation during SSVEP-based BCI, showing statistically significant improvements in upper limb rehabilitation scores among stroke patients [70].

In the domain of bioinspired soft robotics, several notable breakthroughs have been achieved in recent years. The goal of this domain is to develop robotic systems that emulate human behaviors and biological functions [71]. One of the key technologies contributing to this advancement is the use of SMAs [72], which endow robots with a form of mechanical "memory". SMAs are capable of returning to their original shapes through a solid-solid phase transition, a phenomenon in which the atomic structure of the material rearranges while remaining in the solid state throughout the transformation. These transitions can be triggered by various types of external stimuli. A common method for initiating this transformation is through temperature modulation, where thermal stimuli alter the internal structure. As temperature increases or decreases, it produces force that deforms the material into controlled ways, enabling reconfiguration in robotic components.

The evolution of soft robotic actuators is expected to continue, driven by advancements in both materials science and actuation technologies [73]. These improvements will enhance the competitiveness of soft actuators in comparison to traditional rigid alternatives. As actuation strategies mature, their applications will expand across diverse fields, offering increased functionality and adaptability. Simultaneously, the development of novel materials will unlock new opportunities for deployment in environments previously deemed impractical. Material innovation is seen as central to the future of soft robotics, as enhancements in material properties can resolve current limitations while broadening performance capabilities. For instance, hydrogels—a widely used soft material—must overcome challenges such as evaporation to remain viable for long-term use [74]. Continued materials breakthroughs will be essential to maximizing the benefits of soft robots while minimizing reliability concerns.

19.7 Conclusion

In this chapter, we presented a comprehensive overview of the key components that constitute a functional soft robotic system, with a particular emphasis on applications for individuals experiencing motor impairments or paralysis. We began by surveying various actuation mechanisms—ranging from fluid-powered actuators such as pneumatic and hydraulic systems, to shape memory alloys, dielectric elastomer actuators, and magnetically driven systems with their working principles and case studies. Next, we explored brain-computer interfaces, introducing the distinction between invasive and non-invasive modalities. We focused on widely used non-invasive techniques such as EEG, fNIRS, and MEG, while briefly addressing ECoG as a prominent invasive method. We then discussed foundational BCI paradigms, highlighting design considerations and examples including P300, SSVEP, and EEG-based systems. We proceeded to detail control and sensing strategies in soft robotics, including open-loop, closed-loop, and autonomous control architectures. Various sensor technologies—such as resistive, capacitive, and fiber-optic systems—were discussed in the context of their adaptability and application-specific advantages. Following this, we outlined several key application areas, including facial expression recognition, rehabilitation, and lower-limb locomotion, supported by recent experimental findings. The chapter concluded with a discussion on the role of minimally invasive surgery augmented by soft robotics, and a deep dive into how artificial intelligence and machine learning are advancing the cognitive and interactive capabilities of soft robots. Finally, we reflected on recent technological developments, such as shape memory alloys and hydrogel-based actuators, and outlined the future trajectory of materials and intelligent control in soft robotics.

References

1. F. Safavi et al., Emerging frontiers in human–robot interaction. J. Intell. Robot. Syst. Theory Appl. **110**(2), 45 (2024). https://doi.org/10.1007/s10846-024-02074-7
2. F. Safavi, D. Pei, P. Olikkal, R. Vinjamuri, New horizons in human–robot interaction: synergy, cognition, and emotion, in *Discovering the Frontiers of Human-Robot Interaction*, (Springer, Cham, 2024), pp. 103–133. https://doi.org/10.1007/978-3-031-66656-8_5
3. B.S. Armour, E.A. Courtney-Long, M.H. Fox, H. Fredine, A. Cahill, Prevalence and causes of paralysis—United States, 2013. Am. J. Public Health **106**(10), 1855–1857 (2016). https://doi.org/10.2105/AJPH.2016.303270
4. A. Kawala-Sterniuk et al., Summary of over fifty years with brain-computer interfaces—A review. Brain Sci. **11**(1), 43 (2021). https://doi.org/10.3390/brainsci11010043
5. C. Hegde, J. Su, J.M.R. Tan, K. He, X. Chen, S. Magdassi, Sensing in soft robotics. ACS Nano **17**(16), 15277–15307 (2023). https://doi.org/10.1021/acsnano.3c04089
6. M.S. Xavier, A.J. Fleming, Y.K. Yong, Finite element modeling of soft fluidic actuators: Overview and recent developments. Adv. Intell. Syst. **3**(2) (2021). https://doi.org/10.1002/aisy.202000187
7. A. Wilkening, D. Baiden, O. Ivlev, Assistive acting movement therapy devices with pneumatic rotary-type soft actuators. Biomed. Tech. Eng. **57**(6) (2012). https://doi.org/10.1515/bmt-2011-0141
8. H.K. Yap et al., A fully fabric-based bidirectional soft robotic glove for assistance and rehabilitation of hand impaired patients. IEEE Robot. Autom. Lett. **2**(3), 1383–1390 (2017). https://doi.org/10.1109/LRA.2017.2669366
9. X. Wang, Q. Cheng, Z. Wang, Y. Lu, Z. Zhang, X. Zhao, A pneumatic soft glove system based on bidirectional bending functionality for rehabilitation. Biomimetics **10**(3), 129 (2025). https://doi.org/10.3390/biomimetics10030129
10. P. Polygerinos, Z. Wang, K.C. Galloway, R.J. Wood, C.J. Walsh, Soft robotic glove for combined assistance and at-home rehabilitation. Robot. Auton. Syst. **73**, 135 (2015). https://doi.org/10.1016/j.robot.2014.08.014
11. P. Polygerinos, K.C. Galloway, E. Savage, M. Herman, K.O. Donnell, C.J. Walsh, Soft robotic glove for hand rehabilitation and task specific training, in *2015 IEEE International Conference on Robotics and Automation (ICRA)*, Mei 2015, pp. 2913–2919. https://doi.org/10.1109/ICRA.2015.7139597
12. M. Pan et al., Soft actuators and robotic devices for rehabilitation and assistance. Adv. Intell. Syst **4**(4) (2022). https://doi.org/10.1002/aisy.202100140
13. L. Morris, R.S. Diteesawat, N. Rahman, A. Turton, M. Cramp, J. Rossiter, The-state-of-the-art of soft robotics to assist mobility: A review of physiotherapist and patient identified limitations of current lower-limb exoskeletons and the potential soft-robotic solutions. J. Neuroeng. Rehabil. **20**(1), 18 (2023). https://doi.org/10.1186/s12984-022-01122-3
14. S. Chen, H. Xu, F. Haseeb, W. Fan, Q. Wei, A pneumatic–hydraulic hybrid actuator for underwater soft robot swimming and crawling. Sens. Actuators A Phys. **356**, 114284 (2023). https://doi.org/10.1016/j.sna.2023.114284

15. P.S. Lobo, J. Almeida, L. Guerreiro, Shape memory alloys behaviour: A review. Procedia Eng. **114** (2015). https://doi.org/10.1016/j.proeng.2015.08.025
16. M.G. Mataee, M.T. Andani, M. Elahinia, Adaptive ankle-foot orthoses based on superelasticity of shape memory alloys. J. Intell. Mater. Syst. Struct. **26**(6) (2015). https://doi.org/10.1177/1045389X14544145
17. B.B. Park et al., Structure-activity relationships of Baicalein and its analogs as novel TSLP inhibitors. Sci. Rep. **9**(1) (2019). https://doi.org/10.1038/s41598-019-44853-5
18. A. Hadi, K. Alipour, S. Kazeminasab, M. Elahinia, ASR glove: A wearable glove for hand assistance and rehabilitation using shape memory alloys. J. Intell. Mater. Syst. Struct. **29**(8) (2018). https://doi.org/10.1177/1045389X17742729
19. F. Carpi, A. Mannini, D. De Rossi, Elastomeric contractile actuators for hand rehabilitation splints, in *Electroactive Polymer Actuators and Devices (EAPAD) 2008*, vol. 6927, (2008). https://doi.org/10.1117/12.774644
20. S. Pourazadi, S. Ahmadi, C. Menon, Towards the development of active compression bandages using dielectric elastomer actuators. Smart Mater. Struct. **23**(6), 065007 (2014). https://doi.org/10.1088/0964-1726/23/6/065007
21. J.H. Youn et al., Dielectric elastomer actuator for soft robotics applications and challenges. Appl. Sci. **10**(2) (2020). https://doi.org/10.3390/app10020640
22. J. Wang, A. Chortos, Control strategies for soft robot systems. Adv. Intell. Syst. **4**(5), 2100165 (2022). https://doi.org/10.1002/aisy.202100165
23. J. Qu et al., Advanced flexible sensing technologies for soft robots. Adv. Funct. Mater. **34**(29), 2401311 (2024). https://doi.org/10.1002/adfm.202401311
24. M. Cianchetti, C. Laschi, A. Menciassi, P. Dario, Biomedical applications of soft robotics. Nat. Rev. Mater. **3**(6) (2018). https://doi.org/10.1038/s41578-018-0022-y
25. M. Zheng, M. Liu, Y. Cheng, W. Chen, L. Wang, X. Qin, Stimuli-responsive fiber/fabric actuators for intelligent soft robots: From current progress to future opportunities. Nano Energy **129**, 110050 (2024). https://doi.org/10.1016/j.nanoen.2024.110050
26. J. Qu et al., Recent advances on underwater soft robots. Adv. Intell. Syst. **6**(2) (2024). https://doi.org/10.1002/aisy.202300299
27. A. Sarker, T. Ul Islam, M.R. Islam, A review on recent trends of bioinspired soft robotics: Actuators, control methods, materials selection, sensors, challenges, and future prospects. Adv. Intell. Syst. **7, no 3**, 2400414 (2025). https://doi.org/10.1002/aisy.202400414
28. K. Liu, W. Chen, W. Yang, Z. Jiao, Y. Yu, Review of the research progress in soft robots. Appl. Sci. (Switzerland) **13**(1) (2023). https://doi.org/10.3390/app13010120
29. P. Rothemund, N. Kellaris, S.K. Mitchell, E. Acome, C. Keplinger, HASEL artificial muscles for a new generation of lifelike robots—Recent Progress and future opportunities. Adv. Mater. **33**(19), 2003375 (2021). https://doi.org/10.1002/adma.202003375
30. O. Yasa et al., An overview of soft robotics. Annu. Rev. Control Rob. Auton. Syst. **6** (2023). https://doi.org/10.1146/annurev-control-062322-100607
31. J. Peksa, D. Mamchur, State-of-the-art on brain-computer Interface technology. Sensors **23**(13) (2023). https://doi.org/10.3390/s23136001
32. P. Olikkal, D. Pei, B.K. Karri, A. Satyanarayana, N.M. Kakoty, R. Vinjamuri, Biomimetic learning of hand gestures in a humanoid robot. Front Hum Neurosci **18**, 1391531 (2024). https://doi.org/10.3389/fnhum.2024.1391531
33. M.K. Burns, D. Pei, R. Vinjamuri, Myoelectric control of a soft hand exoskeleton using kinematic synergies. IEEE Trans. Biomed. Circuits Syst. **13, no 6**, 1351–1361 (2019). https://doi.org/10.1109/TBCAS.2019.2950145
34. A. Ajendla, M. Patel, P. Olikkal, R.K. Vinjamuri, Mental health management through wearables and AI innovation, in *Smart Healthcare, Clinical Diagnostics, and Bioprinting Solutions for Modern Medicine*, (IGI Global, 2025), pp. 193–212. https://doi.org/10.4018/979-8-3373-0659-9.ch010
35. C.Y. Chu, R.M. Patterson, Soft robotic devices for hand rehabilitation and assistance: A narrative review. J. Neuroeng. Rehabil. **15**(1) (2018). https://doi.org/10.1186/s12984-018-0350-6

36. R. Caminiti, P. Johnson, A. Urbano, Making arm movements within different parts of space: Dynamic aspects in the primate motor cortex. J. Neurosci. **10**(7), 2039–2058 (1990). https://doi.org/10.1523/JNEUROSCI.10-07-02039.1990
37. V. Mondini, R.J. Kobler, A.I. Sburlea, G.R. Müller-Putz, Continuous low-frequency EEG decoding of arm movement for closed-loop, natural control of a robotic arm. J. Neural Eng. **17**(4) (2020). https://doi.org/10.1088/1741-2552/aba6f7
38. P. Tai et al., Brain-computer interface paradigms and neural coding. Front. Neurosci. **17** (2023). https://doi.org/10.3389/fnins.2023.1345961
39. K.C. Galloway, Y. Chen, E. Templeton, B. Rife, I.S. Godage, E.J. Barth, Fiber optic shape sensing for soft robotics. Soft Robot. **6**(5) (2019). https://doi.org/10.1089/soro.2018.0131
40. M. Calisti, G. Picardi, C. Laschi, Fundamentals of soft robot locomotion. J. R. Soc. Interface **14**(130) (2017). https://doi.org/10.1098/rsif.2017.0101
41. P. Olikkal, D. Pei, T. Adali, N. Banerjee, R. Vinjamuri, Data fusion-based musculoskeletal synergies in the grasping hand. Sensors (Basel) **22**(19) (2022). https://doi.org/10.3390/S22197417
42. P. Olikkal, D. Pei, T. Adali, N. Banerjee, R. Vinjamuri, Musculoskeletal synergies in the grasping hand. Proc. Annu. Int. Conf. IEEE Eng. Med. Biol. Soc. EMBS **2022**(July), 3649–3652 (2022). https://doi.org/10.1109/EMBC48229.2022.9871023
43. N. Cheng et al., Brain-computer Interface-based soft robotic glove rehabilitation for stroke. I.E.E.E. Trans. Biomed. Eng. **67**(12) (2020). https://doi.org/10.1109/TBME.2020.2984003
44. X.Q. Shi, H.L. Heung, Z.Q. Tang, Z. Li, K.Y. Tong, Effects of a soft robotic hand for hand rehabilitation in chronic stroke survivors. J. Stroke Cerebrovasc. Dis. **30**(7) (2021). https://doi.org/10.1016/j.jstrokecerebrovasdis.2021.105812
45. J. Kim et al., Soft robotic apparel to avert freezing of gait in Parkinson's disease. Nat. Med. **30**(1) (2024). https://doi.org/10.1038/s41591-023-02731-8
46. R. Mane, T. Chouhan, C. Guan, BCI for stroke rehabilitation: Motor and beyond. J. Neural Eng. **17**(4) (2020). https://doi.org/10.1088/1741-2552/aba162
47. Y.S. Lee, S.H. Bae, S.H. Lee, K.Y. Kim, Neurofeedback training improves the dual-task performance ability in stroke patients. Tohoku J. Exp. Med. **236**(1) (2015). https://doi.org/10.1620/tjem.236.81
48. Y. Tian, T. Kanade, J.F. Cohn, Facial expression recognition, in *Handbook of Face Recognition*, ed. by S.Z. Li, A.K. Jain, (Springer, London, 2011), pp. 487–519. https://doi.org/10.1007/978-0-85729-932-1_19
49. L. Zhang, D. Tjondronegoro, Facial expression recognition using facial movement features. IEEE Trans. Affect. Comput. **2**(4) (2011). https://doi.org/10.1109/T-AFFC.2011.13
50. Z. Yan, Y. Song, R. Zhou, L. Wang, Z. Wang, Z. Dai, Facial expression realization of humanoid robot head and strain-based anthropomorphic evaluation of robot facial expressions. Biomimetics **9**(3) (2024). https://doi.org/10.3390/biomimetics9030122
51. K.W. Kwok, H. Wurdemann, A. Arezzo, A. Menciassi, K. Althoefer, Soft robot-assisted minimally invasive surgery and interventions: Advances and outlook. Proc. IEEE **110**(7) (2022). https://doi.org/10.1109/JPROC.2022.3167931
52. Y. Chi, Y. Zhao, Y. Hong, Y. Li, J. Yin, A perspective on miniature soft robotics: Actuation, fabrication, control, and applications. Adv. Intell. Syst. **6**(2) (2024). https://doi.org/10.1002/aisy.202300063
53. T. George Thuruthel, E. Falotico, L. Beccai, F. Iida, Editorial: Machine learning techniques for soft robots. Front. Robot. AI **8** (2021). https://doi.org/10.3389/frobt.2021.726774
54. Z. Zhao, Q. Wu, J. Wang, B. Zhang, C. Zhong, A.A. Zhilenkov, Exploring embodied intelligence in soft robotics: A review. Biomimetics **9**(4) (2024). https://doi.org/10.3390/biomimetics9040248
55. Z. Ye, L. Zheng, W. Chen, B. Wang, L. Zhang, Recent advances in bioinspired soft robots: Fabrication, actuation, tracking, and applications. Adv. Mater. Technol. **9**(21), 2301862 (2024). https://doi.org/10.1002/admt.202301862
56. Z. Sun et al., Artificial intelligence of things (AIoT) enabled virtual shop applications using self-powered sensor enhanced soft robotic manipulator. Adv. Sci. **8**(14) (2021). https://doi.org/10.1002/advs.202100230

57. T. Čakurda, M. Trojanová, P. Pomin, A. Hošovský, Deep learning methods in soft robotics: Architectures and applications. Adv. Intell. Syst. **7**(5), 2400576 (2025). https://doi.org/10.1002/aisy.202400576
58. K.J. Jyothish, S. Mishra, A survey on robotic prosthetics: Neuroprosthetics, soft actuators, and control strategies. ACM Comput. Surv. **56**(8), 1–44 (2024). https://doi.org/10.1145/3648355
59. N. Ghobadi, N. Sepehri, W. Kinsner, T. Szturm, Beyond human touch: Integrating soft robotics with environmental interaction for advanced applications. Actuators **13**(12), 507 (2024). https://doi.org/10.3390/act13120507
60. N. Robinson, R. Mane, T. Chouhan, C. Guan, Emerging trends in BCI-robotics for motor control and rehabilitation. Curr. Opin. Biomed. Eng. **20**, 100354 (2021). https://doi.org/10.1016/j.cobme.2021.100354
61. M. Alimardani, K. Hiraki, Passive brain-computer interfaces for enhanced human-robot interaction. Front. Robot. AI **7** (2020). https://doi.org/10.3389/frobt.2020.00125
62. J. Kalra et al., How visual stimuli evoked P300 is transforming the brain–computer interface landscape: a PRISMA compliant systematic review. IEEE Trans. Neural Syst. Rehabil. Eng. **31**, 1429–1439 (2023). https://doi.org/10.1109/TNSRE.2023.3246588
63. S. Yang et al., Exploring the use of brain-computer interfaces in stroke neurorehabilitation. Biomed. Res. Int. **2021**(1) (2021). https://doi.org/10.1155/2021/9967348
64. M. Song, H. Jeong, J. Kim, S.-H. Jang, J. Kim, An EEG-based asynchronous MI-BCI system to reduce false positives with a small number of channels for neurorehabilitation: A pilot study. Front. Neurorobot. **16** (2022). https://doi.org/10.3389/fnbot.2022.971547
65. K.M. Hossain, M.A. Islam, S. Hossain, A. Nijholt, M.A.R. Ahad, Status of deep learning for EEG-based brain–computer interface applications. Front. Comput. Neurosci. **16** (2023). https://doi.org/10.3389/fncom.2022.1006763
66. B. Premchand et al., A personalized multimodal BCI–soft robotics system for rehabilitating upper limb function in chronic stroke patients. Biomimetics **10**(2), 94 (2025). https://doi.org/10.3390/biomimetics10020094
67. R. Zhang et al., Hybrid brain-computer Interface controlled soft robotic glove for stroke rehabilitation. IEEE J. Biomed. Heal. Inf. **28**(7), 4194–4203 (2024). https://doi.org/10.1109/JBHI.2024.3392412
68. S. Zare, S.I. Beaber, Y. Sun, NeuroFlex: Feasibility of EEG-based motor imagery control of a soft glove for hand rehabilitation. Sensors **25**(3), 610 (2025). https://doi.org/10.3390/s25030610
69. Y. An, J. Wong, S.H. Ling, Development of real-time brain-computer interface control system for robot. Appl. Soft Comput. **159**, 111648 (2024). https://doi.org/10.1016/j.asoc.2024.111648
70. X. Li, J. Wang, X. Cao, W. Huang, Y. Hu, Soft robotic glove with alpha band brain computer interface for post-stroke hand function rehabilitation, in *2022 14th Biomedical Engineering International Conference (BMEiCON)*, (2022), pp. 1–5. https://doi.org/10.1109/BMEiCON56653.2022.10012103
71. C. Deng, Z. Li, Review: Advanced drive technologies for bionic soft robots. J. Bionic Eng. **22**(2), 419–457 (2025). https://doi.org/10.1007/s42235-025-00664-1
72. M.W. Gifari, H. Naghibi, S. Stramigioli, M. Abayazid, A review on recent advances in soft surgical robots for endoscopic applications. Int. J. Med. Robot. Comput. Assisted Surg. **15**(5) (2019). https://doi.org/10.1002/rcs.2010
73. D. Trivedi, C.D. Rahn, W.M. Kier, I.D. Walker, Soft robotics: Biological inspiration, state of the art, and future research. Appl. Bionics Biomech. **5**(3) (2008). https://doi.org/10.1080/11762320802557865
74. A. López-Díaz, A.S. Vázquez, E. Vázquez, Hydrogels in soft robotics: Past, present, and future. ACS Nano **18**(32), 20817–20826 (2024). https://doi.org/10.1021/acsnano.3c12200

Index

R. Vinjamuri (ed.), *Bridging the Gap between Mind and Machine*, https://doi.org/10.1007/978-3-032-06713-5

MIX
Papier aus verantwortungsvollen Quellen
Paper from responsible sources
FSC® C105338
FSC
www.fsc.org

If you have any concerns about our products,
you can contact us on
ProductSafety@springernature.com

In case Publisher is established outside the EU,
the EU authorized representative is:
Springer Nature Customer Service Center GmbH
Europaplatz 3, 69115 Heidelberg, Germany

Printed by Libri Plureos GmbH
in Hamburg, Germany